Communications in Computer and Information Science 438

Communications in Computer and Information Science

Valeri M. Mladenov Plamen Ch. Ivanov (Eds.)

Nonlinear Dynamics of Electronic Systems

22nd International Conference, NDES 2014
Albena, Bulgaria, July 4-6, 2014
Proceedings

Volume Editors

Valeri M. Mladenov
Technical University Sofia
Faculty of Automation
Department of Theory of Electrical Engineering
Kliment Ohridski Blvd. 8
Sofia, 1000, Bulgaria
E-mail: valerim@tu-sofia.bg

Plamen Ch. Ivanov
Boston University
Physics Department
590 Commonwealth Avenue
Boston, MA 02215, USA
E-mail: plamen@buphy.bu.edu

ISSN 1865-0929 e-ISSN 1865-0937
ISBN 978-3-319-08671-2 e-ISBN 978-3-319-08672-9
DOI 10.1007/978-3-319-08672-9
Springer Cham Heidelberg New York Dordrecht London

Library of Congress Control Number: 2014942072

Typesetting: Camera-ready by author, data conversion by Scientific Publishing Services, Chennai, India

Printed on acid-free paper

Springer is part of Springer Science+Business Media (www.springer.com)

Preface

These proceedings comprise papers presented during the International Conference on Nonlinear Dynamics of Electronic Systems (NDES) 2014. The unique feature of the conference is to bring together theoretical aspects of nonlinear science in electrical engineering, physics, and mathematics, as well as related applications. Nonlinear oscillations, synchronization, chaotic behavior, neural networks, and complex systems together with power electronics, biomedical systems and neurocomputing, networks of infrastructure and social interactions are a few topical examples of the need for expertise in the wide field of nonlinear science.

The NDES conference series was founded in 1993, and through the years has been established as one of the most prominent series of conferences in the field of nonlinear science and its applications. NDES conferences took places in Dresden (1993), Krakow (1994), Dublin (1995), Seville (1996), Moscow (1997), Budapest (1998), Rønne (1999), Catania (2000), Delft (2001), Izmir (2002), Scuol (2003), Vora (2004), Potsdam (2005), Dijon (2006), Tokushima (2007), Nizhniy Novgorod (2008), Rapperswil (2009), Dresden (2010), Kolkata (2011), Wolfenbüttel (2012) and Bari (2013). The organizer of this 22nd issue of NDES was the Technical University of Sofia, Bulgaria. The conference was held during July 4–6, 2014, in Albena, one of the preeminent resorts in Europe located on the Bulgarian Black Sea coast.

During NDES 2014 seven plenary talks were given by some of the most famous researchers in the world of nonlinear science: Vadim Anishchenko from Saratov State University, Russia; Alain Arneodo from Ecole Normale Superiore, Lyon, France; Shlomo Havlin from Bar-Ilan University, Ramat-Gan, Israel; Juergen Kurths from Humboldt-Universität Berlin, Germany; Ruedi Stoop from the University and ETH of Zürich, Switzerland; Wolfgang Mathis from Institut für Theoretische Elektrotechnik, Hannover, Germany; and Ronald Tetzlaff from Technische Universität Dresden, Germany.

In total 65 manuscripts were submitted to NDES 2014. All manuscripts passed a multistep review process before the final decisions. After a thorough peer-review process, the program co-chairs selected 47 papers. The quality of all manuscripts received was high, and it was not possible to include all good papers in the final conference program. The selected papers were divided by subject into the following main topics: nonlinear oscillators, circuits and electronic systems; networks and nonlinear dynamics; and nonlinear phenomena in biological and physiological systems. In parallel to NDES 2014, a satellite workshop on "Electro-physiological Signals in Living Beings: Data and Methods of Nonlinear Analysis" was organized by Prof. Plamen Ch. Ivanov from Boston University, Harvard Medical School, and Bulgarian Academy of Sciences and by Prof. Antonio Scala from the London Institute for Mathematical Sciences and Institute for

Complex Systems at CNR, Italy. The workshop was partially supported by the FET Open EU project PLEASED (Plants Employed as Sensing Devices). Eight lectures, not included in these proceedings, were presented at the workshop.

We thank all NDES 2014 participants for their contributions to the conference program and to these proceedings. Many thanks also go to the Bulgarian organizers for their support and hospitality. We also express our sincere thanks to all reviewers for their help during the manuscript review procedure and for their valuable comments and recommendations to ensure the high quality of the contributions in this proceedings volume.

April 2014

Valeri M. Mladenov
Plamen Ch. Ivanov

Organization

Program Committee Chairs

Valeri M. Mladenov
Plamen Ch. Ivanov

International Program Committee

Alain Arneodo
Alexandar Pisarchik
Alexander Dmitriev
Bert Kappen
Boris Bezruchko
Celso Grebogi
Daniele Marinazzo
Erik Lindberg
Giuliano Scarcelli
Hans Hermann
Ivan Angelov
Jesus M. Cortes
Josè R. Croca
Jürgen Kurths
Kyandoghere Kyamakya
Kristina Kelber
Leon Chua
Maciej Ogorzalek
Marcelo O. Magnasco
Martin Hasler
Ned Corron
Nikos Mastorakis
Oreste Piro
Plamen Ch. Ivanov
Ronald Tetzlaff
Ruedi Stoop
Sebastano Stramaglia
Shlomo Havlin
Syamal K. Dana
Thomas Ott
Toshimichi Saito
Vadim Anishchenko
Valeri Mladenov
Vladimir Nekorkin
Wolfgang Mathis
Wolfgang Schwarz
Yoshifumi Nishio

Local Organizing Committee

Georgi Tsenov
Agata Manolova

Table of Contents

Poincaré Recurrences Near the Critical Point of Feigenbaum Attractor Birth

Yaroslav Boev, Galina Strelkova, and Vadim Anishchenko

Department of Physics, Saratov State University,
Astrakhanskaya str. 83, 410012 Saratov, Russia
boev.yaroslav@gmail.com
{wadim,strelkovagi}@info.sgu.ru
http://chaos.sgu.ru

Abstract. The evolution of the Afraimovich–Pesin dimension of a sequence of Poincaré recurrence times is analyzed when approaching the critical point of Feigenbaum attractor birth. It is shown for two one-dimensional maps that the Afraimovich–Pesin dimension abruptly increases at the critical point. This indicates that this point is singular and requires a special theoretical analysis.

Keywords: Poincaré recurrence, Afraimovich–Pesin dimension.

1 Introduction

Poincaré recurrences are one of the fundamental properties of the temporal evolution of dynamical systems. By now, the mathematical theory of Poincaré recurrences has been rather fully developed. It describes the statistics of return times both in a local neighborhood of a given initial state (the so-called *local approach* [1,2,3]) and in a whole set of the system phase space (the *global approach* [4,5]).

In the framework of the local approach, it has been established that the mean return time is related to the probability (Kac's lemma [2]) and the distribution of recurrence times obeys an exponential law in systems with mixing [3]. In the framework of the global approach, recurrence times are considered in all covering elements of the whole set and their statistics is then studied.

In our paper we apply the global theory of Poincaré recurrences and analyze the features of the *Afraimovich–Pesin dimension* (the AP dimension) in two one-dimensional maps near the critical point of Feigenbaum attractor birth.

2 Problem Statement

The considered set of phase trajectories of a system is covered with balls of size $\varepsilon \ll 1$. A minimal time of the first recurrence $\tau_{\inf}(\varepsilon_i)$ is defined for each covering element ε_i $(i = 1, 2, \ldots, m)$ and its mean value is then found as follows:

$$\langle \tau_{\inf}(\varepsilon) \rangle = \frac{1}{m} \sum_{i=1}^{m} \tau_{\inf}(\varepsilon_i). \tag{1}$$

V.M. Mladenov and P.C. Ivanov (Eds.): NDES 2014, CCIS 438, pp. 1–8, 2014.

It has been shown in [4,5] that in the general case, the mean minimal return time can be estimated by the following expression:

$$\langle \tau_{\mathrm{inf}}(\varepsilon) \rangle \sim \phi^{-1}(\varepsilon^{\frac{d}{\alpha_C}}), \quad \varepsilon \ll 1\,, \tag{2}$$

where d is the fractal dimension of the set under study, α_C is the dimension of a sequence of return times, i.e., the AP dimension, and ϕ is a calibration function that depends on the topological entropy h_T of the system in the considered regime. If the dynamical mode is chaotic, then $h_T > 0$ and the calibration function is $\phi(t) \sim \exp(-t)$. In this case, from (2) we have [4,6]

$$\langle \tau_{\mathrm{inf}}(\varepsilon) \rangle \sim -\frac{d}{\alpha_C} \ln \varepsilon, \quad \varepsilon \ll 1. \tag{3}$$

When $h_T = 0$, the calibration function is $\phi(t) \sim 1/t$ and (2) can be rewritten as follows [4,6]:

$$\ln\langle \tau_{\mathrm{inf}}(\varepsilon) \rangle \sim -\frac{d}{\alpha_C} \ln \varepsilon, \quad \varepsilon \ll 1. \tag{4}$$

The dependence of $\langle \tau_{\mathrm{inf}}(\varepsilon) \rangle$ on $\ln \varepsilon$ given by (3) represents a straight line with slope $k = -d/\alpha_C$. A similar plot can be obtained for the expression (4). Thus, the slope k can be defined numerically from the corresponding plots. Knowing its values and the fractal dimension d we can easily calculate the AP dimension $\alpha_C = d/|k|$.

It has been shown in [7] that the topological entropy can be evaluated by the positive Lyapunov exponent (for one-dimensional maps in a chaotic regime) or the Kolmogorov–Sinai entropy, which equals to the sum of positive Lyapunov exponents for 2- and higher-dimensional systems [8].

Properties of the AP dimension can be analyzed using simple discrete-time systems. We exemplify this with a one-dimensional logistic map in the form:

$$x_{n+1} = r x_n (1 - x_n), \tag{5}$$

where r is the control parameter. For $r > r^* = 3.569946\ldots$, a chaotic attractor is realized in (5) and characterized by the positive Lyapunov exponent $\lambda > 0$. The envelope $\lambda(r)$ is described by the law [9]:

$$\lambda = 0.87(r - r^*)^{0.45}, \; r^* = 3.5699, \tag{6}$$

where r^* is the critical parameter value that corresponds to the birth of the Feigenbaum attractor. We study the statistics of Poincaré recurrences when approaching the critical value $r = r^*$ from above ($r > r^*$). It is expected that (3) is valid when $r > r^*$ and (4) holds at $r = r^*$.

From both mathematical and physical points of view, it is important to analyze the evolution of the AP dimension for the indicated transition, i.e., from the case of $h_T > 0$ to the case of $h_T = 0$. The aim of our paper is to study the features of this transition.

3 Numerical Results

Numerical calculations performed for the system (5) show that the dependence (3) holds to a high accuracy (the error is less than 1%) in the parameter range $r^* < r \leq 4$. This is exemplified by the numerical results shown in Fig. 1 for the parameter value $r = 3.5723$, which is close to r^*.

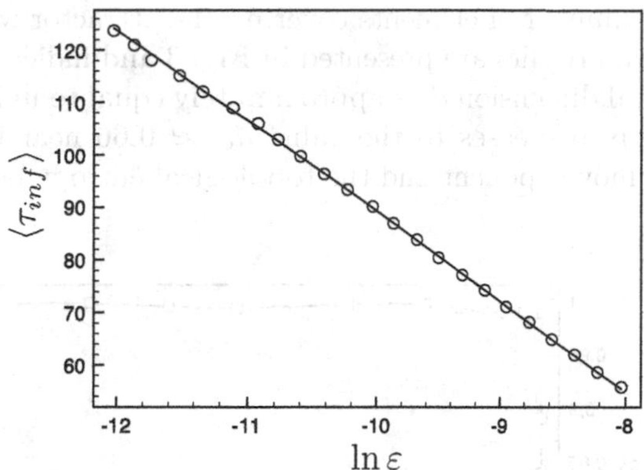

Fig. 1. Mean minimal return time $\langle \tau_{\text{inf}} \rangle$ on ε for $r = 3.5723$

In this case, the slope of the plot is $k = -16.6$. When one approaches the critical value $r = r^*$ from above, the value of $|k|$ dramatically increases. This is illustrated in Fig. 2, where the slope $|k|$ is shown as a function of the parameter r of the system (5).

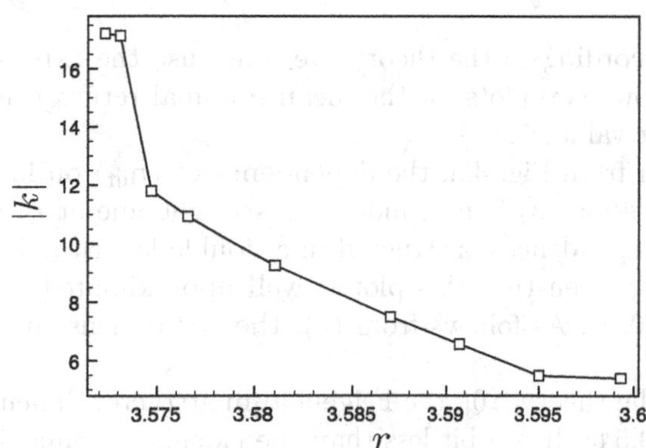

Fig. 2. Slope coefficient $|k|$ of the plot $\langle \tau_{\text{inf}} \rangle$ on $\ln \varepsilon$ for different values of the parameter r of the system (5)

Besides the slope $|k|$, we need to know the value of the fractal dimension d to find the AP dimension α_C using (3). For $r > r^*$, the system (5) has an attractor consisting of a countable set of intervals on a one-dimensional set of x. The fractal dimension of the attractor is $d = 1$. To corroborate this, we calculate the capacity dimension d_C according to the definition [10]:

$$d_C = -\lim_{\varepsilon \to 0} \frac{\ln N(\varepsilon)}{\ln \varepsilon}, \tag{7}$$

where $N(\varepsilon)$ is the number of elements covering the attractor with line segments of size ε. Calculation results are presented in Fig. 3 and indicate that d_C, which evaluates the fractal dimension d, is approximately equal to unity in the interval $r > r*$ and sharply decreases to the value $d_C \approx 0.66$ near the critical point $r \gtrsim r^*$. The Lyapunov exponent and the topological entropy become zero at the

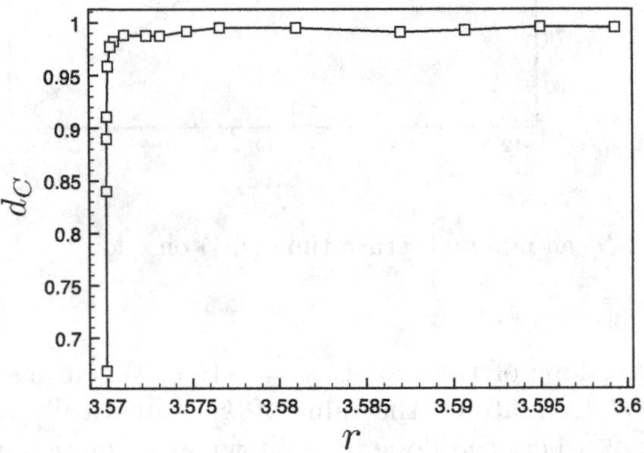

Fig. 3. Capacity dimension d_C of attractors in the system (5) for different values of the control parameter r

critical point. According to the theory, we must use the expression (4) in this case. Figure 4 shows two plots for the mean minimal return time on ε near the critical parameter value r^*.

As can be seen from Fig. 4,a, the dependence of $\langle \tau_{\text{inf}} \rangle$ on $\ln \varepsilon$ plotted in the scale of the expression (3) is not, indeed, a straight line at $r = r^*$. Figure 4,b shows the same dependence constructed in a double logarithmic scale according to (4). It is clearly seen that this plot is well approximated by a straight line with slope $|k| \simeq 0.63$. As follows from (4), the AP dimension is $\alpha_C = d/|k|$ in this case.

According to the theory [10], the Feigenbaum attractor dimension at the critical point is $d \simeq 0.543$. It is a bit less than the calculated value, i.e., $d_C \simeq 0.645$. Our calculations give the AP dimension $\alpha_C = d/|k| \simeq 1.025$. Using the theoretical value $d = 0.543$, we obtain $\alpha_C \simeq 0.86$. The difference between these results can be explained by the calculation error, which grows when we approach the

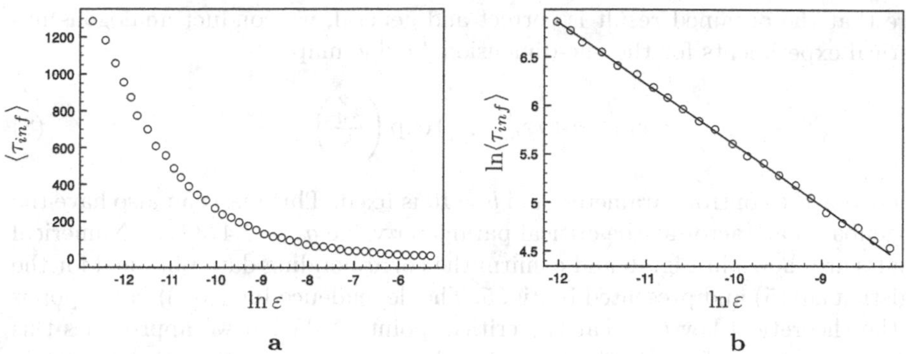

Fig. 4. Mean minimal return time $\langle \tau_{\text{inf}} \rangle$ on ε plotted in different scales for the control parameter $r = 3.569946$ (this is the closest point to the critical value)

critical point. However, the main point to be noted is that the AP dimension α_C is significantly greater than zero, regardless of the calculation error.

The final result is illustrated in Fig. 5 where the Lyapunov exponent λ and the AP dimension α_C are shown as functions of the parameter r for the system (5). As can be seen from the plots, for $r > r^*$, the numerical results are in complete agreement with theoretical ones, i.e., $\alpha_C \approx \lambda$ and tends to zero when one approaches $r = r^*$ from above. In this parameter region, the theoretical expression (3) is confirmed numerically as mentioned earlier in [7,8,11]. However, α_C abruptly increases from almost zero value to unity at the critical point. In addition, the linear dependence in Fig. 4,b corroborates that (4) is suitable to describe the dependence $\tau_{\text{inf}}(\varepsilon)$ at the critical point.

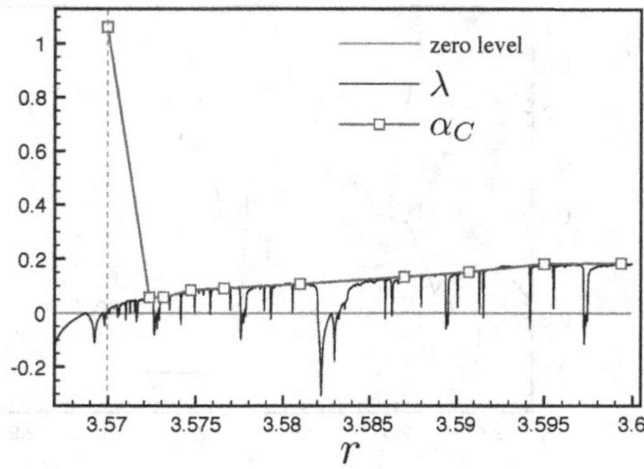

Fig. 5. Lyapunov exponent λ and the AP dimension α_C as functions of the control parameter r of the map (5)

The abrupt increase of the AP dimension at the critical point of Feigenbaum attractor birth is unexpected for us, at least from a physical viewpoint. To make sure that the obtained result is correct and general, we conduct analogous numerical experiments for the one-dimensional cubic map:

$$x_{n+1} = (ax_n - x_n^3)\exp\left(\frac{x_n^2}{b}\right), \tag{8}$$

where a is the control parameter, and $b = 10$ is fixed. This map can also have the Feigenbaum attractor at the critical parameter value $a^* = 2.4738\ldots$. Numerical results are shown in Fig. 6 and confirm the corresponding data obtained for the logistic map (5) and presented in Fig. 5. The dependence $\ln\langle\tau_{\rm inf}(\varepsilon)\rangle$ corresponds to the theoretical law (4) near the critical point a^*. When we approaches this point from above, the AP dimension sharply increases up to the value $\alpha_{\rm C} \approx 0.9$. As for the map (5), for $a > a^*$, the AP dimension and the Lyapunov exponent coincide within the calculation accuracy [11].

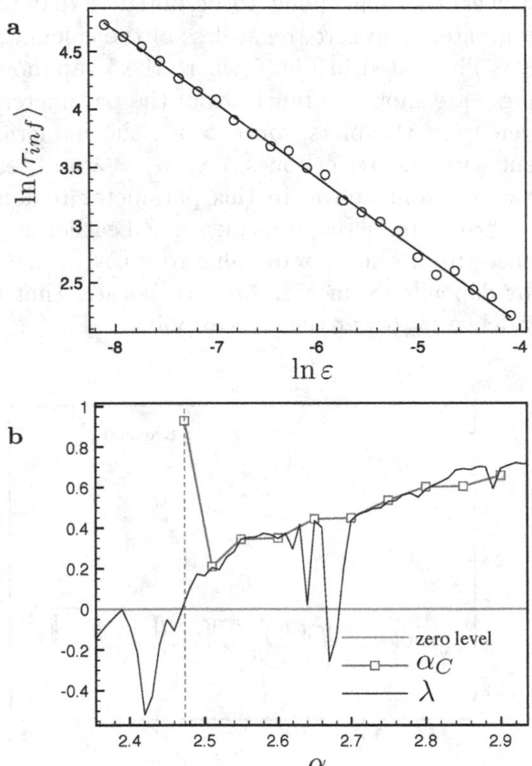

Fig. 6. Numerical results for the map (8). (a) Dependence of $\ln\langle\tau_{\rm inf}\rangle$ on $\ln\varepsilon$ calculated through (4) at the critical point $a^* \approx 2.4738$, and (b) the AP dimension $\alpha_{\rm C}$ and Lyapunov exponent λ as functions of the parameter a.

4 Discussion and Conclusions

Analysis of the numerical results presented in Figs. 5 and 6 enables us to draw the following conclusion. The statistics of Poincaré recurrences is described by the theoretical expression (3) in the parameter range that corresponds to the chaotic dynamics with positive values of the topological entropy h_T and the Lyapunov exponent $\lambda > 0$. Moreover, the relationship $\lambda \simeq \alpha_C$ holds within the accuracy of calculation error. Since α_C corresponds to h_T and λ, the AP dimension in the regime of a chaotic attractor can be treated as a measure of complexity of the system dynamics.

Both the AP dimension and the Lyapunov exponent vanish when a system parameter approaches from above the critical point of Feigenbaum attractor birth. We have $h_T = \lambda = 0$ at the critical point. According to the theory, at this point the statistics of Poincaré recurrences is defined by (4). This fact is corroborated by the numerical data shown in Figs. 4 and 6,a. However, the AP dimension drastically increases at the critical point and takes the value $\alpha_C \approx 1$.

Besides the Feigenbaum attractor, there are another minimal sets having zero topological entropy, for example, shifts on the circle for an irrational winding number. These motions are ergodic but without mixing. One can assume that for such systems, the AP dimension describes different properties of a system. In particular, it is shown that for irrational rotations on the circle, the AP dimension characterizes the rate of approximations of an irrational number [4,5]. The physical meaning of the AP dimension at the critical point of Feigenbaum attractor birth is still unclear and requires further investigations.

Acknowledgments. We are grateful to Prof. V. Afraimovich for fruitful discussions. The reported study was partially supported by RFBR, research project No. 13-02-00216a and by the Russian Ministry of Education and Science (project code 1008).

References

1. P.H.: Sur le Probléme des Trois Corps et les Equations de la Dynamique. Acta Mathematica 13, A3–A270 (1890)
2. Kac, M.: Lectures in Applied Mathematics. Interscience, London (1957)
3. Hirata, M., Saussol, B., Vaienti, S.: Statistics of Return Times: A General Framewok and New Applications. Comm. in Math. Physics 206, 33–55 (1999)
4. Afraimovich, V.: Pesin's Simension for Poincaré Recurrences. Chaos 7, 12–20 (1997)
5. Afraimovich, V., Ugalde, U., Urias, J.: Dimension of Poincaré Recurrences. Elsevier (2006)
6. Afraimovich, V.S., Lin, W.W., Rulkov, N.F.: Fractal Dimension for Poincaré Recurrences as an Indicator of Synchronized Chaotic Regimes. Int. J. Bifurc. Chaos 10, 2323–2337 (2000)
7. Penne, V., Saussol, B., Vaienti, S.: Fractal and Statistical Characteristics of Recurrence Times. In: Talk at the Conference "Disorder and Chaos", Rome, (September 1997), preprint CPT (1997)

8. Anishchenko, V.S., Astakhov, S.V., Boev, Y.I., Biryukova, N.I., Strelkova, G.I.: Statistics of Poincaré Recurrences in Local and Global Approaches. Commun. Nonlinear Sci. Numer. Simulat. 18, 3423–3435 (2013)
9. Huberman, B., Rudnik, J.: Scaling Behavior of Chaotic Flows. Phys. Rev. Lett. 45, 154–156 (1980)
10. Schuster, H.G.: Deterministic Chaos. Physik-Verlag, Weinheim (1984)
11. Anishchenko, V.S., Astakhov, S.V.: Poincaré Recurrence Theory and Its Application to Nonlinear Physics. Physics – Uspekhi 56, 955–972 (2013)

Quasi-periodic Oscillations in the System of Three Chaotic Oscillators

Alexander P. Kuznetsov, Yuliya V. Sedova, and Ludmila V. Turukina

Kotel'nikov Institute of Radio-Engineering and electronics of RAS,
Saratov Branch,
410019, Saratov, Zelyenaya, 38, Russia

Abstract. The dynamics of three coupled chaotic Rössler systems is considered. We discuss scenarios for the evolution of different types of regimes. The possibility of two- and three-frequency quasi-periodicity is shown. We considered the occurrence of resonanses on three-frequency torus, which leads to two-freqiency quasi-periodic regimes. The illustrations in the form of charts of the Lyapunov exponents, phase portraits of attractors plotted in the Poincare section and bifurcation diagrams are presented. We discuss the type of quasi-periodic bifurcation in the system.

Keywords: chaotic oscillations, quasi-periodic oscillations, invariant tori, bifurcation.

1 Introduction

The problem related to oscillations of coupled oscillators of different nature remains the focus of researchers in different fields of physics, chemistry, biology. The examples are radio-electronic oscillators , Josephson contacts, ion traps [1-4], etc. One of the interesting aspects is the problem of synchronization of chaotic systems. The traditional approach in this case is to study the regimes for which the dynamics is chaotic, although it may be both synchronous and asynchronous [4,5]. In the works [4,5] the corresponding structure of the parameter plane (frequency detuning – parameter of coupling) is studied for two coupled Rossler oscillators. Also they pointed to the existence of different windows of periodic regimes. We consider here another situation when the dynamics of coupled chaotic systems becomes quasi-periodic. This is explained by a stabilizing effect of dissipative coupling, which, however, retains some basic oscillatory rhythm of the individual oscillators. We will discuss this problem by the example of three chaotic Rössler oscillators. In this case we found not only a two-frequency quasi-periodic regimes, but also three-frequency quasi-periodic regimes.

2 Three Chaotic Oscillators

Let us consider the system of three coupled Rössler oscillators:

V.M. Mladenov and P.C. Ivanov (Eds.): NDES 2014, CCIS 438, pp. 9–14, 2014.

$$
\begin{aligned}
\dot{x}_1 &= -y_1 - z_1,\\
\dot{y}_1 &= x_1 + py_1 + \mu(y_2 - y_1),\\
\dot{z}_1 &= q + (x_1 - r)z_1,\\
\dot{x}_2 &= -(1-\Delta_1)y_2 - z_2,\\
\dot{y}_2 &= (1-\Delta_1)x_2 + py_2 + \mu(y_1 + y_3 - 2y_2),\\
\dot{z}_2 &= q + (x_2 - r)z_2,\\
\dot{x}_3 &= -(1-\Delta_2)y_3 - z_3,\\
\dot{y}_3 &= (1-\Delta_2)x_3 + py_3 + \mu(y_2 - y_3),\\
\dot{z}_3 &= q + (x_3 - r)z_3.
\end{aligned}
\tag{1}
$$

Here Δ_1 is the frequency detuning between the first and second oscillators and Δ_2 is the frequency detuning between the first and third oscillators. We fix parameters p=0.15, q=0.4 and r=8.5. This corresponds to the chaotic regime in individual subsystems.

Let us disscus the question of how the regimes of different types are embedded in parameter space. For this, we use the method of the charts of Lyapumov exponents [6-10]. We calculate the spectrum of Lyapunov exponents at each grid point on the parameter plane. Then we color these points in accordance with its signature. The corresponding chart is given in Fig.1. It is plotted on the (Δ_1, μ) plane. The periodic regimes lettered by P, two- and three-frequency quasi-periodic regimes T2 and T3 (with one and two zero Lyapunov exponents respectively), regimes of chaos C (with one positive Lyapunov exponent), regimes of hyperchaos HC2 and HC3 (with two and three positive Lyapunov exponents respectively) are marked by different colors. Regime of “amplitude death” AD is responsible for disappearance of oscillations due to their suppression of a dissipative coupling. The color legend is at the right of the figure.

The phase portraits of attractors plotted in the Poincaré section are shown in Fig. 2 (The Poincaré section is defined by relations y=0 and x>0). Two-frequency torus T2 exists for large values of coupling. In this case Poincaré section is the invariant curve close to a circle. Three-frequency torus T3 arises softly from this invariant curve as the parameter of coupling is decreased. One can see a very intricately shaped invariant curve with a further decrease of coupling. This invariant curve corresponds to one of the possible two-frequency resonant tori T_R2. Note that the number of resonance windows is sufficiently large for these values of the frequency detuning. At small coupling the tori are destroyed with the appearance of chaos and hyperchaos.

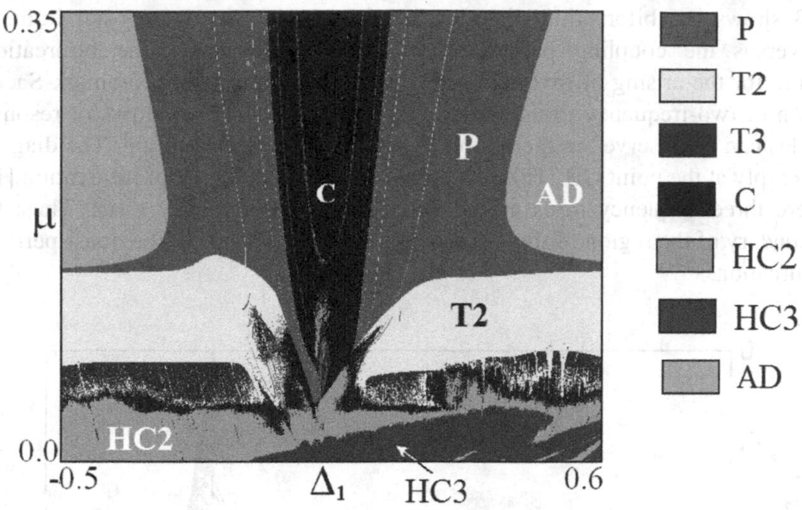

Fig. 1. Chart of Lyapunov exponent for the system (1), Δ_2=0.05

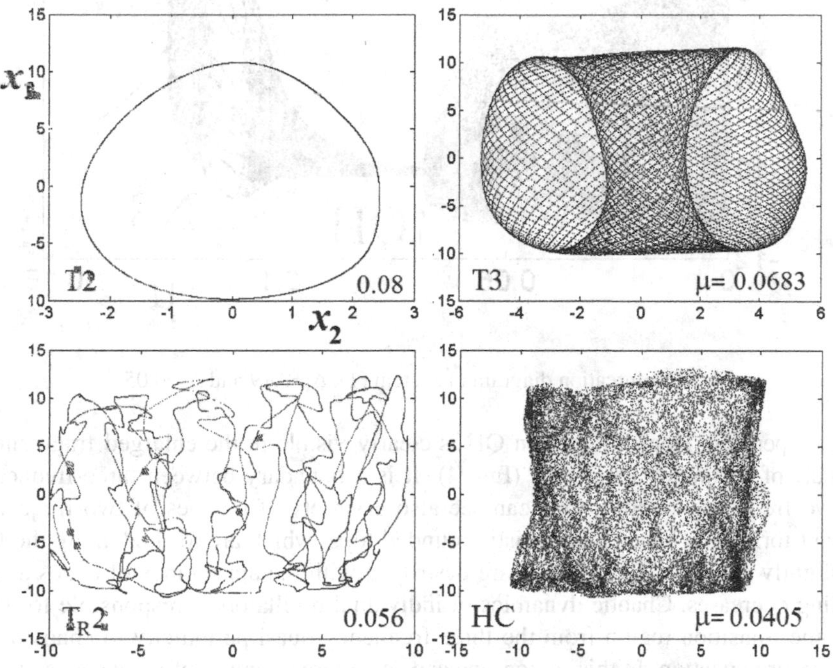

Fig. 2. Phase portraits of attractors at Poincaré section, Δ_1=0.19 and Δ_2=0.05

Fig. 3 shows the bifurcation diagram for the attractor in the chosen Poincare section versus the coupling parameter. This Figure illustrates the bifurcations responsible for the arising of invariant tori of different dimensions. Neumark-Sacker bifurcation of two-frequency torus occurs at the point NS. The windows of resonant limit cycles can be observed in the region of smaller values of coupling. The diagram widens sharply at the point QH. This is a point of quasi-periodic Hopf bifurcation [10-11], where three-frequency torus arises softly from two-frequency torus. Thus, the upper boundary of the region of three-frequency tori corresponds to the quasi-periodic Hopf bifurcation.

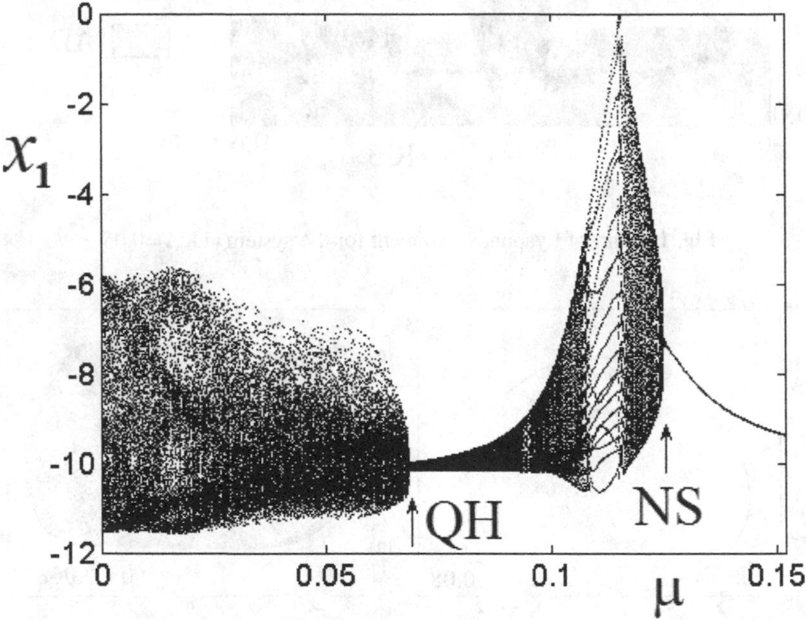

Fig. 3. Bifurcation diagram of system (1), Δ_1=0.19 and Δ_2=0.05

Quasi-periodic Hopf bifurcation QH is clearly visible in the enlarged fragment of the chart of Lyapunov exponents (Fig. 4). It is a boundary between three-frequency and two-frequency regions. One can see also a variety of tongues of two-frequency resonant tori. They have characteristic rounded tops, which are located along the QH line slightly above it. The tongues are destroyed with the appearance of chaos as the coupling decreases. Chaotic dynamics of individual oscillators is responsible for this. Note, the transition region from the three-frequency quasi-periodicity to chaos has a complex organization. In this region one can see a large number of resonances, which are much more numerous than in the case of regular oscillators [9].

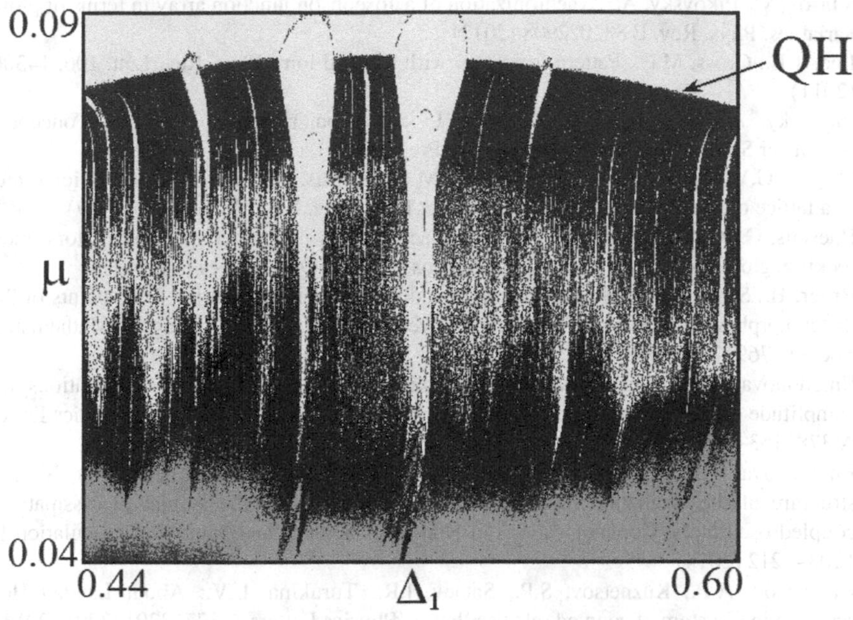

Fig. 4. Enlarged fragment of the parameter plane from Fig. 1, QH is a quasi-periodic Hopf bifurcation

3 Conclude

Thus, the effect of dissipative coupling on the chaotic oscillators can lead not only to the chaotic synchronization and appearance of periodic regimes, but also to the appearance of two- and three-frequency quasi-periodic oscillations. And quasi-periodic Hopf bifurcation is responsible for this. The reason is probably that chaotic regime is characterized by presence of large number of unstable limit cycles [4]. Adding of coupling can stabilize these cycles and this leads to appearance of the set of the resonant tori of different types in the dynamics of the system. With increasing of the number of chaotic oscillators the tori of higher and higher dimensions can be observed. We can expected, that this behavior would be typical for other chaotic systems.

This work was supported by the GrantNSc-1726.2014.2.

References

1. Temirbayev, A., Nalibayev, Y.D., Zhanabaev, Z.Z., Ponomarenko, V.I., Rosenblum, M.: Autonomous and forced dynamics of oscillator ensembles with global nonlinearcoupling: An experimental study. Phys. Rev. E 87, 062917 (2013)

2. Vlasov, V., Pikovsky, A.: Synchronization of a Josephson junction array in terms of global variables. Phys. Rev. E 88, 022908 (2012)
3. Lee, T.E., Cross, M.C.: Pattern formation with trapped ions. Phys. Rev. Lett. 106, 143001 (2011)
4. Pikovsky, A., Rosenblum, M., Kurths, J.: Synchronization: A Universal Concept in Nonlinear Sciences. Cambridge University Press (2001)
5. Osipov, G.V., Pikovsky, A.S., Rosenblum, M.G., Kurths, J.: Phase synchronization effects in a lattice of nonidentical Rossler oscillators. Phys. Rev. E 55, 2353–2361 (1997)
6. Baesens, C., Guckenheimer, J., Kim, S., MacKay, R.S.: Three coupled oscillators: mode locking, global bifurcations and toroidal chaos. Physica D 49, 387–475 (1991)
7. Broer, H., Simó, C., Vitolo, R.: The Hopf-saddle-node bifurcation for fixed points of 3D-diffeomorphisms: The Arnol'd resonance web. Reprint from the Belgian Mathematical Society, 769–787 (2008)
8. Emelianova, Y.P., Kuznetsov, A.P., Turukina, L.V.: Quasi-periodic bifurcations and "amplitude death" in low-dimensional ensemble of van der Pol oscillators. Physics Letters A 378, 153–157 (2014)
9. Emelianova, Y.P., Kuznetsov, A.P., Turukina, L.V., Sataev, I.R., Chernyshov, N.Y.: A structure of the oscillation frequencies parameter space for the system of dissipatively coupled oscillators. Communications in Nonlinear Science and Numerical Simulation 19, 1203–1212 (2014)
10. Kuznetsov, A.P., Kuznetsov, S.P., Sataev, I.R., Turukina, L.V.: About Landau–Hopf scenario in a system of coupled self-oscillators. Physics Letters A 377, 3291–3295 (2013)
11. Broer, H., Simó, C., Vitolo, R.: Quasi-periodic bifurcations of invariant circles in low-dimensional dissipative dynamical systems. Regular and Chaotic Dynamics 16, 154–184 (2011)

Taming Chaos by Using Induced External Signals: Experimental Results

Arturo Buscarino, Carlo Famoso, Luigi Fortuna, and Mattia Frasca

DIEEI, University of Catania, Italy

Abstract. The topic of chaos control is widely studied in literature. In this paper, some new experimental results regarding an innovative strategy for taming chaos are proposed. From the experimental trends, it appears that a wide range of unstable limit cycles are extracted from chaotic electronic circuits like the Chua's circuit and the Lorenz system.

Keywords: Chaos control, nonlinear circuits.

1 Introduction

The concept of taming chaos is related to the stabilization of the unstable limit cycles included in a strange attractor. The essential feature of taming chaos is that the control strategy has to be as less invasive as possible exploiting a low level forcing signal. Numerous strategies have been studied [1,2] including the use of analog devices or digital control systems. In any case the control strategy must be consistent in the sense that the original dynamics of the system must be essentially maintained. This means that the control law must be not destructive with respect to the internal dynamics of the system [3,4,5]. In this paper, it is presented an experimental strategy that allow us to tame chaos by using an impulsive signal transmitted to the controlled circuit by using a transformer. This allowed us to decouple the main circuit from the control signal generator. The obtained results are shown as a gallery of limit cycles that can be extracted from the controlled electronic circuits.

The proposed control strategy can be considered as the introduction of a small perturbation to the bifurcation parameter from a circuital point-of-view. This resembles classical chaos control techniques which are based on parameter perturbation with sinusoidal signal, but in this case the control is performed acting directly on a circuital level.

The study is developed to the Chua's circuit and to the Lorenz system. Moreover it is general. The paper is organized as follows: in Sec. 2 the control strategy is presented, a fundamental section is dedicated to the experimental results. The conclusive remarks will follow.

2 Strategy for Taming Chaos

The new strategy introduced in this paper for taming chaos is based on the introduction of a weak control signal in a given electronic circuit implementing a

V.M. Mladenov and P.C. Ivanov (Eds.): NDES 2014, CCIS 438, pp. 15–21, 2014.

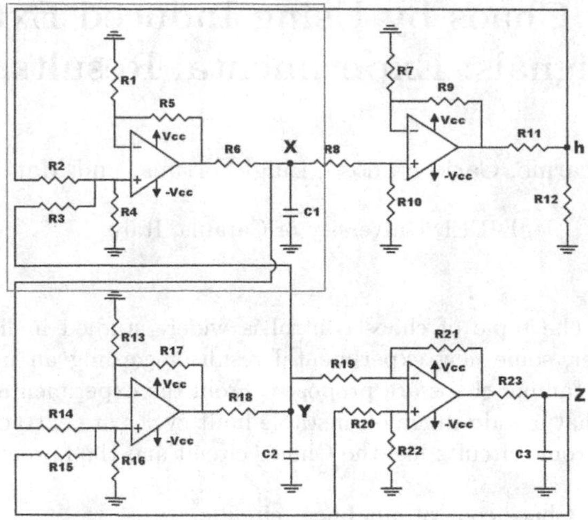

Fig. 1. Circuit implementation of the Chua's circuit. Components: $R_1 = 4k\Omega$, $R_2 = 13.3k\Omega$, $R_3 = 5.6k\Omega$, $R_4 = 20k\Omega$, $R_5 = 20k\Omega$, $R_6 = 380\Omega$ (potentiometer), $R_7 = 112k\Omega$, $R_8 = 112k\Omega$, $R_9 = 1M\Omega$, $R_{10} = 1M\Omega$, $R_{11} = 12.1k\Omega$, $R_{12} = 1k\Omega$, $R_{13} = 51.1k\Omega$, $R_{14} = 100k\Omega$, $R_{15} = 100k\Omega$, $R_{16} = 100k\Omega$, $R_{17} = 100k\Omega$, $R_{18} = 1k\Omega$, $R_{19} = 8.2k\Omega$, $R_{20} = 100k\Omega$, $R_{21} = 100k\Omega$, $R_{22} = 7.8k\Omega$, $R_{23} = 1k\Omega$, $C_1 = C_2 = C_3 = 100nF$, $V_{cc} = 9V$.

chaotic dynamics. We considered two paradigmatic examples of chaotic circuits, i.e. the Chua's circuit and the Lorenz system.

Let us consider the Chua's circuit implementation based on State Controlled Cellular Nonlinear Networks [6], as shown in Fig. 1. This circuit is governed by the following equations:

$$\begin{cases} C_1 R_6 \frac{dX}{d\tau} = -X + \frac{R_5}{R_3} Y + \frac{R_5}{R_2} h \\ C_2 R_{18} \frac{dY}{d\tau} = -Y + \frac{R_{17}}{R_{14}} X + \frac{R_{17}}{R_{15}} Z \\ C_3 R_{23} \frac{dZ}{d\tau} = -Z + \frac{R_{21}}{R_{20}} Z + \frac{R_{21}}{R_{19}} Y \end{cases} \tag{1}$$

where:

$$h = \frac{R_{12}}{R_{11}+R_{12}} \frac{R_9}{R_8} (|X+1| - |X-1|) \tag{2}$$

The component values, reported in the caption of Fig. 1, are chosen in order to match Eqs. (1) with the adimensional equations of the Chua's circuit:

$$\begin{cases} \dot{x} = \alpha(y - h(x)) \\ \dot{y} = x - y + z \\ \dot{z} = -\beta y \end{cases} \tag{3}$$

where $\beta = 14.286$ and $h(x)$ represents the nonlinearity of the system:

$$h(x) = m_1 + \frac{1}{2}(m_0 - m_1)(|x+1| - |x-1|) \tag{4}$$

with $m_0 = -\frac{1}{7}$ and $m_1 = \frac{2}{7}$.

A temporal rescaling $\kappa = \frac{1}{C_2 R_{18}} = \frac{1}{C_3 R_{23}} = 10000$ has been introduced and the different dynamical behavior shown by the Chua's circuit can be observed by varying the single bifurcation parameter α implemented in the circuit by resistor R_6, according to the relation $\alpha = \frac{R_5}{R_3}\frac{R_{18}}{R_6}$.

The external control signal is provided to the circuit introducing a slight modification in the first integrator as reported in Fig. 2: the secondary winding of a transformer with a fixed voltage ratio, which in the following is considered equal to $\frac{1}{10}$, is inserted in series with the resistor R_6, i.e. the resistor directly controlling the bifurcation parameter. The primary winding of the transformer is connected to a signal generator providing a square pulse wave, with an amplitude $A = 1V$ and a 20% duty cycle. Varying the frequency of this control signal, we are able to stabilize the limit cycles included in the chaotic attractor, as the results presented in the following section will clearly demonstrate.

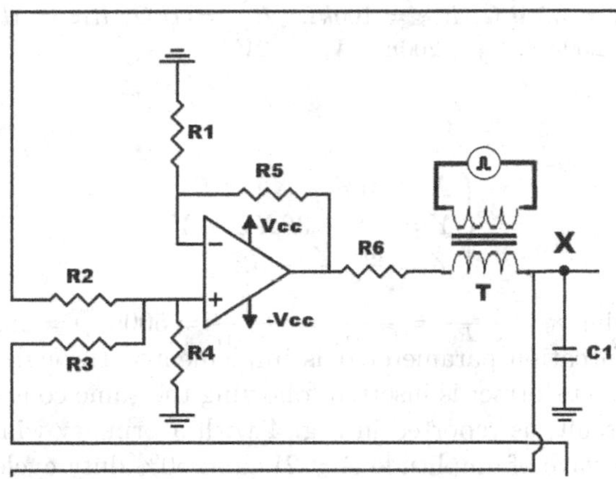

Fig. 2. Circuit scheme for the control strategy obtained including transformer T in the first CNN cell of the Chua's circuit implementation reported in Fig. 1

In order to emphasize the generality of the proposed control approach, a further experiment has been considered starting from the circuital implementation of the well-known Lorenz system. The circuit shown in Fig. 3 obeys to the following equations:

$$\begin{cases} C_1 R_5 \frac{dX}{d\tau} = -X - \frac{R_4}{R_1}X + \frac{R_4}{R_2}X + \frac{R_4}{R_3}Y \\ C_2 R_{11} \frac{dY}{d\tau} = -Y - \frac{R_{10}}{R_7}XZ + \frac{R_{10}}{R_8}X \\ C_3 R_{17} \frac{dZ}{d\tau} = -Z - \frac{R_{16}}{R_{13}}Z + \frac{R_{16}}{R_{14}}XY \end{cases} \quad (5)$$

whose components are chosen according to the original Lorenz dynamics in which an amplitude rescaling has been introduced in order to avoid saturations imposed by voltage supply. In fact the values reported in the caption of Fig. 3 match Eqs. (5) with the following set of dynamical equations:

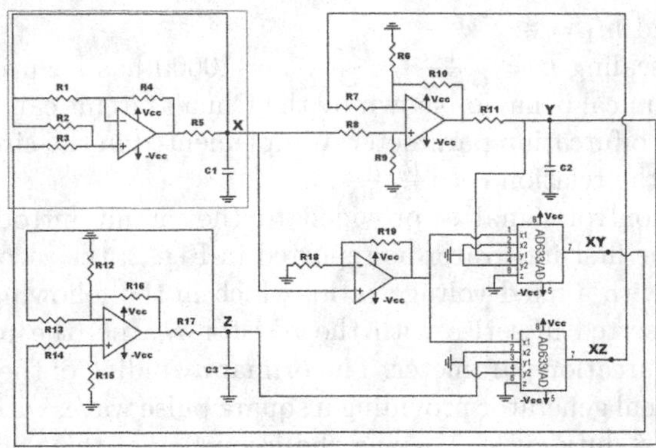

Fig. 3. Circuital implementation of the Lorenz system. Components: $R_1 = 10k\Omega$, $R_2 = 100k\Omega$, $R_3 = 10k\Omega$, $R_4 = 100k\Omega$, $R_5 = 1k\Omega$, $R_6 = 5.6k\Omega$, $R_7 = 3.3k\Omega$, $R_8 = 3.6k\Omega$, $R_9 = 3.19k\Omega$, $R_{10} = 100k\Omega$, $R_{11} = 1k\Omega$, $R_{12} = 3.3k\Omega$, $R_{13} = 37.5k\Omega$, $R_{14} = 3.3k\Omega$, $R_{15} = 3.74k\Omega$, $R_{16} = 100k\Omega$, $R_{17} = 1k\Omega$, $R_{18} = 1k\Omega$, $R_{19} = 9k\Omega$, $C_1 = 200nF$, $C_2 = 200nF$, $C_3 = 200nF$, $V_{cc} = 9V$.

$$\begin{cases} \dot{X} = \alpha(Y - X) \\ \dot{Y} = \rho X - 30XZ - Y \\ \dot{Z} = \frac{100XY}{30} - \beta Z \end{cases} \tag{6}$$

fixing a time scaling $\kappa = \frac{1}{C_1 R_5} = \frac{1}{C_2 R_{11}} = \frac{1}{C_3 R_{17}} = 5000$, $\rho = 28$, and $\beta = \frac{8}{3}$. In this case, the bifurcation parameter α is implemented through represented by resistor R_5. The transformer is inserted following the same considerations made for the Chua's circuit, as reported in Fig. 4 with a primary winding connected to a square pulse wave of amplitude $A = 2V$ and 50% duty cycle.

Initially the bifurcation parameter of the controlled circuit is fixed in order to observe a stable limit cycle, turning off the external control signal. The starting limit cycle is considered as a *gene* from which, varying the frequency of the control signal applied to the primary winding of the transformer, various limit cycles, of different periodicity, can be stabilized. We will now explore the effect of the different frequency on the circuit behavior also with respect to initial genes of different periodicity.

3 Experimental Trends

In the following, a gallery of the different stable limit cycles obtained starting from a given *gene* is reported for the two considered examples.

As concerns the Chua's circuit, whose attractor for $\alpha = 9$ is reported in the first row of Tab. 1, we started from limit cycles with different periodicity selected varying the bifurcation parameters. The effect of the frequency of the control signal is summarized in Tab. 1.

Table 1. Taming chaos in Chua's circuit. Different limit cycles obtained starting from seven genes obtained at the given frequency of the control signal.

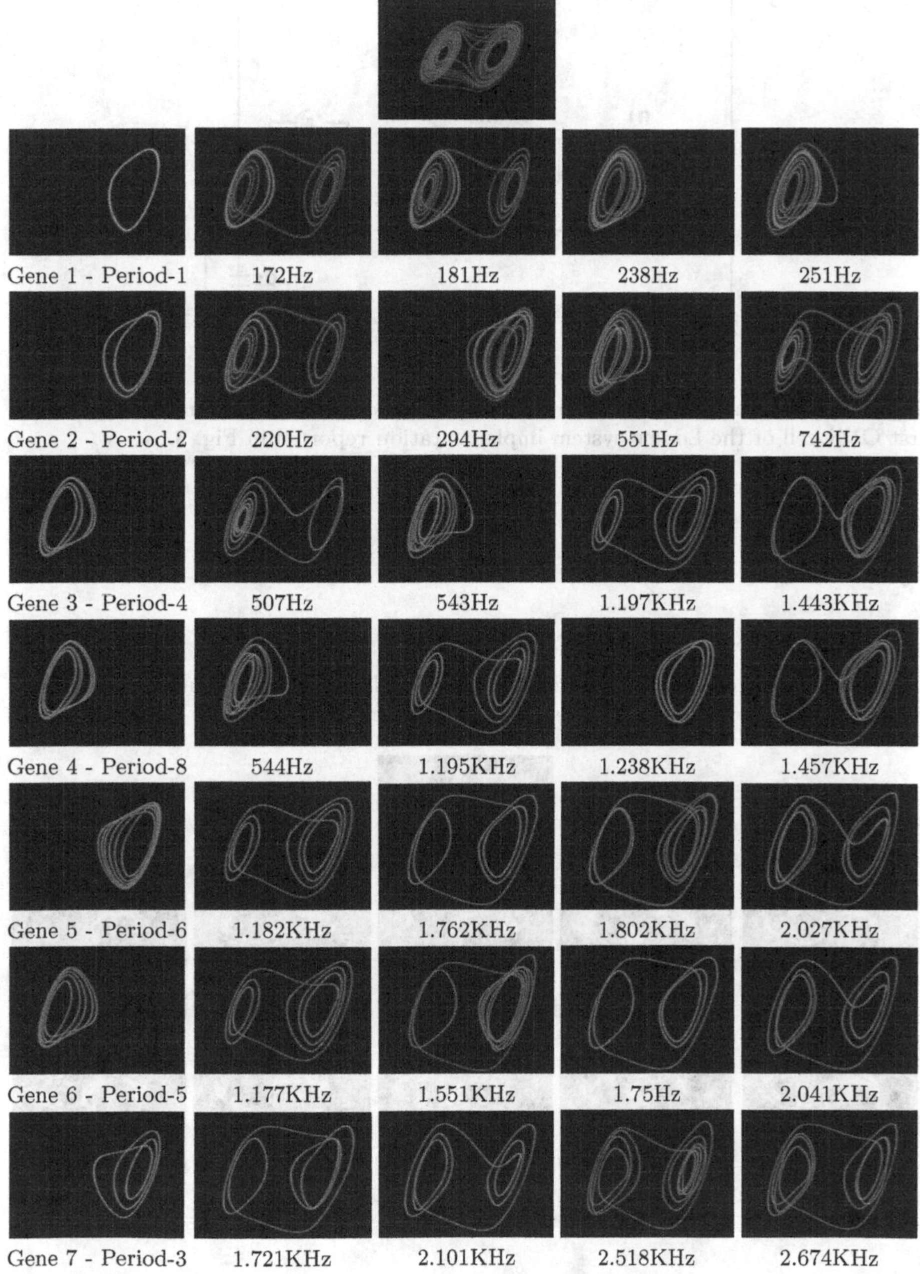

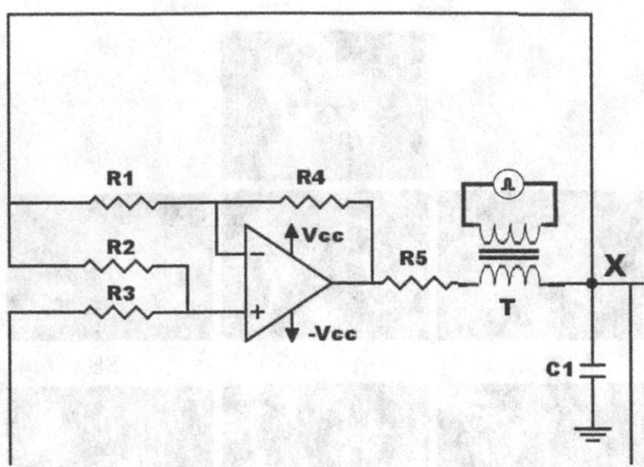

Fig. 4. Circuit scheme for the control strategy obtained including transformer T in the first CNN cell of the Lorenz system implementation reported in Fig. 3

Table 2. Taming chaos in Lorenz circuit. Different limit cycles obtained starting from two genes obtained at the given frequency of the control signal.

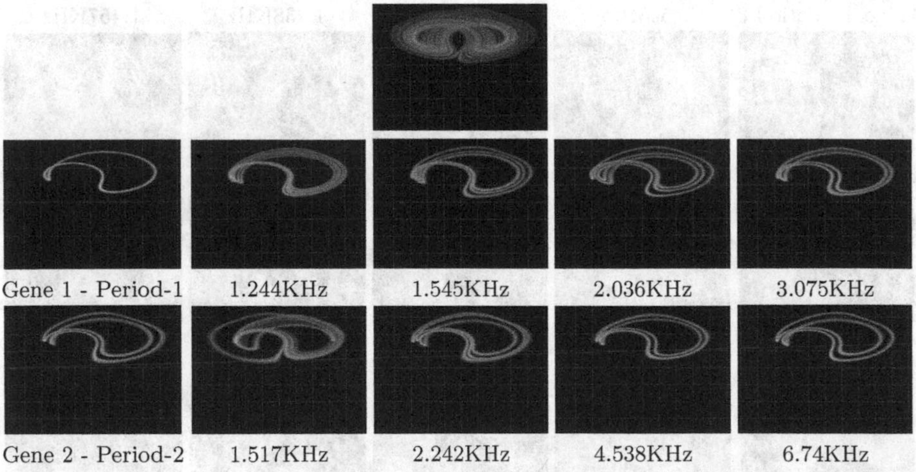

Gene 1 - Period-1	1.244KHz	1.545KHz	2.036KHz	3.075KHz
Gene 2 - Period-2	1.517KHz	2.242KHz	4.538KHz	6.74KHz

The same experiments have been performed for the Lorenz circuit, whose attractor for $\alpha = 10$ is shown in the first row of Tab. 2, starting from two different limit cycles. The stabilized limit cycles at different frequencies of the control signal are reported in Tab. 2.

4 Conclusions

A new strategy for taming chaos based on the introduction of a low level signal in an electronic circuit has been proposed. The strategy exploits a two windings transformer in which the secondary is inserted in the controlled circuit in series with the resistor implementing the bifurcation parameter, while the primary windings is driven by a square pulse wave with a fixed amplitude and duty cycle. Starting from an initial limit cycle, which we called gene cycle, a set of unstable limit cycles included in the chaotic attractor can be stabilized varying the frequency of the control signal. The results obtained in our experiments allows to assess that the proposed approach is effective and general.

Moreover, analyzing the gallery of different limit cycles obtained starting from different genes, it can be noticed that the same limit cycle can be stabilized at different frequencies of the input signal, provided that we start from genes with different periodicity. This latter consideration allows to postulate that there is a coherence between different control laws acting on slightly different systems. A deeper investigation, both numerical and experimental, on the relationship between these two characteristics can be performed, demonstrating the existence of a close connection.

References

1. Chen, G., Yu, X.: Chaos Control, Theory and Applications. Springer (2003)
2. Schuster, H.G.: Handbook of Chaos Control. Wiley (2006)
3. Nijmeijer, H.: Control of chaos and synchronization. Systems & Control Letters 31, 259–262 (1997)
4. Ditto, W.L., Rauseo, S.N., Spano, M.L.: Experimental control of chaos. Physical Review Letters 65, 3211 (1990)
5. Wu, T., Chen, M.: Chaos control of the modified Chua's circuit system. Physica D: Nonlinear Phenomena 164, 53–58 (2002)
6. Fortuna, L., Frasca, M., Xibilia, M.G.: Chua's Circuit Implementations: Yesterday, Today, Tomorrow, World Scientific (2009)

A Study on Performance Improvement of Natural Synchronization Scheme Base on Noise-Induced Synchronization Theory

Hiroyuki Yasuda and Mikio Hasegawa*

Department of Electrical Engineering, Tokyo University of Science,
6-3-1 Niijuku, Katsushika, Tokyo, Japan
hirobacon@haselab.ee.kagu.tus.ac.jp, hasegawa@ee.kagu.tus.ac.jp

Abstract. The phases of nonlinear limit-cycle oscillators can be synchronized by adding identical noise sequence to each of them, which is called noise-induced synchronization. In our previous researches, we have shown that the nonlinear oscillators synchronize even by adding natural environmental noise or fluctuations. Because the natural fluctuations, such as temperature, humidity and sounds, measured by the sensors located in nearby have high cross-correlation, the oscillators installed in independent devices can be synchronized by those natural signals. In this paper, we investigate the synchronization performance with changing the parameters of our proposed method, such as the noise input amplitude and the input interval. Our experimental results show that strong noise is effective for quick synchronization and weak noise is effective for sustention of the synchronization, and the performance of the natural synchronization can be improved by tuning the parameters.

Keywords: Synchronization, Noise-induced synchronization, Environmental noise, Nonlinear oscillator.

1 Introduction

The phases of nonlinear limit cycle oscillators can be synchronized by adding identical noise to each of them. This phenomenon has been called noise-induced phase synchronization[1,2]. Based on this phenomenon, we have proposed a natural synchronization scheme that synchronizes uncoupled devices, on which the nonlinear limit-cycle oscillator is installed. We have already shown feasibility of the proposed scheme using several natural fluctuations, such as the humidity of the air, the temperature, environmental sounds and so on [3,4,5]. Those natural environmental fluctuations obtained at the neighboring devices have high similarity. By adding such similar fluctuations to each device independently, our proposed scheme realizes natural synchronization of those devices, without any interactions or exchanges of the signals between them.

* Corresponding author.

V.M. Mladenov and P.C. Ivanov (Eds.): NDES 2014, CCIS 438, pp. 22–29, 2014.

One of application examples of the proposed synchronization method is time synchronization of the wireless sensor network devices. For the wireless sensor networks, synchronization of the sensor devices is important to reduce power consumption. Because it is hard task to replace the batteries of a large number of battery-powered wireless sensor devices, it is important to develop a low power consumption protocol. One of the approaches is intermittent data transmission with extending sleep time using synchronized sensor nodes. Simplest synchronization schemes are to exchange clock timing information between the sensor devices, to receive GPS time signals, and so on. However, those schemes have overheads in power consumption or requires an additional transmitter for the time signals. In our proposed method, it is possible to remove those overheads, because time synchronization can be achieved by natural environmental fluctuations that are collected at each sensor independently. By experiments using real data, such as the temperature, humidity of the air, we have already shown that synchronization can be achieved by the proposed scheme [3,4]. In real situations, achievement of synchronization should be as fast as possible, with maintaining the synchronization for a long time.

In this paper, we investigate the parameter dependency of the synchronization performance of the proposed scheme, in order to optimize the performance. We change the noise amplitude and the interval of noise input, and investigate the synchronization performance. We use the white Gaussian noise and a real natural data, the humidity and the environmental sound, as the input noises. We investigate synchronization performance by evaluating the success rate of synchronization within a certain period of time and the sustention rate after achieving synchronization.

2 Noise-Induced Phase Synchronization Theory

This section shows the theory of the noise-induced synchronization phenomenon, which is the base of our proposed natural synchronization scheme. We define the dynamics of the nonlinear oscillator using the following an ordinary differential equation,

$$\dot{\mathrm{X}}(t) = \mathrm{F}(\mathrm{X}). \tag{1}$$

The dynamics of its phase, $\theta(t)$ can be defined as follows using the angular frequency ω,

$$\dot{\theta}(t) = \omega. \tag{2}$$

Here, we consider synchronization of two limit cycle oscillators, which have common noise input to both. The dynamics of limit cycle oscillators with the Gaussian white noise $\xi(t)$, as the common noise, are expressed as follows,

$$\dot{\mathrm{X}}_1(t) = \mathrm{F}(\mathrm{X}_1) + \xi(\mathrm{t}), \tag{3}$$

$$\dot{\mathrm{X}}_2(t) = \mathrm{F}(\mathrm{X}_2) + \xi(\mathrm{t}). \tag{4}$$

The phase of these oscillators with the common noise can be expressed as follows,

$$\dot{\theta}_1(t) = \omega + Z(\theta_1)\xi(t) \tag{5}$$

$$\dot{\theta}_2(t) = \omega + Z(\theta_2)\xi(t), \tag{6}$$

where $Z(\theta) = grad_X\theta(X)|_{X=X_0(\theta)}$, which is called the phase sensitivity function. Here, we define the phase difference of these two oscillators as $\phi = \theta_1 - \theta_2$.

The linear growth rate (average Lyapunov exponent) of the phase difference ϕ can be calculated as follows,

$$\begin{aligned} \Lambda &= \left\langle \frac{d}{dt} ln|\phi(t)| \right\rangle = \epsilon^2 \left\langle Z''(\theta(t)) Z(\theta(t)) \right\rangle \\ &\cong \frac{\epsilon^2}{2\pi} \int_0^{2\pi} Z''(\theta) Z(\theta) d\theta \\ &= -\frac{\epsilon^2}{2\pi} \int_0^{2\pi} Z'(\theta)^2 d\theta \leq 0. \end{aligned} \tag{7}$$

Because the linear growth rate of the phase difference is smaller than 0 in average, the phase difference decreases. Thus, the phase of two limit cycle oscillators can be synchronized by adding a common identical noise to both.

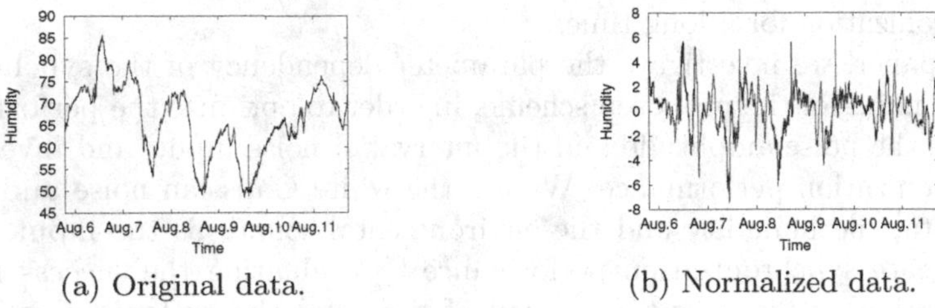

(a) Original data. (b) Normalized data.

Fig. 1. Time series of the humidity data and its normalized data

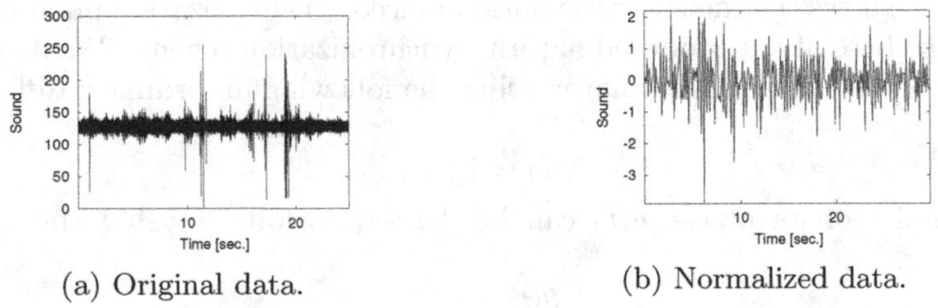

(a) Original data. (b) Normalized data.

Fig. 2. Time series of the environmental sound and its normalized data

3 Time Synchronization Method Using Natural Environmental Fluctuations as the Additive Sequences of Noise-Induced Synchronization

In our proposed scheme, we apply natural fluctuations to the nonlinear oscillators as the common noise of noise-induced synchronization phenomenon. As the natural environmental fluctuations, we use temperature or humidity of the air, environmental sounds and so on. Even if the input noises are not completely same, noise-induced synchronization can be achieved if the noises have high cross-correlation among them. It has shown by real experiments that the natural fluctuations within the neighborhood has high cross-correlation and synchronization can be achieved [3,4].

In the following experiments, we use the real natural environmental fluctuations, the humidity and the environmental sound. In order to collect the humidity data, we use 8 sensor devices of the wireless sensor networks, MICAz [6]. We measured the real natural data at outside corridor of Kudan Building of Tokyo University of Science. The distances between the sensors are around 10 to 30m. To collect the environmental sound, we use a voice recorder. We recorded the sounds at a park in Katsushika, Tokyo. Figs. 1(a) and 2(a) show examples of the measured data.

From Figs. 1(a) and 2(a), we found that the mean, standard deviation, and fluctuation period are different depending on the data. In the theoretical proof of noise-induced synchronization, the input noise has been assumed to have constant average. Therefore, it is necessary to normalize the amplitude of noise with adjusting the sampling frequency of the environmental fluctuations. This normalization must be an online algorithm, which can be run in real systems. In order to keep constant average, we use the time-average and the moving average for humidity. For the environmental sounds, we use the number of zero crossing in a certain period. We use difference of the zero crossing counts for the normalized data. We use normalized data as shown in Figs. 1 and 2, for the humidity and the environmental sound, respectively.

Fig. 3 shows the feasibility of the proposed scheme using natural environmental fluctuations. It shows the differences of the phases of two uncoupled oscillators. We use the two different normalized humidity data collected by different wireless sensor devices at the same time. From Fig. 3, it can be seen that the phase difference between the oscillators is reduced, and synchronization can be achieved by the natural environmental fluctuations.

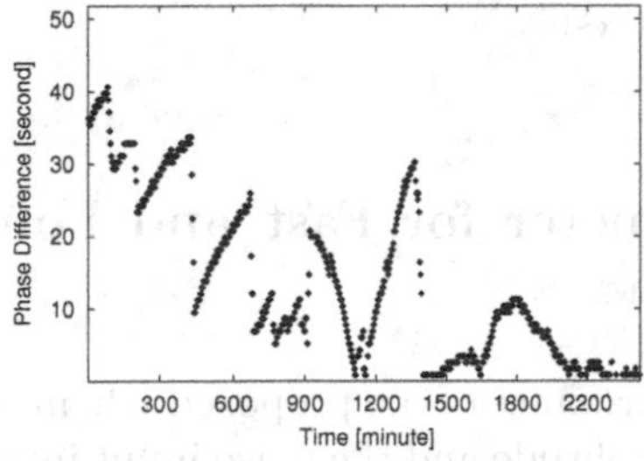

Fig. 3. Time series of the difference of phases of two nonlinear limit-cycle oscillators. The two different normalized humidity data are added independently to each oscillators at the same time.

In our proposed synchronization system, each device runs the nonlinear oscillator in themselves. Each device adds the environmental data collected by their own sensors to the oscillator running on it. If the devices are located in near-by and the environmental data obtained by their sensors have high correlation, the oscillators running on the devices will synchronize by noise-induced synchronization phenomenon. The internal clocks of the devices can be also synchronized by adjusting them according to the phase of the nonlinear oscillator. By using this proposed scheme, time synchronization between the devices becomes possible only by natural signal, without any artificial signal exchange.

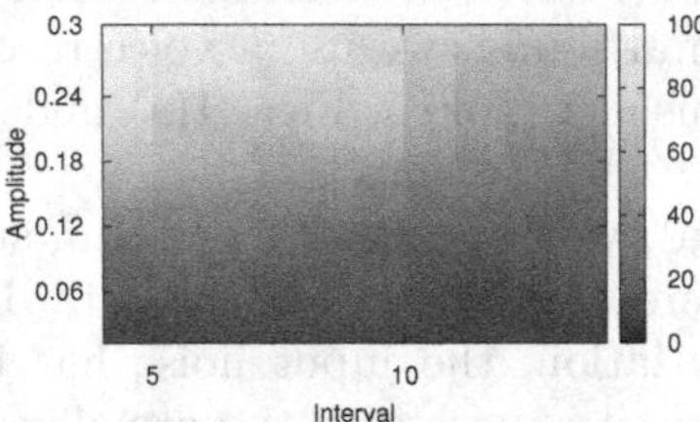

Fig. 4. Dependency of the success rate of synchronization on the noise input interval and the noise amplitude. The white Gaussian noise sequences with high cross-correlation are used for this result.

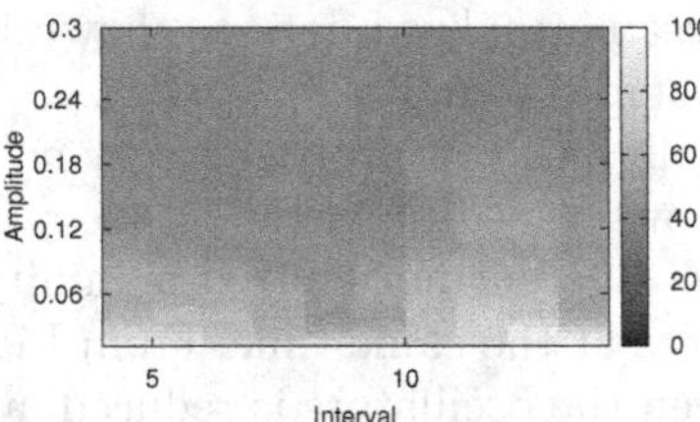

Fig. 5. Dependency of the sustention rate of synchronization on the noise input interval and the noise amplitude. The white Gaussian noise sequences with high cross-correlation are used for this result.

4 Optimal Parameter for Fast and Long-Term Synchronization

The synchronization performance of the proposed scheme depends on the parameters, such as the noise amplitude and the noise input interval. Those parameters have to be optimized to achieve high performance. Here, we analyze the parameter dependency of the synchronization performance and investigate a parameter setting scheme to improve the synchronization performance for given natural fluctuations.

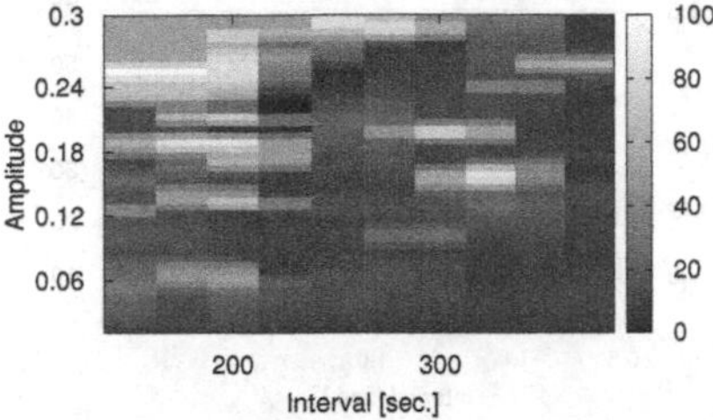

Fig. 6. Dependency of the success rate of synchronization on the noise input interval and the noise amplitude. Natural fluctuation, humidity time series, are used for this result.

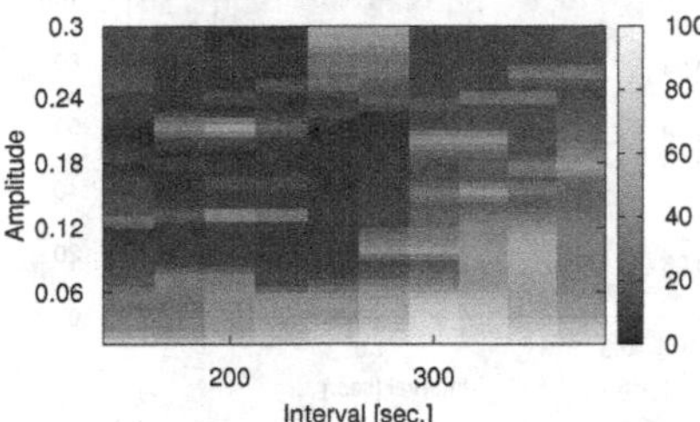

Fig. 7. Dependency of the sustention rate of synchronization on the noise input interval and the noise amplitude. Natural fluctuation, humidity time series, are used for this result.

At first, we investigate dependency of the synchronization performance on the noise input interval and the noise amplitude, with using the white Gaussian noise. We use the FitzHugh-Nagumo oscillators as a limit-cycle nonlinear oscillator. In the following simulations, each state of the oscillators are calculated in a discrete manner by the 4th order Runge-Kutta method. Each natural environmental data is normalized to make their standard deviation same as the oscillator. The achievement of synchronization is defined as that keeping phase difference less than 0.5% of the period of the limit-cycle oscillator. Synchronization performance is evaluated by the success rate and the sustention rate of the synchronization. The success rate is defined as the percentage of achieving synchronization at least once in each run of simulation. Sustention rate is defined as the percentage of time that keeps synchronization.

Figs. 4 and 5 show the dependency of the success rate and the sustention rate on the parameter setting. The vertical axis shows amplitude of the noise and the horizontal axis shows interval of noise input. The color bars of the figures show the success rate and the sustention rate of synchronization. From the Figs. 4 and 5, we can see that the success rate is high when the noise amplitude is strong, and the sustention rate is high when the noise amplitude is weak. Therefore, it is found that there is an optimal parameter set for high achievement rate and high sustention rate.

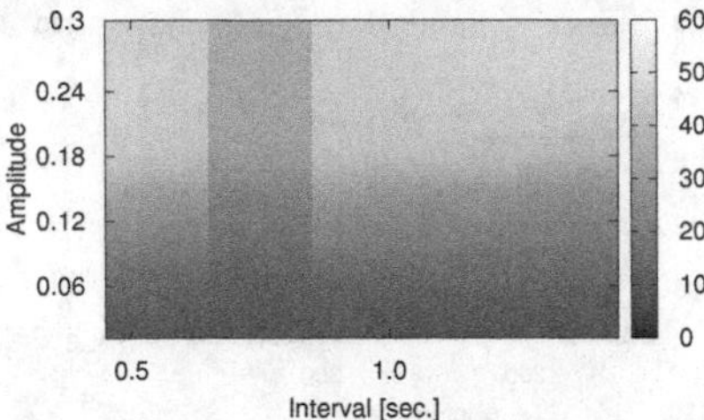

Fig. 8. Dependency of the success rate of synchronization on the noise input interval and the noise amplitude. Natural fluctuation, environmental sound time series, are used for this result.

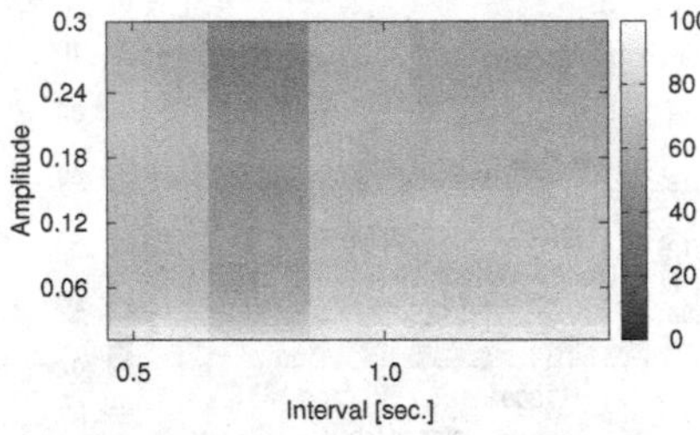

Fig. 9. Dependency of the sustention rate of synchronization on the noise input interval and the noise amplitude. Natural fluctuation, environmental sound time series, are used for this result.

Secondly, we examine the synchronization performance using natural fluctuations, the humidity and the environmental sounds. We use the humidity data of 30 days which was collected by wireless sensor networks that have been placed in 6 outdoor points of Kundan building of Tokyo University of Science. For the environmental sounds, we use 3 hours data, which was recorded at 8 outdoor point in a park in Katsushika, Tokyo.

Figs. 6–9 shows the success rate and sustention rate with changing the interval and amplitude of input noise, respectively. The amplitude of noise input is varied from 0.002 to 0.1 times the standard deviation of the nonlinear oscillator amplitude. The intervals of inputs are varied from 133 to 400 seconds for the case of the humidity, and from 0.5 seconds to 1.5 seconds for the case of the environmental sound. Similar to the case of the white Gaussian noise, the success rate is high when the amplitude is strong, and sustention rate is high when the amplitude is weak. From figs. 6 and 7, achievement rate and sustention rate can be optimized in the case of that the amplitude is 0.1 and the interval is 266 seconds, for the humidity data. Similarly, from fig 8 and 9, synchronization performance can be maximized when the value of amplitude set to 0.1 and the interval set to 1.3 seconds, for the environmental sounds. These results show that the synchronization performances of the proposed natural synchronization scheme can be improved by tuning the parameters.

5 Conclusion

In this paper, we investigate the parameter dependency of synchronization performance of the proposed natural synchronization scheme, which is based on the noise-induced synchronization phenomenon. We examined the dependency of the performance on the amplitude and the interval of the noise inputs. Our results show that there are optimal settings of the parameters of the amplitude and the interval of the inputs, and the performance of the proposed natural synchronization scheme can be improved. In addition, we have also found that the strong noises are effective for the fast success of synchronization, and the weak noise is effective for the long sustention of keeping synchronized state.

As our future work, in order to realize the synchronization with higher accuracy, we will introduce other natural environmental signals that have high cross-correlation and higher frequencies. We also would like to develop the autonomous parameter adjustment method for the oscillators and the natural fluctuations, to further improve the performance of the proposed scheme.

References

1. Teramae, J., Tanaka, D.: Robustness of the Noise-Induced Phase Synchronization in a General Class of Limit Cycle Oscillators. Physical Review Letter 93, 204103 (2004)
2. Nakao, H., Arai, K., Kawamura, Y.: Noise-induced synchronization and clustering in ensembles of uncoupled limit-cycle oscillators. Physical Review Letter 98, 184101 (2007)
3. Yasuda, H., Hasegawa, M.: Natural Synchronization of Wireless Sensor Networks by Noise-Induced Phase Synchronization Phenomenon. IEICE Trans. Commun. E96-B(11), 2749–2755 (2013)
4. Harashima, M., Yasuda, H., Hasegawa, M.: Synchronization of Wireless Sensor Networks using Natural Environmental Signals Based on Noise-Induced Phase Synchronization Phenomenon. In: Proc. IEEE Vehicular Technology Conference (Spring 2012)
5. Honda, Y., Yasuda, H., Hasegawa, M., Nakao, H., Aihara, K.: Time Synchronization Scheme based on Noise-Induced Synchronization using Environmental Sound. In: Proc. RISP International Workshop on NCSP, 5PM1-3-2 (2013)
6. Hill, J., Culler, D.: Mica: A Wireless Platform for Deeply Embedded Networks. IEEE Micro 22, 12–24 (2002)

Experiments on an Ensemble of Globally and Nonlinearly Coupled Oscillators

Amirkhan A. Temirbayev[1], Zeunulla Zh. Zhanabaev[2], Yerkebulan Nalibayev[2], Aisha Zh. Naurzbayeva[2], and Akmaral K. Imanbayeva[2]

[1] Laboratory of Engineering Profile, al-Farabi Kazakh National University, al-Farabi avenue 71, 050040, Almaty, Kazakhstan
[2] Institute of Experimental and Theoretical Physics, al-Farabi Kazakh National University, al-Farabi avenue 71, 050040, Almaty, Kazakhstan
akmaral@physics.kz

Abstract. In this paper we perform an experimental setup of the electronic oscillator ensemble with global and nonlinear coupling. Using the experimental setup with 72 units, we systematically analyze the ensemble dynamics for the cases of linear and nonlinear coupling. With an increase in the coupling strength we first observe formation and then destruction of a synchronous cluster, so that the dependence of the order parameter on the coupling strength is not monotonic. After destruction of the cluster the ensemble remains nevertheless coherent, i.e., it exhibits an oscillatory collective mode (mean field). We show that the system is now in a self-organized quasiperiodic state. In this state, frequencies of all oscillators are smaller than the frequency of the mean field, so that the oscillators are not locked to the mean field they create and their dynamics is quasiperiodic.

Keywords: Synchronization, oscillator ensemble, global coupling, nonlinear coupling.

1 Introduction

Collective dynamics of large ensemble of interacting units received a lot of attention within last decades [1-5]. However, this subject remains in the focus of attention of many researchers, with an emphasis on such aspects as clustering [6], effects of internal delays [7], effects of external forcing [8] or feedback [9], interaction of several populations [10], etc. The main effect of the global coupling is the well-understood emergence of collective synchrony, reflected in the increase in the mean-field amplitude with the interaction strength and often referred to as the Kuramoto transition. Further well-known effects are clustering [11] and chaotization of the mean field [12, 13]. Experiments on globally coupled oscillators have been performed by Hudson, Kiss, and collaborators [14]. Using an ensemble of 64 electrochemical oscillators, they have confirmed most theoretical predictions. In particular, they have demonstrated the Kuramoto transition in ensembles of periodic and chaotic

V.M. Mladenov and P.C. Ivanov (Eds.): NDES 2014, CCIS 438, pp. 30–36, 2014.

oscillators. Other laboratory experiments have been conducted with Josephson junctions, photochemical oscillators, and vibrating motors on a common support [15].

In [16] for the first time predicted very interesting state, so-called self-organized quasiperiodic state (SOQ). In this state, the frequencies of oscillators differ from the frequency of the mean field; the oscillators are not entrained by the field and therefore demonstrate a quasiperiodic dynamics.

The primary goal of this work is experimental verifications of these results. For this purpose, we performed experiments with electronic oscillators, globally coupled via a common feedback loop with a phase-shifting unit. The coupling is nonlinear in the sense that the phase shift depends on the amplitude of the collective oscillation. We demonstrate, with an increase in the strength of the global coupling, a transition from an asynchronous state to collective synchrony and then to SOQ. Typically the tendency to synchrony increases with the coupling strength. However, in some setups the increase of the coupling parameter makes the initially attractive interaction repulsive, leading to the breakup of synchrony. As a result, the system undergoes a transition either to an asynchronous state or to a state of partial synchrony. In the latter case, the system stays at the border between synchrony and asynchrony and exhibits interesting dynamics, in particular, SOQ states.

2 Experimental Setup and Main Results

In this section we describe our setup with 72 globally coupled electronic generators. First we present the implementation of an individual unit. Next, we discuss organization of the linear and the nonlinear global coupling and of the common external forcing.

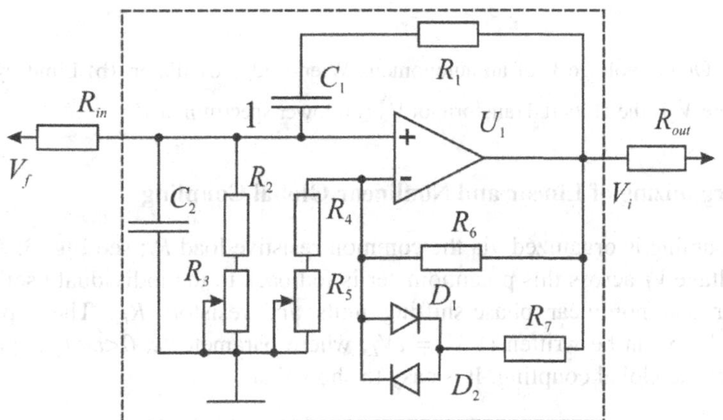

Fig. 1. Wien-bridge oscillator. Here V_i is the output voltage of the *i*-th oscillator and V_f is the output voltage of the global feedback loop (cf. Fig. 3).

2.1 Wien-Bridge Oscillator

A scheme of an individual generator is given in Fig. 1; it represents a nonlinear amplifier with a positive frequency-dependent feedback via the Wien bridge. The amplifier is implemented by the operational amplifier U_1; resistors R_4, R_5, R_6, R_7; and diodes D_1, D_2. The Wien bridge consists of resistors R_1, R_2, R_3 and capacitors C_1, C_2. These elements determine the frequency of the oscillation. Fine frequency tuning is performed by the trimmer resistor R_3, so that all oscillators in the ensemble have close frequencies ≈1.1 kHz. With the help of the trimmer resistor R_5 the amplitudes of all uncoupled oscillators were tuned to approximately same value $V \approx 1.5\ V$; see Fig. 2.

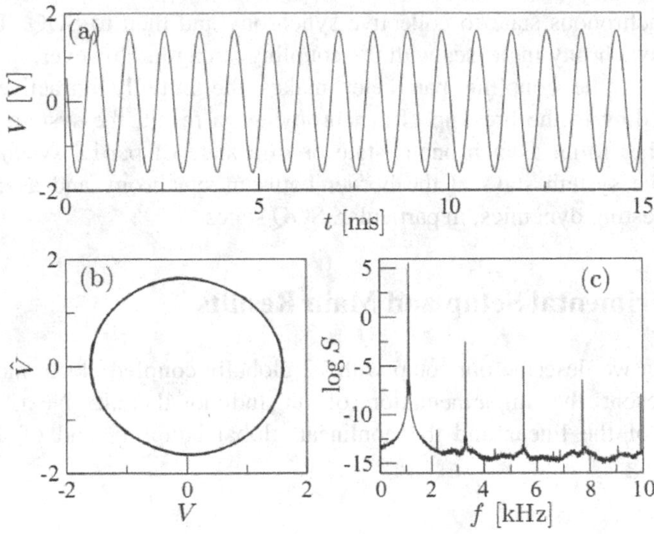

Fig. 2. (a) Output voltage V of an autonomous Wien-bridge oscillator. (b) Limit cycle of the system; here $\hat{V}$ is the Hilbert Transform of V. (c) Power spectrum of V.

2.2 Organizing of Linear and Nonlinear Global Coupling

Global coupling is organized via the common resistive load R_c; see Fig. 3. A fraction of the voltage V_L across this potentiometer is fed back to the individual oscillators via the linear and nonlinear phase-shifting units and resistors R_{in}. The input to the feedback loop can be written as $V_c = \varepsilon V_L$, where parameter ε, $0<\varepsilon<1$, quantifies the strength of the global coupling. It is easy to show that

$$V_c = \varepsilon \frac{\sum_{i=1}^{N} V_i}{N + R_{out}/R_c}, \tag{1}$$

where V_i is the output voltage of the i-th oscillator. Since $R_{out} \ll NR_c$, we have $V_L \approx N^{-1}\sum_i V_i = V_{mf}$, where the subscript mf stands for the mean field. Thus, the $V_C \approx \varepsilon V_{mf}$ is of the mean-field type. The voltage V_c from the common load is fed back to all oscillators via the feedback loop, which includes either linear or both linear and nonlinear phase-shifting units. The linear subunit is an active all-pass filter which shifts the phase of the signal but keeps its amplitude. The nonlinear PSU is implemented by a high-pass first-order filter, where nonlinear properties of diodes provide a dependence of the phase shift between input and output on the amplitude of the input signal.

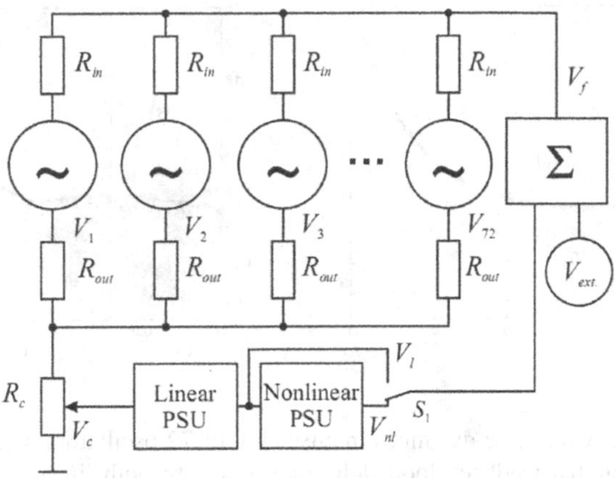

Fig. 3. Scheme of the globally coupled system. Individual generators are shown here by one symbol and a detailed scheme is given in Fig. 1. With the help of the switch, the nonlinear unit can be excluded from the feedback loop. The strength of the feedback is governed by the potentiometer R_c. Common forcing by the external voltage V_{ext} is organized via the summator, results with external forcing was shown in [18].

In our experiments we vary linear phase shift of the linear phase-shifting unit and the strength ε of the global coupling. For each set of parameters we record output voltages, V_i, for all $N = 72$ oscillators and the mean-field voltage, V_{mf}, across R_c. The sampling frequency is f_s = 20 kHz. In each measurement we make five recordings, with $M = 5 \times 10^4$ points per record.

First we perform the experiments with the linear PSU only. In Fig. 4 we present frequencies of individual oscillators and frequency of mean field in dependence on coupling strength ε for the particular settings of the liner phase shift $\approx 0.65\pi$. As expected, Fig.5 shows, that in the linear case we observe a monotonic growth of the order parameter R with the coupling strength ε. Due to the finite size of the ensemble, R is not small in the asynchronous state.

Now we discuss the case when the nonlinear PSU is witched on. The transitions for liner phase shift ≈ 0.65π shown in Figs. 4 and 5 (right side). We see that the system undergoes transitions between five different states. For small coupling system is asynchronous: order parameter is small and frequencies of oscillators remain unchanged. The oscillators synchronize for the coupling $\varepsilon \approx 0.5$, and then synchrony becomes unstable. The slow oscillators leave the synchronous group and the order parameter decreases. For sufficiently large coupling all oscillators are not entrained by the mean field. However, the mean field has a nonzero amplitude; i.e., the SOQ state emerges. The picture quantitatively coincides with the theoretical and numerical result for phase oscillators with uniform frequency distribution [19].

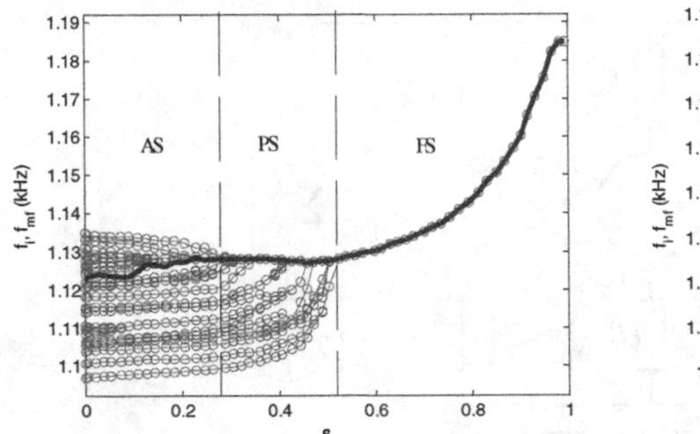

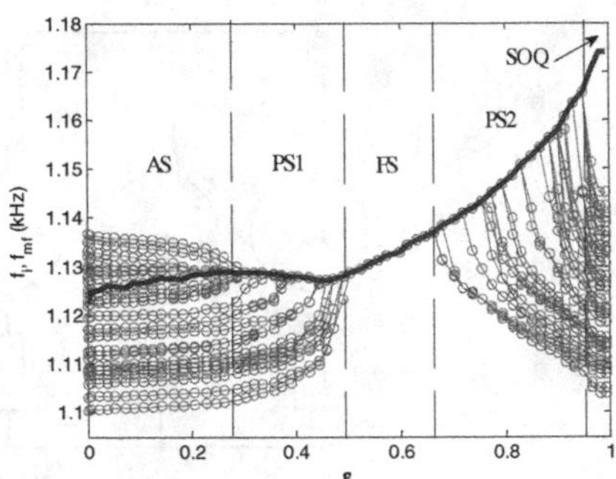

Fig. 4. (Color online) Collective dynamics in ensemble of 72 oscillators with the linear phase-shifting unit in the global feedback loop [left side] and with both linear and nonlinear PSUs [right side]; linear phase shift is $\gamma = 0.65\pi$. Frequencies of individual oscillators (circles, red online) and of the mean field (solid line, blue online). The vertical dotted lines separate different dynamical states: asynchrony (AS), partial synchrony (PS 1,2), full synchrony (FS), and self-organized quasiperiodicity (SOQ).

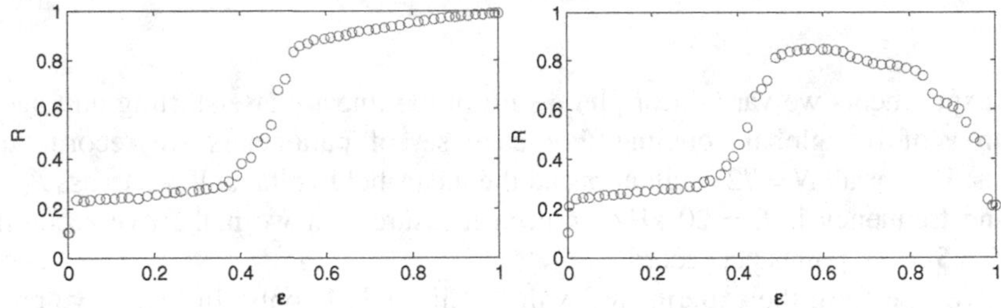

Fig. 5. (Color online) Collective dynamics in ensemble of 72 oscillators with the linear phase-shifting unit in the global feedback loop [left side] and with both linear and nonlinear PSUs [right side]; linear phase shift is $\gamma = 0.65\pi$. Order parameter: In the linear case (left) it grows monotonically with ε, but in the nonlinear case (right) the dependence is not monotonic.

3 Discussion

We have performed experiments with an ensemble of 72 globally coupled van der Pol–like electronic oscillators, treating the cases of linear and nonlinear coupling. The nonlinear coupling was implemented via a circuit with an amplitude dependent phase shift. We have observed synchronous ensemble dynamics, with all elements of the ensemble oscillating with a common frequency. Next, we have shown that in case of nonlinear feedback, increase of coupling results in synchrony breaking, but the ensemble remains in a coherent state. In this state, all oscillators have different frequencies, but, contrary to the simple case of asynchronous dynamics, their phases are distributed nonuniformly and therefore the oscillators produce a nonzero, coherent mean field whose frequency is larger than all oscillator frequencies. Thus, oscillators are not frequency locked to the mean field and therefore exhibit quasiperiodic behavior.

We believe that our results are relevant for investigation of other oscillator populations with amplitude-dependent phase shift or time delay in the global feedback loop. As a possible direction for future experiments we mention investigation of different forms of nonlinear coupling and of different scenarios of transitions from synchrony to asynchronous, though coherent states.

References

1. Kuramoto, Y., Araki, H.: International Symposium on Mathematical Problems in Theoretical Physics. Springer Lecture Notes Phys., vol. 39, p. 420. Springer, New York (1975); Chemical Oscillations, Waves and Turbulence. Springer, Berlin (1984)
2. Pikovsky, A., Rosenblum, M., Kurths, J.: Synchronization: A Universal Concept in Nonlinear Sciences. Cambridge University Press, Cambridge (2001)
3. Strogatz, S.H.: Sync: The Emerging Science of Spontaneous Order. Hyperion, New York (2003)
4. Ott, E.: Chaos in Dynamical Systems, 2nd edn. Cambridge University Press, Cambridge (2002)
5. Acebrón, J.A., Bonilla, L.L., Pérez Vicente, C.J., Ritort, F., Spigler, R.: Rev. Mod. Phys. 77, 137 (2005)
6. Kori, H., Kuramoto, Y.: Phys. Rev. E 63, 46214 (2001); Liu, Z., Lai, Y.-C., Hoppensteadt, F.C.: ibid. 63, 055201 (2001)
7. Niebur, E., Schuster, H.G., Kammen, D.M.: Phys. Rev. Lett. 67, 2753 (1991)
8. Baibolatov, Y., Rosenblum, M., Zhanabaev, Z.Z., Kyzgarina, M., Pikovsky, A.: Phys. Rev. E 80, 46211 (2009)
9. Rosenblum, M.G., Pikovsky, A.S.: Phys. Rev. Lett. 92, 114102 (2004)
10. Okuda, K., Kuramoto, Y.: Prog. Theor. Phys. 86, 1159 (1991)
11. Okuda, K.: Physica D 63, 424 (1993)
12. Matthews, P.C., Strogatz, S.H.: Phys. Rev. Lett. 65, 1701 (1990)
13. Olmi, S., Politi, A., Torcini, A.: Europhys. Lett. 92, 60007 (2010)
14. Wang, W., Kiss, I.Z., Hudson, J.L.: Chaos 10, 248 (2000)
15. Barbara, P., Cawthorne, A.B., Shitov, S.V., Lobb, C.J.: Phys. Rev. Lett. 82, 1963 (1999)
16. Rosenblum, M., Pikovsky, A.: Phys. Rev. Lett. 98, 064101 (2007)

17. Temirbayev, A.A., Zhanabaev, Z.Z., Tarasov, S.B., Ponomarenko, V.I., Rosenblum, M.: Phys. Rev. E 85, 015204 (2012)
18. Temirbayev, A.A., Zhanabaev, Z.Z., Nalibayev, Y.D., Ponomarenko, V.I., Rosenblum, M.: Phys. Rev. E 87, 062917 (2013)
19. Baibolatov, Y., Rosenblum, M., Zhanabaev, Z.Z., Pikovsky, A.: Phys. Rev. E 82, 016212 (2010)

A Stability Analysis Method for Period-1 Solution in Two-Mass Impact Oscillator

Hiroki Amano[1], Hiroyuki Asahara[2], and Takuji Kousaka[1]

[1] Faculty of Engineering, Oita University, 700 Dannoharu, Oita, Japan
amano@bifurcation.jp, takuji@oita-u.ac.jp

[2] Faculty of Engineering, Fukuoka University, 8-19-1 Nanakuma, Fukuoka, Japan
asahara@bifurcation.jp

Abstract. In this paper, we propose a stability analysis method for the period-1 solution in two-mass impact oscillators. First, we describe a dynamical model and its solution. Next, we define the Poincaré map and then we derive derivative of the Poincaré map. In particular, we explain the elements of the Jacobian matrix to perform the stability analysis numerically. Finally, we apply this method to a simple two-mass impact oscillator and confirm its validity.

Keywords: Impact oscillator, Poincaré map, Stability analysis, Numerical integration method.

1 Introduction

An interrupted dynamical system depends on the state and on time [1, 2]. Impact oscillator belongs to the interrupted dynamical system which is observed in the mechanical field. For example, the bouncing ball, the cutting mechanics, the gear-system, the impact-dampers, and a rigid overhead wire-pantograph are typical examples [3–7]. It is known that the bifurcation phenomena are observed in these systems. Thus, it is important to calculate the stability of a periodic solution for understanding the qualitative property of impact oscillator.

For example, the bouncing-ball and the rigid overhead wire-pantograph are categorized as the impact oscillator. These systems oscillate with one-degree-of-freedom and have fixed or periodic boundary. There are many studies, which calculate the impact oscillator [8, 9]. In particular, Ref. [9] discussed the calculation method of the Lyapunov exponents in the impact oscillator. On the other hand, the impact-dampers and gear-system are categorized as the two-mass impact oscillator, which does not have the periodic boundary because a mass mutually affects each other. However, the previous methods do not apply for the two-mass impact oscillator due to its complex of behavior [10, 11]. In this shortfall, we proposed the stability analysis method in Ref. [12]. However, this method applicable only to the two-mass impact oscillator whose motion equation is described by the linearity [13].

In this paper, we propose the stability analysis method for the period-1 orbit in the two-mass impact oscillator with linear motion equation. First, we describe two-mass impact oscillator. Next, we define the Poincaré map and then we derive its derivative, which leads to elements of the Jacobian matrix. After that we construct an algorithm of

V.M. Mladenov and P.C. Ivanov (Eds.): NDES 2014, CCIS 438, pp. 37–44, 2014.

the stability analysis for the two-mass impact oscillator. Finally, we apply this algorithm to the simple model and confirm its validity.

2 Analytical Method

2.1 Description of the Analytical Model

Figure 1 shows the behavior of the solutions in this system. We consider the two-mass impact oscillator described following motion equation.

$$\begin{cases} \dfrac{dx}{dt} = f_x(t, x, u, \boldsymbol{\lambda}_1) \\ \dfrac{du}{dt} = f_u(t, x, u, \boldsymbol{\lambda}_1) \end{cases}, \quad \text{for mass-1,} \tag{1}$$

$$\begin{cases} \dfrac{dy}{dt} = g_y(y, v, \boldsymbol{\lambda}_2) \\ \dfrac{dv}{dt} = g_v(y, v, \boldsymbol{\lambda}_2) \end{cases}, \quad \text{for mass-2,} \tag{2}$$

where $t \in \boldsymbol{R}$, $(\boldsymbol{x}, \boldsymbol{y})^\top = \left((x, u)^\top, (y, v)^\top\right)^\top \in \boldsymbol{R}^4$ is state variable for $(\boldsymbol{f}, \boldsymbol{g})^\top = \left((f_x, f_u)^\top, \left(g_y, g_v\right)^\top\right)^\top \in \boldsymbol{R}^4 \rightarrow \boldsymbol{R}^4$. In addition, $\boldsymbol{\lambda}_1 \in \boldsymbol{R}^\alpha$ and $\boldsymbol{\lambda}_2 \in \boldsymbol{R}^\beta$ are the system parameters. Note that Eqs. (1) and (2) are the non-autonomous and the autonomous system one of which includes an external force with period T. Let the solution of Eqs. (1) and (2) be

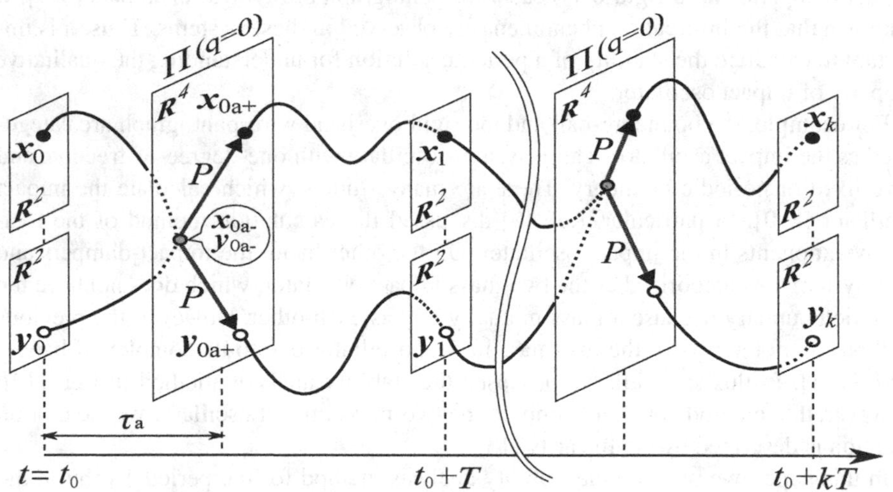

Fig. 1. Behavior of the waveforms

following equation

$$\begin{cases} x(t) = \phi_x(t; t_0, x_0, u_0, \boldsymbol{\lambda}_1) \\ u(t) = \phi_u(t; t_0, x_0, u_0, \boldsymbol{\lambda}_1) \end{cases}, \quad \text{for mass-1}, \tag{3}$$

$$\begin{cases} y(t) = \psi_y(t; t_0, y_0, v_0, \boldsymbol{\lambda}_2) \\ v(t) = \psi_v(t; t_0, y_0, v_0, \boldsymbol{\lambda}_2) \end{cases}, \quad \text{for mass-2}, \tag{4}$$

where $(\phi, \psi)^\top = \left((\phi_x, \phi_u)^\top, \left(\psi_y, \psi_v\right)^\top\right)^\top \in \boldsymbol{R}^4 \to \boldsymbol{R}^4$, x_0, u_0, y_0, and v_0 mean the initial value at $t = t_0$.

Next, we define the local section $\prod$ consists of scalar function q.

$$\prod = \left\{(\boldsymbol{x}, \boldsymbol{y}) \in R^4 : q(x, y, \boldsymbol{\lambda}_q) = 0, q : \boldsymbol{R}^4 \to \boldsymbol{R}\right\}. \tag{5}$$

Now, $\boldsymbol{\lambda}_q \in \boldsymbol{R}^\gamma$ is a parameter. Furthermore, the solution reaches to the local section at $t = t_0 + \tau_a(x_0, y_0)$. Here, we define the jump phenomenon P on the local section as

$$\begin{aligned} P : \prod \; &\to \prod \\ \begin{bmatrix} \boldsymbol{x}_{0a-} \\ \boldsymbol{y}_{0a-} \end{bmatrix} &\mapsto \begin{bmatrix} \boldsymbol{x}_{0a+} \\ \boldsymbol{y}_{0a+} \end{bmatrix} = \begin{bmatrix} \boldsymbol{r}(\boldsymbol{x}_{0a-}, \boldsymbol{y}_{0a-}) \\ \boldsymbol{s}(\boldsymbol{x}_{0a-}, \boldsymbol{y}_{0a-}) \end{bmatrix}, \end{aligned} \tag{6}$$

where $\boldsymbol{r}$ and $\boldsymbol{s}$ are $(\boldsymbol{r}, \boldsymbol{s})^\top \in \boldsymbol{R}^4 \to \boldsymbol{R}^4$. Moreover, $\boldsymbol{x}_{0a-}$, $\boldsymbol{y}_{0a-}$ and $\boldsymbol{x}_{0a+}$, $\boldsymbol{y}_{0a+}$ are the solutions before and after the jumping phenomenon. In addition, note that $\tau_a(\boldsymbol{x}_0, \boldsymbol{y}_0)$ depends on the initial value of two-mass. Moreover, we define the solutions at $t = t_0 + T$ as follows:

$$\boldsymbol{x}_1 = \phi(T; t_0 + \tau_a(\boldsymbol{x}_0, \boldsymbol{y}_0), \boldsymbol{x}_{0a+}, \boldsymbol{\lambda}_1), \tag{7}$$

$$\boldsymbol{y}_1 = \psi(T; t_0 + \tau_a(\boldsymbol{x}_0, \boldsymbol{y}_0), \boldsymbol{y}_{0a+}, \boldsymbol{\lambda}_2). \tag{8}$$

Here, we explain the behavior of this system. We discretize the solution by every interval T. We consider the solution which leaves at $t = t_0$. When the solution reaches to the local section at $t = t_0 + \tau_a(\boldsymbol{x}_0, \boldsymbol{y}_0)$, the solution jump by Eq. (6). After that, the solution repeats above behaviors. Next, we consider the Poincaré map M. Now, we define the maps as follows:

$$\begin{aligned} M_0 : \boldsymbol{R}^4 \; &\to \prod \\ \begin{bmatrix} \boldsymbol{x}_0 \\ \boldsymbol{y}_0 \end{bmatrix} &\mapsto \begin{bmatrix} \boldsymbol{x}_{0a-} \\ \boldsymbol{y}_{0a-} \end{bmatrix}, \end{aligned} \tag{9}$$

$$\begin{aligned} M_1 : \prod \; &\to \boldsymbol{R}^4 \\ \begin{bmatrix} \boldsymbol{x}_{0a+} \\ \boldsymbol{y}_{0a+} \end{bmatrix} &\mapsto \begin{bmatrix} \boldsymbol{x}_1 \\ \boldsymbol{y}_1 \end{bmatrix}. \end{aligned} \tag{10}$$

Consequently, the Poincaré map M is given by

$$\begin{aligned} M : \boldsymbol{R}^4 \; &\to \boldsymbol{R}^4 \\ \begin{bmatrix} \boldsymbol{x}_0 \\ \boldsymbol{y}_0 \end{bmatrix} &\mapsto \begin{bmatrix} \boldsymbol{x}_1 \\ \boldsymbol{y}_1 \end{bmatrix} = M_1 \circ P \circ M_0. \end{aligned} \tag{11}$$

2.2 The Method of Stability Analysis

First, we discuss the stability analysis for period-1 orbit in the two-mass impact oscillator. The characteristic equation of period-1 orbit is described as

$$\begin{aligned}\chi(\mu) &= \det(DM(\boldsymbol{x}_0, \boldsymbol{y}_0) - \boldsymbol{\mu I}_4) \\ &= \det\left(\begin{bmatrix} \frac{\partial M(\boldsymbol{x}_0)}{\partial \boldsymbol{x}_0} & \frac{\partial M(\boldsymbol{x}_0)}{\partial \boldsymbol{y}_0} \\ \frac{\partial M(\boldsymbol{y}_0)}{\partial \boldsymbol{x}_0} & \frac{\partial M(\boldsymbol{y}_0)}{\partial \boldsymbol{y}_0} \end{bmatrix} - \boldsymbol{\mu I}_4\right) = 0,\end{aligned} \tag{12}$$

where $DM(\boldsymbol{x}_0, \boldsymbol{y}_0)$ is derivative for Poincaré map. In addition, μ is the characteristic multiplier. Note that, Eq. (12) consists of a combination of the non-autonomous and the autonomous system.

Next, we consider the elements of Eq. (12) to solve this equation. As an example we describe $\partial M(\boldsymbol{x}_0)/\partial \boldsymbol{x}_0$ and $\partial M(\boldsymbol{y}_0)/\partial \boldsymbol{y}_0$ as follows:

$$\frac{\partial M(\boldsymbol{x}_0)}{\partial \boldsymbol{x}_0} = -\frac{\partial \boldsymbol{\phi}}{\partial \boldsymbol{x}_{0a+}} \left.\frac{\partial \boldsymbol{\phi}}{\partial t}\right|_{t=\tau_a} \frac{\partial \tau_a}{\partial \boldsymbol{x}_0} + \left\{\frac{\partial \boldsymbol{\phi}}{\partial \boldsymbol{x}_{0a+}} \frac{\partial \boldsymbol{r}}{\partial \boldsymbol{x}_0} + \frac{\partial \boldsymbol{\phi}}{\partial \boldsymbol{y}_{0a+}} \frac{\partial \boldsymbol{s}}{\partial \boldsymbol{x}_0}\right\}, \tag{13}$$

$$\frac{\partial M(\boldsymbol{y}_0)}{\partial \boldsymbol{y}_0} = -\left.\frac{\partial \boldsymbol{\psi}}{\partial t}\right|_{t=T-\tau_a} \frac{\partial \tau_a}{\partial \boldsymbol{y}_0} + \left\{\frac{\partial \boldsymbol{\psi}}{\partial \boldsymbol{x}_{0a+}} \frac{\partial \boldsymbol{r}}{\partial \boldsymbol{y}_0} + \frac{\partial \boldsymbol{\psi}}{\partial \boldsymbol{y}_{0a+}} \frac{\partial \boldsymbol{s}}{\partial \boldsymbol{y}_0}\right\}. \tag{14}$$

Here, $\partial \boldsymbol{r}/\partial \boldsymbol{x}_0$, $\partial \boldsymbol{s}/\partial \boldsymbol{x}_0$, $\partial \boldsymbol{r}/\partial \boldsymbol{y}_0$, and $\partial \boldsymbol{s}/\partial \boldsymbol{y}_0$ are given by the following equation

$$\begin{aligned}\frac{\partial \boldsymbol{r}}{\partial \boldsymbol{x}_0} = \left.\frac{\partial \boldsymbol{r}}{\partial t}\right|_{t=\tau_a} \frac{\partial \tau_a}{\partial \boldsymbol{x}_0} &+ \left\{\frac{\partial \boldsymbol{r}}{\partial \boldsymbol{x}} \frac{\partial \boldsymbol{x}_{0a+}}{\partial \boldsymbol{x}_{0a-}} + \frac{\partial \boldsymbol{r}}{\partial \boldsymbol{y}} \frac{\partial \boldsymbol{y}_{0a+}}{\partial \boldsymbol{x}_{0a-}}\right\}\left\{\left.\frac{\partial \boldsymbol{\phi}}{\partial t}\right|_{t=\tau_a} \frac{\partial \tau_a}{\partial \boldsymbol{x}_0} + \frac{\partial \boldsymbol{\phi}}{\partial \boldsymbol{x}_0}\right\} \\ &+ \left\{\frac{\partial \boldsymbol{r}}{\partial \boldsymbol{x}} \frac{\partial \boldsymbol{x}_{0a+}}{\partial \boldsymbol{y}_{0a-}} + \frac{\partial \boldsymbol{r}}{\partial \boldsymbol{y}} \frac{\partial \boldsymbol{y}_{0a+}}{\partial \boldsymbol{y}_{0a-}}\right\}\left\{\left.\frac{\partial \boldsymbol{\psi}}{\partial t}\right|_{t=\tau_a} \frac{\partial \tau_a}{\partial \boldsymbol{x}_0} + \frac{\partial \boldsymbol{\psi}}{\partial \boldsymbol{x}_0}\right\},\end{aligned} \tag{15}$$

$$\begin{aligned}\frac{\partial \boldsymbol{s}}{\partial \boldsymbol{x}_0} = \left.\frac{\partial \boldsymbol{s}}{\partial t}\right|_{t=\tau_a} \frac{\partial \tau_a}{\partial \boldsymbol{x}_0} &+ \left\{\frac{\partial \boldsymbol{s}}{\partial \boldsymbol{x}} \frac{\partial \boldsymbol{x}_{0a+}}{\partial \boldsymbol{x}_{0a-}} + \frac{\partial \boldsymbol{s}}{\partial \boldsymbol{y}} \frac{\partial \boldsymbol{y}_{0a+}}{\partial \boldsymbol{x}_{0a-}}\right\}\left\{\left.\frac{\partial \boldsymbol{\phi}}{\partial t}\right|_{t=\tau_a} \frac{\partial \tau_a}{\partial \boldsymbol{x}_0} + \frac{\partial \boldsymbol{\phi}}{\partial \boldsymbol{x}_0}\right\} \\ &+ \left\{\frac{\partial \boldsymbol{s}}{\partial \boldsymbol{x}} \frac{\partial \boldsymbol{x}_{0a+}}{\partial \boldsymbol{y}_{0a-}} + \frac{\partial \boldsymbol{s}}{\partial \boldsymbol{y}} \frac{\partial \boldsymbol{y}_{0a+}}{\partial \boldsymbol{y}_{0a-}}\right\}\left\{\left.\frac{\partial \boldsymbol{\psi}}{\partial t}\right|_{t=\tau_a} \frac{\partial \tau_a}{\partial \boldsymbol{x}_0} + \frac{\partial \boldsymbol{\psi}}{\partial \boldsymbol{x}_0}\right\},\end{aligned} \tag{16}$$

$$\begin{aligned}\frac{\partial \boldsymbol{r}}{\partial \boldsymbol{y}_0} = \left.\frac{\partial \boldsymbol{r}}{\partial t}\right|_{t=\tau_a} \frac{\partial \tau_a}{\partial \boldsymbol{y}_0} &+ \left\{\frac{\partial \boldsymbol{r}}{\partial \boldsymbol{x}} \frac{\partial \boldsymbol{x}_{0a+}}{\partial \boldsymbol{x}_{0a-}} + \frac{\partial \boldsymbol{r}}{\partial \boldsymbol{y}} \frac{\partial \boldsymbol{y}_{0a+}}{\partial \boldsymbol{x}_{0a-}}\right\}\left\{\left.\frac{\partial \boldsymbol{\phi}}{\partial t}\right|_{t=\tau_a} \frac{\partial \tau_a}{\partial \boldsymbol{y}_0} + \frac{\partial \boldsymbol{\phi}}{\partial \boldsymbol{y}_0}\right\} \\ &+ \left\{\frac{\partial \boldsymbol{r}}{\partial \boldsymbol{x}} \frac{\partial \boldsymbol{x}_{0a+}}{\partial \boldsymbol{y}_{0a-}} + \frac{\partial \boldsymbol{r}}{\partial \boldsymbol{y}} \frac{\partial \boldsymbol{y}_{0a+}}{\partial \boldsymbol{y}_{0a-}}\right\}\left\{\left.\frac{\partial \boldsymbol{\psi}}{\partial t}\right|_{t=\tau_a} \frac{\partial \tau_a}{\partial \boldsymbol{y}_0} + \frac{\partial \boldsymbol{\psi}}{\partial \boldsymbol{y}_0}\right\},\end{aligned} \tag{17}$$

$$\begin{aligned}\frac{\partial \boldsymbol{s}}{\partial \boldsymbol{y}_0} = \left.\frac{\partial \boldsymbol{s}}{\partial t}\right|_{t=\tau_a} \frac{\partial \tau_a}{\partial \boldsymbol{y}_0} &+ \left\{\frac{\partial \boldsymbol{s}}{\partial \boldsymbol{x}} \frac{\partial \boldsymbol{x}_{0a+}}{\partial \boldsymbol{x}_{0a-}} + \frac{\partial \boldsymbol{s}}{\partial \boldsymbol{y}} \frac{\partial \boldsymbol{y}_{0a+}}{\partial \boldsymbol{x}_{0a-}}\right\}\left\{\left.\frac{\partial \boldsymbol{\phi}}{\partial t}\right|_{t=\tau_a} \frac{\partial \tau_a}{\partial \boldsymbol{y}_0} + \frac{\partial \boldsymbol{\phi}}{\partial \boldsymbol{y}_0}\right\} \\ &+ \left\{\frac{\partial \boldsymbol{s}}{\partial \boldsymbol{x}} \frac{\partial \boldsymbol{x}_{0a+}}{\partial \boldsymbol{y}_{0a-}} + \frac{\partial \boldsymbol{s}}{\partial \boldsymbol{y}} \frac{\partial \boldsymbol{y}_{0a+}}{\partial \boldsymbol{y}_{0a-}}\right\}\left\{\left.\frac{\partial \boldsymbol{\psi}}{\partial t}\right|_{t=\tau_a} \frac{\partial \tau_a}{\partial \boldsymbol{y}_0} + \frac{\partial \boldsymbol{\psi}}{\partial \boldsymbol{y}_0}\right\}.\end{aligned} \tag{18}$$

Finally, we obtain the following equation

$$q(\boldsymbol{\phi}(\tau_a, t_0, \boldsymbol{x}_0, \boldsymbol{\lambda}_1), \boldsymbol{\psi}(\tau_a, t_0, \boldsymbol{y}_0, \boldsymbol{\lambda}_2), \boldsymbol{\lambda}_q) = 0. \tag{19}$$

By Eq. (19), we can easily derive $\partial\tau_a/\partial\boldsymbol{x}_0$ and $\partial\tau_a/\partial\boldsymbol{y}_0$

$$\frac{\partial\tau_0}{\partial\boldsymbol{x}_0} = \frac{-\dfrac{\partial q}{\partial\boldsymbol{x}}\dfrac{\partial\boldsymbol{\phi}}{\partial\boldsymbol{x}_0}}{\dfrac{\partial q}{\partial\boldsymbol{x}}\left(\left.\dfrac{\partial\boldsymbol{\phi}}{\partial t}\right|_{t=\tau_a} - \left.\dfrac{\partial\boldsymbol{\psi}}{\partial t}\right|_{t=\tau_a}\right)},$$
$$\frac{\partial\tau_0}{\partial\boldsymbol{y}_0} = \frac{\dfrac{\partial q}{\partial\boldsymbol{y}}\dfrac{\partial\boldsymbol{\psi}}{\partial\boldsymbol{y}_0}}{\dfrac{\partial q}{\partial\boldsymbol{y}}\left(\left.\dfrac{\partial\boldsymbol{\phi}}{\partial t}\right|_{t=\tau_a} - \left.\dfrac{\partial\boldsymbol{\psi}}{\partial t}\right|_{t=\tau_a}\right)}. \tag{20}$$

3 Application

First, we propose the two-mass impact oscillator shown in Fig.2 to confirm the proposed method validity.

$$M_1\ddot{X} + C_1\dot{X} + K_1X = A\cos(\Omega T), \qquad \text{for mass-1}, \tag{21}$$

$$M_2\ddot{Y} + C_2\dot{Y} + K_2Y = 0, \qquad \text{for mass-2}. \tag{22}$$

Here, when the mass reach to the other mass, the velocity of each mass changes as follows:

$$\begin{cases} \dot{X}^+ - \dot{Y}^+ = -\varepsilon\left(\dot{X}^- - \dot{Y}^-\right) \\ M_1\dot{X}^+ + M_2\dot{Y}^+ = M_1\dot{X}^- + M_2\dot{Y}^- \end{cases} \tag{23}$$

Moreover, $\dot{X}^-$, $\dot{Y}^-$ and $\dot{X}^+$, $\dot{Y}^+$ are the velocities before and after the jumping phenomenon. Also, ε is coefficient of restitution between each mass. Then, gap of an equilibrium point is written as follows:

$$\Delta = Y^* - X^*, \tag{24}$$

where X^* and Y^* are equilibrium point of each mass. Here, we define the dimensionless values as

$$m = \frac{M_1}{M_2}, \qquad x = \frac{XK_1}{A}, \qquad y = \frac{YK_1}{A}, \qquad \zeta_i = \frac{C_i}{2\sqrt{K_iM_i}},$$
$$\omega = \Omega\sqrt{\frac{M_1}{K_1}}, \qquad \omega_0 = \sqrt{\frac{K_2m}{K_1}}, \qquad t = T\sqrt{\frac{K_1}{M_1}}, \qquad d = \frac{\Delta K_1}{A}. \tag{25}$$

Where, $i = 1, 2$. The Eqs. (21),(22),(23), and (24) are rewritten as follows using the Eq. (23):

$$\ddot{x} + 2\zeta_1\dot{x} + x = \cos(\omega t), \qquad \text{for mass-1}, \tag{26}$$

$$\ddot{y} + 2\zeta_2\omega_0\dot{y} + \omega_0^2 y = 0, \qquad \text{for mass-2.} \tag{27}$$

$$\begin{cases} \dot{x}^+ - \dot{y}^+ = -\varepsilon(\dot{x}^- - \dot{y}^-) \\ m\dot{x}^+ + \dot{y}^+ = m\dot{x}^- + \dot{y}^- \end{cases}, \tag{28}$$

$$d = y^* - x^*. \tag{29}$$

Now, we define the local section $\prod$ using the scalar function q.

$$\prod = \{q(x, y, d) = y - x - d = 0\}. \tag{30}$$

Next, we explain the behavior of the system. The solution obeys Eqs. (26) and (27). When one mass impacts another mass, the velocity changes as Eq. (28). After that, the solution repeats above behaviors. Here, we set the system parameters as

$$\zeta_1 = 0.05, \ \zeta_2 = 0.10, \ m = 2.0, \ \varepsilon = 0.48, \ \omega_0 = 0.10, \ d = 1.0. \tag{31}$$

Note that ω is changing parameter. The period-doubling bifurcation occurs at $\omega = 0.21061$. In this parameter, $\mu_1 = -0.171$, $\mu_2 = -1.000$, $\mu_3 = 0.183 + 0.065i$, and $\mu_4 = 0.183 - 0.065i$ are the result of stability analysis using the proposed method. Now, $\mu_i (i = 1, 2, 3, 4)$ represents the characteristic multiplier. In addition, Fig.3 is the 1-parameter bifurcation diagram upon varying the parameter ω from $\omega = 0.209$ to $\omega = 0.212$. Also, Fig.4 shows the waveforms corresponding to Fig.3 (*a*) and (*b*). By Fig.3, Fig.4, and the result of the proposed method, the period-doubling bifurcation occurs at the same parameter. Thus, the validity of the proposed method was confirmed.

4 Conclusion

In this paper, we have proposed a stability analysis method for the period-1 orbit of the two-mass impact oscillator. First, we explained the dynamical system of the two-mass

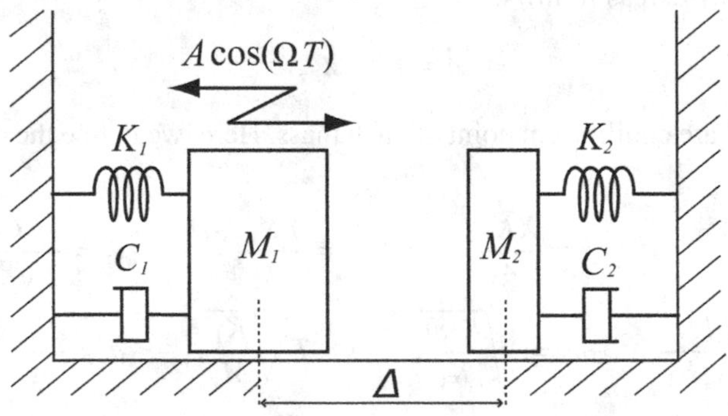

Fig. 2. Two-mass impact oscillator

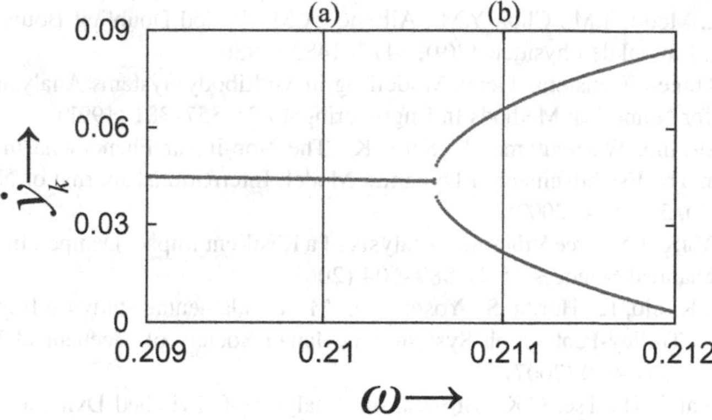

Fig. 3. 1-parameter bifurcation diagram

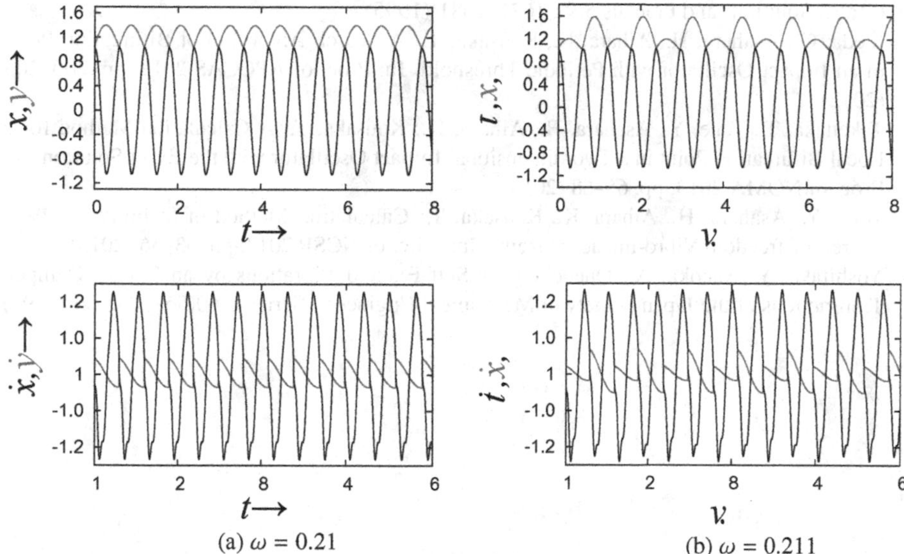

(a) $\omega = 0.21$

(b) $\omega = 0.211$

Fig. 4. Waveforms

impact oscillator and its solution. Next, we defined the Poincaré map and showed the derivative of the map to perform the stability analysis. Finally, we applied this method to the simple two-mass impact oscillator and we could confirm the validity of this method.

References

1. Shaw, S.W., Holmes, P.J.: A Periodically Forced Piecewise Linear Oscillator. Journal of Sound and Vibration 90(1), 129–155 (1983)
2. Izhikevich, E.M.: Simple Model of Spiking Neurons. IEEE Transactions on Neural Networks 14(6), 1569–1572 (2003)

3. Tufillaro, N.B., Mello, T.M., Choi, Y.M., Albano, A.M.: Period Doubling Boundaries of a Bouncing Ball. Journal de Physique 47(9), 1477–1482 (1986)
4. Cardona, A.: Three-dimensional Gears Modelling in Multibody Systems Analysis. International Journal for Numerical Methods In Engineering 40(2), 357–381 (1997)
5. Karube, S., Hoshino, W., Soutome, T., Sato, K.: The Non-linear Phenomena in Vibration Cutting System The Establishment of Dynamic Model. International Journal of Non-Linear Mechanics 37(3), 541–564 (2002)
6. Cheng, C.C., Wang, J.Y.: Free Vibration Analysis of a Resilient Impact Damper. International Journal of Mechanical Sciences 45(4), 589–604 (2003)
7. Kawamura, S., Kitajo, K., Horita, S., Yoshizawa, M.: Fundamental Study on Impact Oscillations of Rigid Trolley-Pantograph System. The Japan Society of Mechanical Engineers, Series C 73(728), 974–980 (2007)
8. Ma, Y., Kawakami, H., Tse, C.K.: Bifurcation Analysis of Switched Dynamical Systems with Periodically Moving Borders. IEEE Transactions on Circuits and Systems-Part I 51(6), 1184–1193 (2004)
9. Muller, P.C.: Calculation of Lyapunov Exponents for Dynamic Systems with Discontinuities. Chaos, Solutions and Fractals 5(9), 1671–1681 (1995)
10. Ikeda, G., Asahara, H., Aihara, K., Kousaka, T.: A Search Algorithm of Bifurcation Point in an Impact Oscillator with Periodic Threshold. In: Proc. of APCCAS 2012, pp. 200–203 (2012)
11. Takenaka, T., Tone, Y., Asahara, H., Aihara, K., Kousaka, T.: A Calculation Method for a Local Bifurcation Point in a Two-dimensional Impact Oscillator with the Exact Solution. In: Proc. of NOMA 2013, pp. 65–68 (2013)
12. Tone, Y., Asahara, H., Aihara, K., Kousaka, T.: Calculation Method of Stability for Two-degree-of-freedom Vibro-impact Systems. In: Proc. of NCSP 2013, pp. 53–56 (2013)
13. Yoshitake, Y., Sueoka, A.: Quenching of Self-Excited Vibrations by an Impact Damper. Transactions of the Japan Society of Mechanical Engineers, Series C 60(569), 50–56 (1994)

Manifold Piecewise Linear Chaotic System on Cylinder and Super Expanding Chaos

Shotaro Suzuki[1], Kazuyuki Kimura[1], Tadashi Tsubone[2], and Toshimichi Saito[1]

[1] Hosei University, Koganei-shi, Tokyo, 184-8584 Japan
tsaito@hosei.ac.jp
[2] Nagaoka University of Technology, Nagaoka-shi, Nigata, 940-2188 Japan

Abstract. This paper presents a novel autonomous chaotic system: the manifold piecewise linear system on the cylinder. This system is defined by second order continuous flow on the cylinder with hysteresis switching and the trajectories do not diverge. This system can exhibit super expanding chaos characterized by very large positive Lyapunov exponent. The dynamics is integrated into a piecewise linear one-dimensional return map and chaos generation is guaranteed theoretically.

Keywords: Chaos, autonomous systems, piecewise linear systems, switched dynamical systems.

1 Introduction

The manifold piecewise linear system (MPL) is an autonomous chaotic system consists of second-order continuous system and hysteresis switching of the equilibrium points [1]-[3]. The MPL exhibits double screw chaotic attractor and has remarkable characteristics. First, the dynamics is integrated into a piecewise linear one-dimensional return map and chaotic generation is guaranteed theoretically. Second, the MPL is realized by a simple electric circuit and chaotic behavior is confirmed in the laboratory. Third, the chaotic behavior is applicable in engineering systems such as chaos-based communication and radar systems [4][5]. Up to the present, a variety of autonomous chaotic systems have been presented and the dynamics have been analyzed: the Lorenz system is the first autonomous chaotic system and the Chua's circuit is the well-known smooth autonomous chaotic circuit [6]-[8]. These systems have been contributed in development of nonlinear dynamical system theory and its engineering applications.

This paper presents a novel autonomous chaotic system: the manifold piecewise linear system on the cylinder (CMPL). The CMPL is defined by second order continuous flow on the cylinder with hysteresis switching and the trajectories do not diverge. The dynamics is simplified into a piecewise linear one-dimensional return map and chaos generation is guaranteed theoretically. Especially, the CMPL can exhibit super expanding chaos characterized by very large positive Lyapunov exponent. Note that, in the MPL, the divergent trajectory is inevitable and the super expanding chaos is impossible. Although this paper studies the most basic version of the CMPL, the results may be developed into a circuit with a variety of chaotic attractors and its engineering applications.

V.M. Mladenov and P.C. Ivanov (Eds.): NDES 2014, CCIS 438, pp. 45–50, 2014.

2 Manifold Piecewise Linear Chaotic System

In this section, we summarize the MPL as preparation to present the CMPL. The MPL is defined by the following second-order piecewise linear system with hysteresis switching:

$$\ddot{x} - 2\delta\dot{x} + x = \begin{cases} +p & (+) \\ -p & (-) \end{cases} \tag{1}$$

Let $\boldsymbol{x} \equiv (x, \dot{x})$. In order to define the switching rule, we divide the x-axis L into two half lines:

$$L = L_+ \cup L_-, \; L_+ \equiv \{\boldsymbol{x} | x \geq X_T, \; \dot{x} = 0\}, \; L_- \equiv \{\boldsymbol{x} | x < X_T, \; \dot{x} = 0\}$$

The switching rule of MPL: The right hand side of Eq. (1) is switched from $(+)$ *to* $(-)$ *if* $\boldsymbol{x}$ *hits* L_- *and is switched from* $(-)$ *to* $(+)$ *if* $\boldsymbol{x}$ *hits* L_+.

The MPL is characterized by three parameters: damping δ, equilibriun point p and the switching threshold X_T. For simplicity, we consider the case

$$0 < \delta < 1 \; (\omega \equiv \sqrt{1 - \delta^2}), \; 0 < p, \; X_T = 0 \tag{2}$$

In this case, the MPL has unstable complex characteristic roots $\delta \pm j\omega$. As shown in Fig. 1, the trajectory rotates divergently around the equilibrium point p $(-p)$. For simplicity, let point on x-axis be represented by its x-component. If the trajectory hits negative x-axis L_- (positive x-axis L_+), the equilibrium point is switched from p to $-p$ ($-p$ to p). Repeating in this manner, the MPL exhibits double-screw chaotic attractor. The dynamics is integrated into the 1D return map. Let x_n denote the n-th intersection of the trajectory and x-axis. Since x_n determines x_{n+1}, we can define the 1-D return map

$$F : L \to L, \; x_n \mapsto x_{n+1}, \; L = \{(x, \dot{x}) \mid \dot{x} = 0\} \tag{3}$$

The map can be described exactly.

$$x_{n+1} = F(x_n) = \begin{cases} -\beta(x_n - p) + p & \text{for } x_n \geq 0 \\ -\beta(x_n + p) - p & \text{for } x_n < 0 \end{cases} \quad \beta \equiv e^{\delta\frac{\pi}{\omega}} > 1 \tag{4}$$

Let $T = F(0) = p(1 + \beta)$ and let $I_T = [-T, \; T]$. If $F(T) > -T$ then I_T is an invariant interval on which the map is expanding.

$$F(I_T) \subset I_T, \; |DF(x)| = \beta > 1 \text{ on } I_T \tag{5}$$

where $DF(x)$ is the slope of F at x. In this case, the map has a positive Lyapunov exponent $\ln \beta$ and chaos generation is guaranteed theoretically [3]. That is, $F(T) \geq T$ is a sufficient condition for chaos generation. In the case of Eq. (2), $F(T) > -T$ is satisfied for $1 < \beta < 2$. Conversely, if $F(T) < -T$ then I_T is not an invariant interval and trajectories diverge.

Fig. 2 shows several examples of trajectories and return maps where β is used as a parameter instead of δ, for convenience.

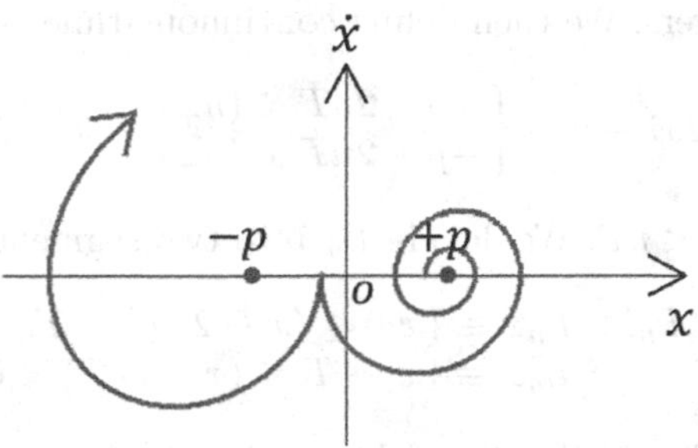

Fig. 1. Switching of MPL

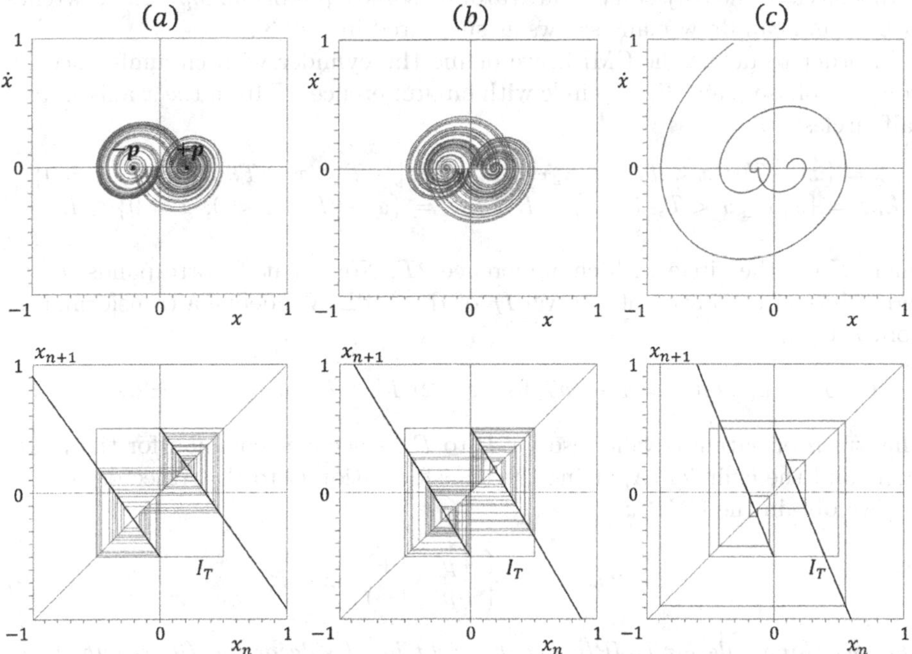

Fig. 2. Trajectories and return maps of the MPL. (a) double screw chaos for $\beta = \sqrt{2}$, (b) double screw chaos for $\beta = 1.7$, (c) divergent trajectory for $\beta = 2.5$

3 Manifold Piecewise Linear System on Cylinder

First, we define an MPL with infinite equilibria. We divide the x-axis L into segments with length $2T$:

$$L = \bigcup_{n=-\infty}^{\infty} L_n, \ L_n = \{\boldsymbol{x}| \ |x - 2nT| < T, \dot{x} = 0\}.$$

where n denotes integers. We then define continuous-time system on L_n:

$$\ddot{x} - 2\delta\dot{x} + x = \begin{cases} +p + 2nT & (n_+) \\ -p + 2nT & (n_-) \end{cases}, \quad \boldsymbol{x} \in L_n \tag{6}$$

where $(n-1)T < np < nT$. We devide L_n into two segments:

$$L_n = L_{n+} \cup L_{n-}, \; L_{n+} \equiv \{\boldsymbol{x}|0 \le (x - 2nT) < T, \; \dot{x} = 0\}$$
$$L_{n-} \equiv \{\boldsymbol{x}| - T \le (x - 2nT) < 0, \; \dot{x} = 0\}$$

The switching rule (Fig. 3): Let the right hand side of Eq. (6) is np where (n_+) is some integer. The right hand side is switched to (n_-) if $\boldsymbol{x}$ hits L_{n-} and is switched to (n_+) if $\boldsymbol{x}$ hits L_{n+}.

In this case, the trajectory rotates around some equilibrium $\pm p + 2nT$, swtches to $L_{\pm n}$ and can draw many screws as suggested in Fig. 3.

In order to define the CMPL, we define the cylinder with circumference $2T$ from the phase plane P, the circle with circumference $2T$ from the x axis L, and half circles:

$$\begin{aligned} &P_c = \{\boldsymbol{x}| - T < x < T, \; x \in S_{2T}\}, \; L_c = \{\boldsymbol{x}| - T < x < T, \; x \in S_{2T}, \; \dot{x} = 0\} \\ &L_{c+} = \{\boldsymbol{x}|0 \le x < T, \; \dot{x} = 0\} \subset L_c, \; L_{c-} = \{\boldsymbol{x}| - T \le x < 0, \; \dot{x} = 0\} \subset L_c \end{aligned} \tag{7}$$

where S_T is the circle with circumference $2T$. Note that T corresponds to the end points of the invariant interval I_T of the MPL. We define a transformation from L to L_c

$$G : L \to L_c : G(x) = x - 2nT \text{ for } |x - 2nT| < T, \quad n = 0, \pm 1, \pm 2, \cdots \tag{8}$$

This transformation is valid also for P to P_c, since $\dot{x}$ is common for the phase plane and the cylinder, Applying the transformation G to the trajectory of Eq. (6), we obtain the CMPL:

$$\ddot{x} - 2\delta\dot{x} + x = \begin{cases} +p & (+) \\ -p & (-) \end{cases}, \boldsymbol{x} \in P_c \tag{9}$$

The switching rule for CMPL: Let the right hand side of Eq. (9) is either (+) or (-). If trajectory hits the border $x = T$ (respectively $x = -T$) then it jumps to $x = -T$ (respectively $x = T$) holding $\dot{x}$ constant. The right hand side of Eq. (9) is switched from (+) to (-) if $\boldsymbol{x}$ hits L_{c-} and is switched from ($-$) to (+) if $\boldsymbol{x}$ hits L_{c+}.

Note that the CMPL is characterized by three parameters: damping δ, x-component of the equilibrium p and the circumference $2T$. The CMPL exhibits chaotic trajectory as shown in Fig. 4. Note that the trajectory does not diverge because it is defined on the cylinder.

After similar consideration as the MPL, the dynamics of CMPL can be integrated into a piecewise linear 1D return map f from L_c to itself:

$$x_{n+1} = f(x_n) = \begin{cases} G(-\beta(x_n - p) + p) & \text{for } x_n \in L_{c+} \\ G(-\beta(x_n + p) - p) & \text{for } x_n \in L_{c-} \end{cases} \tag{10}$$

where $\beta \equiv e^{\delta \frac{\pi}{\omega}} > 1$. For simplicity, we focus on the case

$$0 < \delta < 1, \quad p(\beta + 1) = T = 0.5 \tag{11}$$

For convenience, we use β as a parameter instead of δ. Several examples of chaotic attractor and return maps are shown in Fig. 4. We can see that the orbit does not diverge for all finite β. Since $\beta > 1$, map is expanding, has positice lyapunov exponent $\ln \beta$ and chaos generation is guaranteed theoretically.

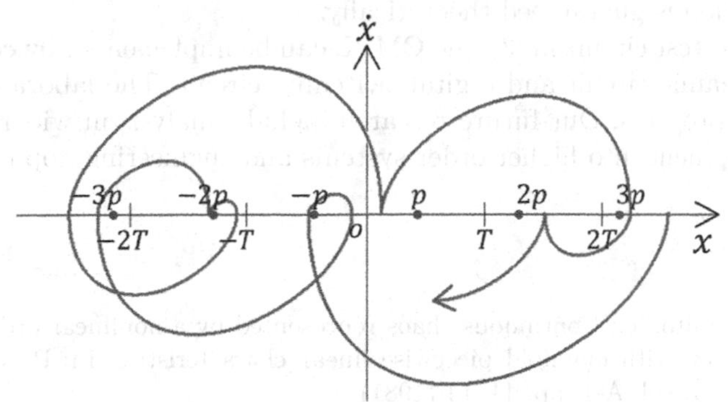

Fig. 3. Switching of MPL with infinite equilibria

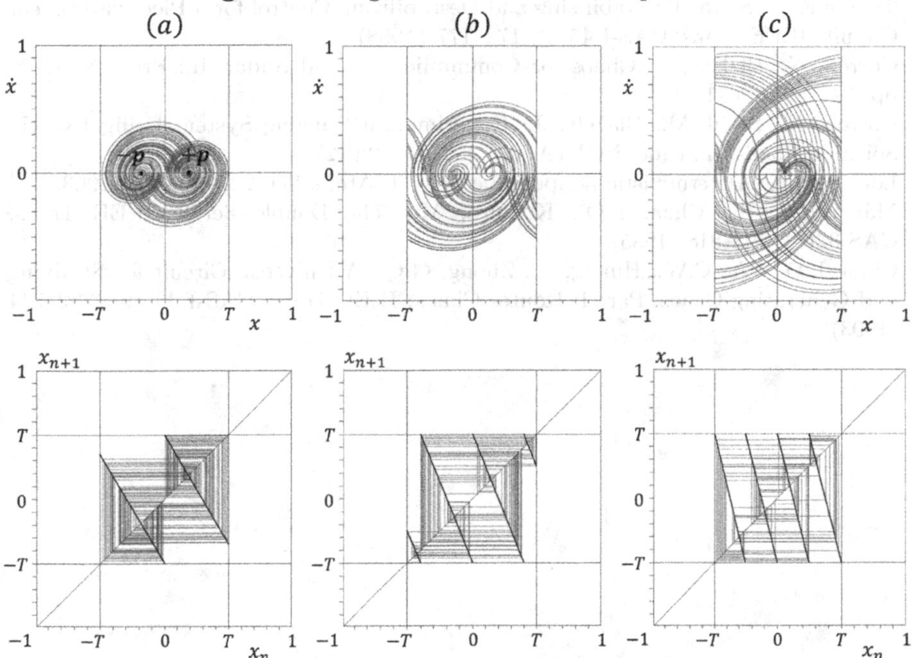

Fig. 4. Trajectories and return maps of the CMPL. (a) double screw chaos for $\beta = 1.7$, (b) super expanding chaos for $\beta = 2.5$, (c) super expanding chaos for $\beta = 4$

Note hat the CMPL can exhibit chaos with Lyapunov exponent $\ln \beta > \ln 2$. Since such a chaotic behavior is impossible in the MPL, we refer to chaos for $\ln \beta > \ln 2$ as a super expanding chaos.

4 Conclusions

CMPL is presented in this paper. It exhibits chaos and no divergence trajectory exists. The dynamics is simplified into a piecewise linear 1D return map and chaos generation is guaranteed theoretically.

Referring to test circuit in [3]. the CMPL can be implemented by combination of analog dynamic circuit and digital switching circuit. The laboratory experiments are in progress, Our future research include analysis in wider parameter range, development into higher order systems and engineering applications.

References

1. Fujuta, H., Saito, T.: Continuous chaos represented by a nonlinear ordinary differential equation with manifold piecewise linear characteristics. In: Proc. Int. Wiss. Koll., Ilmenau, vol. A-1, pp. 11–14 (1981)
2. Saito, T., Fujuta, H.: Chaos in a Manifold Piecewise Liner System. Trans. IECE 64-A(10), 827–834 (1981)
3. Tsubone, T., Saito, T.: Stabilizing and Destabilizing Control for a Piecewise Linear Circuit. IEEE Trans. CAS-I 45(2), 172–177 (1998)
4. Corron, N., Blakely, J.: Chaos for Communication and Radar. In: Proc. NOLTA, pp. 322–325 (2011)
5. Corron, N., Stahl, M., Blakely, J.: Experimental Ranging System Using Exactly Solvable Chaos. In: Proc. NOLTA, pp. 454–457 (2012)
6. Lorenz, E.N.: Deterministic nonperiodic flow. J. Atom. Sci. 20, 130–141 (1963)
7. Matsumoto, T., Chua, L.O., Komuro, M.: The Double Scroll. IEEE Trans. CAS 32(8), 798–818 (1985)
8. Chua, L.O., Wu, C.W., Huang, A., Zhong, G.Q.: A Universal Circuit for Studying and Generating Chaos. Part I: Routes Chaos: IEEE Trans. CAS-I 40(10), 732–744 (1993)

Spectrum Sensor Based on a Self-Organizing Feature Map

Artūras Serackis* and Liudas Stašionis

Department of Electronic Systems, Vilnius Gediminas Technical University, Naugarduko 41-413, Vilnius, LT-03227, Lithuania
Arturas.Serackis@vgtu.lt, Liudas.Stasionis@dok.vgtu.lt

Abstract. The paper focuses on the problem of burst signal detection in noisy radio environment. The ability to detect a burst signal in environment with low signal-to-noise level is crucial for cognitive radio solutions for which the detection of primary user transmission should be performed quickly and in energy efficient way. While the currently used spectrum sensing techniques based on the analysis of signal spectrum energy are simple to implement and are efficient, the sensitivity of these sensors in particular environment highly depends on the threshold estimation technique used in implementation. The classifier based on the self-organizing feature map proposed in this paper, takes the decision of the primary user signal presence in the measured environment. An experimental investigation was performed in 25 MHz wide frequency band on 949 MHz central frequency. The proposed spectrum sensing technique was compared with the alternatively proposed semi-adaptive threshold setting techniques for energy based spectrum sensors.

Keywords: self-organizing feature map, cognitive radio, spectrum sensor, energy detector.

1 Introduction

The increasing demand on the mobile data transmission puts additional challenges to the radio spectrum sharing by GSM and LTE users. The shifting of the terrestrial digital television to lower frequency channels in order to occupy the bandwidth for mobile data transmission does not solve the spectrum-sharing problem. However, the current spectrum sharing techniques leaves gaps in time domain, which we can use for concurrent – secondary user – signal transmission.

The sensitivity of the sensor, which detects the gap in the spectrum, directly depends on the algorithm used in cognitive radio solution [1]. In addition, the influence of continuously changing environment should be taken into account by the sensor. Therefore, the spectrum sensor should have a possibility to adapt to current environment state.

* Corresponding author.

V.M. Mladenov and P.C. Ivanov (Eds.): NDES 2014, CCIS 438, pp. 51–58, 2014.

The currently used spectrum sensing methods may be divided into five classes [2]: waveform-based sensing, cyclostationary-based sensing, radio identification based sensing, matching filter based sensing and energy detector based sensing methods. The simplest methods are based on the analysis of the spectrum energy changes [3], [4], [5]. The energy detector based spectrum sensors does not require initial knowledge of the primary user signal characteristics, but uses advanced decision threshold estimation algorithms which should be able to adapt to particular, currently present, environmental noise. Many theoretical suggestions for automatic threshold estimation are suggested in literature [3], [4], [5], [6], [7]. The stationary or simplified models of the environment used to estimate the efficiency of these proposed algorithms still requires manually setting the probability of false alarm and estimating the standard deviation of the noise [8]. While these two parameters plays important role for sensitivity of the spectrum sensor, there is no suggested method to set them adaptively.

An artificial neural network based solutions are suggested to predict channel status from previous channel observations [9]. The experimental investigations in previous authors work has shown the possibility to improve primary user detection for low SNR situations [10]. The use of multilayer perceptron for channel state prediction [9] or single perceptron to set the threshold for decision of the presence of primary user signal [10] requires a supervised training algorithm implemented in the sensor node. In both solutions, we need the training data with known examples of free and occupied channel. It limits the possibility of the sensor to manually retrain itself if the noise conditions or primary user signal type have been changed. The application of self-organizing feature map (SOM) to classify different primary user signals has been already analyzed with four spectral and seven temporal features used as an input [11]. We propose an unsupervised training of the classifier, based on a SOM for detection of unoccupied spectrum channels with the energy and standard deviation of the analyzed spectrum sub-band, used as features of the SOM.

We compare the newly introduced spectrum sensor with alternative low power spectrum sensing solutions, based on the energy detector and estimation of the modified standard deviation [10]. We have performed an experimental investigation on the real environment data collected using low cost software defined radio (SDR). In this paper, we cover three main questions:

- How the fully automatic SOM based spectrum sensor performs on a real radio spectrum data?
- Do it is needed to perform an update of SOM neurons – neighboring neurons of the winner neuron – to achieve acceptable spectrum sensing?
- What is the influence of SOM topology selection (hexagonal, rectangular or triangular) to the detection capabilities of the spectrum sensor?

1.1 The Structure of Proposed Spectrum Sensor

The model of the proposed spectrum sensor consists of three main parts: signal spectrum estimation, feature extraction of selected sub-band and SOM based noise detection (see Fig. 1). In the first part of spectrum sensor given in Fig. 1, we use the SDR as a receiver.

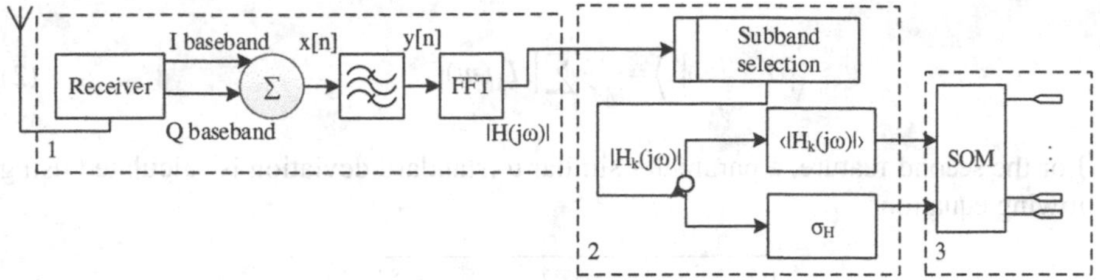

Fig. 1. Structure of the spectrum sensor based on Self-Organizing Feature Map

The receiver gives I and Q baseband nodes as an output, which are added together in order to convert I/Q data into the plane signal for band-pass filtering:

$$x[n] = I[n] + jQ[n]. \tag{1}$$

For selected windowed signal, the sensor estimates the fast Fourier transform (FFT). The signals received in the analyzed frequency band has different spectral characteristics (continuous spectrum, burst signals) so the spectrum sensors are tested for various types of primary user signals (see Fig. 2).

Fig. 2. Illustration of the analyzed frequency spectrum occupancy

Analysis algorithm divides the signal spectrum into narrow channels and estimates two feature vectors for each one. The average value of the signal energy in analyzed sub-band is estimated using following equation:

$$\left\langle |H_k(j\omega)|^2 \right\rangle = \frac{1}{M} \sum_{m=0}^{M-1} |H_k(m)|^2 . \tag{2}$$

For the second feature, a parameter similar to standard deviation is calculated using following equation:

$$\sigma_{\mathrm{H}} = \sqrt{\frac{1}{M} \sum_{m=0}^{M-1} \left(H_k(m) - \left\langle |H_k(j\omega)|^2 \right\rangle \right)^2} . \tag{3}$$

We made the modification of standard deviation formula in order to reduce the computational load of the algorithm and perform in parallel to energy calculation [10]. The changes in standard deviation formula lead to the value differences (in comparison to the results with classical expression) of about 0.07 % and do not affect the precision of spectrum sensor.

The decision block in the spectrum sensor, proposed in this paper, is based on SOM. The main idea of SOM application in this task is to classify the sub-bands into having no other signals but noise and sub-bands with possible presence of the primary user signal. The main task of the SOM block is to find the sub-bands with only noise present in it in order to mark this sub-band as a gap in the signal spectrogram.

2 Experimental Investigation of the Spectrum Sensor

In order to test the performance of the proposed spectrum sensing solution, we performed an experimental investigation in real environment.

2.1 Experimental Data

We have collected the real experimental data on the central frequency of 949 MHz with the bandwidth of 25 MHz. An 8 s view (duration of one column is 2 ms) of analyzed frequency band occupancy by active data transmission is given in Fig. 2.

The experimental data measurements were made by collecting 10 000 000 time windows in selected frequency range in order to classify them into free (only the noise is present) and occupied by data transmission channels (with primary user signal). During provided experimental investigation 3 817 328 channels of 20 kHz width were occupied by primary user and 6 182 672 channels were free for transmission. These numbers shows the theoretical possibility to transmit nearly twice-higher amount of data through the same frequency channel.

2.2 Experimental Setup

In order to test the influence of selection of neighboring SOM neurons to update their weights during self-organization phase, we have tested different two-dimensional SOM topology combinations: rectangular, hexagonal and diamond. The selected topology defines the number of neighboring neurons selected for updating their weights. The number of neurons in the first neighboring set for rectangular structure is eight, for hexagonal structure is six and for diamond structure is four. The second neighboring set has respectively 16, 12 and 8. One fired neuron we select to indicate the channel free from primary user signals. The firing of the rest SOM neurons we treat as an indicator of particular (unknown) type of primary user signal. The more neurons are used in SOM topology, the more different types of primary user signals possibly might be detected. For the experimental investigation, we selected 40 neurons in total (five rows with eight neurons in each).

Additionally we tested the performance of the proposed sensor by giving different feature vectors to the SOM input: energy estimate only, standard deviation estimate only and both estimates at once.

The success of SOM unsupervised training depends on the selected number of epochs N, learning rate η and selected neighboring function. In our experimental investigation, we updated the weights according to the following expression:

$$\omega(n+1)=\omega(n)+\eta\sigma\exp(-2n/N). \tag{4}$$

Here $\sigma\exp(-2n/N)$ is a neighboring function, which controls, how strongly the weights of the neighboring to the fired one neuron are updated. For the experimental investigation, we selected the following SOM parameters: $N=420$, $\eta=0.1$, $\sigma_1=0.5$ (for the first neighboring set) and $\sigma_2=0.3$ (for the second neighboring set).

2.3 Results of the Experimental Investigation

The goal of the spectrum sensor is to detect a gap in monitored frequency band. In an idealized situation the spectrum sensor should detect situation when no primary user emission is present also should fire when the emission is present. Therefore, we analyzed the results of experimental investigation in order to estimate the number of detected frequency gaps and the number of not detected emissions in the monitored frequency band. We have tested different size and topology of SOM in order to maximize the number of detected gaps in frequency band and minimize the number of not detected emissions.

Not all detected by SOM based spectrum sensor signal frequency gaps were free from primary user. It is a situation when the neuron, responsive to the free sub-band detection did fire even if primary user signals are present. The percentages of not detected emissions are shown in charts given in Fig. 3.

The simplest in implementation SOM topology is based on a linear topology of SOM neurons. During the experimental investigation, we made an additional simplification for the one of the topologies – only the weights of fired neuron were updated. In this situation, no topological structure is estimated for SOM. However looking at the results of experimental investigation it is seen in Fig. 3 that the amount of detected emissions mostly does not reach 80 %. In addition, the detection rate varies in wide range from 40 % to 90 %. It shows that the performance of such spectrum sensor highly depends on the number of neurons used but also in Fig. 3 we can see that the dependence is not smooth and seems to be chaotic.

Three alternative SOM topologies were analyzed in order to ensure that the topological structure estimation might increase the efficiency of the spectrum sensor.

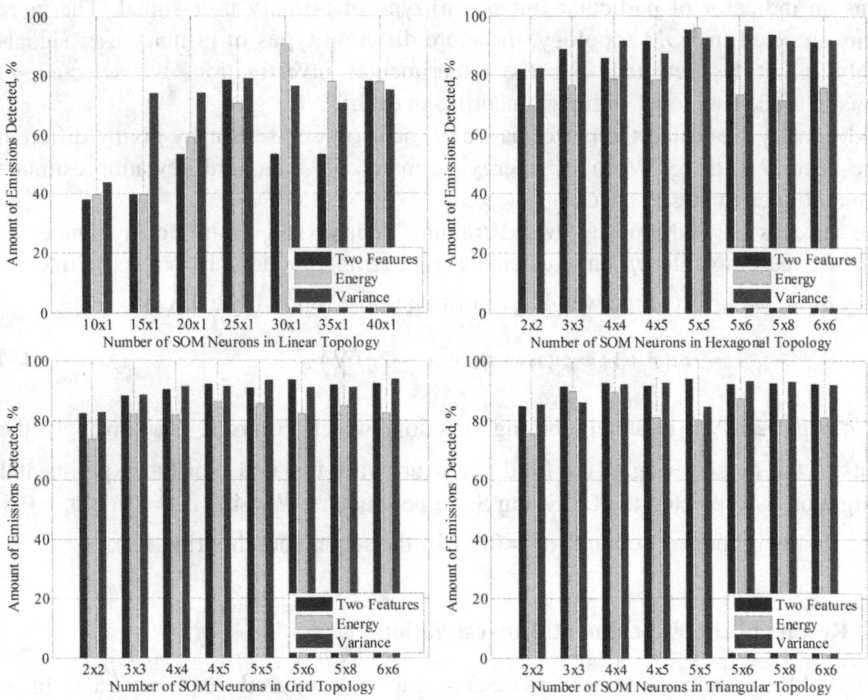

Fig. 3. Comparison of spectrum sensor performance for different feature vectors when: only the weights of fired neuron are updated; weights are updated in hexagonal topology; weight are updated in grid topology and weights are updated in triangular topology

Table 1. Comparison of different SOM topology by estimation of gap detection performance

SOM Topology	Number of False Alarm Indications (Rate in %)		
	Minimum	Average	Maximum
Single neuron weights update	0	580 625 (9.39 %)	5 804 664 (93.89 %)
Hexagonal Topology	0	194 136 (3.14 %)	824 535 (13.34 %)
Grid Topology	0	178 087 (2.88 %)	487 333 (7.88 %)
Triangular Topology	0	187 025 (3.02 %)	428 846 (6.94 %)

The weights update for the neighboring neurons (to the one, which fires) is giving higher rate of primary user emissions estimation (see Fig. 3). During the experimental investigation, we have selected the neighboring neurons for weights update accordingly to three alternative topologies: hexagonal, grid based (rectangular) and diamond. The number of neighboring neurons are selected by slightly increasing this number from four (2 × 2) to 36 (6 × 6) and testing the constructed spectrum sensor on the same data.

The comparison of received results has shown that in order to get emission estimation rate higher than 90 % at least nine neighboring neurons should be used. The hexagonal structure gave worst results from three structures analyzed; however, the received emission detection rate was still higher comparing to non-topological case. The emission detection rate shows the risk of starting the secondary user transmission when the emission of the primary user (not detected by sensor) is present in the sub-band.

Table 1 shows the results of frequency gap detection in the analyzed 25 MHz band. The false alarm ratio shows the amount of frequency gaps that were not detected by spectrum sensor. The sensing results vary depending on SOM topology and number of neurons used. All proposed spectrum sensor types (one or several topology realizations of each type) are able to find all present frequency gaps (by identifying a sub-band as having no emission when no emission is present). However we should take into account that together with correct frequency gaps found there are additionally not detected emissions (percentage of correctly detected emissions is given in Fig. 3) – additional amount of sub-bands marked as a gap, but with emission present. In average, the least number of mistakes is received using grid based topology (see Table 1). The number of mistakes in average for the triangular topology is close to the grid based one but the maximum number of mistakes is lower. Taking into account that the number of gap detection mistakes should be minimized, the spectrum sensor based on triangular SOM topology should be considered as the most efficient.

To compare the received spectrum sensing results a previous results received by authors using energy detector and modified standard deviation based spectrum sensors are used [1], [12]. The experimental investigations in previous research were made also using different type of signals. The performance of these detectors depends on the manually selected probability of false alarm ratio. This makes a relation between number of not detected emissions and false alarm ratio. By increasing the false alarm probability, we may decrease the number of not detected emissions and vice versa. Although there is hard to compare precisely the performance of SOM based spectrum sensor, which performs fully automatically, and previously investigated sensors for which the optimal probability of false alarm could be manually selected, the range of received values still could be compared. The best performance of energy detector based spectrum sensor was received with false alarm ratio in the range from 2 % to 5 % (is different for different types of signals) and amount of not detected emissions were in the range from 12 % to 14%.

The spectrum sensing results, received using SOM based spectrum sensor with triangular topology performs with average false alarm ratio close to 3 % and amount of not detected emissions is in the range from 6 % to 15 %.

3 Conclusions

The SOM based spectrum sensor proposed in this paper is able to automatically adapt to the unknown environment and use one firing neuron as indicator of frequency gap in sub-band that could be used for secondary user signal transmission.

The performance of SOM based spectrum sensor analyzed or real radio data. Four different topologies were tested in order to find the optimal topology to maximize the number of detected emissions and minimize the false alarm ratio. The experimental investigation results (false alarm ratio: 3 %, number of not detected emissions: 6 % – 15 %) compared to the spectrum sensing performance of energy detector based spectrum sensor (false alarm ratio: 2 % – 5 %, number of not detected emissions: 12 % – 14 %). These results shows the ability to replace the semi-adaptive threshold setting technique form energy detector with SOM keeping similar spectrum sensing performance without the need of setting any parameters manually and adding the ability of spectrum sensor to adapt to the environment changes.

References

1. Stašionis, L., Serackis, A.: Experimental study of spectrum sensing algorithm with low cost SDR. In: 22nd International Conference on Electromagnetic Disturbances, pp. 117–120. Technika, Vilnius (2012)
2. Yucek, T., Arslan, H.: A survey of spectrum sensing algorithms for cognitive radio applications. Communications Surveys & Tutorials 11(1), 116–130 (2009)
3. Nair, P.R., Vinod, A.P., Krishna, A.K.: An adaptive threshold based energy detector for spectrum sensing in cognitive radios at low SNR. In: IEEE International Conference on Communication Systems, pp. 574–578 (2010)
4. Zhang, W., Mallik, R.K., Letaief, K.: Optimization of cooperative spectrum sensing with energy detection in cognitive radio networks. IEEE Transactions on Wireless Communications 8(12), 5761–5766 (2009)
5. Joshi, D.R., Popescu, D.C., Dobre, O.A.: Gradient-Based Threshold Adaptation for Energy Detector in Cognitive Radio Systems. IEEE Communications Letters 15(1), 19–21 (2011)
6. Xie, S., Liu, Y., Zhang, Y., Yu, R.: A Parallel Cooperative Spectrum Sensing in Cognitive Radio Networks. IEEE Transactions on Vehicular Technology 59(8), 4079–4092 (2010)
7. Kim, H., Shin, K.G.: In-Band Spectrum Sensing in IEEE 802. 22 WRANs for Incumbent Protection. IEEE Transactions on Mobile Computing 9(12), 1766–1779 (2010)
8. Kim, K., Xin, Y.: Rangarajan, S.: Energy Detection Based Spectrum Sensing for Cognitive Radio: An Experimental Study. In: Global Telecommunications Conference, pp. 1–5. IEEE Press (2010)
9. Tumuluru, V.K., Wang, P., Niyato, D.: A Neural Network Based Spectrum Prediction Scheme for Cognitive Radio. In: IEEE International Conference on Communications, pp. 1–5 (2010)
10. Stasionis, L., Serackis, A.: Burst Signal Detector based on Signal Energy and Standard Deviation. Electronics and Electrical Engineering 30(2), 48–51 (2014)
11. Newman, T.R., Clancy, T.C.: Security Threats to Cognitive Radio Signal Classifiers. In: 4th International Conference on Cognitive Radio Oriented Wireless Networks and Communications, pp. 1–6 (2009)
12. Stasionis, L., Serackis, A.: A new approach for spectrum sensing in wideband. In: EUROCON, pp. 125–132. IEEE Press (2013)

A Collision PSO for Search of Periodic Points

Kazuki Maruyama and Toshimichi Saito

Hosei University, Koganei, Tokyo, 184–8584 Japan
tsaito@hosei.ac.jp

Abstract. This paper studies application of PSO to search of periodic points of the Hénon map. The search problem is translated into discrete multi-solution problem evaluated by plural cost function with logical operation. The PSO can use inter-particle collision that can be effective to avoid trapping into partial/local solutions. Performing basic numerical experiments, we investigate the algorithm capability.

Keywords: Particle swarm optimizer (PSO), multi-solution problem (MSP), multi-objective problem (MOP), Hénon map.

1 Introduction

The particle swarm optimizer is a population-based paradigm for solving optimization problems inspired by flocking behavior of living beings [1] [2]. The particles correspond to potential solutions and construct a swarm. Referring to their past history, the particles communicate to each other and try to find an optimal solution of an objective problem. The PSO is simple in concept, is easy to implement and has been applied to optimization problems in various systems, e.g., signal processors, artificial neural networks, and power electronics [3]-[8].

This paper studies the collision particle swarm optimizer (CPSO) and its application to search of periodic points of the Hénon map. The search problem is translated into a discrete multi-solution problem (DMSP) that aims at finding plural discrete solution regions (DSRs). The DSRs correspond to solutions of analog multi-solution problems (AMSPs [13]-[15]). The DMSP is evaluated in a multi-objective problem (MOP) consisting of plural cost functions and logical operation. The evaluation is different form Ref. [18] that uses one cost function. Since the DMSPs are different from the AMSPs, the existing methods for the ASMs [13]-[17] cannot be applied directly to the DMSP.

The dynamics of the CPSO is governed by a deterministic difference equation defined on a discrete search space [11][18]. Such deterministic systems are useful from viewpoints of reproducibility and simplicity. The CPSO uses the artificial inter-particle collision that can be effective to avoid trapping into local optima. In standard PSOs, such trapping causes stagnation of the search and obstructs to find solutions. The discrete search space is convenient to define the collision. Performing basic numerical experiments, we have confirmed that the artificial collision can be effective to avoid trapping into partial/local solutions. There exist various methods for search of periodic points of dynamical systems, however,

V.M. Mladenov and P.C. Ivanov (Eds.): NDES 2014, CCIS 438, pp. 59–67, 2014.

they are developed for local search. The CPSO does not require differentiability of the objective function and is available for both global and local searches.

2 The Discrete Multi-solution Problem

Here we define the DMSP for search of periodic points of the Hénon map [12]:

$$\begin{cases} x_1(n+1) = F_1(x_1(n), x_2(n)) \equiv 1 - ax_1^2 + x_2 \\ x_2(n+1) = F_2(x_1(n), x_2(n)) \equiv bx_1 \end{cases} ab.\ \boldsymbol{x}(n+1) = \boldsymbol{F}(\boldsymbol{x}(n)) \quad (1)$$

where n is a discrete time and $\boldsymbol{x} \equiv (x_1, x_2)$ is the 2-dimensional state variable. As parameters (a, b) vary, this 2-dimensional map can exhibit various chaotic/periodic phenomena [12]. We define periodic points and the DMSP for this map. The definition is available for general discrete dynamical system by replacing (x_1, x_2) with D-dimensional state vector.

A point $\boldsymbol{p}$ is said to be a periodic point with period m if $\boldsymbol{p} = \boldsymbol{F}^m(\boldsymbol{p})$ and $\boldsymbol{p} \neq \boldsymbol{F}^l(\boldsymbol{p})$ for $0 \leq l < m$ where $\boldsymbol{F}^m$ is the m-fold composition of $\boldsymbol{F}$. A periodic point with period 1 is said to be a fixed point.

Exploring desired periodic points is a basic problem to analyze the dynamics. For the analysis, define the basic function $G_m(\boldsymbol{x})$.

$$G_m(\boldsymbol{x}) \equiv \|\boldsymbol{F}^m(\boldsymbol{x}) - \boldsymbol{x}\| \geq 0,\ \boldsymbol{x} \in S_A \equiv \{\boldsymbol{x} \mid x_j \in [X_{Lj},\ X_{Rj}]\},\ j = 1, 2. \quad (2)$$

S_A is the analog search space. j is the dimension. $\|\cdot\|$ denotes the Euclidean distance is used in this paper. If m is not a prime number, it is described by the prime factorization: $m = m_1^{n_1} \times \cdots \times m_K^{n_K}$ where $m_k \geq 2$ is the k-th submultiple and $k = 1, \cdots, K$. The m-periodic point is given by the solution of

$$G_m(\boldsymbol{x}) = 0,\ G_{m/m_k}(\boldsymbol{x}) \neq 0,\ \boldsymbol{x} \in S_A \quad (3)$$

Let $\boldsymbol{x}_s$ be solutions of this equation. In general, the number of solutions is at least m and is a multiple of m. Equation (3) has multiple-solution $\boldsymbol{x}_{s1}, \cdots, \boldsymbol{x}_{sN_s}$ where N_s is the number of solutions. If m is a prime number then the periodic point is given by $G_m(\boldsymbol{x}) = 0$, $G_1 \neq 0$ for $\boldsymbol{x} \in S_A$. Discretizing the S_A onto M^D lattice points, Eq. (3) is transformed into the DMSP:

$$G_m(\boldsymbol{x}) \leq T_m,\ G_{m/m_k}(\boldsymbol{x}) \geq T_{m/m_k},\ \boldsymbol{x} \in S_D, \quad (4)$$

$$S_D = \{\boldsymbol{x} \mid x_j \in \{d_{0j}, \cdots, d_{Mj}\}\},\ d_{nj} = X_{Lj} + n(X_{Rj} - X_{Lj})/M$$

where $n = 0, \cdots, M$. M is the number of lattice points per one dimension and T_m is a criteria for the DMSP. The DMSP intends to find N_s disjoint subsets $\text{DSR}_i \in S_D$ whose elements satisfy Eq. (4).

For simplicity, this paper considers a basic problem: search of periodic points with period 4.

$$G_4(\boldsymbol{x}) = 0,\ G_2(\boldsymbol{x}) \neq 0, \boldsymbol{x} \in S_A \quad (5)$$

Figure 1 (a) shows contour maps of G_4 with solutions. The objective periodic points correspond to four solutions of Eq. (5). Discretizing this, we obtain a DMSP.

$$G_4(\boldsymbol{x}_i) \le T_4,\ G_2(\boldsymbol{x}_i) \ge T_2,\ \boldsymbol{x} \in S_D \equiv \{\boldsymbol{x} \mid x_j \in \{d_{0j}, \cdots, d_{Mj}\}\} \quad (6)$$

where $d_{nj} = X_{Lj} + n(X_{Rj} - X_{Lj})/M$, $j = 1, 2$. The DMSP aims at finding 4 disjoint subsets DSR_1 to DSR_4 whose elements satisfy Eq. (6).

3 The CPSO Algorithm

In order to define the CPSO, we introduce several notations. Let t be a time step and let $\boldsymbol{P}^t \equiv (P_1^t, \cdots, P_N^t)$ denote the particle swarm at step t. N denotes the number of particles. Let $P_i^t \equiv (\boldsymbol{x}_i^t, \boldsymbol{v}_i^t)$ be the i–th particle, $\boldsymbol{x}_i^t \equiv (x_{i1}^t, \cdots, x_{iD}^t)$ be its position and let $\boldsymbol{v}_i^t \equiv (v_{i1}^t, \cdots, v_{iD}^t)$ be its velocity where $i = 1, \cdots, N$. The position $\boldsymbol{x}_i^t$ is a potential solution. The update of the particle is based on the personal best position ($\boldsymbol{pbest}_i^t$) and local best position ($\boldsymbol{lbest}_i^t$). The personal best gives the best value in the past history and the best value is given in the algorithm. The local best is the best of the personal bests in a neighbor of P_i^t. The neighbor particles are determined depending on the topology of the particle swarm [2] [16]. For convenience, we use the ring topology where both sides particles are the neighbors of a particle (Fig 2 (a)).

Step 1: Initialization of the particle swarm $\boldsymbol{P}^t$. Let $t = 0$. $\boldsymbol{x}_i^0$ is assigned randomly in S_D and $\boldsymbol{v}_i^0$ is assigned randomly in $S_V = \{\boldsymbol{v} \mid -d_{Mj}/2 \le v_j \le d_{Mj}/2\}$.

Step 2: If some particle satisfies Eq. (6) then particle is declared as a solution candidate.

Step 3: Update of personal best. Depending on the $\boldsymbol{pbest}_i^t$, either of the following two cases is applied:

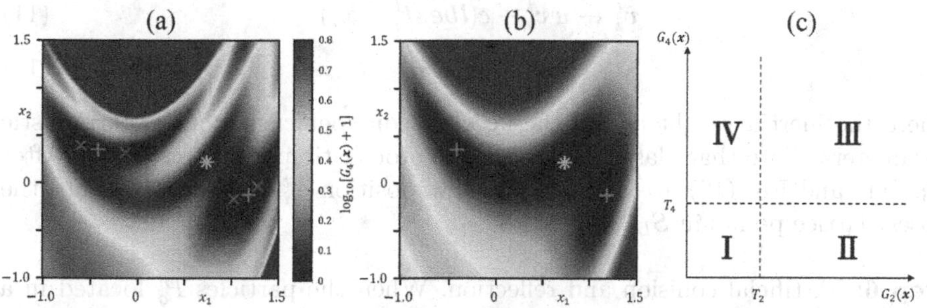

Fig. 1. Hénon map for $a = 0.96875$, $b = 0.3$. (a) Contour map of G_4. Red x, green + and orange *-marks denote periodic point with period 4, period 2 and period 1, respectively. (b) Contour map of G_2. (c) G_2–G_4 plane.

Case 1: $G_4(\boldsymbol{pbest}_i^t) \leq T_4$ AND $G_2(\boldsymbol{pbest}_i^t) < T_2$

$$\boldsymbol{pbest}_i^t = \begin{cases} \boldsymbol{x}_i^t & \text{if } G_2(\boldsymbol{x}_i^t) > G_2(\boldsymbol{pbest}_i^t) \\ \boldsymbol{pbest}_i^t & \text{otherwise} \end{cases} \tag{7}$$

Case 2: $G_4(\boldsymbol{pbest}_i^t) > T_4$ OR $G_2(\boldsymbol{pbest}_i^t) \geq T_2$

$$\boldsymbol{pbest}_i^t = \begin{cases} \boldsymbol{x}_i^t & \text{if } G_4(\boldsymbol{x}_i^t) < G_4(\boldsymbol{pbest}_i^t) \\ \boldsymbol{pbest}_i^t & \text{otherwise} \end{cases} \tag{8}$$

Figure 1 (c) illustrates the G_4 versus G_2 space where Case 1 corresponds to region I and Case 2 corresponds to region II ∪ III ∪ IV. Note that, in Case 1, the G_2 is the object to increase and the function G_4 can increase. In Case 2, G_4 is the object to decrease and G_2 can decrease. This is one method for target Eq. (6) that includes plural basic functions G_4 and G_2.

Step 4: Update of local best.
Case 1: $G_4(\boldsymbol{pbest}_i^t) \leq T_4$ AND $G_2(\boldsymbol{pbest}_i^t) < T_2$

$$\boldsymbol{lbest}_i^t = \begin{cases} \boldsymbol{pbest}_c^t & \text{if } G_2(\boldsymbol{pbest}_c^t) \text{ is the maximum value in } N_i \\ & \quad \text{and } G_2(\boldsymbol{pbest}_i^t) > G_2(\boldsymbol{lbest}_i^t) \\ \boldsymbol{lbest}_i^t & \text{otherwise} \end{cases} \tag{9}$$

where $N_i \equiv \{i-1, i, i+1\}$ is the neighbor of the i-th particle in the ring topology (see Fig. 2 (a)).
Case 2: $G_4(\boldsymbol{pbest}_i^t) > T_4$ OR $G_2(\boldsymbol{pbest}_i^t) \geq T_2$

$$\boldsymbol{lbest}_i^t = \begin{cases} \boldsymbol{pbest}_c^t & \text{if } G_4(\boldsymbol{pbest}_c^t) \text{ is the minimum value in } N_i \\ & \quad \text{and } G_4(\boldsymbol{pbest}_i^t) < G_4(\boldsymbol{lbest}_i^t) \\ \boldsymbol{lbest}_i^t & \text{otherwise} \end{cases} \tag{10}$$

Step 5: The velocity and position of the particles are updated based on the $\boldsymbol{lbest}_i^t$.

$$\boldsymbol{v}_i^t \leftarrow w\boldsymbol{v}_i^t + c(\boldsymbol{lbest}_i^t - \boldsymbol{x}_i^t) \tag{11}$$

$$\boldsymbol{x}_i^t \leftarrow \boldsymbol{x}_i^t + \boldsymbol{v}_i^t \tag{12}$$

where the inertia weight w and the acceleration coefficient c are deterministic parameters. Note that classic PSOs include stochastic parameters [1] [2]. After Eq. (11) and Eq. (12) are applied, the new position $\boldsymbol{x}_i^t$ is discretized onto the closest lattice points in S_D.

Step 6: Artificial collision and reflection. When the particles P_q^t located in a vicinity area of the particles P_p^t, the artificial collision is applied.

$$\begin{matrix} \boldsymbol{v}_p^t \leftarrow ((\boldsymbol{v}_q^t - \boldsymbol{v}_p^t) \cdot \hat{\boldsymbol{c}})\hat{\boldsymbol{c}} + \boldsymbol{v}_p^t \\ \boldsymbol{v}_q^t \leftarrow ((\boldsymbol{v}_p^t - \boldsymbol{v}_q^t) \cdot \hat{\boldsymbol{c}})\hat{\boldsymbol{c}} + \boldsymbol{v}_q^t \end{matrix}, \quad \hat{\boldsymbol{c}} = \frac{\boldsymbol{x}_q^t - \boldsymbol{x}_p^t}{|\boldsymbol{x}_q^t - \boldsymbol{x}_p^t|} \tag{13}$$

where "·" denotes the inner product and $\hat{c}$ is a unit vector derived from position of the two particles. For simplicity, we have constructed the vicinity area of the i-th particle by its joint 8 lattice points as shown in Fig. 2(b). Using this velocity, the positions are changed:

$$\boldsymbol{x}_p^t \leftarrow \boldsymbol{x}_p^t + \boldsymbol{v}_p^t,\ \boldsymbol{x}_q^t \leftarrow \boldsymbol{x}_q^t + \boldsymbol{v}_q^t \tag{14}$$

For simplicity, the collision is assumed to occur only between two particles. The collision detection is performed from the t-th particle to the $t+N$-th particle modulus N in order to avoid unfair collision detection. Note that the discrete search space is convenient to define the collision. When some particle P_i^t hits the border of S_D, the particle is reflected at the border:

$$\begin{cases} v_{ij}^t \leftarrow v_{ij}^t - 2(x_{ij}^t - d_{0j}),\ \ x_{ij}^t \leftarrow x_{ij}^t - 2(x_{ij}^t - d_{0j}) & \text{if } x_{ij} \text{ hits } X_{Lj} \\ v_{ij}^t \leftarrow v_{ij}^t - 2(x_{ij}^t - d_{Mj}),\ x_{ij}^t \leftarrow x_{ij}^t - 2(x_{ij}^t - d_{Mj}) & \text{if } x_{ij} \text{ hits } X_{Rj} \end{cases} \tag{15}$$

After the artificial collision and reflection, the updated particle position is discretized onto the closest lattice points in S_D.

Step 7: Let $t \rightarrow t+1$, go to **Step 2**, and repeat until $t = t_{max}$.

After this algorithm, we obtain several sets of candidate particles extracted in **Step 2**. If we can construct desired DSRs from the candidate particles as shown in Fig. 3 (f), the DMSP is said to be solved successfully.

4 Numerical Experiments

We have applied the CPSO to the DSM defined in Eq. (6): search of 4 DSRs that correspond to periodic points with period 4 of the Hénon map. Since it

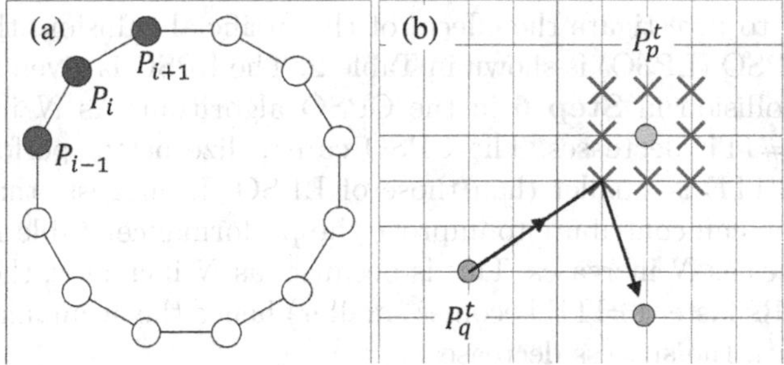

Fig. 2. Neighbor and vicinity. (a) Ring topology. Both sides particles (P_{i-1} and P_{i+1}, (blue)) are the neighbors of the particle P_i (red). (b) Vicinity for collision. Blue x-mark lattice points construct vicinity of particle P_p^t (yellow). If a P_q^t (green) moves into the vicinity then it is judged to collide with P_p^t.

is hard to investigate effects of all the parameters, we have fix the parameter values after trial-and-errors: $N = 30$, $t_{max} = 50$, $M = 128$, $T_4 = T_2 = 0.05$, $w = 0.7$ and $c = 1.4$. Note that the brute force require $M \times M = 2^{14}$ iterations and this CPSO executes at most $N \times t_{max} = 1500$ iterations in total. Although 2^{14} iterations are possible and all the DSRs must be found in this situation, our purpose is not only finding all the DSRs. Our purpose is also to investigate basic capability of the CPSO and provide basic information for developments of efficient search algorithm of various systems including the Hénon map. As M and/or D increases, the brute force is impossible.

Applying the CPSO, desired solution candidates can be found favorably. At $t = t_{max}$, we have obtained four subsets of the solution candidates and have marked DSR_1 to DSR_4 manually as shown in Fig. 3 (f). The DSR_1 to DSR_4 include the four periodic points of period 4 of the Hénon map. Note that automatic marking of the DSRs is not easy and is remain in future problems.

Figure 3 shows snapshots of personal bests on the G_2 versus G_4 plain. After the classification of the DRSs, four solution candidates are indicated in the figure. The particles are colored by classification of the four DSRs in Fig. 3 (f). We can see that DSR_1 (brown), DSR_2 (green) and DSR_3 (blue) are found until $t = 7$ and DSR_4 (magenta) is found until $t = 24$. The results suggest that the particle position and personal best do not concentrate near certain DSRs and that the particle swarm preserve adequate diversity until $t = t_{max}$. In usual PSO without the artificial collision, particles tend to be trapped into certain DSRs and hard to escape from the trap [18].

We have performed 100 trials of the numerical experiments for various N, other parameters are fixed. The results are summarized in Table 1 where the following three feature quantities are used:

SR: the average rate of successful runs.
#ITE: the average number of iterations until success.
#COL: the average number of collisions per one particle until success.

where the success means that at least one solution candidate is found in every DSR. In order to investigate the effects of the artificial collision, the results of the local-best PSO (LPSO) is shown in Table 2. The LPSO is given by removing the artificial collision in **Step 6** in the CPSO algorithm. As N increases, SR increases and #ITE decreases. The CPSO can realize better performance: SR is larger and #ITE is smaller than those of LPSO. It suggests that the inter-particle collision can contribute to improve the performance. Table 1 shows that #COL decreases as N increases. This is because, as N increases, the CPSO can find all the DSRs faster (#ITE becomes smaller) hence the accumulated number of collisions until the success decreases.

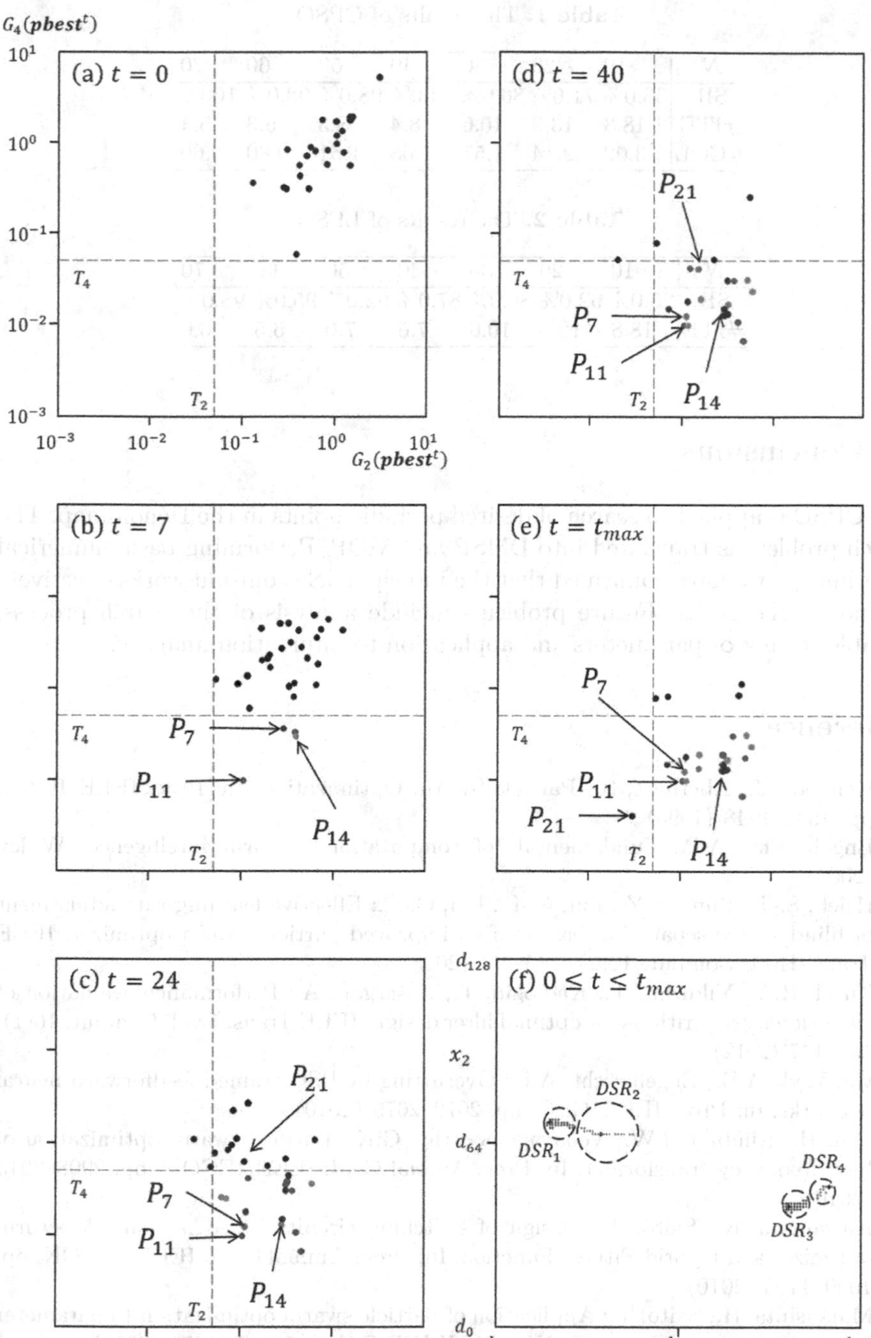

Fig. 3. Personal best on the G_2-G_4 plane in the search process. (a) to (e) Snapshots. Colored particles denote solution candidates in each DSRs. The P_7,P_{11},P_{14} and P_{21} are candidate particles that represent the four DSRs. (f) The color classification of four DSRs.

Table 1. The results of CPSO

N	10	20	30	40	50	60	70
SR	33.0%	71.0%	86.0%	94.0%	98.0%	98.0%	100%
#ITE	18.3	13.9	10.6	8.4	8.3	6.3	5.4
#COL	3.02	2.14	1.57	1.08	1.61	0.80	0.60

Table 2. The results of LPSO

N	10	20	30	40	50	60	70
SR	26.0%	62.0%	80.0%	87.0%	92.0%	95.0%	98.0%
#ITE	18.8	12.0	10.6	7.6	7.0	6.5	6.0

5 Conclusions

The CPSO is applied to search of desired periodic points in the Hénon map. The search problem is translated into DMSP and MOP. Performing basic numerical experiments, we have confirmed that the inter-particle collision works effectively to find all the DSRs. Future problems include analysis of the search process, suitable setting of parameters and application to bifurcation analysis.

References

1. Kennedy, J., Eberhart, R.: Particle Swarm Optimization. In: Proc. IEEE/ICNN, pp. 1942–1948 (1995)
2. Engelbrecht, A.P.: Fundamentals of computational swarm intelligence. Willey (2005)
3. Hsieh, S.-T., Sun, T.-Y., Lin, C.-L., Liu, C.-C.: Effective learning rate adjustment of blind source separation based on an improved particle swarm optimizer. IEEE Trans. Evol. Comput. 12(2), 242–251 (2008)
4. Vural, R.A., Yildirim, T., Kadioglu, T., Basargan, A.: Performance evaluation of evolutionary algorithms for optimal filter design. IEEE Trans. Evol. Comput. 16(1), 135–147 (2012)
5. van Wyk, A.B., Engelbrecht, A.P.: Overfitting by PSO trained feedforward neural networks. In: Proc. IEEE/CEC, pp. 2672–2679 (2010)
6. Qin, H., Kimball, J.W., Venayagamoorthy, G.K.: Particle swarm optimization of high-frequency transformer. In: Proc. Annual Conf. IEEE/IECON, pp. 2908–2913 (2010)
7. Kawamura, K., Saito, T.: Design of switching circuits based on particle swarm optimizer and Hybrid Fitness Function. In: Proc. Annual Conf. IEEE/IECON, pp. 1099–1103 (2010)
8. Matsushita, H., Saito, T.: Application of particle swarm optimization to parameter search in dynamical systems. NOLTA, IEICE E94-N, 10, 458–471 (2011)
9. Sevkli, Z., Sevilgen, F.E.: Discrete particle swarm optimization for the orienteering problem. In: Proc. IEEE/CEC, pp. 1937–1944 (2010)
10. Lee, K.-B., Kim, J.-H.: Mass-spring-damper motion dynamics-based particle swarm optimization. In: Proc. IEEE/CEC, pp. 2348–2353 (2008)

11. Jin'no, K., Shindo, T.: Analysis of dynamical characteristic of canonical deterministic PSO. In: Proc. IEEE/CEC, pp. 1105–1110 (2010)
12. Ott, E.: Chaos in Dynamical Systems. Cambridge Univ. Press (1993)
13. Parsopoulos, K.E., Vrahatis, M.N.: On the computation of all global minimizers through particle swarm optimization. IEEE Trans. Evol. Comput. 8(3), 211–224 (2004)
14. Parrott, D., Li, X.: Locating and tracking multiple dynamic optima by a particle swarm model using speciation. IEEE Trans. Evol. Comput. 10(4), 440–458 (2006)
15. Li, X.: Niching without niching parameters: Particle swarm optimization using a ring topology. IEEE Trans. Evol. Comput. 14(1), 150–169 (2010)
16. Saito, T., Miyagawa, E.: Growing-tree particle swarm optimizer with simple tabu search function. In: Proc. NOLTA, pp. 376–379 (2009)
17. Sano, R., Shindo, T., Jin'no, K., Saito, T.: PSO-based multiple optima search systems with switched topology. In: Proc. IEEE/CEC, pp. 3301–3307 (2012)
18. Maruyama, K., Saito, T.: Deterministic Particle Swarm Optimizers with Collision for Discrete Multi-solution Problems. In: Proc. IEEE/SMC, pp. 1335–1340 (2013)

Minimum Description Length Principle for Fat-Tailed Distributions

Bono Nonchev

Faculty of Mathematics and Informatics, Sofia University, Bulgaria
bono.nonchev@gmail.com

Abstract. While in theory many processes should have normal distribution or at least show asymptotic normality, in practice nonlinear systems often exhibit fat-tail distributions. Non-normality is hard to model, but also hard to detect (model selection). The problem is exacerbated by the varying complexity of the models, i.e. their propensity to overfit.

The Minimum Description Length principle applies Shannon's information theory in statistical enquiry to balance between goodness of fit and model complexity. More specifically, the Normalized Maximum Likelihood (NML) model, stochastic distribution complexity are discussed.

Prior research has shown the distribution complexity for spherical distributions (uncorrelated identically distributed samples) in closed form.

The purpose of this paper is to extend the MDL framework to cover the independent samples case. A general and optimized numerical method for the calculation of the distribution complexity and the stochastic complexity is presented, with results shown for the Student-T distribution.

Keywords: MDL, Model Selection, Complexity, Fat-Tails.

1 Introduction

The problem discussed in this paper is problem of determining the distribution of a sample using the Minimum Description Length principle (MDL). The elements of the sample are assumed to be independent and identically distributed, with a pre-specified marginal distribution.

This problem of model selection is one of the classical problems in statistics. Using a naïve approach to selection using the Neyman-Pearson lemma runs into problems as soon as the simple exact two distribution test (dubbed point hypothesis) is extended to a continuum of hypotheses.

The more sophisticated Bayes factors of [1] yields more convincing results and cast the problem into a Bayesian framework, however there is something deeply unsatisfying in assigning subjective prior probabilities. Using Jeffreys' objective priors instead turns out to be very closely connected to the MDL principle.

The Minimum Description Length principle (MDL) in its most basic form states that the more we can compress the data generated by a process, the more we know about it. This simple idea has some very interesting applications and will be presented briefly in Sect. 2.

V.M. Mladenov and P.C. Ivanov (Eds.): NDES 2014, CCIS 438, pp. 68–75, 2014.

The mathematically simpler case of spherical distributions have been discussed in [2]. Spherical distributions exhibit uncorrelatedness, but not independence, except in the case of normal distributions. Moreover, in a very important sense all spherical distributions have identical descriptive power (goodness of fit *plus* complexity) hence cannot be distinguished by the MDL principle.

Section 2 discusses some basic motivation for the MDL principle and defines NML codes. Section 3 shows previous results for with the scale-location distirbution families. The main result is described in Sect. 4, with details on numerical calculation in Sect. 5. Results, conclusions and future work is presented in Sect. 6.

2 Minimum Description Length Principle

A classic example problem used in the inspirational paper of Kolmogorov [3] is that if you are charged with transmitting three sequences of a million symbols 0 or 1 each, e.g.

- 010101010101010101010101010101010...
- 110110011111101111110110011111111111...
- 101010100011101000111010001110101110...

you can certainly do better than transmitting the whole sequence bit by bit, if you exploit regularities in the data.

The first sequence is just 01 repeated, so sending this instruction instead is quite a lot faster. The second has about 9 ones for each zero, so we can encode long strings of ones with shorter codes, thus transmit shorter codes (on average). For the third there is not much we can do, as it is random with equal probabilities of 0 and 1.

In each case knowledge of the patterns in the data allows us to compress it, which is why the MDL principle equates knowledge with compression.

However, allowing any code to be used renders the problem of finding the shortest codes uncomputable. The main insight of Rissanen for the MDL principle is to restrict the set of codes to those corresponding to probability distributions. In addition we only have to use the distribution as a description method and do not have to assume that the modelled process is really generated by it.

Suppose we have a random variable X with distribution f. There is an optimal code called Shannon-Fano code for X that encodes an obvservation x with a codeword with length

$$L(x) = -\ln f(x) + \Delta \quad (1)$$

where Δ is a constant dependent only on the precision of x we want to achieve, so we skip it in the equations below.

Most research in the area is focused on extending this and finding suitable coding schemes when there are many possible distributions for X. In the literature a set of distributions with some defining characteristic (e.g. the set of all normal distributions) is called a model. For each distribution there is an optimal code, namely the Shannon-Fano code. A single distribution that approximates the codelength for all distributions in a model "well" is called an universal model and the main line of research on the MDL principle is the discovery of those models.

More general overview of the MDL principle can be found in [4] and one focused on statistical modelling in [5].

In this paper the Normalized Maximum Likelihood model is used, first introduced in [6] and subsequently thoroughly explored in various settings. Suppose we have a sample that we want to model using a parametric family with parameter θ and have its MLE $\hat{\theta}(x)$. A natural idea is to construct the distribution

$$f_{NML}(x) = \frac{f\left(x|\hat{\mu}(x), \hat{\sigma}(x)\right)}{\int f\left(y|\hat{\mu}(y), \hat{\sigma}(y)\right) dy} \tag{2}$$

or equivalently use code with length

$$L_{NML}(x) = -\ln f(x|\hat{\theta}(x)) + \ln \int f\left(y|\hat{\mu}(y), \hat{\sigma}(y)\right) dy = -\ln f(x|\hat{\theta}) + COMP_n\left(f\right) .$$

The second term is called the complexity of the model defined and is defined only when $COMP_n\left(f\right) < \infty$.

This is the basis for the stochastic complexity (SC) criterion for model selection: having a finite number of competing models, encode the sample using the NML distribution for each model and choose the one having the smallest codelength $L_{NML}^{\mathcal{M}}(x)$. The chosen model is the best description of the data at hand.

The main result in this paper is the practical formula for the Monte-Carlo computation of the model complexity for i.i.d. samples.

3 Scale-Location Families

In this section some basic definitions and previous results from [2,7] are provided.

As it turns out for scale-location families the model complexity is infinite. There are several ways to deal with the infinities, most notable of which are the *renormalization* by complexity conditional on the data space as presented in [8] and the usage of complexity conditional on the parameter space in [9]. The first approach allows comparison between different distributions, but the renormalization step is not really needed to make sense of the complexity.

A scale-location family is a family of distributions having p.d.f. $f\left(\mathbf{x}^n|\mu, \sigma\right)$ for which a function $g(\mathbf{y}^n)$ exists satisfying

$$f\left(\mathbf{x}^n|\mu, \sigma\right) = \sigma^{-n} g\left(\frac{\mathbf{x}^n - \mu}{\sigma}\right) .$$

The proof for the following trivial property of the MLEs for scale-location families can be found in [2]:

Lemma 1. If $\hat{\mu}(\mathbf{x}^n)$ and $\hat{\sigma}(\mathbf{x}^n)$ are MLE for a scale-location family (i.e. they exist and are unique), then $\hat{\mu}(\sigma\mathbf{y}^n+\mu)=\sigma\hat{\mu}(\mathbf{y}^n)+\mu$ and $\hat{\sigma}(\sigma\mathbf{y}^n+\mu)=\sigma\hat{\mu}(\mathbf{y}^n)$.

The model complexity for many parametric families, including the above, turns out to be infinite, so a natural choice is to use the complexity conditional on $\mathbf{x}^n$. This has been studied in [2,7] and the following important decomposition applies (*[7], Theorem 1*, pp. 109):

$$COMP_n\left(\mathcal{M}|\left\{-R\leq\hat{\mu}\leq R, D\leq\hat{\sigma}\right\}\right)=\ln 2RD^{-1}+\ln DC_n\left(\mathcal{M}\right) . \quad (3)$$

The last term is called the distribution complexity, because it does not depend on the restriction of $\mathbf{x}^n$, freeing the model comparison procedure of the arbitrary bounds R and D.

The deficiency is that the above is only suitable for model selection when the MLEs $\hat{\mu}$ and $\hat{\sigma}$ are comparable accross distributions. This necessitates careful parameterization and the case where the estimates are the sample mean and sample standard deviation, which is the case of spherical distributions, has been thoroughly explored in [2].

4 Model Complexity for Independent and Identically Distributed Samples

In this paper we extend the distribution complexity and alter it slightly in order to cover the independent samples case. The MLEs are no longer the sample mean and standard deviation, so it makes sense to define the range in (3) on sample mean and variance. In this case $DC_n(\mathcal{M})$ is defined as follows:

Definition 1. The distribution complexity is defined as

$$\begin{aligned}DC_n\left(\mathcal{M}\right)&=\mathbb{E}_{\mathbf{Y}^n}\left[s_{\mathbf{Y}^n}\delta\left(\hat{\mu}\left(\mathbf{Y}^n\right)\left(1-\hat{\sigma}\left(\mathbf{Y}^n\right)\right)\right)\right]\\&=\int s_{\mathbf{y}^n}\delta\left(\mu(\mathbf{y}^n)\right)\delta\left(1-\sigma(\mathbf{y}^n)\right)g(\mathbf{y}^n)d\mathbf{y}^n . \quad (4)\end{aligned}$$

In case of spherical distributions with proper parameterization $s_{y^n}=\hat{\sigma}$ and the delta function restricts $\hat{\sigma}=1$, so both definitions coincide. [1]

Noteing the difference in defitions of the model complexity here, the following theorem applies:

Theorem 1. *For a scale-location family the conditional complexity can be decomposed as*

$$COMP_n\left(\mathcal{M}|\mathcal{A}\right)=\ln 2RD^{-1}+\ln DC_n\left(\mathcal{M}\right)$$

where $\mathcal{A}=\{-R\leq\overline{x}\leq R, D\leq s_{\mathbf{x}^n}\}$.

[1] Moreover, the argument in [2] can be simplified by the definition above and remove the reparameterization required to obtain distribution complexity comparable across distributions.

Proof. Most of the proof follows [2] and below we note the different steps. The first step is to rewrite the integral using the standard density $g(\mathbf{x}^n)$:

$$
\begin{aligned}
COMP &= \ln \int_{\mathbf{x}^n \in \mathcal{A}} f\left(\mathbf{x}^n | \mu = \hat{\mu}\left(\mathbf{x}^n\right), \sigma = \hat{\sigma}\left(\mathbf{x}^n\right)\right) d\mathbf{x}^n \\
&= \ln \int I_{\mathcal{A}} \delta\left(\mu - \hat{\mu}\left(\mathbf{x}^n\right)\right) \delta\left(\sigma - \hat{\sigma}\left(\mathbf{x}^n\right)\right) \sigma^{-n} g\left(\frac{\mathbf{x}^n - \mu}{\sigma}\right) d\mu\, d\sigma\, d\mathbf{x}^n .
\end{aligned}
$$

where $I_{\mathcal{A}}$ represents the boundaries. Then the substitution $\mathbf{y}^n = \frac{\mathbf{x}^n - \mu}{\sigma}$ is made, having $|J| = \sigma^n$. An alternative form of $I_{\mathcal{A}}$ can be used to simplify the integrand

$$
\begin{aligned}
\mathcal{A}(\mathbf{x}^n) &= \mathcal{A}(\mathbf{y}^n, \mu, \sigma) \\
&= \left\{ \mathbf{y}^n | \overline{y} \in \left[\frac{-R - \mu}{\sigma}, \frac{R - \mu}{\sigma}\right], s_{\mathbf{y}^n} \in \left[\frac{D}{\sigma}, \infty\right) \right\} \\
&= \left\{ \mathbf{y}^n | \mu \in \left[-R - \sigma\overline{y}, R - \sigma\overline{y}\right], \sigma \geq \frac{D}{s_{\mathbf{y}^n}} \right\} .
\end{aligned}
$$

Making use of Lem. 1 and the basic properties of the delta function we get

$$
\begin{aligned}
&COMP_n\left(\mathcal{M}|\mathcal{A}\right) = \\
&= \ln \int \mathcal{A}(y^n, \mu, \sigma)\sigma^{-2} \delta\left(\hat{\mu}\left(\mathbf{y}^n\right)\right) \delta\left(1 - \hat{\sigma}\left(\mathbf{y}^n\right)\right) \sigma^{-n} g\left(\mathbf{y}^n\right) \sigma^n d\mu\, d\sigma\, d\mathbf{y}^n .
\end{aligned}
$$

Now the only dependence to σ and μ in the integrand remains $\mathcal{A}(y^n, \mu, \sigma)\sigma^{-2}$ so we integrate it as

$$
\begin{aligned}
\int\int \mathcal{A}(y^n, \mu, \sigma)\sigma^{-2} d\sigma d\mu &= \int_{-R-\sigma\overline{y}}^{R-\sigma\overline{y}} \int_{s_{\mathbf{y}^n}}^{\infty} \sigma^{-2} d\sigma\, d\mu \\
&= 2RD^{-1} s_{\mathbf{y}^n} .
\end{aligned}
$$

Then the complexity is expressed as

$$
\begin{aligned}
COMP_n\left(\mathcal{M}|\mathcal{A}\right) &= \ln \int 2RD^{-1} s_{\mathbf{y}^n} \delta\left(\hat{\mu}\left(\mathbf{y}^n\right)\right) \delta\left(1 - \hat{\sigma}\left(\mathbf{y}^n\right)\right) g\left(\mathbf{y}^n\right) d\mathbf{y}^n \\
&= \ln 2RD^{-1} + \ln \int s_{\mathbf{y}^n} \delta\left(\hat{\mu}\left(\mathbf{y}^n\right)\right) \delta\left(1 - \hat{\sigma}\left(\mathbf{y}^n\right)\right) dG(\mathbf{y}^n) .
\end{aligned}
$$

Thus a formula for the calculation of the model complexity for an i.i.d. sample have been obtained. Its usage is discussed in the next section.

5 Numerical Calculation and Challenges

The formulas in this section are calculated with the assumption that the marginal distribution is absolutely continuous, unimodal and the log-likelihood is differentiable with respect to the parameters and the data.

The first challenge in the calculation of the integral is the dimensionality, and as remarked in [7], the formula 4 successfully solves the problem. Even though the integral is high-dimensional (dimenstion is equal to the sample size), it can be efficiently calculated by simulating from an n-dimensional distribution.

Another problem is the delta functions under the integral. A change of variables is done to eliminate the delta function, i.e. $\mathbf{y}^n \to \mathbf{y}^{n-2}, m = \hat{\mu}(\mathbf{y}^n), s = \hat{\sigma}(\mathbf{y}^n)$. Then

1. allow $\mathbf{y}^{n-2}$ to vary independently, i.e. simulate from $G(\mathbf{y}^{n-2})$;
2. solve for y_{n-1} and y_n
3. average to obtain the Monte-Carlo estimate of the integral.

This is the approach applied to obtain the results below. First we change the variables to remove the delta function

$$\begin{aligned} DC_n(\mathcal{M}) &= \int s_{\mathbf{y}^n} \delta\left(\hat{\mu}\left(\mathbf{y}^n\right)\right) \delta\left(1-\hat{\sigma}\left(\mathbf{y}^n\right)\right) dG(\mathbf{y}^n) \\ &= \int s_{\mathbf{y}^n} \delta(m)\, \delta(1-s)\, |J|\, g(y_{n-1}, y_n) g(\mathbf{y}^{n-2}) d\mathbf{y}^{n-2}\, dm\, ds \\ &= \int s_{\mathbf{y}^n_*} |J^*|\, g(y^*_{n-1}, y^*_n) dG(\mathbf{y}^{n-2}) \end{aligned}$$

where, for fixed $\mathbf{y}^{n-2}$, $s_{\mathbf{y}^n_*}$ is the sample standard deviation, J^* is the Jacobian and y^*_{n-1} and y^*_n are solution of

$$\begin{cases} \hat{\mu}(\mathbf{y}^n) &= 0 \\ \hat{\sigma}(\mathbf{y}^n) &= 1 \end{cases} . \tag{5}$$

For some values of $\mathbf{y}^{n-2}$ there is no solution of 5 - for them the integrand is 0. Let T be the set of $\mathbf{y}^{n-2}$ for which there is a solution of 5.

Because the solution is not unique due to the symmetry of y_{n-1} and y_n, we can make it so by enforcing $y_{n-1} \le y_n$ and splitting the integration in two, by symmetry having

$$DC_n(\mathcal{M}) = 2 \int_{T \cap \{y^*_{n-1} \le y^*_n\}} s_{\mathbf{y}^n_*} |J^*|\, g(y^*_{n-1}, y^*_n) dG(\mathbf{y}^{n-2}) .$$

5.1 Jacobian

A more serious problem is the behaviour of $|J^*|$ near to the boundary of T, namely that $|J^*| \to \infty$ when $y^*_{n-1} \to y^*_n$. This is a significant problem if the determinant is calculated using finite differences. Fortunately it is possible to obtain those partial derivatives using the partial derivatives of the log-likelihood.

Writing log-likelihood as $h(\mathbf{y}^n|\mu,\sigma) = \ln f(\mathbf{y}^n|\mu,\sigma)$, y^*_{n-1} and y^*_n can be defined for the interior of T as solution of

$$\frac{\partial h(\mathbf{y}^n|\mu,\sigma)}{\partial \mu} = \frac{\partial h(\mathbf{y}^n|\mu,\sigma)}{\partial \sigma} = 0$$

The total derivatives with respect to m and s will be also equal to zero:

$$0 = \frac{d}{dm}\left[\frac{\partial h(\mathbf{y}^n|\mu,\sigma)}{\partial \mu}\right] = \frac{\partial^2 h(\mathbf{y}^n|\mu,\sigma)}{\partial \mu^2} + \sum_{i=n-1}^{n} \frac{\partial^2 h(\mathbf{y}^n|\mu,\sigma)}{\partial y_i \partial \mu}\frac{dy_i}{dm}.$$

Having this with respect to $\frac{d}{dm}$ and $\frac{d}{ds}$ for both $\frac{\partial h(\mathbf{y}^n|\mu,\sigma)}{\partial \mu}$ and $\frac{\partial h(\mathbf{y}^n|\mu,\sigma)}{\partial \sigma}$ can be rewritten as four linear equations:

$$\begin{pmatrix} \frac{\partial y_{n-1}}{\partial m} & \frac{\partial y_{n-1}}{\partial s} \\ \frac{\partial y_n}{\partial m} & \frac{\partial y_n}{\partial s} \end{pmatrix} = -\begin{pmatrix} \frac{\partial^2 h(\mathbf{y}^n|\mu,\sigma)}{\partial \mu^2} & \frac{\partial^2 h(\mathbf{y}^n|\mu,\sigma)}{\partial \mu \partial \sigma} \\ \frac{\partial^2 h(\mathbf{y}^n|\mu,\sigma)}{\partial \sigma \partial \mu} & \frac{\partial^2 h(\mathbf{y}^n|\mu,\sigma)}{\partial \sigma^2} \end{pmatrix} \begin{pmatrix} \frac{\partial^2 h(\mathbf{y}^n|\mu,\sigma)}{\partial y_{n-1} \partial \mu} & \frac{\partial^2 h(\mathbf{y}^n|\mu,\sigma)}{\partial y_n \partial \mu} \\ \frac{\partial^2 h(\mathbf{y}^n|\mu,\sigma)}{\partial y_{n-1} \partial \sigma} & \frac{\partial^2 h(\mathbf{y}^n|\mu,\sigma)}{\partial y_n \partial \sigma} \end{pmatrix}^{-1}.$$

Since the log-likelihood is analytically differentiable in the case of Student-T distribution, this provides a closed-form solution for the Jacobian. It is precise enough so that the numerical integration can converge.

It is obvious that the divisor is the reason why for the boundary of T the Jacobian becomes infinite when $y^*_{n-1} \to y^*_n$ - the matrix becomes badly conditioned.

5.2 Numerical Optimization

In order to find y^*_{n-1} and y^*_nwe sill have to perform numerical optimization. This is by far the slowest part of the calculation and there is really no way arround it.

To speed up the calculations and since a lot of independent optimizations are performed, an optimized massively parallel algorithm for particle swarm optimization is implemented in C++ that solves for y^*_{n-1} and y^*_n on a GPU.

6 Results and Future Work

Using the above numerical tricks results were computed for three degrees of freedom for the Student-T distribution, as well as the analytic formula for the Normal distribution. Figure 1 summarize and compare their distributional complexity.

It seems that the model complexity for the Student-T distribution is smaller the fatter the tails get. This was also the case for spherical distributions in [2], but here there is no curvature caused by the dependence of the marginals which is very pronounced for some spherical distributions.

The application of the notion of distribution complexity in a more complex model like linear regression or GARCH is left as future work.

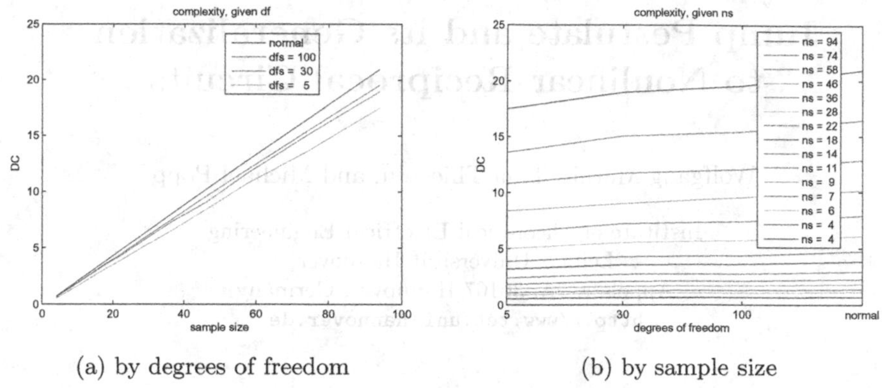

(a) by degrees of freedom (b) by sample size

Fig. 1. The distribution complexity shown in two ways. On the left, the distribution complexity shown for different degrees of freedom. On the right, for different sample sizes.

Acknowledgements. I am particularly grateful to my PhD advisor prof. Plamen Mateev for his continuing support and inspiration.

This work was supported by the European Social Fund through the Human Resource Development Operational Programme under contract BG051PO001-3.3.06-0052 (2012/2014).

The research is partially supported by appropriated state fund for research allocated to Sofia University (contract № 111/2013), Bulgaria.

References

1. Kass, R., Raftery, A.: Bayes factors. Journal of the American Statistical Association 90(430), 773–795 (1995)
2. Nonchev, B.: Minimum Description Length Principle and Distribution Complexity of Spherical Distributions. In: Proceedings of the 18th European Young Statisticians Meeting (2013)
3. Kolmogorov, A.N.: On Tables of Random Numbers. Sankhya: The Indian Journal of Statistics, Series A 25, 369–376 (1963)
4. Grünwald, P.: The Minimum Description Length Principle. MIT Press (2007)
5. Rissanen, J.: Information and Complexity in Statistical Modeling (Information Science and Statistics). Springer (January 2007)
6. Shtarkov, Y.: Universal Sequential Coding of Single Messages. Problems of Information Transmission 23, 175–186 (1987)
7. Nonchev, B.: Minimum Description Length Principle in Discriminating Marginal Distributions. Pliska Studia Mathematica Bulgarica 22(125), 101–114 (2013)
8. Rissanen, J.: MDL Denoising. IEEE Transactions on Information Theory 46(7), 2537–2543 (2000)
9. Stine, R., Foster, D.: The Competitive Complexity Ratio. In: Proceedings of the 2001 Conference on Information Sciences and Systems. WP8, pp. 1–6 (2001)

Historical Remarks on Andronov-Witt's Jump Postulate and its Generalization to Nonlinear Reciprocal Circuits

Wolfgang Mathis, Tina Thiessen, and Michael Popp

Institute of Theoretical Electrical Engineering
Leibniz Universität Hannover
Appelstr. 9A, 30167 Hannover, Germany
`http://www.tet.uni-hannover.de`

Abstract. In this paper we consider circuit equations where so-called impasse points arise such that solutions cannot be continued. The jump postulate was formulated for degenerated linear circuits by Andronov and Witt in 1930. Based on the Brayton-Moser potential of reciprocal and topological complete nonlinear circuits, we develop a method in order to obtain unique trajectories for this class of circuits with switching behavior without regularization. Our approach is illustrated with an example including so-called resonance tunnel diodes.

Keywords: Nonlinear circuits, jump effect, reciprocity, Brayton-Moser potential.

1 Electronic Circuits with Jump Behavior

Since the beginning of the electromagnetic age, switching systems were of fundamental importance. For example, Morse signals of telegraph communication systems were generated by a switch that is human controlled and also telephone systems need switches or electrical relays as essential parts. In these cases, switches were realized by electromechanical systems such as relays. Switching behavior of physical systems with an electrical arc was already observed by Blondel [6]; cf. [14]. However, with the appearance of electronic tubes, engineers tried to realize switching functionality using these new electronic devices. A first tube circuit was published by Abraham and Bloch [1] in 1917 where a periodic switching signal was generated with their "multivibrator", but because of the first world war the corresponding journal paper was published in 1919. Nearly at the same time, Eccles and Jordan [10] presented their "ionic relay". In dependence of a electrical control signal this tube circuit switched between two circuit states such that it is called later on as "flip-flop" circuit. In 1934 Schmitt developed his "thermionic trigger circuit" and published it in 1938 [31]. Although these circuits are different in its details there are two essential similarities:

1. As functions of time, currents and voltages of the circuits are assumed to have only two values (including a tolerance) – two-valued circuits – where a changing from one value to the other occurs in a very fast switching process.

V.M. Mladenov and P.C. Ivanov (Eds.): NDES 2014, CCIS 438, pp. 76–83, 2014.

2. For creating a fast switching behavior, the positive feedback principle will be used. Otherwise, a nonlinear current-voltage characteristic with a negative section is needed.

Since two-valued circuits are basic for the realization of digital systems, several other devices and circuits were constructed with a bivalent behavior.

In the late 1960ies tunnel diodes were developed by Esaki [11], which were intended for use in memory cells. However, these semiconductors were not suitable for integrated circuits, which began at the same time. More recently, Esaki presented resonant tunnel diodes (RTD) [12]. These semiconductor devices are realized using techniques from nanoelectronics and have similar current-voltage characteristic as Esaki's tunnel diodes. Although there are several suggestions for RTD-circuits, this circuit technology is rather in an experimental status; cf. [5].

In 1926 the Dutch physicist van der Pol developed a first mathematical concept for the analysis of multivibrators. He showed that this circuit can be described by van der Pol's famous differential equation [43]

$$\ddot{v} - \varepsilon(1 - v^2)\,\dot{v} + v = 0, \tag{1}$$

if $\varepsilon \gg 0$. Another approach was developed by the Russian scientists Andronov and Witt [2]. These authors formulated "jump conditions" based on physical properties of electronic circuits and derived discontinuous solutions [3]. Further details about relaxation oscillators can be found in a paper of Ginoux and Letellier [14].

Around 1950 Dorodnitsin, Volosov, Tichonov and Gradshtein developed an asymptotic theory of differential equations with a small parameter attached to the higher derivatives. Further results were presented by Pontryagin, Mishchenko, Bogoliubov and Mitropolski; cf. [22], [23], [26]. A more recent monograph is presented by Grasman [16]. Other approaches related to the theory of singular perturbation are the methods of multiple time scales [30] and regularization theory [18], where additional circuit elements are included such that circuit equations have smooth solutions.

Although the analytic approaches of dynamic systems with switching behavior became very powerful, researchers expected additional insights by using a geometrical approach. First ideas were presented in the late 1950ies by the Russian school of geometric differential systems as well as mathematicians from France and US. Fundamental impacts were a paper of Smale on differentiable dynamical systems [33] and a monograph of Arnol'd published in 1971 about the geometrical theory of ordinary differential equations [4]. Moreover, a geometric theory of singular perturbation was presented by Fenichel [13]. A corner stone for circuit theory was Smale's publication [34] in 1972. In his celebrated paper Smale gave a sketch about a new geometric setting of circuit theory and formulated several open questions. The fundamental aspects of this theory were already constructed ten years ago by Moser and Brayton [24], [25] studying tunnel diode circuits. Later on, some open questions of Smale were discussed and the concept was generalized by the Japanese School of Ishiraku and Matsumoto [20], [21] as well as the Dutch mathematician Takens in direction of constrained dynamical systems and researchers from US; e. g. [35], Ihring [17]. In collaboration with many researchers Chua from the UC-Berkeley showed that the geometric theory of nonlinear circuits is suitable to serve as a solid basis of nonlinear circuit theory [8].

Also the dynamical circuits with switching behavior ("jumps") were discussed in a geometrical setting by Tchizawa [36], Sastry and Desoer [30]. Closely related to jump effects, so-called impasse points were studied by many authors (e. g. Reissig [28]), but these exceptional points became also very interesting in circuit simulation and therefore in numerical analysis of differentiable algebraic equations (DAE); for recent details cf. Riaza [29]. However, there are some open questions that will be addressed in this paper.

It was already mentioned by Andronov and Witt [3] (p. 163) that "we are also unable to determine the manner in which the representative point reaches the phase curve" and they pointed out that "the condition of jump, which in the case of linear systems settles this question (see page 37), fails here". Later on Chua and Alexander use the condition of jump for linear systems [9]. We will generalized the condition of jump to reciprocal nonlinear circuits where a Brayton-Moser potential function exists; cf. Weiss et al. [44].

At first we will give a short overview of our setting and discuss the corresponding results; cf. Thiessen et al. [37], [38], [39], [40]. In addition we present some recent results.

2 Geometric Circuit Theory and Jump Effects

At first, we formulate the describing equations of a circuit in a coordinate-free manner that is decomposed into a passive lossless n-port and circuit elements at the ports. For this purpose, we construct the state space $\mathcal{S}$ using Kirchhoff's current and voltage laws and the constitutive current-voltage relationships of the circuit elements. We assume that the Kirchhoff space $\mathcal{K}$ is a n-dimensional linear manifold whereas the Ohm space $\mathcal{O}$ is a smooth manifold. With respect to the currents and voltages both manifolds can be embedded into a $\mathbb{R}^{2n}$. An admissible state space is intersection $\mathcal{K}\cap\mathcal{O}$ where $\mathcal{K}$ and $\mathcal{O}$ are transversal [20]. Furthermore, we assume that we have a 2-form $G\colon \mathbb{R}^\gamma \times \mathbb{R}^\lambda \to T(\mathbb{R}^\gamma \times \mathbb{R}^\lambda) \otimes T(\mathbb{R}^\gamma \times \mathbb{R}^\lambda)$ that is constructed by the voltage of the γ capacitors and the currents of the λ inductors and a 1-form Ω that is zero on $\mathcal{K}$ [20]. Using a pullback π^* of $\pi\colon \mathbb{R}^\gamma \times \mathbb{R}^\lambda \to \mathcal{S}$, we obtain the forms

$$g := \pi^* \circ G, \quad \omega := \pi^* \circ \Omega. \tag{2}$$

Now, we are able to formulate an equation that determines the dynamical vector field X of a circuit

$$g(X, Y) = \omega(Y), \tag{3}$$

for all vector fields $Y\colon \mathcal{S} \to T\mathcal{S}$, and the equation of motion

$$\dot{\xi} = X \circ \xi. \tag{4}$$

It is known from the analysis, that if ω is a closed 1-form, a scalar function $P: \mathcal{S} \to I\!R$ exists with $\omega = dP$ where d is the exterior derivative. In circuit theory ω is a closed 1-form for reciprocal and topological complete circuits [25], [7], [44] and P is called mixed potential function. Then, Brayton-Moser's equation of motion can be derived in local coordinates $\mathbf{i} = (i_1, \ldots, i_\lambda)^T$, $\mathbf{u} = (u_{\lambda+1}, \ldots, u_{\lambda+\gamma})^T$

$$L_\rho \frac{di_\rho}{dt} = \frac{\partial P(\mathbf{i}, \mathbf{u})}{\partial i_\rho}, \tag{5}$$

$$C_\sigma \frac{du_\sigma}{dt} = -\frac{\partial P(\mathbf{i}, \mathbf{u})}{\partial u_\sigma}, \tag{6}$$

with $\rho = 1, \ldots, \lambda$ and $\sigma = \lambda + 1, \ldots, \lambda + \gamma$. Under certain conditions the vector

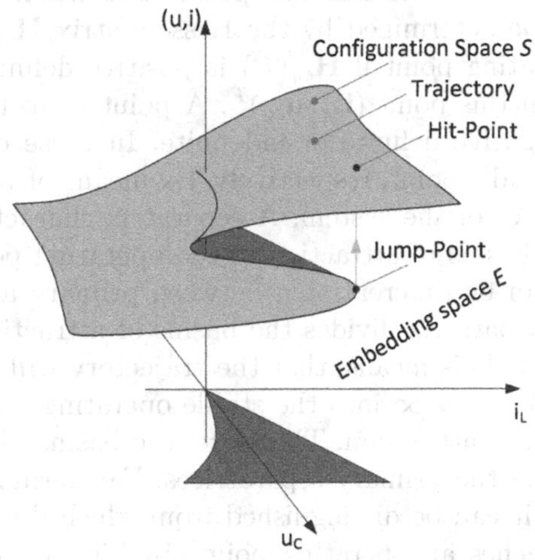

Fig. 1. State Space with Jump

field X is singular and possesses impasse points; cf. [28]. Such cases can be illustrated in Fig. 1 where the state space $\mathcal{S}$ has a fold with respect to the coordinate plane of local coordinates. Using linearization techniques, we can show that the dynamics is stable within the light (blue) areas of $\mathcal{S}$ and unstable in the dark (blue) area; cf. [40].

A trajectory is a solution of (5), (6) in each stable point of $\mathcal{S}$, but if it reaches the edge of the fold the trajectory encounters an impasse point or "jump point". Then, the local equations (5), (6) are not valid anymore and the trajectory cannot be continued. In such a case there are two possibilities: 1) Regularization with additional circuit elements. 2) Application of Andronov-Witt's condition of jump.

After a regularization of the circuit equations the dynamics is well-defined [9] but Tikhonov's theorem has to be taken into account, cf. [42], [23]. From a numerical point of view we obtain highly stiff differential equations where numerical difficulties can arise; cf. Knorrenschild [19]. Therefore, we consider the second possibility. Andronov and Witt formulated their jump postulate for degenerated linear circuits in the following manner [2]: *The energy of the system cannot undergo a jump, and hence the system can jump from one state to another only if it possesses the same energy in the initial and in the final state.* It was already known by Andronov and Witt [2] that the hit-point – the "next" point on the state space $\mathcal{S}$ – is in general not unique in nonlinear circuits. This statement was rediscovered several times but still Andronov-Witt's simple rule was recommended; cf. e. g. [9]. We have shown that even in simple nonlinear circuits more than one candidate for a hit-point arises; cf. [27].

According to Moser [24], an equilibrium point $(\mathbf{i}_{eq}, \mathbf{u}_{eq})^T$, i.e. an operating point, of a circuit, can be specified as point in which the mixed potential function $P(\mathbf{i}, \mathbf{u})$ has an extremum or a saddle point. The stability of an operating point $(\mathbf{i}_{eq}, \mathbf{u}_{eq})^T$ can be determined by the Hesse-matrix $\mathbf{H}_{eq}(P)$ of $P(\mathbf{i}, \mathbf{u})$. A point is a stable operating point if $\mathbf{H}_{eq}(P)$ is positive definite. In this case P exhibits a minimum in this point $(\mathbf{i}_{eq}, \mathbf{u}_{eq})^T$. A point is an unstable operating point if $\mathbf{H}_{eq}(P)$ is negative definite or indefinite. In these cases P exhibits a local maximum or a saddle point, respectively. By means of P it is also possible to construct separatrices of the system. A separatrix characterizes the boundary between different basins of attraction of the operating points, i.e. different system behavior. It can be differentiated between primary and secondary separatrices. A primary separatrix divides the basins of attraction of the different stable operating points. This means, that the trajectory will proceed for different initial conditions for $t \to \infty$ into the stable operating point which belongs to the respective basin of attraction. Therefore, the basins of attraction can be determined by knowing the primary separatrices. Furthermore, by means of a secondary separatrix, it can be distinguished from which direction a trajectory asymptotically approaches an operating point. In Fig. 3 the mixed potential

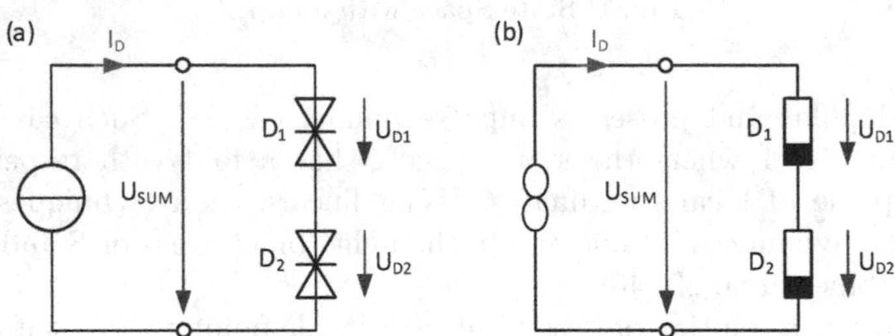

Fig. 2. Series connection of RTDs with a current source; (a) Circuit, (b) Circuit model

function is shown for a series connection of two RTDs (Fig. 2) where the current through the series circuit is chosen as $I_D = 0.8\,I_{\text{peak}}$. As one can see, P exhibits 9 different extrema and 4 minima for this current. Just like a ball in a hilly area the state of the circuit will be guided by the mixed potential function to the proper hit-point. Obviously, the mixed potential function and with it the extrema change in dependence of time. With respect to the current state of the circuit we obtain a unique trajectory.

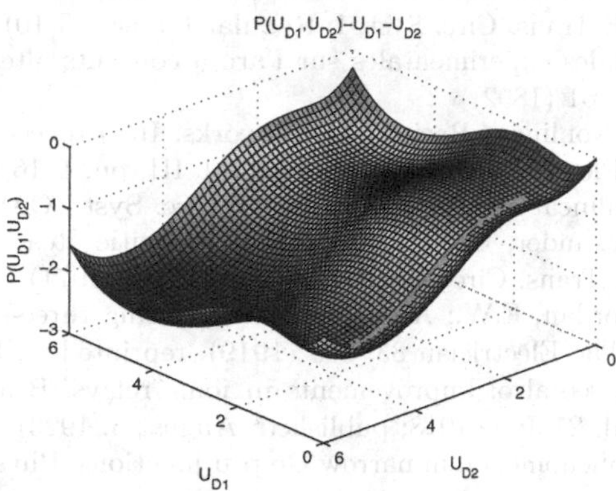

Fig. 3. Mixed potential function for series connected RTDs ($I_D = 0.8\, I_{\text{peak}}$)

Based on our algorithm for calculation a "jump point" along a trajectory [39] we are able to calculate the corresponding hit-point [40]. Moreover, we were able to explain the switching behavior of a so-called MOBILE circuit which consists of RTD devices in such a manner that it is consistent with the experimental results; cf. Thiessen et al. [41] and Popp et al. [27].

3 Conclusions

In our paper we discuss electronic circuits that possess jumping behavior and the history of the conditions of jump. Then, we present a coordinate-free description of circuits and show that impasse points and states where a jump can be possible are equivalent. Furthermore, we present an algorithm for calculating a unique hit-point for nonlinear reciprocal circuits and illustrate it by a RTD circuit.

Acknowledgment. We thank Dr. Dr. Jean-Marc Ginoux, Université de Toulon, France, for very helpful advices with respect to the historical part of this paper.

References

1. Abraham, H., Bloch, E.: Mesure en valeur absolue des périodes des oscillations électriques de haute fréquence. J. Phys. Theor. Appl. 9(1), 211–222 (1919)
2. Andronov, A.A., Witt, A.A.: Discontinuous periodic solutions and the theory of the Abraham-Bloch multivibrator. Dokl. Akad. Nauk SSSR (8), 189 (1930) (in Russian)

3. Andronov, A.A., Witt, A.A.: Theory of Oscillations. Princeton University Press, Princeton (1949)
4. Arnol'd, V.I.: Ordinary Differential Equations. Izd. Nauka Moskva (1971,1972) (in Russian)
5. Ramesh, A., Park, S.-Y., Berger, P.R.: 90 nm 32x32 bit Tunneling SRAM Memory Array With 0.5 ns Write Access Time, 1 ns Read Access Time and 0.5 V Operation. IEEE Trans. Circ. Syst.-I: Regular Papers 58(10), 2432–2445 (2011)
6. Blondel, A.: Études experimentales sur l'arc a courants alternatifs. La Lumiére électrique 43, 51–61 (1892)
7. Brayton, R.K.: Nonlinear Reciprocal Networks. In: Proceedings of SIAM-AMS Symposium on Electrical Network Theory, vol. III, pp. 1–16 (1971)
8. Chua, L.O.: Nonlinear circuits. IEEE Trans. Circ. Syst. CAS-31, 69–87 (1984)
9. Chua, L.O., Alexander, G.: The Effects of Parasitic Reactances on Nonlinear Networks. IEEE Trans. Circ. Syst. CT-18, 520–532 (1971)
10. Eccles, W.H., Jordan, F.W.: A trigger relay utilizing three-electrode thermionic vacuum tubes. The Electrician 83, 298 (1919), reprinted in: Radio Review, 1(3), 143–146 (1919), see also: Improvements in ionic relays, British patent number: GB 148582 (filed: 21 June 1918; published: August 5, 1920)
11. Esaki, L.: New phenomenon in narrow Ge p-u junctions. Phys. Rev. 109, 603–604 (1958)
12. Chang, L.L., Esaki, L., Tsu, R.: Resonant Tunneling in Semiconductor Double Barriers. Appl. Phys. Lett. 24, 593 (1974)
13. Fenichel, N.: Geometric singular perturbation theory for ordinary differential equations. J. Differential Equations 31, 53–98 (1979)
14. Ginoux, J.-M., Letellier, C.: Van der Pol and the history of relaxation oscillations: Toward the emergence of a concept. Chaos 22, 023120 1–15 (2012)
15. Goto, E., Murata, K., Nakazawa, K., Nakagawa, K., Moto-Oka, T., Matsuoka, Y., Ishibashi, Y., Ishida, H., Soma, T., Wada, E.: Esaki Diode High Speed Logical Circuits. IRE Transactions on Electronic Computers EC-9, 25–29 (1960)
16. Grasman, J.: Asymptotic Methods for Relaxation Oscillations and Applications. Springer, Berlin (1987)
17. Ihring, E.: The Regularization of Nonlinear Electrical Circuits. Proc. American Math. Society 47(1), 179–183 (1975)
18. Ikegami, G.: On Regularization of Nonlinear Electrical Circuits. SIAM Journ. Appl. Math. 49(5), 1296–1309 (1989)
19. Knorrenschild, M.: Differential/algebraic Equations as Stiff Ordinary Differential Equations. SIAM Journ. Numerical Anal. 29(6), 1694–1715 (1992)
20. Matsumoto, T.: On the Dynamics of Electrical Networks. Diff. Equations 21, 179–196 (1976)
21. Matsumoto, T., Chua, L.O., Kawakami, H., Ichiraku, S.: Geometric properties of dynamic nonlinear networks: transversality, local-solvability and eventual passivity. IEEE Trans. Circuits and Systems CAS-28, 406–428 (1981)
22. Mishchenko, E., Pontryagin, L.: Differential Equations with a Small Parameter Attached to the Highest Derivatives and Some Problems in the Theory of Oscillations. IRE Trans. Circuit Theory 7, 527–535 (1960)
23. Mishchenko, E., Rozov, N.: Differential Equations with Small Parameters and Relaxation Oscillations. Plenum Press, New York (1980)
24. Moser, J.K.: Bistable Systems of Differential Equations with Applications to Tunnel Diode Circuits. IBM Journ. Res. Dev. 5, 226–240 (1961)
25. Brayton, R., Moser, J.K.: A theory of non-linear networks. I, II, Quart. Appl. Math. 22, 1–33, 81–104 (1964)

26. O'Malley, R.E.: Introduction to Singular Perturbations. Academic Press, New York (1974)
27. Popp, M., Thiessen, T.: Zorn, Ch., Mathis, W.: On the Functionality of RTD MOBILE Circuits - A Geometric Approach. In: Proc. IEEE ISCAS 2014, Melbourne, Australia (2014) (accepted)
28. Reissig, G.: Differential-algebraic equations and impasse points. IEEE Trans. Circuits and Systems, Part I CAS-43, 122–133 (1996)
29. Riaza, R.: Differential-Algebraic Systems - Analytical Aspects and Circuit Applications. World Scientific Publ., New Jersey (2008)
30. Sastry, S.S., Desoer, C.A.: Jump Behavior of Circuits and Systems. IEEE Trans. Circ. Syst. CAS-28(12), 1109–1124 (1981)
31. Schmitt, O.H.: A Thermionic Trigger. Journal of Scientific Instruments 15, 24–26 (1938)
32. Silva-Madriz, R., Sastry, S.S.: Multiple Time Scale for Nonlinear Systems. Circ. Systems Signal Process 5(1), 153–169 (1986)
33. Smale, S.: Differentiable dynamical systems. Bull. Amer. Math. Soc. 73, 747–817 (1967)
34. Smale, S.: On the mathematical foundations of electrical circuit theory. J. Diff. Geom. 7, 193–210 (1972)
35. Takens, F.: Constrained equations: A study of implicit differential equations and their discontinuous solutions. In: Hilton, P. (ed.) SSD 1991. LNCS, vol. 525, pp. 143–234. Springer, Heidelberg (1991)
36. Tchizawa, K.: An Analysis of Nonlinear Systems with Respect to Jump. Yokohama Math. Journ. 32, 203–214 (1984)
37. Thiessen, T., Mathis, W.: Geometric Dynamics of Nonlinear Circuits and Jump Effects, COMPEL: The Intern. Journ. Comp. Mathem. Electr. Electronic Eng. 30(4), 1307–1318 (2011)
38. Sarangapani, P., Thiessen, T., Mathis, W.: Differential Algebraic Equations of MOS Circuits and Jump Behavior. Adv. Radio Sci. 10, 221–327 (2012)
39. Thiessen, T., Mathis, W.: Simulating Circuits with Impasse Points. Adv. Radio Sci. 11, 113–118 (2013)
40. Thiessen, T., Plönnigs, S., Mathis, W.: Fast Switching Behavior in Nonlinear Electronic Circuits: A Geometric Approach. In: Kyamakya, K., Halang, W.A., Mathis, W., Chedjou, J.C., Li, Z. (eds.) Selected Topics in Nonlinear Dynamics. SCI, vol. 459, pp. 99–116. Springer, Heidelberg (2013)
41. Thiessen, T., Zorn, C., Mathis, W.: On the Analysis of Series Connected Tunneling Diodes. In: Proc. NDES 2012, Wolfenbütte, pp. 1–4 (2012)
42. Tikhonov, A.M.: Systems of differential equations containing small parameters for the derivatives. Mat. Sb. 31, 575–586 (1952)
43. van der Pol, B.: On "relaxation-oscillations". Phil. Mag., Series 7, 2, 978–992 (1926)
44. Weiss, L., Mathis, W., Trajkovic, L.: A Generalization of Brayton & Moser's Mixed Potential Function. IEEE Trans. Circ. Syst. CAS-45, 423–427 (1998)

Note on Lossless Transmission Lines Terminated by Triode Generator Circuits

Vasil Angelov

Department of Mathematics, University of Mining and Geology "St. I. Rilski"
1700 Sofia, Bulgaria
angelov@mgu.bg

Abstract. This note is a preliminary announcement of results concerning lossless transmission lines terminated by a circuit equivalent to a triode generator. The circuit consists of *L*-load parallel to in series connected *CR*-loads. Using Kirchhoff's law we formulate boundary conditions corresponding to nonlinear loads. Then we reduce the mixed problem for hyperbolic (Telegrapher) system to an initial value problem for a neutral system on the boundary. We introduce an operator in a suitable function space whose fixed point is a periodic solution of the neutral system. The obtained conditions are easily verifiable on Numerical example.

Keywords: Kirchhoff's law, Lossless transmission line, Mixed problem for hyperbolic system, Neutral equation, Periodic solution, Fixed point theorem.

1 Introduction

The main purpose of the present note is to consider transmission lines terminated by circuits shown on Fig. 1. Such configurations present a simplified scheme of a triode generator circuit (cf. [1]). Various applications one can found in [2]-[6]. The cases of in series and parallel connected elements in [7] and [8] are investigated. The general method for investigation for transmission lines in [9] is proposed. Here we follow this method and first derive boundary conditions and formulate the mixed problem for Telegrapher system in lossless case. Then we reduce the mixed problem to an initial value problem on the boundary. The main difficulty is that every transmission line terminated by a prescribed circuit should be considered individually. So in this case it is not possible to exclude some intermediate voltage functions and one obtains a system of 4 equations for 4 unknown functions. The system obtained is a neutral one (cf. [9]) and we obtain existence of periodic solutions by fixed point method [9].

2 Derivation of the Boundary Conditions by Kirchhoff's Law

We proceed from a lossless transmission line terminated by circuits with nonlinear loads as shown on Fig. 1. Such a configuration is a simplified scheme of triode generator circuit (cf. [1]-[6]). In order to derive boundary conditions generated by nonlinear circuits we use Kirchhoff's law (cf. Fig. 1).

V.M. Mladenov and P.C. Ivanov (Eds.): NDES 2014, CCIS 438, pp. 84–91, 2014.

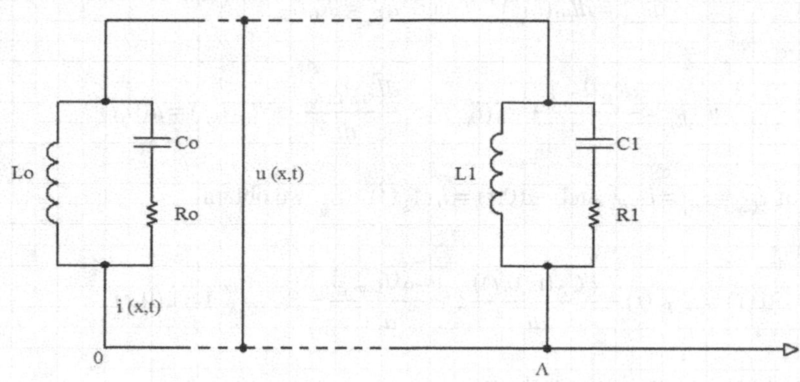

Fig. 1. Lossless transmission line terminated by triode generator circuits

If Λ is the length of the transmission line then propagation time is $T=\Lambda/(1/\sqrt{LC})=\Lambda\sqrt{LC}$, where L is per unit-length inductance and C – per unit-length capacitance, that is, these are specific parameters of the line.

In accordance with Kirchhoff's law (cf. Fig. 1) we have to sum the voltages of the elements C_0 and R_0 and then to sum the current of C_0R_0 with the current of L_0.

The second end of the line is terminated by the same configuration (cf. Fig. 1) and we proceed in the same way. Assume that R_p, L_p and $C_p\,(p=0,1)$ are nonlinear elements, that is, $R_p=R_p(i)$, $L_p=L_p(i)$ and $C_p=C_p(u)$ are prescribed nonlinear characteristics.

Introducing denotations $\tilde{C}_p(u_{C_p})=C_p(u_{C_p}).u_{C_p}$, $\tilde{L}_p(i_{L_p})=L_p(i_{L_p}).i_{L_p}$ we have

$$u_{R_p}=R_p(i),\ u_{\Psi_p}=\frac{d\Psi_p}{dt}=\frac{d\tilde{L}_p(i)}{dt}\equiv\frac{d\left(L_p(i)i\right)}{dt}=\left[i\frac{dL_p(i)}{di}+L_p(i)\right]\frac{di}{dt}$$

and

$$u_{C_pR_p}=R_p(i_{C_pR_p})+\frac{d\tilde{L}_0(i_{C_pR_p})}{di_{C_pR_p}}\frac{di_{C_pR_p}}{dt}\equiv R_p(i_{C_pR_p})+\left(i_{C_pR_p}\frac{dL_p(i_{C_pR_p})}{di_{C_pR_p}}+L_p(i_{C_pR_p})\right)\frac{di_{C_pR_p}}{dt}\,(p=0,1).$$

Then Kirchhoff's law yields $-i(0,t)=i_{C_0R_0}+i_{L_0}$. But $i_{L0}=d\tilde{C}_0(u_{L0})/dt$ and $u_{L_0}=u(0,t)$. Therefore

$$-i(0,t)=i_{C_0R_0}(t)+d\tilde{C}_0(u_{L0})/dt \Leftrightarrow -i(0,t)=i_{C_0R_0}(t)+d\tilde{C}_0(u(0,t))/dt.$$

On the other hand $u_{C_0}+u_{R_0}=u_{C_0R_0}=u(0,t)=u_{L_0}$.

But

$$u_{C_0} = d\tilde{L}_0(i_{C_0})/dt \text{ and } u_{R_0} = R_0(i_{R_0})$$

and therefore

$$u_{C_0R_0} = \frac{d\tilde{L}_0(i_{C_0})}{dt} + R_0(i_{R_0}) \Rightarrow \frac{d\tilde{L}_0(i_{C_0R_0})}{dt} + R_0(i_{R_0}) = u(0,t).$$

In view of $i_{C_0} = i_{R_0} = i_{C_0R_0}$ and $-i(0,t) = i_{C_0R_0}(t) + i_{L_0}$ we obtain

$$-i(0,t) = i_{C_0R_0}(t) + \frac{d\tilde{C}_0(u(0,t))}{dt}; \quad \frac{d\tilde{L}_0(i_{C_0R_0})}{dt} + R_0(i_{C_0R_0}) = u(0,t).$$

For the right end respectively we have:

$$i(\Lambda,t) = i_{C_1R_1}(t) + \frac{d\tilde{C}_1(u(\Lambda,t))}{dt}, \quad \frac{d\tilde{L}_1(i_{C_1R_1})}{dt} + R_1(i_{C_1R_1}) = u(\Lambda,t),$$

where $\tilde{C}_p(u) = uC_p(u) = c_p\sqrt[h]{\Phi_p}\, u/\sqrt[h]{\Phi_p - u}$ and

$d\tilde{C}_p(u)/du = C_p(u) + u\left(dC_p(u)/du\right) \Rightarrow d\tilde{C}_p(u)/dt = \left[C_p(u) + u\left(dC_p(u)/du\right)\right]du/dt$

Remark. When the loads are linear and L_p, C_p, R_p are constants then $u_{\Psi_p} = L_p.di/dt$, $i_{C_p} = C_p.du/dt$, $u_{C_pR_p} = R_p.i_{C_pR_p}$.

Now we are able to formulate the mixed problem for the hyperbolic transmission line system: to find a solution $(u(x,t),i(x,t))$ of the lossless (ideal) transmission line system

$$\frac{\partial u(x,t)}{\partial x} + L\frac{\partial i(x,t)}{\partial t} = 0, \quad \frac{\partial i(x,t)}{\partial x} + C\frac{\partial u(x,t)}{\partial t} = 0$$

for $(x,t) \in \Pi = \{(x,t) \in R^2 : 0 \le x \le \Lambda,\ t \ge 0\}$, satisfying the initial conditions

$$u(x,0) = u_0(x),\ i(x,0) = i_0(x) \text{for } x \in [0,\Lambda]$$

and boundary conditions for $x = 0$:

$$\left[i_{C_0R_0}(t)\left(dL_0(i_{C_0R_0}(t))/di_{C_0R_0}\right) + L_0(i_{C_0R_0}(t))\right]\left(di_{C_0R_0}(t)/dt\right) = -R_0(i_{C_0R_0}(t)) + u(0,t),$$

$$\left[u(0,t)\left(dC_0(u(0,t))/du\right) + C_0(u(0,t))\right]\left(du(0,t)/dt\right) = -i_{C_0R_0}(t) - i(0,t)$$

and for $x = \Lambda$

$$\left[i_{C_1R_1}(t)\left(dL_1(i_{C_1R_1}(t))/di_{C_1R_1}\right)+L_1(i_{C_1R_1}(t))\right]\left(di_{C_1R_1}(t)/dt\right)=-R_1(i_{C_1R_1}(t))+u(\Lambda,t),$$

$$\left[u(\Lambda,t)\left(dC_1(u(\Lambda,t))/du\right)+C_1(u(\Lambda,t))\right]\left(du(\Lambda,t)/dt\right)=-i_{C_1R_1}(t)+i(\Lambda,t).$$

3 Reducing the Mixed Problem to a Periodic Initial Value Problem for Delay Differential Equation on the Boundary

Proceeding from the transformation

$$u(x,t)=U(x,t)/2+I(x,t)/2,\ i(x,t)=U(x,t)/(2Z_0)-I(x,t)/(2Z_0)$$

we reach the system

$$\left(i_{C_0R_0}(t)dL_0(i_{C_0R_0}(t))/di_{C_0R_0}+L_0(i_{C_0R_0}(t))\right)di_{C_0R_0}(t)/dt=-R_0(i_{C_0R_0}(t))+U(0,t)/2+I(0,t)/2$$

$$\left(\frac{U(0,t)+I(0,t)}{2}\frac{dC_0((U(0,t)+I(0,t))/2)}{du}+C_0\left(\frac{U(0,t)+I(0,t)}{2}\right)\right)\frac{d}{dt}\left(\frac{U(0,t)+I(0,t)}{2}\right)=$$
$$=-i_{C_0R_0}(t)-\left(U(0,t)+I(0,t)\right)/2Z_0,$$

$$\left(i_{C_1R_1}(t)dL_1(i_{C_1R_1}(t))/di_{C_1R_1}+L_1(i_{C_1R_1}(t))\right)di_{C_1R_1}(t)/dt=-R_1(i_{C_1R_1}(t))+U(\Lambda,t)/2+I(\Lambda,t)/2$$

,

$$\left(\frac{U(\Lambda,t)-I(\Lambda,t)}{2Z_0}\frac{dC_1((U(\Lambda,t)-I(\Lambda,t))/2Z_0)}{du}+C_1\left(\frac{U(\Lambda,t)-I(\Lambda,t)}{2Z_0}\right)\right)\frac{du(\Lambda,t)}{dt}=$$
$$=-i_{C_1R_1}(t)+\left(U(\Lambda,t)-I(\Lambda,t)\right)/2Z_0.$$

An integration along the characteristics yields

$$U(0,t)=U(\Lambda,t+T),\ I(0,t+T)=I(\Lambda,t).$$

We assume that the unknown functions are $U(0,t)\equiv U(t),\ I(\Lambda,t)\equiv I(t)$ and then in view of the last relations obtain the following neutral system:

$$\frac{di_{C_0R_0}(t)}{dt}=\frac{U(t)+I(t-T)-2R_0(i_{C_0R_0}(t))}{2d\tilde{L}_0(i_{C_0R_0}(t))/di_{C_0R_0}},$$

$$\frac{dU(t)}{dt}=-\frac{dI(t-T)}{dt}+\frac{1}{Z_0}\frac{-U(t)+I(t-T)-2Z_0i_{C_0R_0}(t)}{d\tilde{C}_0((U(t)+I(t-T))/2)/du},$$

$$\frac{di_{C_1R_1}(t)}{dt}=\frac{U(t-T)+I(t)-2R_1(i_{C_1R_1}(t))}{2d\tilde{L}_1(i_{C_1R_1}(t))/di_{C_1R_1}}, \tag{3.1}$$

$$\frac{dI(t)}{dt}=-\frac{dU(t-T)}{dt}+\frac{1}{Z_0}\frac{U(t-T)-I(t)-2Z_0i_{C_1R_1}(t)}{d\tilde{C}_1((U(t-T)+I(t))/2)/du},$$

$$U(t)=\tilde{U}_0(t),\ dU(t)/dt=d\tilde{U}_0(t)/dt, t\in[0,T], \tilde{U}_0(T)=0,$$

$$I(t)=\tilde{I}_0(t),\ dI(t)/dt=d\tilde{I}_0(t)/dt,\ t\in[0,T],\ \tilde{I}_0(T)=0,\ i_{C_0R_0}(T)=0,\ i_{C_1R_1}(T)=0$$

where

$$R_p(i_{C_pR_p})=\sum_{n=1}^{m}r_n^{(p)}(i_{C_pR_p})^n, (p=0,1);\ L_p(i)=\sum_{n=1}^{m}l_n^{(p)}i^n;\ \tilde{L}_p(i)=i\,.L_p(i)$$

and $\tilde{U}_0(t)$, $\tilde{I}_0(t)$ are prescribed periodic functions.

We estimate the nonlinear characteristics of the capacitive elements: $C_p(u)=c_p/\sqrt[h]{1-(u/\Phi_p)}$, where $c_p>0, \Phi_p>0,\ h\in[2,3]$ are constants and $|u|\le\phi_0<\hat{\Phi}=\min\{\Phi_0,\Phi_1\}$. If $u\in[-\phi_0,\phi_0]$ then $\tilde{C}_p(u)=C_p(u).u$ and $d\tilde{C}_p(u)/du$ have strictly positive lower bounds.

Indeed we have

$$\frac{d\tilde{C}_p(u)}{du}=c_p\sqrt[h]{\Phi_p}\frac{\Phi_p-((h-1)/h)u}{(\Phi_p-u)^{(1/h)+1}},\quad \frac{d^2\tilde{C}_p(u)}{du^2}=\frac{2c_p\sqrt[h]{\Phi_p}}{(\Phi_p-u)^{(1/h)+2}}\frac{\Phi_p-((2h+1)/h)u}{h}.$$

If we choose $\phi_0<\hat{\Phi}\min\{h/(1+2h),h/(h-1)\}$ it follows $d\tilde{C}_p(u)/du>0$ and $d^2\tilde{C}_p(u)/du^2>0$ for $u\in[-\phi_0,\phi_0]$ and therefore

$$\min\{\tilde{C}_p(u): u\in[-\phi_0,\phi_0]\}=\tilde{C}_p(-\phi_0)\ge\frac{2c_p\sqrt[h]{\Phi_p}}{(\Phi_p+\phi_0)^{(1/h)+2}}\frac{\Phi_p-(2+(1/h))\phi_0}{h}=\hat{C}_p>0,$$

$$\min\left\{\frac{d\tilde{C}_p(u)}{du}:|u|\le\phi_0\right\}=\frac{d\tilde{C}_p(-\phi_0)}{du}=c_p\sqrt[h]{\Phi_p}\frac{\Phi_p+((h-1)/h)\phi_0}{(\Phi_p+\phi_0)^{(1/h)+1}}=\hat{C}_p^1>0,(p=0,1).$$

We need also the estimates

$$\left|\frac{d\widetilde{C}_p(u)}{du}\right| \le \frac{2c_p \sqrt[h]{\Phi_p}}{\left(\Phi_p-\phi_0\right)^{(1/h)+2}} \frac{\Phi_p+\left((2h+1)/h\right)\phi_0}{h} \equiv M_p;$$

$$\left|\frac{d^2\widetilde{C}_p(u)}{du^2}\right| \le \frac{2c_p \sqrt[h]{\Phi_p}\left(h\Phi_p+\phi_0\right)}{h^2 \sqrt[h]{\left(\Phi_p-\phi_0\right)^{1+2h}}} = H_p \ (p=0,1).$$

For $\widetilde{L}_p(i)$ we get

$$d\widetilde{L}_p(i)/di = i\left(dL_p(i)/di\right)+L_p(i) = \sum_{n=1}^{m}(n+1)l_n^{(p)}i^n; \quad d^2\widetilde{L}_p(i)/di^2 = \sum_{n=1}^{m}(n+1)nl_n^{(p)}i^{n-1}.$$

Assumption (L): $|i(t)| \le i_0 \Rightarrow d\widetilde{L}_p(i(t))/di = \sum_{n=1}^{m}(n+1)l_n^{(p)}\left(i(t)\right)^n \ge \hat{L}_p > 0 \ (p=0,1).$

Assumption (IN): $U_0(.),\ I_0(.) \in C_{T_0}^1[0,T], \left|\widetilde{U}_0(t)\right| \le U_0 e^{\mu(t+T-kT_0)}, \left|\widetilde{I}_0(t)\right| \le U_0 e^{\mu(t+T-kT_0)};$ $(k=0,1,2,\ldots,m-1)$.

Assumption (U): $e^{\mu T_0}\left(U_0+I_0\right)/2 \le \phi_0$.

We introduce the following sets for the unknown functions:

$$M_p = \left\{ i_{C_pR_p}(t) \in C_{T_0}^1[T,2T] : \left|i_{C_pR_p}(t)\right| \le I_{C_p} e^{\mu(t-T-kT_0)}, t \in [T+kT_0, T+(k+1)T_0] \right\},$$

$(p=0,1)$

$$M_U = \left\{ U \in C_{T_0}^1[T,2T] : |U(t)| \le U_0 e^{\mu(t-T-kT_0)}, t \in [T+kT_0, T+(k+1)T_0] \right\},$$

$$M_I = \left\{ I \in C_{T_0}^1[T,2T] : |I(t)| \le I_0 e^{\mu(t-T-kT_0)}, t \in [T+kT_0, T+(k+1)T_0] \right\},$$

$(k=0,1,2,\ldots,m-1)$, where $C_{T_0}^1[T,2T]$ is the set of all continuously differentiable T_0-periodic functions, $I_{C_0}, U_0, I_{C_1}, I_0, T_0, \mu > 0$ are constants and $\mu T_0 = \mu_0 = \text{const}$. By a suitable metric the set $M_0 \times M_U \times M_1 \times M_I$ turns out into a metric space.

Now we formulate the main problem: to find a T_0-periodic solution $\left(i_{C_0R_0}(t), U(t), i_{C_1R_1}(t), I(t)\right)$ of (3.1) on $[T,2T]$.

We define an operator $B = (B_0(t), B_U(t), B_1(t), B_I(t))$ by the formulas:

$$B_p^{(k)}(i_{C_0R_0},U,i_{C_1R_1},I)(t):=\int\limits_{T+kT_0}^{t}I_{C_p}(i_{C_0R_0},U,i_{C_1R_1},I)(s)ds-\frac{t-T-kT_0}{T_0}\int\limits_{T+kT_0}^{T+(k+1)T_0}I_{C_p}(i_{C_0R_0},U,i_{C_1R_1},I)(s)ds,(p=0,1)$$

$$B_U^{(k)}(i_{C_0R_0},U,i_{C_1R_1},I)(t):=\int\limits_{T+kT_0}^{t}V(i_{C_0R_0},U,i_{C_1R_1},I)(s)ds-\frac{t-T-kT_0}{T_0}\int\limits_{T+kT_0}^{T+(k+1)T_0}V(i_{C_0R_0},U,i_{C_1R_1},I)(s)ds,$$

$$B_I^{(k)}(i_{C_0R_0},U,i_{C_1R_1},I)(t):=\int\limits_{T+kT_0}^{t}J(i_{C_0R_0},U,i_{C_1R_1},I)(s)ds-\frac{t-T-kT_0}{T_0}\int\limits_{T+kT_0}^{T+(k+1)T_0}J(i_{C_0R_0},U,i_{C_1R_1},I)(s)ds,$$

on every $[T+kT_0,T+(k+1)T_0]$ (k=0,1,…) whose fixed points are periodic solutions.

Theorem 1. Let the assumptions **(U)**, **(L)** and **(IN)** be fulfilled. Then there exists a unique T_0-periodic solution of (3.1).

4 Numerical Example

Here we consider all inequalities guaranteeing an existence-uniqueness result:

$$e^{\mu_0}(U_0+I_0)/2\le\phi_0;\ \ \phi_0<\hat{\Phi}\min\{h/(1+2h),h/(h-1)\};$$

$$\frac{e^{\mu_0}}{\mu\hat{L}_p}\left(\frac{U_0+e^{-\beta}I_0}{2}+\sum_{n=1}^{m}\left|r_n^{(p)}\right|I_{C_p}^n e^{(n-1)\mu_0}\right)\le I_{C_p}\,(p=0,1).;$$

$$I_0e^{-\beta}+e^{\mu_0}\left(U_0+I_0e^{-\beta}+4Z_0I_{C_0}\right)/\,\mu\hat{C}_0^1Z_0\le U_0;$$

$$U_0e^{-\beta}+e^{\mu_0}\left(U_0e^{-\beta}+I_0+2Z_0I_{C_1}\right)/\,\mu Z_0\hat{C}_1^1\le I_0;$$

$$\dot{K}_p=\left(1+\left(e^{\mu_0}-1\right)/\,\mu_0\right)\left(1/2\mu\hat{L}_p\right)$$

$$\times\left\{\left(1/\hat{L}_p\right)\left[\left(\sum_{n=1}^{m}n\left|r_n^{(p)}\right|I_{C_p}^{n-1}e^{(n-1)\mu_0}\right)\cdot\left(\sum_{n=1}^{m}(n+1)\left|l_n^{(p)}\right|I_{C_p}^{n}e^{n\mu_0}\right)+\right.\right.$$

$$\left.\left.+\left(U_0e^{\mu_0}+I_0e^{\mu_0}+\sum_{n=1}^{m}\left|r_n^{(p)}\right|I_{C_p}^{n}e^{n\mu_0}\right)\sum_{n=1}^{m}(n+1)n\left|l_n^{(p)}\right|I_{C_p}^{n-1}e^{(n-1)\mu_0}\right]+1\right\}<1,$$

$$\hat{\hat{K}}_p=\left(1+\left(e^{\mu_0}-1\right)/\,\mu_0\right)\left(1/\mu\hat{C}_p^1\right)\left(2+(1/Z_0)+e^{\mu_0}\left(U_0+I_0+4Z_0I_{C_p}\right)\left(H_p/2Z_0\hat{C}_p^1\right)\right)<1;$$

For a transmission line with length $\Lambda=100m$, $L=0,45\,\mu H/m$, $C=80\,pF/m$, $v=1/\sqrt{LC}=1/\left(6.10^{-9}\right)$; $Z_0=\sqrt{L/C}=75\Omega$. Then $T=\Lambda\sqrt{LC}=6.10^{-7}$ sec .

Consider a propagation of waves with $\lambda_0=(1/6)10^{-3}m$. We have $T_0=1/\left(\lambda_0\sqrt{LC}\right)=10^{-12}$ .Choose $\mu=10^{12}$ and then $\mu T_0=\mu_0=1$. If

$R_0(i) = R_1(i) = 0{,}028i - 0{,}125i^3$ and $\tilde{L}_0(i) = 6i - (1/3)i^3$ then for $i_0 = 1$ one obtains $6i - (1/3)i^3 > 6 - (1/3) = 17/3$, i.e. $1/\hat{L}_0 = 3/17$. For $h = 2$ we have $C_0(u) = C_1(u) = c_0\sqrt{\Phi_0}/\sqrt{\Phi_0 - u}$. For $\phi_0 = 0{,}15$; $c_0 = c_1 = 5.10^{-11}F$; $\Phi_0 = \Phi_1 = 0{,}4\,V \Rightarrow U_0 < \phi_0 < 0{,}4$; $0{,}15 < 0{,}4\min\{2/5\,;2\} = 0{,}16$; $\hat{C}_0^1 = \hat{C}_1^1 \approx 3{,}7.10^{-11}$; $H_0 = H_1 \approx 46{,}7.10^{-11}$; $H_0/\hat{C}_0^1 = 8{,}8$. The above inequalities for $h = 2$, $U_0 = I_0 = 10^{-3}$; $I_{C_0} = I_{C_1} = 10^{-3}$ become

$e.10^{-3} \le 0{,}15$; $0{,}48.10^{-12}\left((1 + e^{-\beta})/2 + 0{,}028\right) \le 1$; $e^{-\beta} + e(1 + e^{-\beta} + 300)/2760 \le 1$.

Since $K_p, \hat{K}_p$ are of order $1/\mu^2$ we compute $\dot{K}_p \le 4.10^{-15} < 1$ and $\hat{\dot{K}}_p \approx 0{,}153 < 1$.

Conclusion. We notice that in contrast of the classical schemes [9] here we proceed in a different way of derivation of the boundary conditions and respectively this leads to different type of delay differential equations. Our fixed point method, however, is again applicable and introducing a suitable operator we obtain existence-uniqueness of periodic solution for lossless lines using easily verifiable inequalities.

We point out that out method to time-varying specific parameters might be applied.

References

1. Weinstein, L.A., Solntzev, W.A.: Lectures on RF Electronics. Sovetskoe Radio, Moscow (1981) (in Russian)
2. Ramo, S., Whinnery, J.R., Duzer, T.: Fields and Waves in Communication Electronics. John Wiley & Sons, Inc., New York (1994)
3. Rosenstark, S.: Transmission Lines in Computer Engineering. Mc Grow-Hill, New York (1994)
4. Maas, S.A.: Nonlinear Microwave and RF Circuits, 2nd edn. Artech House, Inc., Boston (2003)
5. Misra, D.K.: Radio-Frequency and Microwave Communication Circuits, Analysis and Design, 2nd edn. University of Wisconsin-Milwaukee. John Wiley & Sons, Inc., Publication (2004)
6. Miano, G., Maffucci, A.: Transmission Lines and Lumped Circuits, 2nd edn. Academic Press, New York (2010)
7. Angelov, V.G., Hristov, M.: Distortionless Lossy Transmission Lines Terminated by in Series Connected RCL-Loads. Circuits and Systems 2, 297–310 (2011)
8. Angelov, V.G.: Periodic Regimes for Distortionless Lossy Transmission Lines Terminated by Parallel Connected RCL-loads. In: Transmission Lines: Theory, Types and Applications, pp. 259–293. Nova Science Publishers, Inc. (2011)
9. Angelov, V.G.: A Method for Analysis of Transmission Lines Terminated by Nonlinear Loads. Nova Science, New York (2014)

Using the Transfer Entropy to Build Secure Communication Systems

Fabiano Alan Serafim Ferrari[1], Ricardo Luiz Viana[1], and Sandro Ely de Souza Pinto[2]

[1] Federal University of Parana, Physics Department, post office box: 19044, Curitiba, Brazil

[2] State University of Ponta Grossa, Physics Department, Secretariat - Bloco L - Room 115, Av. Carlos Cavalcanti, 4746, Campus Uvaranas, Ponta Grossa, PR, Zip code: 84.030-900

{ferrari,viana}@fisica.ufpr.br,
desouzapinto@gmail.com

Abstract. We are livining in times in which is very hard to keep in secret documents and data. For this reason communication systems more safe and fast have been required and in this sense we proposed in this work some ways to build secure communication systems using transfer entropy in coupled map lattices.

Keywords: Transfer entropy, coupled map lattice, secure communication systems.

1 Introduction

There are five basic elements to stablish a communication: the information source, where the message will be selected; the emitter, responsible to modify the message in such way that it can be transmitted over a channel; the channel, that is responsible to transport the message until the receiver; the receiver, that is responsible to modify the signal in a message to be readed by the addressee and the addressee that receive the sent message[1]. However to build secure communication systems are necessary other elements: the encipherer, that change the message in a cryptogram; a key, that allows the cryptogram be deciphered, and the decypher that using the key will recovery the message[2].

The possibility to use chaotic systems to develop secure communication systems starts with Pecora and Caroll's work in which they show the criterious to chaotic systems reach the synchronization[3]. After that Cuomo and Oppenheim proposed a circuit exploring the synchronization effect and also applied it in communication[4]. Nowadays there are a huge number of models using chaotic systems to develop communication systems[5,6,7,8].

An intersting method to transmit information is the use of transfer entropy. The transfer entropy measures the transport of information and statiscal coherence between systems[9]. Using this idea Hung and Hu develop a system to transmit binary messages with unidirectional coupled map lattice in which the

V.M. Mladenov and P.C. Ivanov (Eds.): NDES 2014, CCIS 438, pp. 92–99, 2014.

coupling direction (clockwise and counterclockwise) represents the message[10]. Based on this idea in this work we will show how to expand the message size using the transfer entropy and replica of networks[11].

In the section II we show how to use the transfer entropy to detect coupling directions, then in the section III we show how to use this technique to build secure communication systems and in the section IV we present our conclusions.

2 Detecting Coupling Directions with Transfer Entropy

Be I and J two variables described by a chaotic map with a constant domain. We can divide this domain in P partitions with same size. So we define i_n and j_n the respective states in the time n such $i, j = 0, 1, 2, \ldots, (P-1)$. Then the transfer entropy can be defined as

$$T_{J \to I} = \sum p(i_{n+1}, i_n, j_n) \log \frac{p(i_{n+1}|i_n, j_n)}{p(i_{n+1}|i_n)}, \tag{1}$$

in which $p(,), p(|)$ are joint probability and conditional joint probablity respectively and the sum is over all the possible states[9].

If we have two variables X and Y described by

$$x_{n+1} = (1-\varepsilon)f(x_n) + \varepsilon f(y_n), \tag{2}$$
$$y_{n+1} = (1-\delta)f(y_n) + \delta f(x_n), \tag{3}$$

where $f(x)$ and $f(y)$ are the local dynamic and ε, δ are the coupling strength. Then we have three possible situations for coupling: (i) the unidirectional coupling in which the X variable affects the Y variable but the opposite doesn't occur, i.e., $\varepsilon = 0.0$ and $\delta \neq 0$ as described by the figure 1 (a); (ii) the unidirectinal coupling in which the Y variable affects X variable but the opposite doesn't occur, i.e., $\varepsilon \neq 0$ and $\delta = 0.0$ as described by the figure 1 (b); (iii) the bidirectional coupling in which both variables affect each other, $\varepsilon \neq 0$ and $\delta \neq 0$ as described by the figure 1 (c).

For a system described by the equations (2) and (3) we can use as local dynamic the tent map,

$$f(t) = \begin{cases} 2t & \text{if } 0 < t < 0.5 \\ 2(1-t) & \text{if } 0.5 \leq t < 1.0 \end{cases}, \tag{4}$$

and the time evolution for the unidirectional cases and bidirectional case will be undistinguishable as we can see in the figures 2 (a),(b) and (c). But if we evaluate the transfer entropy in both senses, $T_{X \to Y}$ and $T_{Y \to X}$, then will be possible to distinguish the cases because for each case we will have a different measure for the transfer entropy:

Figure 2 (a) $T_{X \to Y} = 0.0017$ and $T_{Y \to X} = 0.0001$ what indicates the coupling strength in one direction is bigger than other.

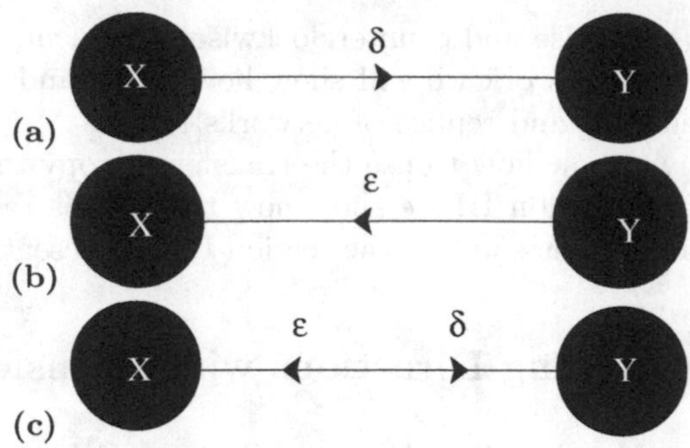

Fig. 1. Three possible coupling configurations: (a) unidirectional coupling $X \to Y$; (b) unidirectional coupling $Y \to X$ and (c) bidirectional coupling

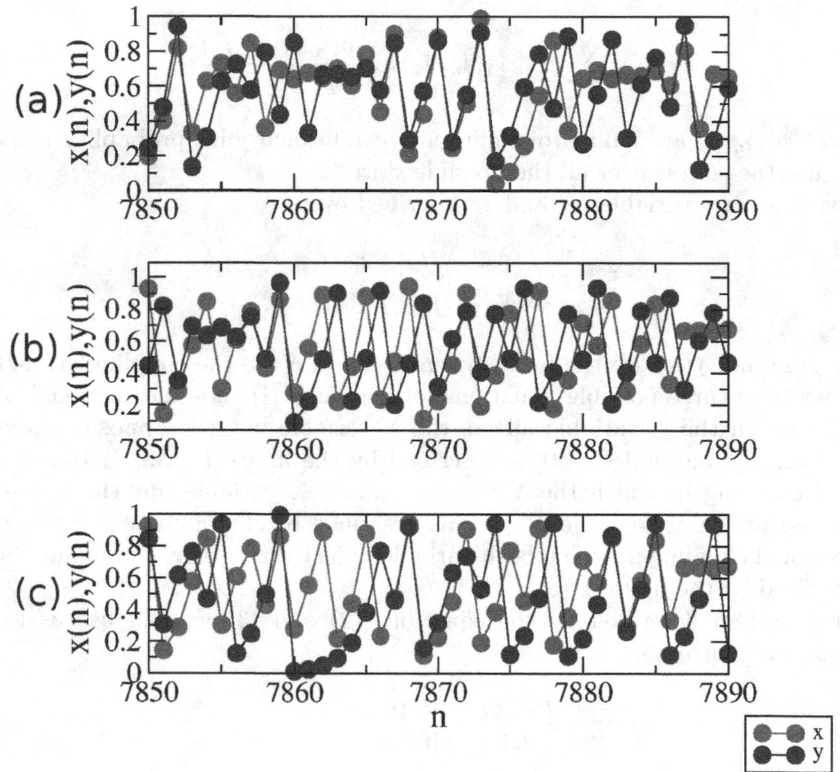

Fig. 2. Time evolution for three different coupling configurations: (a) $\delta = 0.1$ and $\varepsilon = 0.0$; (b)$\delta = 0.1$ and $\varepsilon = 0.1$; (c) $\delta = 0.0$ and $\varepsilon = 0.0$. The color red represents the x variable and the color blue the y variable.

Figure 2 (b) $T_{X \to Y} = 0.0072$ and $T_{Y \to X} = 0.0072$ what indicates the coupling strenghth is the same in both directions.

Figure 2 (c) $T_{X \to Y} = 0.0000$ and $T_{Y \to X} = 0.0000$ what indicates there isn't coupling between the varibles.

In this way, according with the coupling structure we can define a coupling direction and then explore this to build a communication system.

3 Using the Coupling Direction to Build Secure Communication Systems

3.1 Hung and Hu Method

The Hung and Hu model consist in build a coupled map lattice with a local and unidirection coupling[10]. In this model the message is transmitted according with coupling direction. If the coupling is in clockwise direction then the transmitted message is 1, as show in the figure 3 (a) and if the coupling is in counterclokwise direction then the transmitted message is 0, as show in the figure 3 (b).

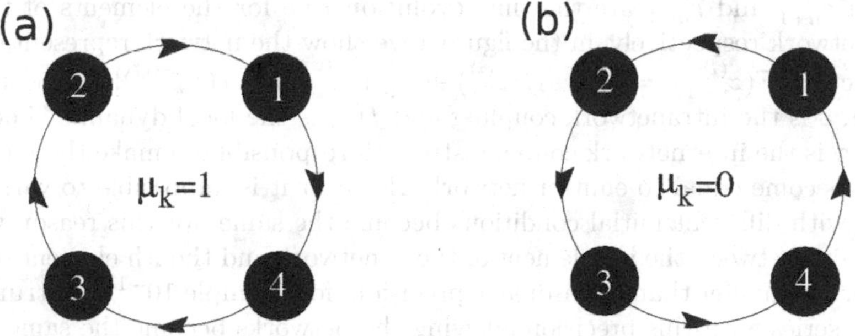

Fig. 3. Description of the message according with the direction: (a) clockwise direction and message=1 and (b) counterclockwise direction and message=0

In a general form the coupled map lattice can be described as

$$x_{n+1}^{(i)} = (1 - \varepsilon) f(x_n^{(i)}) + \varepsilon f(x_n^{(i \pm 1)}) \tag{5}$$

where $\pm$ vary according with the message, $+$ for clockwise direction, $\mu = 1$ and $-$ for counterclockwise direction, $\mu = 0$. The function $f(x)$ is the local dynamic, the tent map is a good option. We have observed that continuous maps work well.

In this model the message is recovery using the transfer entropy to discovery the coupling direction. The security of this system consist in protect the time series of the network elements with a cryptogram then an eavesdropper can't determine the transfer entropy and then can't discovery the hidden message.

3.2 Network Replica Method

While the Hung and Hu method allow us to send only binary messages we present here another method in which we create a network and then a replica of this network what enable us to send the same number of bits as the network size[11]. This method is divided in two stages: the pre-synchronization stage in which we create the replica network and the post-synchronization stage in which we send the message in a safe way.

Pre-synchronization Stage. First of all we assume that the emitter and the receiver are networks. In the pre-synchronization stage the objective is make the emitter and receiver network become the same. To do that we define between the two networks a master-slave coupling where the emitter network (E) is the master and the receiver network (R) is the slave, so the networks can be described as

$$e_{n+1}^{(i)} = F(e_n^{(i)}) \tag{6}$$

$$r_{n+1}^{(i)} = (1-\gamma)F(r_n^{(i)}) + \gamma F(e_n^{(i)}) \tag{7}$$

in which $e_{n+1}^{(i)}$ and $r_{n+1}^{(i)}$ are the time evolution rule for the elements of the E and R network respectively, in the figure 4 we show the network representation. The function $F(z_{n+1}^{(i)}) = (1-\varepsilon)f(z_n^{(i)}) + \frac{\varepsilon}{2}[f(z_n^{(i-1)}) + f(z_n^{(i+1)})]$ is the kernel coupling, ε is the intranetwork coupling and $f(z)$ is the local dynamic. The parameter γ is the internetwork coupling strength responsible to make the receiver network become equal to emitter network. However it is impossible to variables starting with different initial conditions become the same, for this reason when the equality between the ith element of the E network and the ith element of the R network is smaller than a reasonable precision, for example 10^{-14}, we truncate the time series with this precision allowing the networks become the same.

When the network R become a replica of the E network then we turn off the coupling and due to the truncation they won't be different anymore. If for the equations (6) and (7) we choose the tent map and $\gamma = 0.5$ and $\varepsilon = 0.1$ then the replica network can be build. When the networks become the same the second stage starts.

Post Synchronization Stage. In this stage each element of the E network, $e^{(i)}$, will transmit one bit of information. To do that we define a new variable S described by the equation

$$s_{n+1} = (1-\beta)s_n + \frac{\beta}{\eta}\sum_{i=1}^{N} e_n^{(i)} m^{(i)}, \tag{8}$$

in which $\beta \equiv (N-1)/N$, η is the sum of connected sites with S and $m^{(i)}$ is the binary message sent by the ith element of the E network. When the ith element of the E network intend to send the message 1 then this element is connected

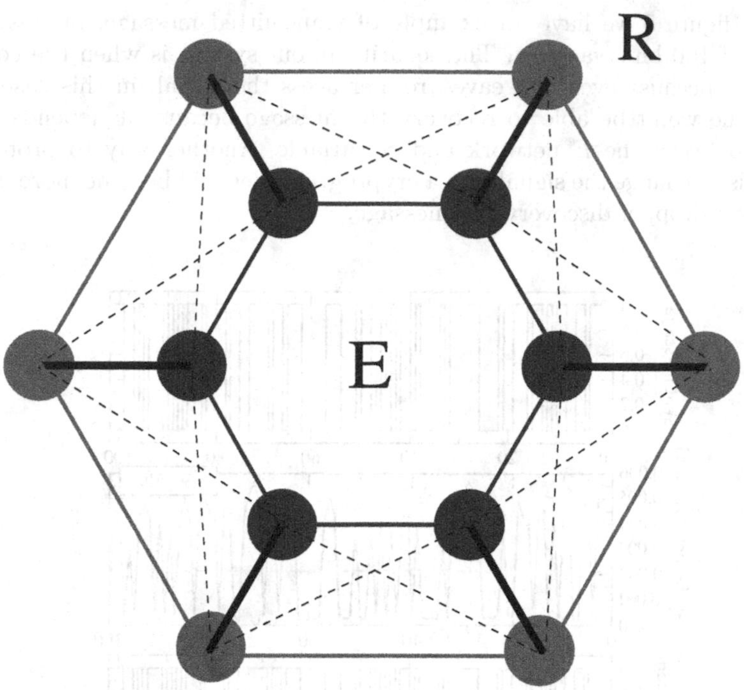

Fig. 4. Network representation. In the color green we have the receiver network (R) and in blue we have the emitter network (E). The lines in the colors green and blue represent the intranetwork coupling (ε). The lines in black represent the internetwork coupling, the coupling strength in the solid lines are equal to $\gamma(1-\varepsilon)$ while in the dashed lines are equal to $\gamma\frac{\varepsilon}{2}$.

with the S variable and $m^{(i)} = 1$ and when the ith element of the E network send the message 0 it is not connected with S and $m^{(i)} = 0$. The message is recovery when the transfer entropy between the ith elements of the R network and the S variable is measured. How the E and R network are the same then the transfer entropy will say if the ith element of the emitter sent or not the message. To differenciate the elements that are or not connected is necessary insert a filter to measure the transfer entropy such that when the measure is above the filter then the result is 1 and the site is connected and when the measure is below then the result is 0 and the site disconnected. In this case the filter is the standard deviation, $\sigma(T)$, related to whole transfer entropy measures. So the recovery message will be

$$m'^{(i)} = \Theta(T_{r_n^{(i)} \to s_{n+1}} - \sigma(T)), \tag{9}$$

in which Θ is the Heaviside step function such

$$\Theta(x) = \begin{cases} 0 \text{ if } x < 0, \\ 1 \text{ if } x \geq 0. \end{cases} \tag{10}$$

In the figure 5 we have an example of transmitted message, in this case a message of 100 bits was sent. The security in our system is when the coupling is turn off because even if a eavesdropper acess the signal, in this case the S variable, he won't be able to recovery the message because it depends on the relation between the E network and S variable. Another way to protect the message is to change the signal into a cryptogram then will become more difficult for an eavesdropper discovery the message.

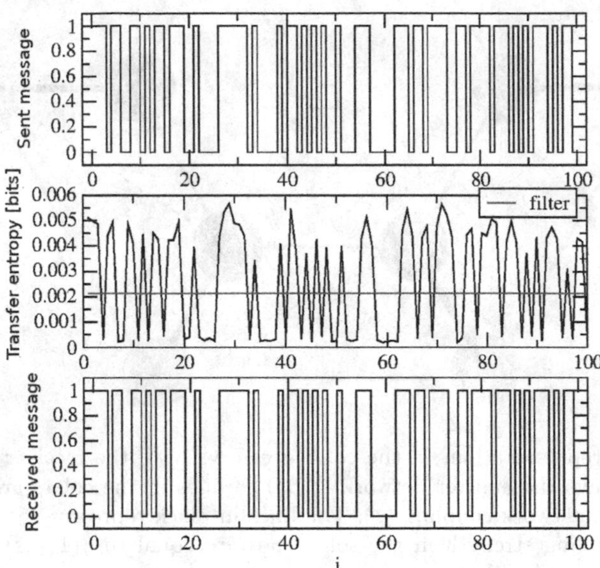

Fig. 5. Example of a message of 100 bits using the replica network method. Above we have the sent message, in the middle the transfer entropy measured between all the elements of the R network and the S variable ($T_{r_n^{(i)} \to s_{n+1}}$) and below we have the received message. In the x-axis i is the index related to each network element. Here the local dynamic is the tent map, $\gamma = 0.5$ and $\varepsilon = 0.1$.

4 Conclusions

The transfer entropy allow us to determine coupling directions in coupled map lattices and this propertie can be used to build secure communication systems. However even with modifications in the Hung and Hu method the system is not fast enough but is safe. In our research we found that continuous map are better to develop this kind of communication systems. We also believe that mechanisms in which the security of the system is depedent of the relation between the emitter, receiver and the signal are important to keep a message in secret.

Acknowledgments. This work was made possible through partial financial support from the following Brazilian research agencies: CNPq, CAPES, and Fundação Araucária. We acknowledge Romeu M. Szmoski for the relevant discussions.

References

1. Shannon, C.E.: A mathematical theory of communication. The Bell System Technical Journal 27(3), 379–423 (1948)
2. Shannon, C.E.: Communication theory of secrecy systems. The Bell System Technical Journal 28(4), 656–715 (1949)
3. Pecora, L.M., Carroll, T.L.: Synchronization in chaotic systems. Physical Review Letters 64(8), 821–826 (1990)
4. Cuomo, K.M., Oppenheim, A.V.: Circuit implementation of synchronized chaos with applications to communications. Physical Review Letters 71(1), 65–69 (1993)
5. Eisencraft, M., Fanganiello, R.D., Grzybowski, Soriano, D.C., Attux, R., Batista, A.M., Macau, E.E.N., Monteiro, L.H.A., Romano, J.M.T., Suyama, R., Yoneyama, T.: Chaos-based communication systems in non-ideal channels. Communications in Nonlinear Science and Numerical Simulation 17(12), 4707–4718 (2012)
6. Jimnez-Rodrgrue, M., Jaimes-Reategui, R., Psarchik, A. N.: Secure communication based on chaotic cipher and chaos synchronization. Discontinuity, Nonlinearity and Complexity 1(1), 57–68 (2012)
7. Ren, H.P., Baptista, M.S., Grebogi, C.: Wireless communication with chaos. Physical Review Letters 110(18), 184101(5) (2013)
8. Szmoski, R.M., Ferrari, F.A.S., Pinto, S.E.S., Viana, R.L., Baptista, M.S.: Cryptography based on chaotic and unsynchronized elements of a network. CRC Press (2013)
9. Schreiber, T.: Measuring information transfer. Physical Review Letters 85, 461–464 (2000)
10. Hung, Y.C., Hu, C.K.: Chaotic communication via temporal transfer entropy. Physical Review Letters 101, 244102(4) (2008)
11. Szmoski, R.M., Ferrari, F.A.S., Pinto, S.E.S., Baptista, M.S., Viana, R.L.: Secure information transfer based on computing reservoir. Physics Letters A 377, 760–765 (2013)

A PROMETHEE – Based Approach for Multiple Objective Voltage Regulator Optimization

Galia Marinova[1] and Vassil Guliashki[2]

[1] Technical University of Sofia, Faculty of Telecommunications, 1000 Sofia, Bul. "Kliment Ohridski", № 8, Bulgaria
gim@tu-sofia.bg

[2] Institute of Information and Communication Technologies – BAS, Section Information processes and decision support systems, 1113 Sofia, "Acad. G. Bonchev" str. Bl. 2, Bulgaria
vggul@yahoo.com

Abstract. An approach to solve the multiple objectives problem for optimal choice of circuit elements parameters among a set of possible alternatives is proposed in this paper. The presented approach is based on the PROMETHEE I method. The considered combinatorial problem is decomposed in sub-problems to reduce considerably the number of investigated alternatives. The approach is illustrated on two voltage regulator circuits. The results obtained are encouraging and the efficiency of the approach increases when the number of circuit elements and the alternatives set cardinality is larger.

Keywords: Voltage Regulator Design and Optimization, Multiple Objective Combinatorial Optimization, PROMETHEE I Method.

1 Introduction

A task in voltage regulator circuits design is to choose the circuit elements in such a manner, that several criteria obtain optimal values. In practice the criteria are often in conflict and multiple objective optimization methods are used to solve this task.

Voltage regulator circuits are seriously investigated and online platforms as PowerEsim [9] and WEBENCH DESIGN CENTER - TI [10] propose automatic multisolution synthesis of switch mode power supply (SMPS) circuits. Both platforms propose an optimization option which includes efficiency, cost and some other criteria, which priority is indicated by the designer. The optimal choice of circuit's elements – transistors (BJT, JFET, and MOSFET), diodes, resistors, capacitors, inductors, transformers, etc. is a combinatorial optimization task, multi-objective in the general case. The full search is time consuming, so different approaches are proposed to solve the optimization task with less calculations. A genetic optimization for SMPS design with smart scan is proposed in [7] and it's illustrated for a circuit in PowerEsim platform with an optimization criterion for maximal efficiency.

A PSpice–based multisolution synthesis of voltage regulator circuits is also proposed in [5]. On the other side statistical optimization methods of voltage regulator circuits design are presented in [3, 4].

V.M. Mladenov and P.C. Ivanov (Eds.): NDES 2014, CCIS 438, pp. 100–113, 2014.

On that basis a multiobjective combinatorial optimization approach is proposed in the paper, applying the PROMETHEE method. Two illustrative examples are given – linear voltage regulator and forward converter circuits. Both circuits are designed for 15 V stabilized output voltage with 1% maximal ripple.

Four criteria are defined:

- Stabilized output voltage to be as close as possible to 15V → **Min | Vst-15 |**
- Minimal output voltage ripple → **Min (Vstripple)**
- Minimal total power dissipation → **Min (P)**
- Maximal efficiency → **Max (η).**

In case the multiple objectives problem for optimal choice of circuit elements parameters among a set of possible alternatives should be solved, the PROMETHEE methods I and II (see [1, 6]) may be very useful. In this paper is proposed a PROMETHEE I – based approach for multiple objective voltage regulator circuit optimizations. The optimization problem can be defined as:

$$\textbf{Min} \quad F(p) = \{F_1(p), F_2(p), \ldots, F_k(p)\} \tag{1}$$

$$\textbf{subject to:} \quad p \in T, \tag{2}$$

where F_i, i=1,…,k; are the optimization objectives, p are the circuit elements parameters, T is a finite set of possible alternatives parameters. The standard circuit elements parameters, permissible in the production, are usually specified in a parameter tables. Below **Vstripple** is denoted by **ΔV**.

Maximization of an objective can easily be transformed in the formulation (1)-(2), because max f(x) = –min (–f(x)). The problem (1)-(2) is a combinatorial one and belongs to the class of NP-hard optimization problems (see [2]). For this reason the statistical optimization methods solving this problem could be very time consuming. Usually the computation time grows exponentially with the problem size, i.e. with the increasing of number of circuit elements parameters (in the concrete case). The PROMETHEE I and II methods [1, 7] are very popular in solving multiple objective problems having as feasible domain a finite set of possible alternatives. Unfortunately they could be also very time consuming tool when the total number of all possible alternatives has to be explored. For this reason an optimization approach combining the problem decomposition with the PROMETHEE I method ranking the explored subset of alternatives is proposed in this paper.

The paper is organized as follows: In section 2 is presented the PROMETHEE – based optimization approach. Section 3 is devoted to the illustrative test instances, which are real circuits. Analysis of the obtained results and some conclusions are given in section 4.

2 The PROMETHEE – Based Optimization Approach

The proposed approach is designed to solve the multiple objective optimization problems interactively with the active participation of the person using the corresponding software (the so called Decision Maker or DM). Before to consider how the investigated subset of alternatives is evaluated and preordered by means of the POMETHEE I method the idea of the problem decomposition will be explained.

2.1 Decomposition of the Optimization Problem

A decomposition of the optimization problem (1)-(2) is proposed here to avoid the full enumeration of the total number of alternatives. This is important, because it could lead to drastic reducing the computational efforts without worsening the quality of the obtained solutions.

The decomposition of the original combinatorial problem is organized in the following manner:

- The DM chooses initially in the interactive dialog regime an important circuit parameter, having as many as possible alternatives. Let it be the parameter t, having the permissible values t_1, t_2, …, t_r; correspondingly. (Also a combination of two circuit parameters may be chosen eventually.)
- Then the set of alternatives T in the problem (1)-(2) is divided into r subsets T_j, j = 1,…,r; each of them containing all possible alternatives for one concrete t_j value. The decomposed problem is formulated as follows:

$$\textbf{Min}\ \ F(p_j) = \{F_1(p_j), F_2(p_j), \ldots, F_k(p_j)\} \tag{3}$$

$$\textbf{subject to:}\ \ p_j \in T_j,\quad \mathbf{j = 1,\ldots,r;} \tag{4}$$

There are r optimization sub-problems for j = 1,…,r;.

2.2 The PROMETHEE – Based Estimation of Alternatives

In this paper the PROMETHEE I method is used to estimate and reorder the different alternatives. It consists in pairwise comparisons. There are three possible relations between the alternatives **a** and **b**, where a ∈ T and b ∈ T. These relations are: preference, indifference and incomparability. They will be denoted by *Pr*, *Ind* and *Inc* respectively. Let be considered the particular criterion f(.), which has to be maximized. The effective choice between the alternatives is made interactively by the DM or by an analyst, according to their feeling of the intensities of preference. In this connection the following parameters have to be fixed:

q - a threshold defining an indifference area,

v - a threshold, defining a strict preference area,

s - a parameter, which value lies between q and v.

Let d = f(a) – f(b). The preference function Pref(a,b) will be considered. It gives the intensity of preference of a over b in function of the deviation d.

A generalized criterion H(d) is associated to each criterion. Six different types of H(d) are defined in the PROMETHEE I method. In this paper is used a generalized criterion H(d) of type 5, known as V-shape with indifference area. It is formulated as follows:

$$H(d) = \begin{cases} 0 & |d| \le q \\ (|d|-q)/(v-q) & q < |d| \le v \\ 1 & |d| > v. \end{cases} \tag{5}$$

This criterion has been often used. For this type of H(d) the intensity of preference increases linearly between q and v.

Let the preference index $\pi(a,b)$ of **a** over **b** over all the criteria be defined in the form:

$$\pi(a, b) = \sum_{l=1}^{k} w_l F_l(a,b), \quad (6)$$

where w_l, l=1,...,k are weights associated to each criterion and $\sum_{l=1}^{k} w_l = 1$.

In this paper is assumed that the weights are equal. In this case $\pi(a,b)$ is simply the arithmetic mean of all the intensities of preference $Pref_j(a,b)$, j = 1,…,k;.
For each pair (a,b) the values $\pi(a,b)$ are calculated. Then for each alternative $a \in T$ the positive outranking flow $\boldsymbol{\Phi}^+(a)$ is calculated as follows:

$$\boldsymbol{\Phi}^+(a) = \sum_{x \in T} \pi(a,x). \quad (7)$$

In our problem the preference index $\pi(b, a) = -\pi(a, b)$. The corresponding negative outranking flow in the problem (1)-(2) is symmetric to $\boldsymbol{\Phi}^+(a)$ and has the opposite direction:

$$\boldsymbol{\Phi}^-(a) = -\boldsymbol{\Phi}^+(a). \quad (8)$$

Hence the positive outranking flow is enough to expresses how each alternative is outranking all the others. The higher $\boldsymbol{\Phi}^+(a)$, the better is the alternative. $\boldsymbol{\Phi}^+(a)$ represents the power of a, it gives its outranking character.

To arrange all explored alternatives in an order according their preference over all the criteria (all the objectives) in the problem (1)-(2) the positive outranking flow $\boldsymbol{\Phi}^+(a)$ is calculated for each explored alternative $a \in T$. Then the alternatives are arranged in a non increasing order of their $\boldsymbol{\Phi}^+$ values.

2.3 The Optimization Approach

- Optimization stage I. Only a part of alternatives in each subset T_j , j=1,...,r; will be explored. The correspondent alternatives are chosen as follows:
  - As first alternative is chosen the alternative, where all parameters $p_j \neq t_j$ are set to their lowest values (the lower bounds of the parameters).
  - The other alternatives included in the subset for exploration are generated by means of the first determined alternative – by changing only one of its parameters to his greatest permissible value consecutively (to the upper bound of the parameter). After the estimation of alternatives the conclusion can be drawn whether the lower or the upper bound value of this parameter gives a better alternative from the point of view of the DM.

  In this way the set of explored alternatives and the computational efforts are reduced considerably. Each alternative of the described subset is estimated by means of PROMETHEE I method.

- When all subset alternatives in all sub-problems are estimated and ordered, only one sub-problem will be separated for further exploration. Here the mean values of each objective for the alternatives in each subset are compared, and the subset, corresponding to the sub-problem with the best mean objective values for the most of the criteria is separated. In case the automatic separation is impossible the DM makes a choice which sub-problem will remain at the next optimization step.
- Optimization stage II. For the separated sub-problem DM determines which parameters should be extra investigated generating alternatives by means of their corresponding intermediate values (between the lower and the upper bound). The generated additional alternatives are estimated by means of PROMETHEE I method.
- STOP-criterion: In case the DM is satisfied with the performed explorations, the optimization procedure terminates. Otherwise the optimization continues repeating the last two steps, where DM determines which parameters should be extra investigated, in order new alternatives to be generated. After that the next optimization stage is executed and the STOP criterion is checked again.

3 Illustrative Test Instances

The PROMETHEE - based approach will be demonstrated on two voltage regulator circuits. The choice of circuit's parameters should optimize the criteria: $|\mathbf{Vst} - 15|$ (the voltage), $\mathbf{\Delta V}$ (the voltage ripple), **P** (the power dissipation) and η (the efficiency). Only η has to be maximized and the other three criteria have to be minimized.

The optimization task is to choose the optimal transistor types (PNP and/or NPN) and the optimal values of the resistors and capacitors in the voltage regulator circuits from a set of standard values.

The voltage regulator circuits are edited in ORCAD/Capture [8] and then simulated with PSpice simulator. The values for the 4 criteria are estimated through PSpice simulations at each optimization step.

3.1 Circuit 1

The first example is the linear voltage regulator circuit presented on Fig. 1. Preliminary simulations have shown that the diode D1 type isn't critical for circuits behavior, so only nine elements are considered in the circuit: the transistors Q1 (NPN) and Q2 (PNP), the 6 resistors R1, R2, R3, R4, R5, R6, and the capacitor C1. The corresponding parameters have the following types and permissible standard values:

$Q1 \in \{$ Q2N2222, BC548A $\}$, $Q2 \in \{$ Q2N 3906, BC559A $\}$,
$R1 \in \{ 1.6k, 1.8k, 2.0k \}$, $R2 \in \{ 110\Omega, 100\Omega, 91\Omega \}$,
$R3 \in \{ 430\Omega, 470\Omega, 510\Omega \}$, $R4 \in \{ 750\Omega, 910\Omega \}$,
$R5 \in \{ 750\Omega, 910\Omega \}$, $R6 \in \{ 1.3k, 1.5k \}$, $C1 \in \{ 4.7\mu, 5.6\mu \}$;

or 1728 alternatives in total.

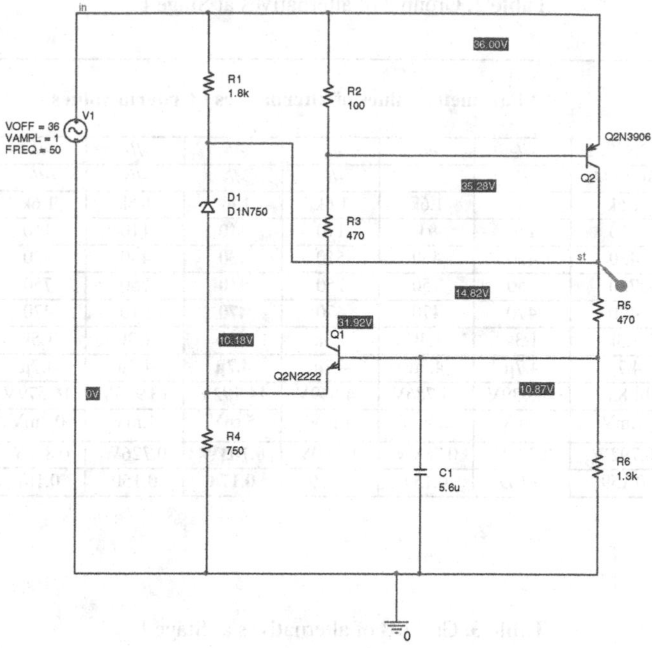

Fig. 1. Linear voltage regulator

Initially DM chooses the combination of transistor Q1 and Q2 to decompose the optimization problem. According the proposed approach the experiment at the first stage of optimization process is performed on the four groups of alternatives, presented in Table 1, Table 2, Table 3 and Table 4 correspondingly:

Table 1. Group 1 of alternatives at Stage I

Elements / Criteria	**Parameter values of alternatives / Criteria values**							
Q1	Q2N2222	-//-	-//-	-//-	**-//-**	-//-	-//-	-//-
Q2	Q2N3906	-//-	-//-	-//-	**-//-**	-//-	-//-	-//-
R1	1.6k	2k	1.6k	1.6k	**1.6k**	1.6k	1.6k	1.6k
R2	110	110	91	110	**110**	110	110	110
R3	430	430	430	510	**430**	430	430	430
R4	750	750	750	750	**910**	750	750	750
R5	470	470	470	470	**470**	510	470	470
R6	1.3k	1.3k	1.3k	1.3k	**1.3k**	1.3k	1.5k	1.3k
C1	4.7μ	4.7μ	4.7μ	4.7μ	**4.7μ**	4.7μ	4.7μ	5.6μ
Vst	14.86V	14.85V	14.796V	14.859V	**14.799V**	13.99V	16.588V	14.86V
ΔV	2.5mV	2.6mV	2.2mV	1.mV	**4.5mV**	2.8mV	0.7mV	1.mV
P	0.792W	0.792W	0.788W	0.792W	**0.703W**	0.727W	0.876W	0.792W
η	0.160	0.160	0.158	0.159	**0.178**	0.150	0.161	0.159

Table 2. Group 2 of alternatives at Stage I

Elements / Criteria	**Parameter values of alternatives / Criteria values**							
Q1	Q2N2222	-//-	-//-	-//-	**-//-**	-//-	-//-	-//-
Q2	BC559A	-//-	-//-	-//-	**-//-**	-//-	-//-	-//-
R1	1.6k	2k	1.6k	1.6k	**1.6k**	1.6k	1.6k	1.6k
R2	110	110	91	110	**110**	110	110	110
R3	430	430	430	510	**430**	430	430	430
R4	750	750	750	750	**910**	750	750	750
R5	470	470	470	470	**470**	510	470	470
R6	1.3k	1.3k	1.3k	1.3k	**1.3k**	1.3k	1.5k	1.3k
C1	4.7μ	4.7μ	4.7μ	4.7μ	**4.7μ**	4.7μ	4.7μ	5.6μ
Vst	14.85V	14.839V	14.786V	14.849V	**14.792V**	13.983V	16.579V	14.85V
ΔV	1.mV	1.mV	2.3mV	1.mV	**5.mV**	4.mV	0.7mV	1.mV
P	0.792W	0.791W	0.788W	0.792W	**0.703W**	0.726W	0.875W	0.792W
η	0.159	0.159	0.159	0.159	**0.178**	0.150	0.161	0.159

Table 3. Group 3 of alternatives at Stage I

Elements / Criteria	**Parameter values of alternatives / Criteria values**							
Q1	BC548A	-//-	-//-	-//-	**-//-**	-//-	-//-	-//-
Q2	BC559A	-//-	-//-	-//-	**-//-**	-//-	-//-	-//-
R1	1.6k	2k	1.6k	1.6k	**1.6k**	1.6k	1.6k	1.6k
R2	110	110	91	110	**110**	110	110	110
R3	430	430	430	510	**430**	430	430	430
R4	750	750	750	750	**910**	750	750	750
R5	470	470	470	470	**470**	510	470	470
R6	1.3k	1.3k	1.3k	1.3k	**1.3k**	1.3k	1.5k	1.3k
C1	4.7μ	4.7μ	4.7μ	4.7μ	**4.7μ**	4.7μ	4.7μ	5.6μ
Vst	14.767V	14.755V	14.676V	14.755V	**14.715V**	13.91V	16.487V	16.767V
ΔV	1.mV	0.5mV	1.mV	0.5mV	**5.mV**	5.mV	0.4mV	1.mV
P	0.786W	0.785W	0.786W	0.785W	**0.698W**	0.721W	0.869W	0.786W
η	0.159	0.158	0.158	0.158	**0.176**	0.150	0.161	0.158

According the mean criteria values in each group of explored alternatives the last three criteria in Group 3 have best values. For this reason the sub-problem 3 having the transistors combination Q1 = BC548A, Q2 = BC559A is chosen by the DM for further optimization.

Table 4. Group 4 of alternatives at Stage 1

Elements/ Criteria	**Parameter values of alternatives / Criteria values**							
Q1	BC548A	-//-	-//-	-//-	**-//-**	-//-	-//-	-//-
Q2	Q2N3906	-//-	-//-	-//-	**-//-**	-//-	-//-	-//-
R1	1.6k	2k	1.6k	1.6k	**1.6k**	1.6k	1.6k	1.6k
R2	110	110	91	110	**110**	110	110	110
R3	430	430	430	510	**430**	430	430	430
R4	750	750	750	750	**910**	750	750	750
R5	470	470	470	470	**470**	510	470	470
R6	1.3k	1.3k	1.3k	1.3k	**1.3k**	1.3k	1.5k	1.3k
C1	4.7μ	4.7μ	4.7μ	4.7μ	**4.7μ**	4.7μ	4.7μ	5.6μ
Vst	14.776V	14.766V	14.714V	14.776V	**14.720V**	13.916V	16.497V	14.777V
ΔV	1.mV	0.5mV	2.mV	1.mV	**5.mV**	3.mV	0.4mV	1mV
P	0.787W	0.786W	0.783W	0.787W	**0.699W**	0.722W	0.869W	0.787W
η	0.159	0.159	0.158	0.159	**0.177**	0.150	0.161	0.159

After the estimation of the alternatives by means of PROMETHEE I method, 5-th H(d) criterion - V-shape with an indifference area the results, presented in Table 5. about the preorder of alternatives in each group are obtained:

Table 5. Preorder of alternatives in each group at Stage I

Preorder of alternatives		1	2	3	4	5	6	7	8
Alternative number	Group 1	5	8	4	3	1	2	6	7
	Group 2	5	6	4	1	2	8	3	7
	Group 3	5	4	2	1	3	8	7	6
	Group 4	5	2	8	4	1	3	6	7

Since the best alternative is the alternative № 5 in each group, followed by the alternatives № 4 and № 2, it could be assumed that corresponding parameter values, which have been changed towards the values in alternative № 1, could be fixed to their upper bound during the further search for the optimal solution.

Hence the following parameters are fixed at the second optimization stage:

Q1 = BC548A, Q2 = BC559A, R1 = 2k, R3 = 510Ω, R4 = 910Ω.

Table 6. Alternatives explored at Stage II

R2	110	100	91	110	100	91	110	100	91	110	100	91	110	100	91	110	100	**91**
R5	470	470	470	470	470	470	510	510	510	510	510	510	510	510	510	510	510	**510**
R6	1.3	1.3	1.3	1.3	1.3	1.3	1.3	1.3	1.3	1.5	1.5	1.5	1.5	1.5	1.5	1.5	1.5	**1.5**
C1	4.7	4.7	4.7	5.6	5.6	5.6	4.7	4.7	4.7	4.7	4.7	4.7	4.7	4.7	4.7	5.6	5.6	**5.6**
Vst	14.664	14.615	14.560	14.664	14.615	14.560	13.706	13.775	13.77	16.395	16.35	16.298	15.378	15.333	15.280	15.378	15.332	**15.280**
ΔV	1	1.5	3	1	2	3	3	6	10	0.7	0.7	1	1	1.4	2	1	1	**1.8**
P	0.696	0.693	0.690	0.696	0.693	0.690	0.638	0.636	0.636	0.765	0.762	0.760	0.701	0.698	0.696	0.701	0.698	**0.696**
η	0.177	0.176	0.175	0.177	0.176	0.176	0.167	0.167	0.167	0.181	0.180	0.180	0.170	0.170	0.169	0.170	0.170	**0.169**

The alternatives presented in Table 6 are explored at the second optimization stage. The alternatives with № 18 and № 15 are estimated as the best at this stage.

DM wishes to perform also a third stage of optimization varying the values of R1, R3 and C. The following parameters are fixed at the second optimization stage:

Q1 = BC548A, Q2 = BC559A, R2 = 91, R4 = 910Ω, R5 = 510Ω, R6 = 1.5k.

The alternatives in Table 7. are estimated and the following results are obtained:

Table 7. Alternatives explored at Stage III

R1	1.6k	2k	1.6k	1.6k	**2k**	1.6k
R3	510	430	430	510	**430**	430
C1	4.7	4.7	4.7	5.6	**5.6**	5.6
V	15.325	15.282	15.384	15.384	**15.281**	15.384
ΔV	6mV	2.2mV	6mV	6mV	**2.2mV**	6mV
P	0.701W	0.696W	0.701W	0.701W	**0.696W**	0.701W
η	0.170	0.169	0.170	0.170	**0.169**	0.170

The alternatives with № 5 and № 2 are estimated as the best at this stage.

The DM is satisfied with the obtained results. At the three optimization stages are explored 32+18+6 = 56 alternatives in total among 1728 possible alternatives.

The best obtained 20 alternatives, including alternatives №№ 1,4,5,8 from Group 1, stage I, №№ 1,4,5,8 from Group 2, stage I, №№ 1,2,4,5 from Group 3, stage I, №№ 2,4,5,8 from Group 4, stage I, as well as №№ 18, 15, stage II, and №№ 5, 2, stage III, are arranged by means of PROMETHEE I method. The 10 alternatives, classified as the best are presented in Table 8.

Table 8. The obtained ten best alternatives

Elements / Criteria	**Parameter values of alternatives / Criteria values**									
Q1	**BC548A**	-//-	-//-	-//-	Q2N2222	Q2N2222	BC548A	BC548A	-//-	-//-
Q2	**BC559A**	-//-	-//-	-//-	Q2N3906	BC559A	BC559A	Q2N3906	-//-	-//-
R1	**2k**	2k	2k	2k	1.6k	1.6k	1.6k	1.6k	1.6k	2k
R2	**91**	91	91	91	110	110	110	110	110	110
R3	**510**	510	510	510	430	430	430	430	510	430
R4	**910**	910	910	910	910	910	910	750	750	750
R5	**510**	510	510	510	470	470	470	470	470	470
R6	**1.5k**	1.5k	1.5k	1.5k	1.3k	1.3k	1.3k	1.3k	1.3k	1.3k
C1	**5.6μ**	4.7μ	5.6μ	4.7μ	4.7μ	4.7μ	4.7μ	5.6μ	4.7μ	4.7μ
V	**15.280V**	15.280V	15.281V	15.282V	14.799V	14.792V	14.715V	14.777V	14.776V	14.766V
ΔV	**1.8mV**	2.mV	2.2mV	2.2mV	4.5mV	5.mV	5.mV	1.mV	1.mV	0.5mV
P	**0.696W**	0. 696W	0. 696W	0. 696W	0.703W	0.703W	0.698W	0.786W	0.787W	0.786W
η	**0.169**	0.169	0.169	0.169	0.178	0.178	0.176	0.159	0.159	0.159
Preorder of alternatives	**1**	2	3	4	5	6	7	8	9	10

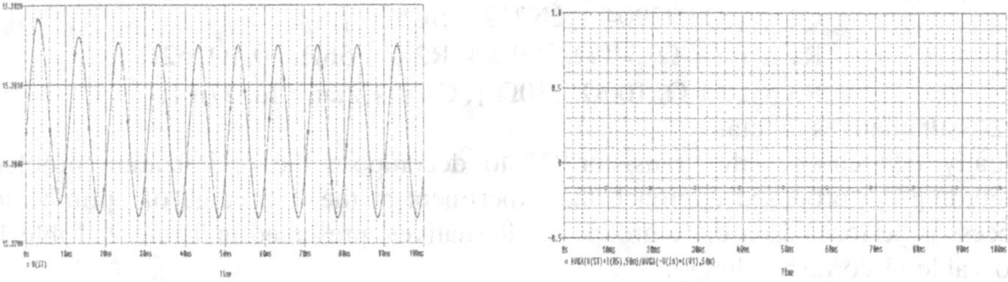

Fig. 2. Stabilized output voltage **Fig. 3.** Regulator efficiency

The DM or the corresponding expert can make a choice which alternative is the best one for a series production. In this test instance only 3,24% of all possible alternatives have been explored and estimated. For test instances with larger number of alternatives this percent decreases.

The PSpice simulation results of the optimal alternative for the circuit from Fig. 1 is presented on the Fig. 2 for the stabilized output voltage and on Fig. 3 for the regulator efficiency. The linear voltage regulator provides very low voltage ripple, which is its main advantage but low regulator efficiency, which is its main drawback.

3.2 Circuit 2

The second example, which will be investigated by means of the proposed approach, is a forward SMPS converter, presented on Fig. 4.

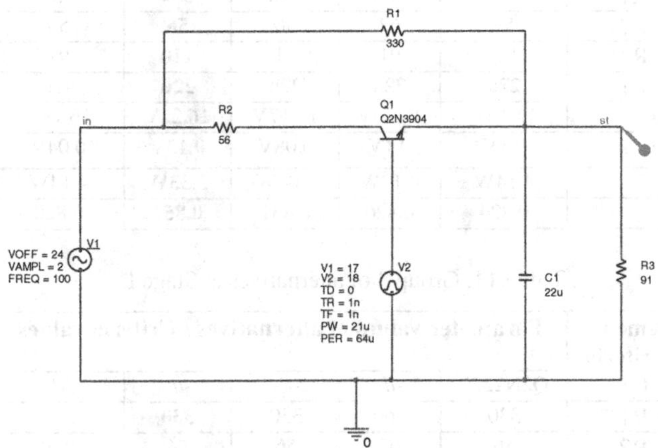

Fig. 4. Forward converter

This circuit consists of five elements: the transistor Q1 (NPN), the 3 resistors R1, R2, R3, and the capacitor C1. The corresponding parameters have the following types and permissible standard values:

Q1 ∈ { Q2N3904, Q2N2222, BC548 },
R1 ∈ { 330Ω, 340Ω, 360Ω }, R2 ∈ { 56Ω, 60.4Ω, 62Ω },
R3 ∈ { 91 Ω, 100Ω, 110Ω }, C1 ∈ { 22u, 32u, 47u };
or 243 alternatives in total.

Initially DM chooses the transistor Q1 to decompose the optimization problem. According the proposed approach the experiment at the first stage of optimization process is performed on three groups of alternatives, presented in Table 9, Table 10, and Table 11 correspondingly:

Table 9. Group 1 of alternatives at Stage I

Elements / Criteria	**Parameter values of alternatives / Criteria values**				
Q1	Q2N3904	-//-	-//-	-//-	-//-
R1	330	360	330	**330**	330
R2	56	62	56	**56**	56
R3	91	91	91	**110**	91
C1	22u	22u	22u	**22u**	47u
Vst	16.1V	16.1V	16.1V	**16.15V**	16.14V
ΔV	0.1V	0.1V	0.09V	**0.12V**	0.04V
P	4.13W	4.12W	4.03W	**3.52W**	4.13W
η	0.826	0.820	0.835	**0.840**	0.821

Table 10. Group 2 of alternatives at Stage I

Elements / Criteria	**Parameter values of alternatives / Criteria values**				
Q1	Q2N2222	-//-	-//-	-//-	-//-
R1	330	360	330	**330**	330
R2	56	62	56	**56**	56
R3	91	91	91	**110**	91
C1	22u	22u	22u	**22u**	47u
Vst	16.16V	16.16V	16.17V	**16.22V**	16.19V
ΔV	0.1V	0.1V	0.08V	**0.13V**	0.04V
P	4.14W	4.12W	4.04W	**3.53W**	4.14W
η	0.824	0.826	0.831	**0.852**	0.820

Table 11. Group 3 of alternatives at Stage I

Elements / Criteria	**Parameter values of alternatives / Criteria values**				
Q1	Q2N2222	-//-	-//-	-//-	-//-
R1	330	360	330	**330**	330
R2	56	62	56	**56**	56
R3	91	91	91	**110**	91
C1	22u	22u	22u	**22u**	47u
Vst	16.85V	16.80V	16.80V	**16.90V**	17.V
ΔV	0.3V	0.3V	0.3V	**0.25V**	0.15V
P	4.14W	4.13W	4.05W	**3.52W**	4.14W
η	0.90	0.92	0.95	**0.85**	0.90

After the estimation of the alternatives by means of PROMETHEE I method, 5-th H(d) criterion - V-shape with an indifference area the results, presented in Table 12. about the preorder of alternatives in each group are obtained:

Table 12. Preorder of alternatives in each group at Stage I

Preorder of alternatives		1	2	3	4	5
Alternative	Group 1	4	3	5	2	1
	Group 2	4	3	5	2	1
	Group 3	4	5	3	2	1

According the mean criteria values in each group of explored alternatives the first and the third criterion in Group 1 have best values and the second criterion has equal mean value in comparison to Group 2 and better mean value in comparison to Group 3. For this reason the sub-problem 1 having the transistor Q1 = Q2N3904 is chosen by the DM for further optimization.

Since the best alternative is the alternative № 4 in each group, followed by the alternatives № 3 and № 5, it could be assumed that corresponding parameter values, which have been changed towards the values in alternative № 1, could be fixed to their upper bound during the further search for the optimal solution.

Hence the following parameters are fixed at the second optimization stage:

Q1 = Q2N3904, R2 = 62Ω, R3 = 110Ω, C1 = 47u.

Also some additional alternatives have been chosen for estimation by the DM.

The alternatives presented in Table 13 are explored at the second stage:

Table 13. Alternatives explored at Stage II

Q1	Q2N3904	-//-	-//-	-//-	-//-	-//-	-//-
R1	**330**	360	360	330	360	360	360
R2	**62**	62	56	62	62	62	56
R3	**110**	110	91	100	100	91	110
C1	**47u**	47u	47u	47u	47u	47u	47u
Vst	**16.2V**	16.2V	16.0V	16.1V	16.1V	16.0V	16.2V
ΔV	**0.1V**	0.12V	0.2V	0.2V	0.2V	0.3V	0.1V
P	**3.51**	3.50	4.12	3.77	3.75	4.02	3.52
η	**0.864**	0.866	0.853	0.862	0.865	0.874	0.860

The alternatives with № 1 and № 6 are estimated as the best at this stage.

The DM is satisfied with the obtained results. At the both optimization stages are explored 15+7 = 22 alternatives in total among 243 possible alternatives, or 9.053% of all alternatives.

The best obtained 5 alternatives are presented in Table 14.

Table 14. The five obtained best alternatives

Elements / Criteria	Parameter values of alternatives / Criteria values				
Q1	**Q2N3904**	-//-	-//-	-//-	-//-
R1	**330**	360	360	330	330
R2	**62**	56	62	56	56
R3	**110**	110	110	110	91
C1	**47u**	47u	47u	22u	47u
Vst	**16.2V**	16.2V	16.2V	16.15V	16.14V
ΔV	**0.1V**	0.1V	0.12V	0.12V	0.04V
P	**3.51W**	3.52W	3.50W	3.52W	4.13W
η	**0.864**	0.860	0.866	0.840	0.891
Preorder of alternatives	**1**	2	3	4	5

At the end the DM can make a choice which alternative is the best one for a series production.

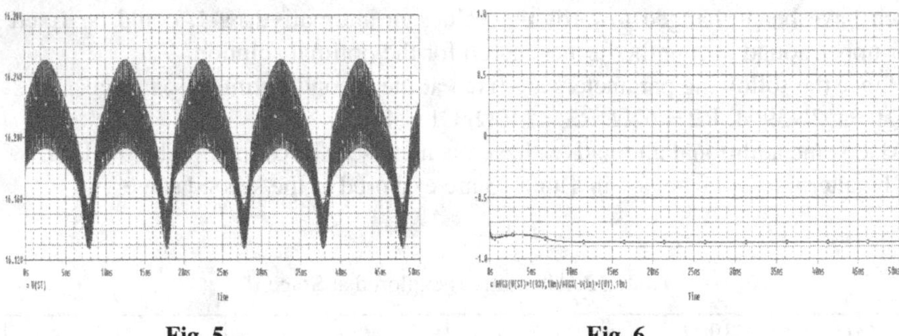

Fig. 5. **Fig. 6.**

The PSpice simulation results of the optimal alternative for the circuit from Fig. 4 is presented on the Fig. 5 for the stabilized output voltage and on Fig. 6 for the regulator efficiency. This Forward converter SMPS circuit provides a large voltage ripple, which is its main drawback and high regulator efficiency, which is its main advantage.

The stabilized voltage of Fig. 1 is closer to 15 V than that one of Fig. 4 and the efficiency of Fig. 4 is higher than that one of Fig. 1.

4 Conclusions

The experimental results obtained and illustrated in the paper confirm that the PROMETHEE I method is a good solution of the task for multiple objective optimization of voltage regulator circuits.

Compared to the genetic algorithm proposed in [7] the PROMETHEE method I in combination with the decomposition approach proposed in this paper finds out the solution by estimating 3,24 % of the total number of alternatives. The genetic algorithm proposed in [7] estimates 3,85% of the total number of alternatives (which are 9600) for a voltage regulator circuit with close number of components, but it is applied to solve single objective problem in contrast to the PROMETHEE I – based approach proposed here, which solves multiple objective optimization problems (the used illustrative test instances have four criteria). When the number of alternatives increases the efficiency of the approach proposed here increases too.

The application of the method proposed allows to optimize the circuits outcoming from multisolution synthesis of voltage regulators with a same specification as a first step and then to select the best solution for a concrete application, or first to select one solution and then to optimize it.

The method proposed is promising for integration in CAD tools for voltage regulator circuits deign.

Acknowledgements. The research work reported in the paper is partly supported by the project AComIn "Advanced Computing for Innovation", grant 316087, funded by the FP7 Capacity Programme (Research Potential of Convergence Regions).

References

1. Bana, E., Costa, C.A. (eds.): Readings in Multiple Criteria Decision Aid. Springer, Berlin (1990)
2. Garey, M.R., Johnson, D.S.: Computers Intractability: A Guide to the Theory of NP-Completeness. W. H. Freeman, San Francisco (1979)
3. Marinova, G.I., Dimitrov, D.I.: Statistical analysis and optimization of voltage regulator circuit using IESD and ORCAD environment. In: ICEST 2003, Sofia, Bulgaria, October 16-18, pp. 478–482 (2003)
4. Marinova, G.I., Guliashki, V.: Improved design centering in a reduced search space for electronic circuits optimization. In: Proceedings of Papers, XL International Scientific Conference on Information, Communication and Energy Systems and Technologies, ICEST 2005, Nis, Serbia and Monte Negro, June 29-July 1, pp. 166–169 (2005)
5. Marinova, G., Dimitrov, D.: Learning Optimal Synthesis of Voltage Regulators using PSpice. COMPEL, Int. J. for Computation and Mathematics in Electrical and Electronic Design 30(4), 1433–1448 (2011)
6. Vincke, P.: Multiple Criteria Decision Aid. John Wiley& Sons, New York (1992)
7. Yeung, H., Poon, N.K., Lai, S.L.: Generic Optimization for SMPS design with Smart Scan and Genetic Algorithm. In: Proceeding of CES/IEEE 5th International Power Electronics and Motion Control Conference, IPEMC 2006, vol. 3 (2006)
8. Cadence ORCAD Design Suit 16.6, Users' guide (2013)
9. PowerEsim, `http://www.poweresim.com`
10. Webench Design Center, Texas Instruments, `http://www.ti.com/lsds/ti/analog/webench/overview.page`

A Design Procedure for Stable High Order, High Performance Sigma-Delta Modulator Loopfilters

Georgi Tsenov and Valeri Mladenov

Dept. of Theoretical Electrical Engineering, Technical University of Sofia, 8, Kliment Ohridski St., Sofia 1000, Bulgaria
{gogotzenov,valerim}@tu-sofia.bg

Abstract. In this paper we present the ideas for design of stable high order single bit sigma-delta modulator loopfilter transfer functions that provide high signal-to-noise ratio. The procedure is backed up with example and results made for third order loopfilter sigma-delta modulators and give the performance impact on them when varying the poles of the noise transfer function, when using optimized zeroes. This loopfilter function design is computed with fast theoretical calculation of the signal to noise ratio with mathematical formula, instead of approximation based on simulations and combined with theory that presents approximated value for modulator's maximal stable DC input signal, resulting in design without the need of simulations of the modulator output bitstream.

Keywords: Sigma-delta modulators, digital signal processing, stability, analog-to-digital conversion, signal-to-noise ratio.

1 Introduction

From decades ago Sigma-Delta modulators are the standard for analog to digital conversion nowadays. When using high oversampling ratios sigma-delta modulators (SDM) can achieve very high signal to noise ratio (SNR) even with small number of quantization levels [1]. They shape the noise and push it to frequencies higher than the operational band of interest. Thanks to its simplicity, single bit code shaping SDM are of greatest interest, because the provided performance is influenced mostly by the loopfilter transfer function and the modulator's oversampling ratio (OSR) and the modulator output is encoded into a bitstream. In the practice the modulator's maximal stable DC input signal range and its SNR are determined by simulations, which also leave a zone of uncertainty. Furthermore a lot of engineers experiment with the loopfilter coefficients in order to achieve higher SNR, but up to date there is still no such a thing as optimal loopfilter transfer function for specific modulator order that provides both high performance and stable modulator behavior. Also, the realistic high performance loopflter transfer functions have the poles grouped into a complex conjugate pairs and one real pole when having odd modulator order and complex conjugate pairs for even loopfilter orders. In order to increase modulator performance

V.M. Mladenov and P.C. Ivanov (Eds.): NDES 2014, CCIS 438, pp. 114–124, 2014.

some authors move one of the complex conjugate pair of poles [3] or the real pole [4] a little bit outside of the unit circle, while keeping the other poles inside resulting in increased SNR and reduced stability limit for maximal DC input signal amplitude beyond which the modulator becomes unstable. Moving a single pole or complex pair a little bit outside the unit circle does not necessarily makes the SDM unstable, as it is a nonlinear system, which makes the SDM behavior analysis harder for those cases.

This paper is presenting a design approach for a higher odd order SDM taking into account the SDM stability and SNR performance. The approach includes variation of the zero positions on the unit circle of sigma-delta modulator loopfilter transfer function with optimized poles, with pole positions given by the delta sigma toolbox. For that type of analysis a parallel decomposition form of the loopfilter given in [2] is used, because with it there is a theory that can predict the stability range. The results presented in [2] allows approximation of the maximal stable DC input signal value without the need of simulations for single bit quantizer modulators with this particular filter form and they are used in the design procedure. For faster SNR calculation a derivation of it from the loopfilter noise transfer function is used, because in this case there is no need of modulator's output bitstream resulting in no need of SDM simulations [5], resulting in reduced number of calculations.

The paper is organized as follows. In the next chapter a theoretical background necessary for understanding the approach is given and description of the method for determination of SDM stability and the method for fast SNR approximation. Then in the third chapter is presented the design procedure targeted to obtain SDM loopfilter transfer functions providing both decent performance and guaranteed stable modulator behavior with presented the procedure results, Then in fourth chapter final conclusion remarks are given.

2 Theoretical Background

For better understanding the design approach here we will remind briefly the results in [2] that are used. The well-known basic structure of an SDM is shown in Fig.1, and consists of a filter with transfer function G(z) followed by a one-bit quantizer in a feedback loop.

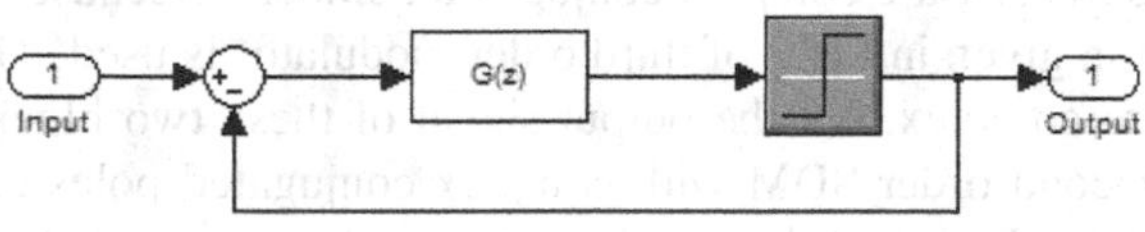

Fig. 1. Basic sigma delta modulator structure

The system operates in discrete time and the input to the loop is a discrete-time sequence $u(n)\in[-1, 1]$, appearing in quantized form at the output. The discrete-time sequence $x(n)$ is output of the filter and quantizer input. Quantizer producing an output of +1 when its input is positive and –1 when its input is negative (single-bit), will not provide a good approximation to its input signal and for that reason an feedback loop is used, acting in such a way as to shift this quantization noise away from a certain frequency band. If an input signal from within this frequency band is applied to the loop, most of the noise imposed by the quantization process will be moved outside of the frequency band of interest and can subsequently be filtered out, leaving a good approximation to the input signal. This process is called noise shaping.

The modulator stability can be obtained without simulations. In [2] authors consider a Nth order modulator with a loop filter transfer function of the form

$$G(z)=\frac{a_1 z^{-1}+...+a_N z^{-N}}{1+d_1 z^{-1}+d_2 z^{-2}+...+d_N z^{-N}} \tag{1}$$

In the general case the loop filter transfer function have complex conjugated roots. Without loss of generality we will consider only one pair of complex conjugated roots. In this case (1) becomes

$$G(z)=\frac{b_1 z^{-1}}{1-\lambda_1 z^{-1}}+...+G_2(z)=\frac{b_1 z^{-1}}{1-\lambda_1 z^{-1}}+...\frac{B_{N-1} z^{-1}+B_N z^{-2}}{1-d_1 z^{-1}-d_2 z^{-2}} \tag{2}$$

where the coefficients b_i, i=1,2,…,N of the fractional components can be found easily using the well-known formula $b_i=\left.\frac{(1-\lambda_i z^{-1})}{z^{-1}}G(z)\right|_{z=\lambda_i}$.

The denominator of the last part of (2) has a complex conjugated pair of roots and therefore (2) becomes:

$$G(z)=\frac{b_1 z^{-1}}{1-\lambda_1 z^{-1}}+...+\frac{b_{N-1} z^{-1}}{1-\lambda_{N-1} z^{-1}}+\frac{b_N z^{-1}}{1-\lambda_N z^{-1}} \tag{3}$$

where

$$\begin{gathered}\lambda_{N-1}=\alpha+j\beta,\ \lambda_N=\alpha-j\beta\\ b_{N-1}=\delta-j\gamma,\ b_N=\delta+j\gamma\end{gathered} \tag{4}$$

i.e. λ_{N-1}, λ_N and b_{N-1}, b_N are complex conjugated numbers. Because of this in [2] the parallel presentation given in Fig.2 of third order modulator is used. The values of the last two blocks are complex, but the output signal of these two blocks is real. They correspond to a second order SDM with complex conjugated poles of the loop filter transfer function $G(z)$. Both signals x_2 and x_3 are complex conjugated, namely

$$\begin{gathered}x_2(k+1)=m(k+1)+jn(k+1)\\ x_3(k+1)=m(k+1)-jn(k+1)\end{gathered} \tag{5}$$

Because of this the input of the quantizer is real i.e.

$$(\delta-j\gamma)x_2(k)+(\delta+j\gamma)x_3(k)=2\delta m(k)+2\gamma n(k) \tag{6}$$

The modulator could be considered as three first order modulators interacting only through the quantizer function. The connected signals with two modulators are complex, but the input and output signals (u and y) are the "true" signals of the modulator. As it is stressed in [2] both modulators work cooperative, because their signals are conjugated. These modulators do not exist in the real SDM and they are introduced to help the analysis of the behavior of the whole system.

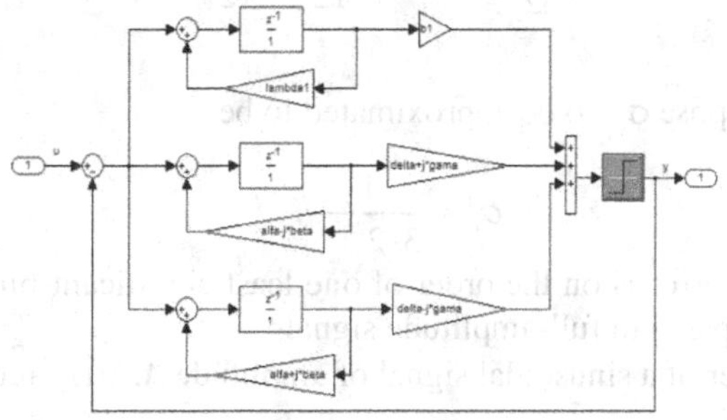

Fig. 2. Block diagram of higher order SDM with parallel loopfilter form

The benefit of this modulator representation is because we can determine whenever the modulator is stable or not by this criterion [2]:

$$\frac{(2-\lambda_1)}{\lambda_1}\frac{b_1}{(\lambda_1-1)} > -\sum_{i=2}^{N-2}\frac{|b_i|}{\lambda_i-1}+\frac{2\,|\,\delta(1-\alpha)+\gamma\beta\,|}{(1-\alpha)^2+\beta^2} \tag{7}$$

Additionally we can also determine the maximal range of input signal ensuring the stability expressed by Δu:

$$\Delta u < \frac{\sum_{i=2}^{N-2}\frac{|b_i|}{\lambda_i-1}-\frac{2|\delta(1-\alpha)+\gamma\beta|}{(1-\alpha)^2+\beta^2}+\frac{b_1(2-\lambda_1)}{\lambda_1(\lambda_1-1)}}{\frac{b_1}{\lambda_1-1}-\sum_{i=2}^{N-2}\frac{|b_i|}{\lambda_i-1}+\frac{2|\delta(1-\alpha)+\gamma\beta|}{(1-\alpha)^2+\beta^2}} \tag{8}$$

The fast approximation formula that will be used for fast Signal to Noise Ratio (SNR) calculations is derived in the following way in [5].

The theory of quantization and the corresponding noise is well-established. The distance between 2 successive quantization levels is called the quantization step size, Q. For a quantizer with a specified number of bits covering the range from +1 to -1 there are 2^{bits} quantization levels and the width of each quantization step is:

$$Q=\frac{2}{(2^{bits}-1)} \tag{9}$$

The quantizer assigns each input sample $u(n)$ to the nearest quantization level. The quantization error is simply the difference between the input and output to the quantizer, e_q=y(u)-u, and is bounded by

$$-\frac{Q}{2} \leq e_q(n) \leq \frac{Q}{2} . \tag{10}$$

The quantization noise power is given by

$$\sigma_e^2 = \frac{1}{Q} \int_{-\frac{Q}{2}}^{\frac{Q}{2}} e_q{}^2 de_q = \frac{Q^2}{12} = \frac{1}{3(2^{bits}-1)^2} . \tag{11}$$

Many authors propose σ_e^2 to be approximated to be

$$\sigma_e^2 \approx \frac{1}{3 \cdot 2^{2bits}} . \tag{12}$$

This quantization error is on the order of one least-significant-bit in amplitude and it is quite small compared to full-amplitude signals.

The average power of a sinusoidal signal of amplitude A, $x(t)=A\cos(2\pi t/T)$, is

$$\sigma_x^2 = \frac{1}{T} \int_0^T (A\cos(2\pi t/T))^2 dt = \frac{A^2}{2} \tag{13}$$

If we assume that the signal is oversampled, then rather than acquiring the signal at the Nyquist rate, $2f_B$, the actual sampling rate is $f_s=2^{r+1}f_B$, with oversampling ratio $OSR=2^r=f_s/2f_B$. The quantization noise is spread over a larger frequency range yet we are still primarily concerned with the noise below the Nyquist frequency.

Most of the noise power is now located outside the signal band. The quantization noise power within the band of interest has decreased by a factor *OSR*. The signal power occurs over the signal band only, so it remains unchanged and is given by (13).

Many authors use the linear SDM model for analysis. The linear model contains two inputs: the input signal *X(z)* and the quantization error *E(z)*. As the basic model shown on Fig.1 a filter is placed in front of the quantizer, known as the 'loop filter' and the output of quantization is fed back and subtracted from the input signal, as shown on Fig. 3.

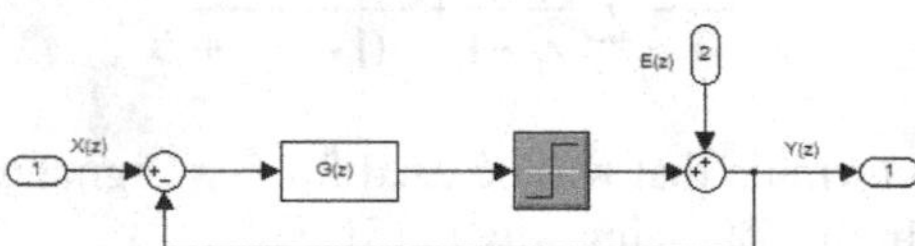

Fig. 3. Representation of a sigma-delta modulator using the linear model

This may be represented by transfer functions applied to both the input signal and the quantization noise. The *Z*-domain output may be represented as

$$Y(z) = STF(z)X(z) + NTF(z)E(z) , \tag{14}$$

where the *STF* is the *Signal Transfer Function* and the *NTF* is the *Noise Transfer Function.* The input to the loop filter is $X(z)$-$E(z)$ so that $Y(z)=G(z)[X(z)-Y(z)]+E(z)$. Rearranging terms, we have:

$$STF(z)=\frac{G(z)}{1+G(z)},\quad NTF(z)=\frac{1}{1+G(z)}\ . \tag{15}$$

Utilizing the linear model of the SDM the noise shaping in the SDM implies a variable noise power in the baseband:

$$\sigma_n^2=\int_{-f_B}^{f_B} S_e^{\,2}(f)\,|\,NTF(f)|^2\,df \tag{16}$$

where $S_e^{\,2}(f)=\sigma_e^{\,2}/f_s$ is the power spectral density of the unshaped quantization noise. The total noise power, $\sigma_e^{\,2}$, remains unchanged, but appropriate choice of the $NTF(z)$ pushes the noise up to the high frequencies.

By definition the SNR is calculated on basis of:

$$\text{SNR(dB)}=10\log_{10}\frac{\sigma_x^2}{\sigma_n^2} \tag{17}$$

We can take the $\sigma_e^{\,2}$, $\sigma_x^{\,2}$ and $\sigma_n^{\,2}$ terms to substitute them to get the general formula for the SNR of any sigma-delta modulator.

$$\text{SNR(dB)}=10\log_{10}\frac{\sigma_x^2}{\sigma_n^2}=10\log_{10}\frac{\frac{A^2}{2}}{\int_{-f_B}^{f_B} S_e^{\,2}(f)\,|\,NTF(f)|^2\,df}=10\log_{10}\frac{A^2\cdot f_s}{2\sigma_e^2\int_{-f_B}^{f_B}|\,NTF(f)|^2\,df} \tag{18}$$

Applying approximation of $\sigma_e^{\,2}$ we get the formula:

$$\text{SNR(dB)}\approx10\log_{10}\frac{3\cdot2^{2bits}\cdot A^2\cdot f_s}{2\int_{-f_B}^{f_B}|\,NTF(f)|^2\,df}\approx10\log_{10}3\cdot2^{2bits}A^2 f_s-10\log_{10}2\int_{-f_B}^{f_B}|\,NTF(f)|^2\,df \tag{19}$$

Using numerical integration this equation can be solved for a SDM with arbitrary noise transfer function, oversampling rate and bit length. These calculations for the SNR approximation, when done with computer, are very fast and precise and are computed much faster than the SNR approximate estimation when using modulator output bitstream simulations. If using a loopfilter transfer function of odd order, using the function coefficients with Eq.8 the stability can be calculated without simulations, while with Eq.19 the SNR of the SDM can be computed without simulations. This provides the tools for SDM analysis without the need of SDM model simulations.

The practical relation of Δu and modulator stability depending on its signal value is given for one example on Figure 4 with usage of loopfilter transfer function that produces loss of stability for input signals with amplitude values lesser than 1 if 1 is the scaled maximal input signal. Usually with appliance of higher input signal the SNR rises, but here when rising the test sine wave amplitude we can observe that at some point there is an SNR decrease and eventually loss of stability. This means that the modulator can be stable for input signals with value higher than that of Δu, but then we can't guarantee its stability (Δu=0.68 in this example).

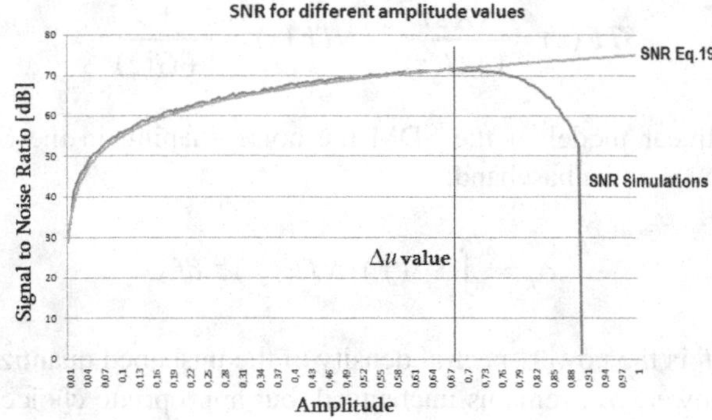

Fig. 4. Relation between Δu and the value of the input signal

3 Design Approach

Some authors [6], claims that if the NTF zeroes (loopfilter poles) are placed not on the DC, but spread in the baseband, the SNR is higher and the quantization noise is spread evenly in the baseband as shown on Figure 5.

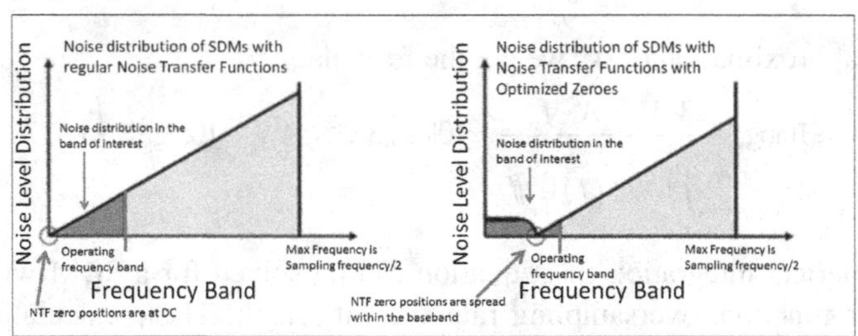

Fig. 5. SDM noise distribution comparison with and without optimized NTF zeroes

The Delta Sigma Toolbox [7] for MATLAB, created by this author, can generate such NTFs of arbitrary order. This Toolbox was used to generate an exemplar noise transfer function with optimized zeroes for third order SDM filter order, which is:

$$NTF(z) = \frac{z^3 - 2.999z^2 + 2.999z - 1}{z^3 - 2.1992z^2 + 1.6876z - 0.4441} \tag{20}$$

For this type of noise transfer function with Eq. (8) we can determine the SDM stability ranges even when moving the real pole outside the unit circle and we can find the SNR very fast with the following constrained optimization procedure:

$$R = \max(SNR)$$
$$NTF = \frac{(1-zero\_ntf_1 * z^{-1})*(1-zero\_ntf_2 * z^{-1})*(1-zero\_ntf_3 * z^{-1})}{(1-pole_1 * z^{-1})*(1-pole_2 * z^{-1})*(1-pole_3 * z^{-1})}$$
$$NTF(z) = \frac{1}{1+G(z)};\ G(z) = \frac{b_1 z^{-1}}{1-zero\_ntf_1 z^{-1}} + \frac{b_2 z^{-1}}{1-zero\_ntf_2 z^{-1}} + \frac{b_3 z^{-1}}{1-zero\_ntf_3 z^{-1}}$$
$$zero\_ntf_2 = \alpha + j\beta,\ zero\_ntf_3 = \alpha - j\beta,\ b_2 = \delta - j\gamma,\ b_3 = \delta + j\gamma$$
$$\mathrm{SNR} = 10\log_{10} 3\cdot 2^{2bits} A^2 f_s - 10\log_{10} 2\int_{-f_B}^{f_B} |NTF(f)|^2\, df\ \ [dB]$$
$$\Delta u < \frac{\sum_{i=2}^{N-2} \frac{|b_i|}{zero\_ntf_i - 1} - \frac{2|\delta(1-\alpha)+\gamma\beta|}{(1-\alpha)^2+\beta^2} + \frac{b_1(2-zero\_ntf_1)}{zero\_ntf_1(zero\_ntf_1 - 1)}}{\frac{b_1}{zero\_ntf_1 - 1} - \sum_{i=2}^{N-2}\frac{|b_i|}{zero\_ntf_i - 1} + \frac{2|\delta(1-\alpha)+\gamma\beta|}{(1-\alpha)^2+\beta^2}}$$
$$zero\_ntf_2 = a * cos(b) + j * c * sin(d),\ zero\_ntf_3 = a * cos(b) - j * c * sin(d)$$
$$pole_2 = conj(pole_3) = const,\ pole_3 = conj(pole_2) = const$$
$$def(zero\_ntf_1) = [1, 1.5],\ def(pole_1) = [0, 1]$$
$$def(a) = [0, 1],\ def(c) = [0, 1]$$
$$def(b) = [0^0, 90^0],\ def(d) = [0^0, 90^0] \tag{21}$$

With R we are going to have the maximal SNR for third order SDMs, while keeping the modulator stability. For this reason NTF from Eq. 20 was used as a starting point for NTF pole variation (or loopfilter zero variation Eq. 15), while varying the poles inside and on the unit circle in various angles in certain increments with constraints specified in SDM literature. The zero on the real axis is varied in increments starting from 1 up to 1.5 values. This variation is done as a max function, with maximum being with respect to the SNR and a constraint optimization with respect to NTF zero placement and pole placement values. So, for the third order modulator with the realistic form of the loopfilter (third order transfer function with one real and one pair of complex conjugate poles) a pole variation is performed (Figure 6) by moving the complex conjugate pair inside the unit circle and the real zero on the unit circle and away from it. In order to have a valid result obtained from stability condition (7) and Δu (8), these equations allows only one of the poles to be outside of the unit circle and this is the reason why the complex conjugate pairs are not moved outside the unit circle. Some authors prefer to move the complex conjugate pair outside of the unit circle, because they end up with higher SNR, but with decreased range of maximal acceptable input signal [4]. In these cases at some point when increasing the input signal amplitude the modulator behavior becomes unstable and maximum input signal clipping limits determined by simulations are used in order to keep the modulator stable. On the other hand moving only the real pole still produces higher SNR (but not as high as when moving complex pairs), while keeping the maximal range of input signal almost equal to 1 [3]. Moving the real zero of the NTF too far away from the unit circle produces decrease of SNR and loss of stability. The modulus of all the variations of one of the complex pole positions thus varies from zero to one. Values equal to almost 1 means that the complex pair is very close and placed almost on the border of unit circle.

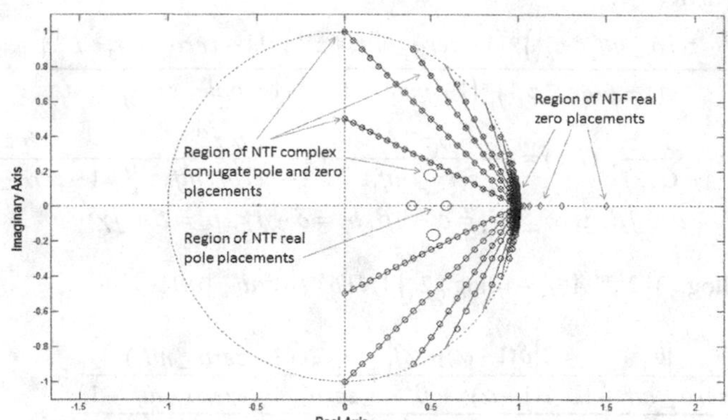

Fig. 6. Position placements of the loopfilter's NTF poles and zeroes

4 Simulation Results

We have implemented the above mentioned in the previous chapter Sigma-Delta Modulator design procedure in MATLAB. In order to verify the correctness of the results from the implementation of the design approach procedure with respect to maximization of the SNR we have tested the resulting candidate NTFs with simulations on a Sigma-Delta Modulator models implemented also in MATLAB.

On Figure 7 are presented the SNR results for one loopfilter case, with the variation of the mentioned parameters. The results for SNR value are given calculated for 64 times oversampling ratio (OSR), input sine wave with scaled amplitude value 0.5 and frequency 2/3 of our band of interest. In that figure is shown SNR values that are calculated and theoretically derived from loopfilter transfer functions [5] that are formed at every iteration step.

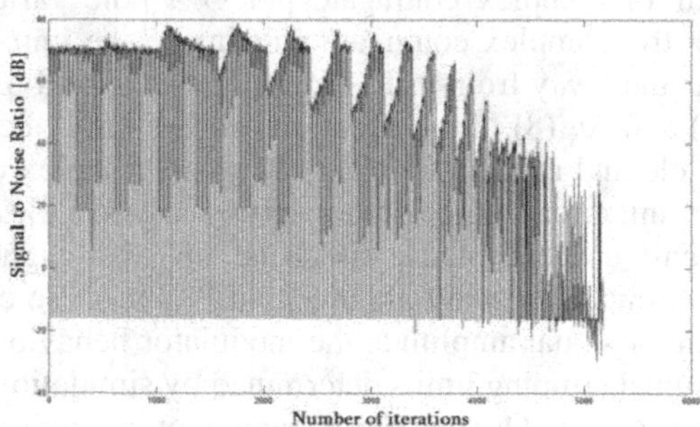

Fig. 7. SNR results calculated on derivations from the loopfiler transfer function for fixed zeros combination and variation of the real and complex conjugate pair poles

For almost all of these transfer functions Δu stayed with values close to 1.

The candidate R Noise Transfer Function that provides max SNR from Eq. 21 is the following (with numbers rounded up to the third digit):

$$NTF(z)= \frac{z^3 - 2.997z^2 + 2.996z - 0.999}{z^3 - 1.997z^2 + 1.535z - 0.467} \tag{22}$$

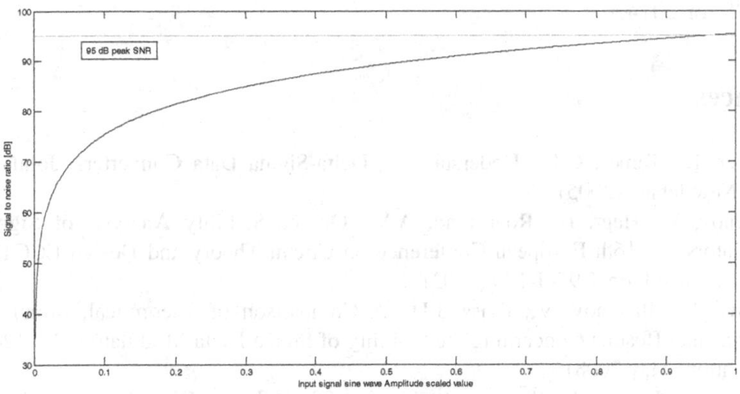

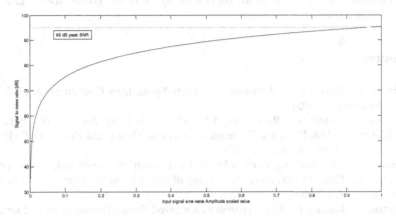

Fig. 8. Input signal and resulting SNR for SDMs when using NTF(z) from Eq. 22

One example of modulator power spectrum shape obtained after simulations, scaled with respect to sampling frequency (Fs=1), when using 64 time Oversampling ratio and test sine wave with scaled amplitude of 0.99 for Sigma-Delta Modulator with Noise transfer function from Eq. (22), is shown on Figure9. In this case the obtained effective SNR is 95dB.

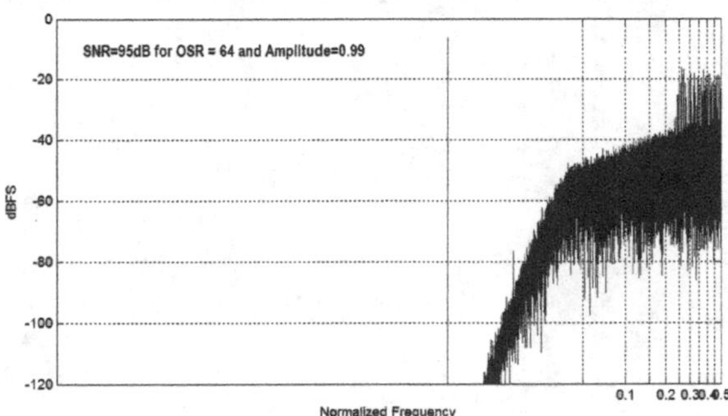

Fig. 9. SDM spectrum of simulated SDM output for the third order transfer function

5 Conclusion

In the paper a design approach for stable high performance sigma-delta modulator is presented. The approach is based on improvement on NTF zero optimized transfer functions with respect to stability and signal to noise ratio. Based on this approach a third order stable SDM with reasonable performance in sense of SNR and stable DC input signal range is obtained. The approach can be generalized for higher order modulators when using odd loopfilter orders.

Acknowledgements. This work is supported by N.W.O. visitor travel grant Nr. 040.11.425 for 2014.

References

[1] Schreier, R., Temes, G.C.: Understanding Delta-Sigma Data Converters. John Wilet & Sons, New Jersey (2005)

[2] Mladenov, V., Hegt, H., Roermund, A.V.: On the Stability Analysis of Sigma-Delta Modulators. In: 16th European Conference on Circuit Theory and Design ECCTD 2003, Cracow, Poland, pp. I-97–I-100 (2003)

[3] Tsenov, G., Mladenov, V., Reiss, J.D.: A Comparison of Theoretical, Simulated, and Experimental Results Concerning the Stability of Sigma Delta Modulators. In: 124th AES Convention (May 2008)

[4] Reefman, D., Janssen, E.: Signal processing for Direct Stream Digital: A tutorial for digital Sigma Delta modulation and 1-bit digital audio processing. Philips Research, Eindhoven, White Paper (December 18, 2002)

[5] Mladenov, V., Karampelas, P., Tsenov, G., Vita, V.: Approximation Formula for Easy Calculation of Signal-to-Noise Ratio of Sigma-Delta Modulators. ISRN Signal Processing, Article ID 731989 (2011) doi:10.5402/2011/731989

[6] Schreier, R.: An Empirical Study of High-Order Single-Bit Delta-Sigma Modulators. IEEE Transactions on Circuits and Systems-11: Analog and Digital Signal Processing 40(8) (August 1993)

[7] Schreier, R.: The delta sigma toolbox, http://www.mathworks.com/

Application of Generalized Instantaneous Reactive/ Non-active Power Theories in the Control of Shunt Active Power Line Conditioners: Practical Evaluation under Nonideal Voltage and Unbalanced Load

Mihaela Popescu, Cristina Alexandra Pătraşcu, and Mircea Dobriceanu

University of Craiova, Faculty of Electrical Engineering, Decebal Bd. 107, 200440 Craiova, Romania
{mpopescu,apatrascu,mdobriceanu}@em.ucv.ro

Abstract. This paper is focused on the practical evaluation of two generalized theories of powers in phase coordinate system, namely generalized instantaneous reactive power and generalized instantaneous non-active power, by their implementation in the real time control of a three-phase three-wire shunt active power line conditioner through a dSPACE-based platform. Based on each theory concepts, specific blocks for reference current generation to achieve the global compensation were conceived first. Then, experimental tests were conducted to prove the ability of the active filtering system to compensate a nonlinear distorted and unbalanced load under nonideal voltage conditions. The good dynamics behaviour of the compensating system is illustrated too.

Keywords: Active power line conditioner, Generalized instantaneous non-active power theory, Generalized instantaneous reactive power theory, Nonlinear load.

1 Introduction

For quite a long time, the compensation of the nonlinear and distorted load in electric power systems is an important topic in the field of power quality improvement due to the increasingly use of such loads.

Clearly, recent advances in power electronics devices and control make the so-called shunt active power line conditioners (APLCs) or active power filters (APFs) the more flexible and efficient solution to eliminate the current harmonic distortion and to compensate both the reactive power and load unbalance.

In order to obtain unity power factor or perfect harmonic cancellation after compensation irrespective of the supply voltage waveform, many methods of reference compensating current generation have been adopted until now. Most of them are time domain based and provide either the current to be compensated or the desired supply current after compensation.

While the most common approaches involve the transformation from phase coordinate system to stationary or rotating two-phase system in order to apply the p-q

V.M. Mladenov and P.C. Ivanov (Eds.): NDES 2014, CCIS 438, pp. 125–133, 2014.

theory of the instantaneous reactive power concepts [1], [2] or the id-iq method [3], [4], there are different approaches whose implementation does not require any reference frame transformation. This last set refers to the theories such as Fryze-Buchholz-Depenbrock (FBD) [5], the generalized instantaneous reactive power [6] and generalized instantaneous non-active power [7].

The attention in this paper is directed to the two above mentioned generalized theories and their practical implementation for total compensation in a shunt compensator through a dSPACE based control system operating together with Matlab/Simulink software. After a brief description of the compensating system, the reference current generation algorithms and the associated developed Simulink blocks are presented. Section 4 refers to the experimental setup and the results achieved under nonideal voltage and unbalanced load, in order to prove the high performance of the developed APLC system. At the end, some conclusions are pointed out.

2 APLC System Structure

A shunt APLC including its control system has been developed for experimental testing. As depicted in Fig. 1, the voltage source inverter is connected to the point of common coupling (PCC) by an inductive filter.

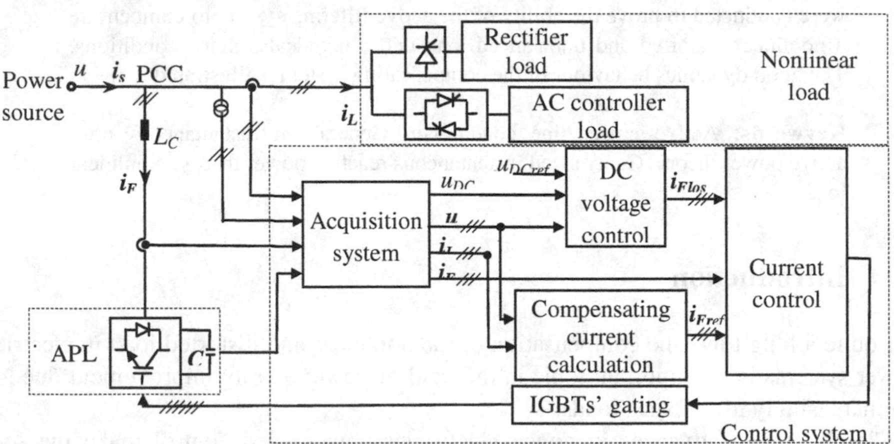

Fig. 1. Single-phase block diagram of shunt APLC system

Based on the sensed load currents and supply voltages, the compensating current calculation block generates the reference currents (i_{Fref}) by the real time implementation of the adopted algorithm. To keep the DC-voltage at its prescribed value in order to cover the power system losses, the additional compensating current i_{Floss}, which is an active current, is generated by the DC-voltage control block. The ability of the current controller to ensure the accurate tracking of the resulting reference current gives the compensation system efficiency.

3 Reference Current Generation

The compensation goal taken into consideration in the reference current generation is to eliminate the load generated harmonics, the load unbalance, as well as the reactive power. Depending on the current decomposition method, the reference compensating current supplied to the current controller (i_{Fref}) can be provided either directly from the expression of load current decomposition, or by subtracting the load current (i_L) from the desired (reference) supply current (i_{sref}), as follows in the vectorial writing:

$$\boldsymbol{i}_{Fref} = \boldsymbol{i}_{sref} - \boldsymbol{i}_L, \tag{1}$$

where the vector of the line currents is defined as:

$$\boldsymbol{i} = [i_a(t) \quad i_b(t) \quad i_c(t)]^T. \tag{2}$$

3.1 Generalized Instantaneous Reactive Power Theory-Based Approach

The foundation of the generalized instantaneous reactive power (GIRP) theory for three-phase power systems, which was introduced by Peng and Lai in 1996, is the decomposition of the current vector into the instantaneous active component ($\boldsymbol{i}_p$) and the instantaneous reactive component ($\boldsymbol{i}_q$) [6].

When used to decompose the load current vector ($\boldsymbol{i}_L$), the associated expression is:

$$\boldsymbol{i}_L = \boldsymbol{i}_{Lp} + \boldsymbol{i}_{Lq}. \tag{3}$$

Following the GIRP's theory concepts, the active and reactive current vectors are expressed by using the instantaneous active power (p_L) and the instantaneous reactive power vector ($\boldsymbol{q}_L$), whose definitions make use of the dot product ($\cdot$) and cross product (x) of voltage and current vectors, i.e.

$$\boldsymbol{i}_{Lp} = p_L/(\boldsymbol{u}\cdot\boldsymbol{u})\cdot\boldsymbol{u}; \qquad \boldsymbol{i}_{Lq} = (\boldsymbol{q}_L \times \boldsymbol{u})/(\boldsymbol{u}\cdot\boldsymbol{u}); \tag{4}$$

$$p_L = \boldsymbol{u}\cdot\boldsymbol{i}_L; \qquad \boldsymbol{q}_L = \boldsymbol{u}\times\boldsymbol{i}_L. \tag{5}$$

The vector $\boldsymbol{u}$ in (4) and (5) corresponds to the three-phase supply voltages system,

$$\boldsymbol{u} = [u_a(t) \quad u_b(t) \quad u_c(t)]^T. \tag{6}$$

After making evident the average (P_L) and oscillatory ($p_{L\sim}$) components of p_L, expression (3) of the load current can be written as follows:

$$\boldsymbol{i}_L = P_L/(\boldsymbol{u}\cdot\boldsymbol{u})\cdot\boldsymbol{u} - [-p_{L\sim}/(\boldsymbol{u}\cdot\boldsymbol{u})\cdot\boldsymbol{u} - (\boldsymbol{q}_L\times\boldsymbol{u})/(\boldsymbol{u}\cdot\boldsymbol{u})]. \tag{7}$$

Thus, for total compensation, the reference current vector to be extracted from PCC is:

$$\boldsymbol{i}_{Fref} = -(p_L - P_L)/(\boldsymbol{u}\cdot\boldsymbol{u})\cdot\boldsymbol{u} - (\boldsymbol{q}_L \times \boldsymbol{u})/(\boldsymbol{u}\cdot\boldsymbol{u}). \quad (8)$$

The associated block diagram shown in Fig. 2 was created in Matlab/Simulink and further used for the experimental implementation on a dSPACE-based platform.

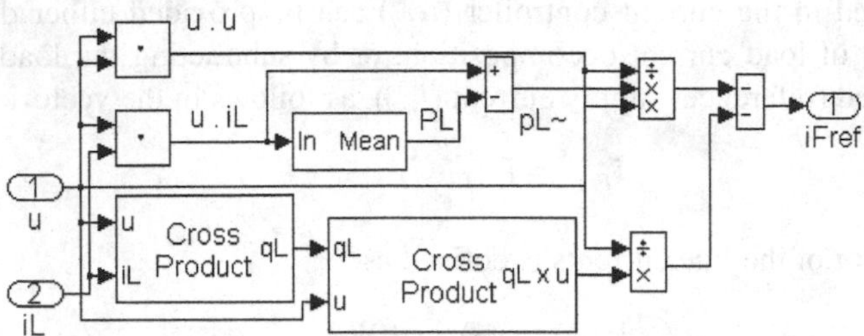

Fig. 2. Block diagram for total compensation strategy based on the GIRP theory concepts

As the quantity in the denominator of the active current given in (4) is not constant when the voltage waveform is distorted, it is expected that the supply current waveform will be more different compared to the voltage waveform as the voltage distortion is higher.

3.2 Generalized Instantaneous Non-active Power Theory-Based Approach

The proposal of a generalized decomposition of the load current vector in poly-phase circuits into the instantaneous active component ($\boldsymbol{i}_{Lp}$) and the so-called instantaneous non-active component ($\boldsymbol{i}_{Lq}$) belongs to Peng and Tolbert [7]. From the very beginning, the applicability in shunt compensation was envisaged.

Neglecting the compensator power losses, the active power at the supply side (P) during an averaging interval T_C is equal to the load active power (P_L).

As only the active current expression is actually defined and the remaining current is the non-active component, the calculation of the reference compensating current is performed by imposing the desired supply current, as required by (1).

The flexibility in implementing the generalized instantaneous non-active power (GINAP) theory for shunt compensation of the load current comes from the general expression of the supply active current,

$$\boldsymbol{i}_{sref} = P/U_P^2 \cdot \boldsymbol{u}_p = \left(\frac{1}{T_C}\int_{-T_C} \boldsymbol{u}(\tau)\cdot\boldsymbol{i}_L(\tau)d\tau\right)\Big/\left(\frac{1}{T_C}\int_{-T_C} \boldsymbol{u}_p(\tau)\cdot\boldsymbol{u}_p(\tau)d\tau\right)\cdot\boldsymbol{u}_p, \quad (9)$$

where the imposed reference voltages in vector $\boldsymbol{u}_p$ give the resulting waveforms of the supply line currents.

Thus, when unity power factor (UPF) is the compensation goal, $\boldsymbol{u}_p$ must be the voltage vector itself. But, in order to obtain sinusoidal supply currents and unity displacement power factor, $\boldsymbol{u}_p$ must contain the fundamental components of $\boldsymbol{u}$.

In the associated Simulink block diagram shown in Fig. 3, the unity power factor strategy is implemented and the equivalent conductance is highlighted.

Moreover, by imposing a proper averaging interval T_C in (9) in relation to the fundamental period of the supply voltage (T), both periodic and non-periodic currents can be compensated [8], [9].

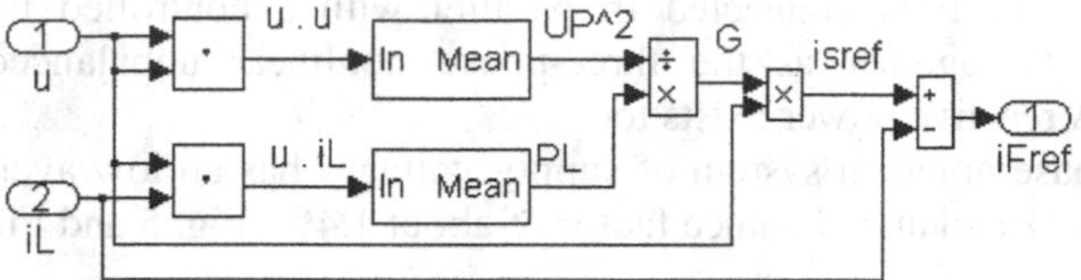

Fig. 3. Block diagram for UPF compensation strategy based on the GINAP theory concepts

4 Experimental Setup and Results

The experimental tests were conducted on a three-phase 15 kVA laboratory prototype consisting of an IGBT-based voltage source inverter with a DC-link capacitor of 1100 μF and an inductive filter of 4.4 mH on the AC side. Based on dSPACE DS1103 PPC controller board with comprehensive I/O, the real-time control system was implemented via Matlab/Simulink environment. The conceived Simulink model of the control system illustrates the analog to digital conversion, the generation of the prescribed currents, the DC-voltage and current control, the digital to analog conversion and the output signals transfer to the digital I/O channels (Fig. 4). The start-up process and the required protections are also taken into consideration.

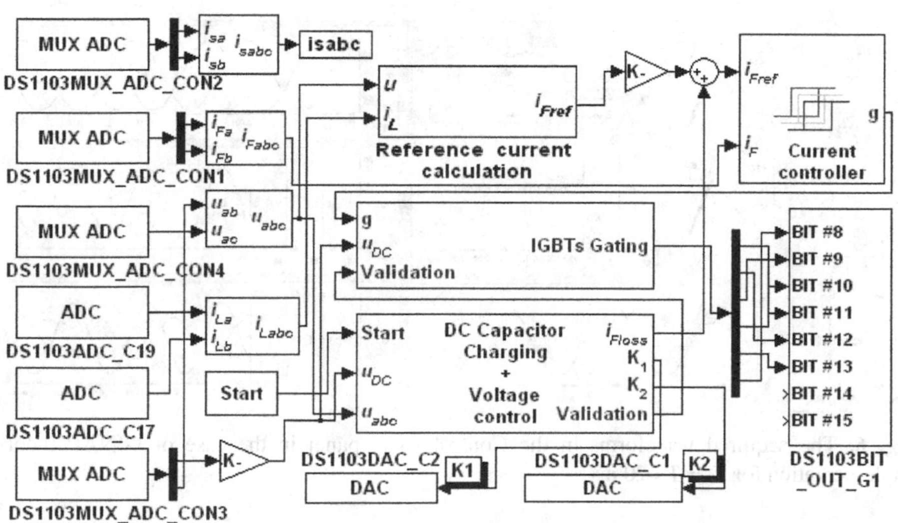

Fig. 4. Compiled Simulink model for the real time control through dSPACE platform

The voltage controller of PI type was designed in accordance with the principle of Modulus Optimum criterion [10], [11]. The DC-voltage prescribed value is 700 V.

By adopting a sampling time of 20 μs and a hysteresis band of 0.4 A for the current controller, the IGBTs' switching frequency was kept below their capability.

One of the nonlinear loads is an AC voltage controller manufactured by Nokian Capacitors Ltd. and especially aimed for testing, which allows producing an unbalanced current. It is connected in parallel with a controlled thyristor-bridge rectifier and acts together as the three-phase nonlinear unbalanced load to be compensated. A reactive power exists too.

The three-phase nonideal system of supply voltages has an low average harmonic distortion of 2.4 % and an unbalance factor of about 1.4% (Fig. 5 and Fig. 6).

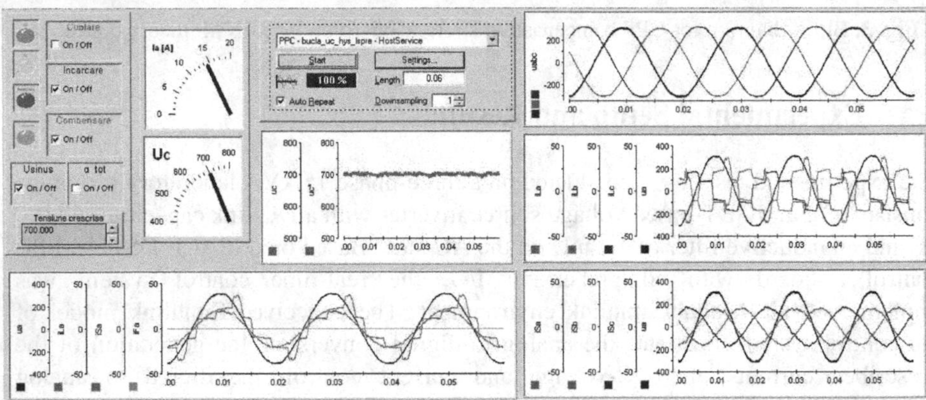

Fig. 5. The acquired waveforms in the Control Desk panel in the case of GIRP theory implementation

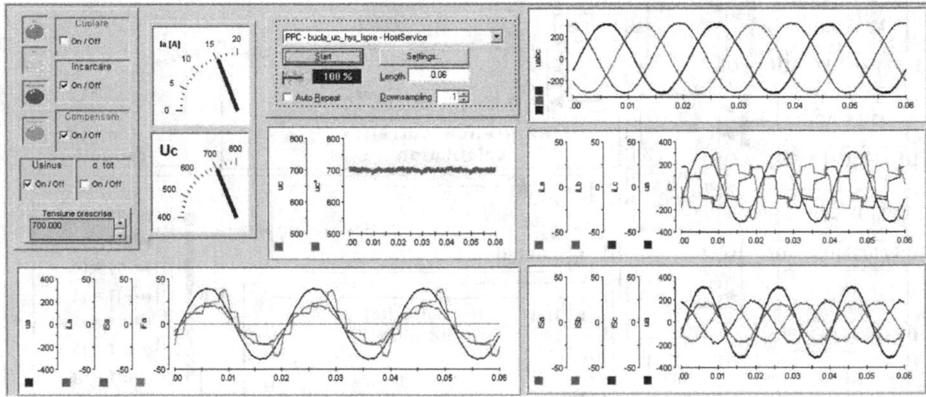

Fig. 6. The acquired waveforms in the Control Desk panel in the case of GINAP theory implementation for $T_C = T$ =20 ms

Table 1 summarizes the results of the experimental tests conducted for both GIRP and GINAP based methods of reference current generation. As subharmonic and interharmonic components exist in the current drawn by the line-commutated loads, three values of the averaging interval were taken into consideration (i.e. $T_C = T/2$, $T_C = T$ and $T_C = 2T$) for the GINAP based method.

To quantify the compensation performance, besides the total harmonic distortion factor (*THD*) on each phase and its average value (THD_e), the three-phase rms value of the line currents (I_e), the unbalance factor (*UF*) according to IEC definition,. the global power factor (*PF*) and the displacement power factor (*DPF*) were calculated.

As it can be seen in Table 1 and results displayed in the conceived Control Desk panel (Fig. 5 and Fig. 6), the compensating system is able to significantly improve the power quality at the supply side in all experimental tests.

The good dynamics behaviour of the filtering system is shown in Fig. 7. The APLC is charged for UPF compensation through the reference current calculation according to GINAP theory for T_C = 20 µs. The steady state regime is rapidly established in both situations of partial coupling of the load and suddenly coupling of the filter.

Table 1. Summary of the compensation performance

	Load side											
	I_{La} (A)	I_{Lb} (A)	I_{Lc} (A)	I_{Le} (A)	P_L (W)	THD_{La} (%)	THD_{Lb} (%)	THD_{Lc} (%)	THD_{Le} (%)	PF_L	DPF_L	UF_L (%)
GIRP	19.3	17.5	14.3	17.1	8659.0	33.52	31.22	24.37	29.70	0.7775	0.8140	15.60
GINAP - *T*/2	19.4	17.7	14.4	17.3	8789.1	33.68	30.91	24.63	29.74	0.7768	0.8131	15.61
GINAP - *T*	18.8	17.2	14.2	16.8	8506.9	33.41	31.08	24.35	29.61	0.7769	0.8135	15.35
GINAP - 2*T*	19.4	17.6	14.6	17.3	8779.3	33.04	30.57	24.67	29.43	0.7827	0.8185	15.08
	Supply side after compensation											
	I_{Sa} (A)	I_{Sb} (A)	I_{Sc} (A)	I_{se} (A)	P_S (W)	THD_{Sa} (%)	THD_{Sb} (%)	THD_{Sc} (%)	THD_{Se} (%)	PF_S	DPF_S	UF_S (%)
GIRP	14.0	14.1	14.4	14.2	9165.4	4.92	5.69	5.58	5.40	0.9965	0.9990	1.83
GINAP - *T*/2	14.3	13.8	14.6	14.2	9281.1	5.29	6.21	5.44	5.65	0.9967	0.9987	2.93
GINAP - *T*	13.9	13.5	14.3	13.9	8998.1	4.69	6.53	5.16	5.46	0.9971	0.9989	2.95
GINAP - 2*T*	14.3	14.1	14.7	14.4	9267.3	5.13	6.38	5.56	5.69	0.9971	0.9989	2.90

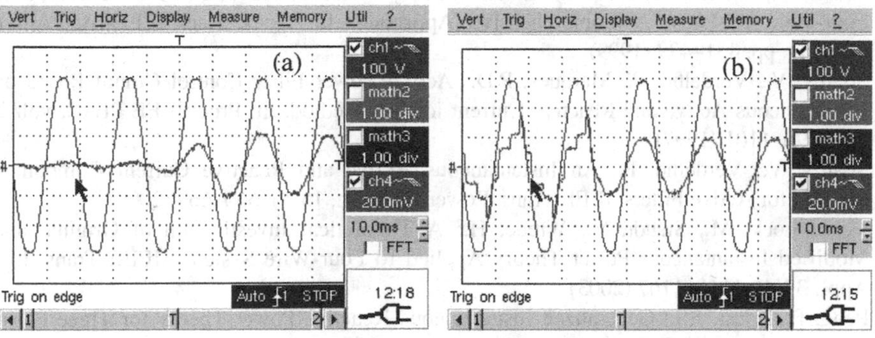

Fig. 7. Experimental phase voltage and supply for GINAP theory implementation when $T_C = T$ =20 ms: (a) APLC compensates the AC controller current and the rectifier is suddenly connected; (b) APLC is suddenly connected to compensate the global load.

As the existing power supply in the laboratory provides a low degree of distortion in the voltage waveform, there is little difference between the results related to GIRP theory, which was conceived for sinusoidal voltage conditions, and those related to GINAP theory applied for UPF strategy under distorted voltage conditions. The implementation of the two strategies gets the supply currents to be almost sinusoidal and balanced, with a power factor over 0.996. The filtering efficiency, in terms of ratio of average harmonic distortion factors at the load and supply sides, the highest value of 5.5 corresponds to GIRP theory, whereas the lowest value (about 5.17) corresponds to GINAP theory in case of $T_C = 2T$. The unbalance level of the supply current is of about 5 times lower by implementing the GINAP theory are even over 8 times lower through GIRP theory implementation (Table 1). Though a small degree of disturbance is identified in the electric power system, the choice of an averaging interval other then the fundamental cycle does not improve the compensation quality.

5 Conclusions

The GIRP and GINAP theories, both of them associated with the phase coordinate system, provide the necessary foundation to develop appropriate strategies for the reference supply current generation in three-phase shunt active line conditioners, so that the total compensation goal is achieved. The developed algorithms were successfully tested by experimental tests on a dSPACE DS1103 platform in the case of distorted and unbalanced load. The good dynamics during compensation is highlighted too.

References

1. Akagi, H., Kanazawa, Y., Nabae, A.: Instantaneous Reactive Power Compensators Comprising Switching Devices without Energy Storage Components. IEEE Trans. Ind. Appl. IA-20(3), 625–630 (1984)
2. Akagi, H., Ogasawara, S., Kim, H.: The Theory of Instantaneous Power in Three-Phase Four-Wire Systems: A Comprehensive Approach. In: Conf. Rec. IEEE-IAS Annu. Meeting, pp. 431–439 (1999)
3. Soares, V., Verdelho, P., Marques, P.D.: Active Power Filter Control Circuit Based on Instantaneous Active and Reactive Current id – iq Method. In: Proc. IEEE PESC, vol. 2, pp. 1096–1101 (1997)
4. Soares, V., Verdelho, P.: An Instantaneous Active and Reactive Current Component Method for Active Filters. IEEE Trans. Power Electron. 15(4), 660–669 (2000)
5. Depenbrock, M., Staudt, V., Wrede, H.: A Theoretical Investigation of Original and Modified Instantaneous Power Theory Applied to Four-Wire Systems. IEEE Trans. Ind. Appl. 39(4), 1160–1167 (2003)
6. Peng, F.Z., Lai, J.S.: Generalized Instantaneous Reactive Power Theory for Three-Phase Power Systems. IEEE Trans. Instrum. Meas. 45(1), 293–297 (1996)
7. Peng, F.Z., Tolbert, L.M.: Compensation of Nonactive Current in Power Systems-Definitions from a Compensation Standpoint, pp. 983–987. IEEE Power Eng. Society Summer Meeting, Seattle (2000)
8. Xu, Y., Tolbert, L.M., Peng, F.Z., Chiasson, J.N., Chen, J.: Compensation-Based Non-active Power Definition. IEEE Power Electronic Letters 1(2), 45–50 (2003)

9. Tlusty, J., Svec, J., Sendra, J.B., Valouch, V.: Analysis of Generalized Non-active Power Theory for Compensation of Non-Periodic Disturbances. In: International Conference on Renewable Energies and Power Quality, Santiago de Compostela (2012)
10. Popescu, M., Bitoleanu, A.: Control Loops Design and Harmonic Distortion Minimization in Active Filtering-Based Compensation Power Systems. Internat. Review Modelling and Simulations 3(4), 581–589 (2010)
11. Popescu, M., Bitoleanu, A., Suru, V.: A DSP-based implementation of the p-q theory in active power filtering under nonideal voltage conditions. IEEE Trans. Ind. Informat. 9(2), 880–889 (2013)

Maximizing Power Transfer in Induction Heating System with Voltage Source Inverter

Alexandru Bitoleanu, Mihaela Popescu, and Vlad Suru

University of Craiova, Romania

{alex.bitoleanu,mpopescu,vsuru}@em.ucv.ro

Abstract. This paper investigates the effects of the matching inductance on the quality of the power transfer in the induction heating system with voltage source inverter and parallel resonance. It is shown that there is an optimal value of the matching inductance that maximizes the fundamental harmonic of the inverter and inductor currents and minimizes the third-order harmonic. Experimental results sustain this conclusion and shown that the active power transferred to the equivalent inductor is also maximized.

Keywords: Induction heating, Resonant voltage source inverter, Matching inductance, Optimal design.

1 Introduction

Induction heating is a non-contact heating process which uses high frequency electricity to heat materials that are electrically conductive. Since it is non-contact, the heating process does not contaminate the material being heated. It is also very efficient since the heat is actually generated inside the work piece.

The work coil is made to resonate at the intended operating frequency by means of a capacitor placed in parallel or series with it. In the same time, the parallel resonance has a very important advantage because magnifies the current through the work coil, which is much higher than the output current capability of the inverter [1], [2].

In the high power design, it is common to use a full-bridge (H-bridge) of four or even more switching devices. Most commonly, the current source inverters are used in combination with parallel resonance and the voltage source inverters are used in combination with series resonance [3], [4], [5], [6], [7].

The authors proposed the voltage source inverters used in combination with parallel resonance because, for induction heating of the pipes, over Curie temperature, this structure have high energetically performances [8]. In this case, a matching inductance placed in series by parallel resonance circuit is mandatory.

The inverter output voltage is strongly distorted, such as, the value of the matching inductance influences, both quantitatively and qualitatively, the power transfer to the heated body. The matching coil size adjustment must take into account the effect on the current harmonic spectrum, especially on the third-order harmonic value, which is the most important. It is mentioned that the entire system is nonlinear, because the

V.M. Mladenov and P.C. Ivanov (Eds.): NDES 2014, CCIS 438, pp. 134–141, 2014.

inverter, the matching coil and the equivalent circuit of heating coil-heated body are nonlinear.

After a short introduction, the analysis of the influence of matching inductance value on the harmonics of orders 1 and 3 is presented in Section II. The analysis is done based on the harmonic equivalent circuit. It is shown that there is an optimal value of the matching inductance that maximizes the fundamental harmonic of the inductor current and minimizes the harmonic of order three.

Section III presents the experimental structure and gives some information about the experimental setup and conducted tests. For each configuration of the matching inductance (i.e. 5, 10, 15, 20, and 25 turns), the RMS current and the inductor active power are calculated. It is shown that, for 20 turns of matching inductance, the active power and the RMS current of the inductor are maximized.

2 Harmonic Currents Analysis

The analysis of the harmonic currents is based on the harmonic equivalent circuit and on the parameters that take into account the frequency of each harmonic (Fig. 1).

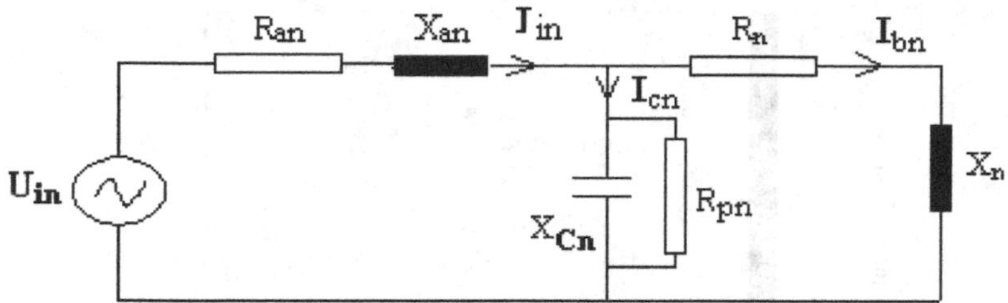

Fig. 1. N harmonic equivalent circuit of the induction-pipe-parallel compensation capacitor system, supplied by the voltage inverter

It is emphasized that only certain parameters are independent of the order harmonics. It is about the capacitance and resistance of the compensation capacitor (C_n=C and R_{pn}=R_p) and the inductivity and the resistance of the matching inductance (L_{an}=R_a and R_{an}=R_a) [9], [10].

The complex electrical quantities were used, by assuming the inverter output voltage harmonics as phase origin. The matching inductance resistance was taken into account and the following assumptions were made.

1. If the inverter transistors are closed during π radians in one period and the DC circuit average voltage is 660 V, then, the RMS value of nth order harmonic voltage is [6], [7],

$$U_{in} = \frac{4 \cdot 660}{n\pi}. \tag{1}$$

2. The complex impedances of the matching inductance, equivalent inductor and compensation capacitor are:

$$\underline{Z}_{an} = R_a + j \cdot n \cdot \omega \cdot L_a \quad (2)$$

$$\underline{Z}_n = R_n + j \cdot n \cdot \omega \cdot L_n \quad (3)$$

$$\underline{Z}_{pcn} = R_p / (1 + j \cdot n \cdot \omega \cdot R_p C) \quad (4)$$

3. The RMS values of the inverter current and inductor current are given by:

$$I_{in} = \frac{U_{in}}{|\underline{Z}_{en}|}; I_{bn} = \frac{U_{in}}{|\underline{Z}_{bn}|} \quad (5)$$

The impedances in (5) have the following expressions:

$$\underline{Z}_{en} = \underline{Z}_{an} + \underline{Z}_{pn}; \ \underline{Z}_{pn} = \frac{\underline{Z}_{pcn} \cdot \underline{Z}_n}{\underline{Z}_{pcn} + \underline{Z}_n}; \ \underline{Z}_{bn} = \underline{Z}_n + \underline{Z}_{an}(1 + \underline{Z}_n / \underline{Z}_{pcn}). \quad (6)$$

Next, the spectra of the first 15 harmonics of the inductor current for different values of the matching inductance are shown (Fig. 2 and Fig. 3).

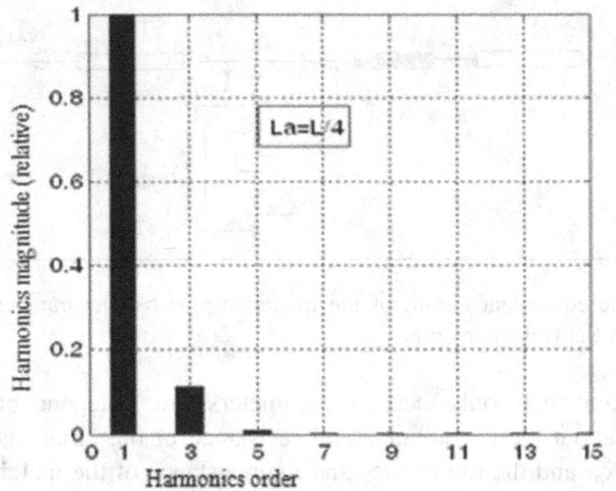

Fig. 2. Harmonic spectra of current through the equivalent inductor for $L_a = 1/4\ L$

The most relevant effect for small values of the matching inductance is on the 3rd harmonic. Thus, for $L_a = 1/4\ L$, the 3rd harmonic is of about 10% of the fundamental harmonic (Fig. 2) and, for $L_a = 2L$, the 3rd harmonic is of about 1.5% of the fundamental (Fig. 3). The 5th harmonic is much smaller that the fundamental, even when the matching inductance value is lower than the inductance of the equivalent inductor (Fig. 3). The 3rd harmonic decreases quickly when the value of the matching inductance increases (Fig. 3).

The harmonics weight of the inductor current is always reduced. This can be explained, on the one hand, by the higher fundamental values and, on the other hand, by the effect of the compensation capacitor.

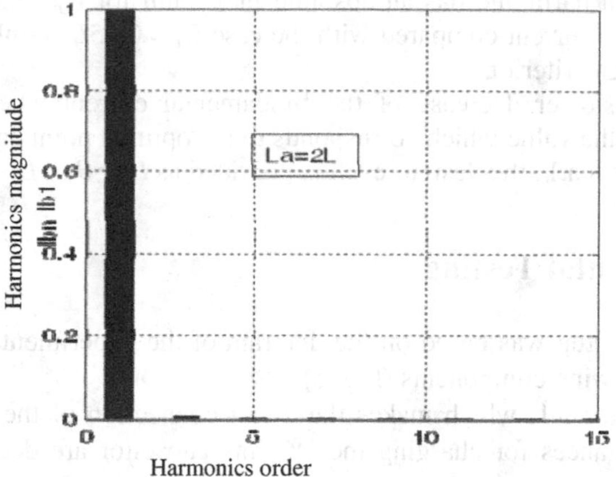

Fig. 3. Harmonic spectra of current through the equivalent inductor for $L_a = 2L$

It was found that, for values of the matching inductance greater than or equal to the corresponding equivalent inductor, the 3rd harmonic weight is under 4%.

As regards the current through the inverter, the effect of values of the matching inductance is more pronounced. For example, the 3rd harmonic becomes lower than the fundamental value for a matching inductance over 1.5 L [10]. Further, the percentage of the 3rd harmonic decreases as L_a increases. There are two reasons for the distortion of the inverter current must be reduced [11]:

1. The inverter current influences the waveform of the supply current;
2. If the inverter current is much distorted, it is more difficult to implement a control algorithm to dynamically adapt the inverter frequency.

More valuable information is provided by the dependences of the fundamental and 3th harmonics in the equivalent inductor current (Fig. 4) on the inductances ratio.

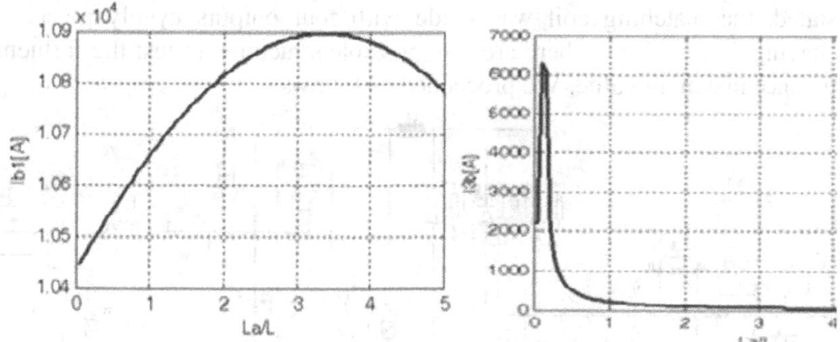

Fig. 4. Fundamental and 3th order harmonic inductor current versus ratio L_a / L

The 3rd harmonic variation confirms the previous results for matching inductance values lower than the equivalent inductor inductance. In addition, it shows the existence of a very high maximum value when $L_a \approx 0.15L$.

The fundamental harmonic has an absolute maximum for $L_a \approx 3.35L$. Even if the increase in the peak current compared with the case $L_a = 0.05L$ is only of 4%, it may be an optimal design criterion.

By noting the slower decrease of the fundamental current when the matching inductance is over the value which corresponds to the optimal point and by correlating with the previous remark, the design result is provided as $L_a \geq 3.35L$.

3 Experimental Testing

The experimental setup was based on the diagram of the experimental model, which highlights the following components (Fig. 5):

1. Phase contactor K1, which makes the direct connection of the static converter (i.e. the resistances for charging the DC-link capacitor are decoupled by short circuit);
2. Phase contactor K3, which makes the DC-link capacitor charging;
3. Three-phase switching coil 0.08mH/350A (L_1);
4. The static converter, which consists of a three-phase diode rectifier and the single-phase voltage source inverter with IGBTs (1200A, 1700V);
5. Compensation capacitors C=64μF, U=3000V, I=11.6kA, f=10kHz;
6. The matching coil La (Fig. 6): number of turns - 25 (four intermediate sockets of five turns each), inner diameter -270 mm, conductor of copper tube - Ø 24 mm, length - 81 cm;
7. The inductor (cooled by water provided by an independent equipment): number of turns – 8, inner diameter -365 mm, conductor of copper tube - Ø 24 mm, length - 42 cm (Fig. 6);
8. Carbon steel heated pipe ≈ 1%: outer diameter - 168 cm, wall thickness - 32 mm;
9. The required voltage and current transducers.

As stated, the matching coil was made with four outputs evenly placed, each section having 5 turns. Thus, there are five possible structures to test the influence of the inductance matching value. We proceeded as follows:

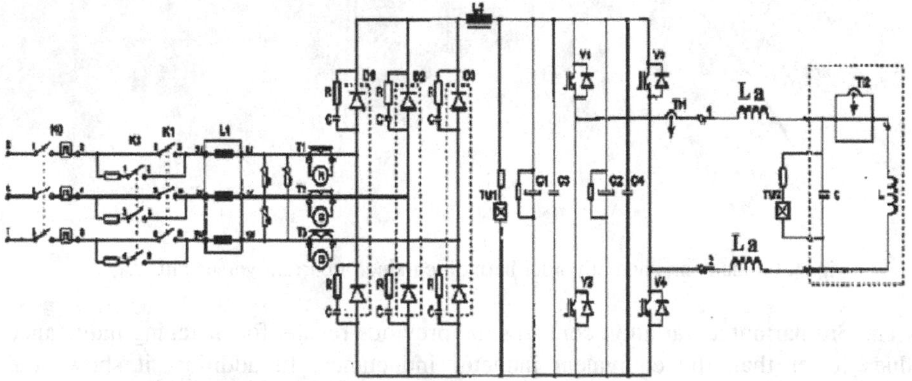

Fig. 5. The electric scheme of the induction heating converter for experimental setup

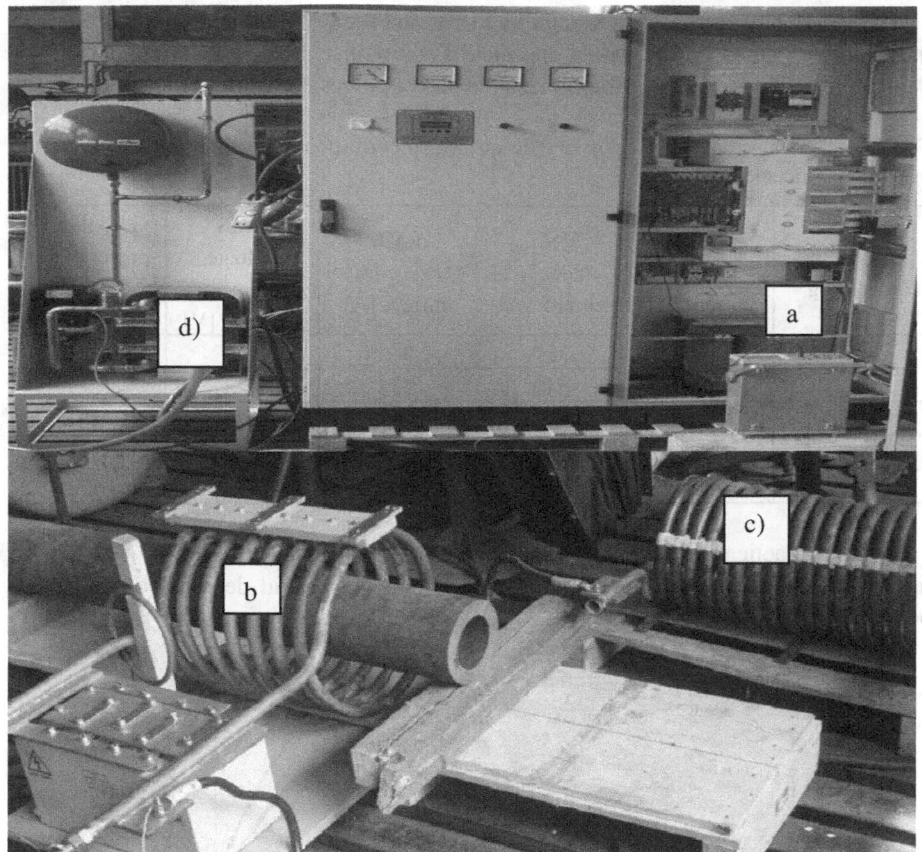

Fig. 6. Pictures of the experimental setup: a) inverter; b) inductor; c) matching coil; d) cooling equipment

- We worked with capacitors of 64 μF and already mentioned inductor and pipe;
- Frequency was prescribed in open circuit at a constant value of 6000 Hz;
- Inverter input voltage the was set at 350 V, by the appropriate control of the rectifier;
- For each of the four configurations of the matching coil, the waveforms of the inverter current, voltage across the inductor and the inductor current were recorded by Metrix OX 7042 oscilloscope with current probe-M HHM811;
- The current through the compensation capacitor was calculated, and then the RMS value of all quantities and the active power transmitted to the inductor (Table I).

In order to calculate the RMS value (X) of the instantaneous quantity x and the active power (P), the following expressions were used:

$$X = \sqrt{1/T\int_{t-T}^{t} x^2 dt}\ ;\ P = 1/T\int_{t-T}^{t} u \cdot i \cdot dt\,. \tag{7}$$

In the expressions above, the significances of the quantities are: *x* - the instantaneous value; *u*, *i* - the instantaneous values of voltage and current; *t* – the observation time; *T* – the period of *x*, *u* and *i*.

The active power transmitted to the inductor and the RMS inductor current confirm the results obtained on the model and clearly show that there is an optimal value of the matching inductance that maximizes the active power transmitted to the load.

Table 1. The values of the main energetic quantities corresponding to the five values of the matching inductance

Number of turns of the matching coil	RMS inverter current [A]	RMS inductor current [A]	RMS capacitor current [A]	RMS voltage across inductor [V]	Active power of the inductor [KW]
5	164	386	566	362	6
10	154	540	775	428	8,75
15	180,6	808	1154	523	13
20	230	1322	1890	670	21
25	153	418	605	376,5	6,5

From a practical standpoint, it is very important that the use of a matching inductance different from the optimum value leads to a rapid decrease of the active power (Fig. 7).

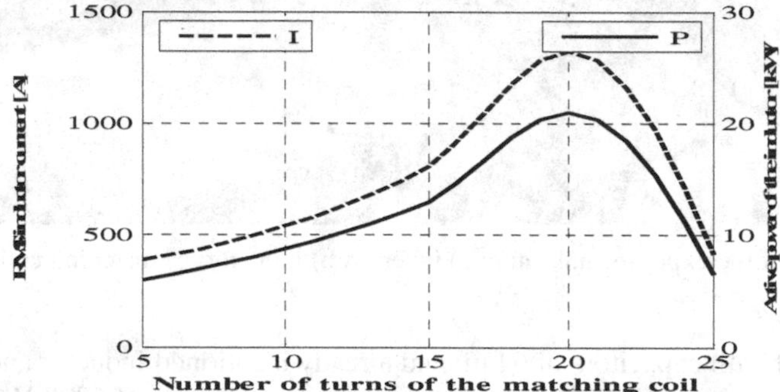

Fig. 7. The dependencies of the RMS current and active power of the inductor versus number of turns of the matching coil

Thus, if the matching inductance value increases or decreases by 6.25% (the number of turns increases or decreases by 25%) compared with the optimum value, the active power received by the inductor decreases by 69% and 38% respectively.

4 Conclusions

The first part of this paper presents a detailed analysis of the influence of matching inductance value on the harmonics of orders 1 and 3 of the currents through the equivalent inductor and through the voltage source inverter.

The experimental setup developed by authors and the performed experimental measurements are presented in the second part.

The original contributions of the paper are:

1. It is shown that there is an absolute maximum of the fundamental currents through the equivalent inductor and through the inverter.
2. It is also shown that, over a certain value of the matching inductance, the amplitude of the third-order harmonic current decreases continuously when the inductivity value increases.
3. An optimal matching inductance value was found, which is dependent on the inductivity of the equivalent inductor.
4. The theoretical results are verified by the performed experimental measurements on the induction heating system developed by authors.
5. It is demonstrated, on the experimental way, that there is an optimum value of the matching inductance which maximizes the power transferred to the heated body.

References

1. Suresh, A., Rama, R.S.: Parallel Resonance Based Current Source Inverter for Induction Heating. European Journal of Scientific Research 58(2), 148–155 (2011)
2. Rudnev, V., Loveless, D., Cook, R., Black, M.: Handbook of Induction Heating. Marcel Dekker, NY (2003)
3. Dieckerhoff, S., Ruan, M.J., De Doncker, R.W.: Design of an IGBT-based LCL-resonant inverter for high-frequency induction heating. In: The 34th Industry Applications Conference, vol. 3, pp. 2039–2045 (1999)
4. Esteve, V., Pardo, J., Jordan, J., Dede, E., Sanchis-Kilders, E., Maset, E.: High Power Resonant Inverter with Simultaneous Dual-frequency Output. In: 36th IEEE Power Electronics Specialists Conference, pp. 1278–1281 (2005)
5. Fujita, H., Uchida, N., Ozaki, K.: Zone controlled induction heating (ZCIH) A new concept in induction heating. In: Power Conversion Conference, Nagoya, Japan, pp. 1498–1504 (2007)
6. Jain, P.K., Espinoza, J.R., Dewa, S.B.: Self-Started Voltage-Source Series-Resonant Converter for High-Power Induction Heating and Melting Applications. IEEE Trans. on Industry Applications 34(3), 518–525 (1998)
7. Kazimierczuk, M.K., Czarkowski, D.: Resonant Power Converters. John Wiley & Sons (2011)
8. Suru, V., Popescu, M., Bitoleanu, A.: Energetic Performances of Induction Heating Systems with Voltage Resonant Inverter. In: Proceedings of International Symposium on Electrical and Electronics Engineering, Galaţi, România, October 11-13, pp. 978–971 (2013) ISBN 978-1-4799-2441-7
9. Popescu, M., Bitoleanu, A.: Power Control System Design in Induction Heating with Resonant Voltage Inverter. Journal of Automation and Control Engineering 2(2), 195–198 (2014)
10. Popescu, M., Bitoleanu, A., Dobriceanu, M.: Analysis and optimal design of matching inductance for induction Heating system with voltage inverter. In: The 8th International Symposium on Advanced Topics in Electrical Engineering (ATEE 2013), Bucharest, Romania, May 23-25, pp. 978–971 (2013) Print ISBN: 978-1-4673-5979-5
11. Callebaut, J.: Induction Heating, Power Quality & Utilisation Guide, Section 7: Energy Efficiency (February 2007)

General Solutions of Nonlinear Equations for the Buck Converter

Yuri Tanovitski, Gennady Kobzev, and Danil Savin

Tomsk State University of Control Systems and Radioelectronics, Tomsk, Russia
{Yuri.Tanovitski,tyn}@ie.tusur.ru

Abstract. To analyze the dynamic properties and set the feedback ratio of the converter, the authors suggest using numerical general solutions of differential equations. This paper presents the analysis of such solutions for various feedback ratios. The paper shows that the parameters that ensure stability and fast transient processes in the small, also ensure stability and fast transient processes in the large.

Keywords: general solution, buck voltage converter, stability.

1 Introduction

The buck converter follows nonlinear differential equations with discontinuous switching functions. It is known that, should improper parameters be used, the converter will have a full range of "complex dynamic" properties (e.g. see [1-2]) – loss of stability of its basic operation mode, emergence of undesired modes, bifurcations, and chaos. The parameters that enable high quality values (stabilization factor) are sometimes nearing bifurcation boundaries. External parameters (load and input voltage) (which, if changed, can result in bifurcations) can vary greatly. Some of the well-known numerical/analytical and numerical methods of analysis (e.g., SPICE family packages) rely on specific solutions of the Cauchy problem $X(t,X_0)$ where X is the vector of state variables, t is time and X_0 is the initial conditions. However, in order to assess performance of the stabilizer in the context of nonlinear dynamics, one needs to analyze a multitude of solutions of $X(t,X_0)$, where X_0 belongs to a set D_0. This analysis must be applied to all permissible variations of the parameters. Over time the set of initial conditions D_0 will evolve into the set $D(t,D_0)$. We will treat the set $D(t,D_0)$ as the general solution of the system of equations, since, unlike a specific solution where one trajectory corresponds to one point of initial conditions, in this case a set of solutions will correspond to a set of initial conditions. Unfortunately, general solutions are known primarily for linear differential equation systems. They are usually presented in the form of composite functions which will meet the solution of the linear system of equations with any initial conditions. In nonlinear systems, where there are no analytical solutions, $D(t,D_0)$ can be observed to transform with the growth of t using numerical methods. This approach was discussed by the authors in [3], where it was used to analyze the stability of the generally

V.M. Mladenov and P.C. Ivanov (Eds.): NDES 2014, CCIS 438, pp. 142–147, 2014.

stabilized voltage converters with PWM. This paper focuses on finding "the best" feedback ratios ensuring the fastest transient processes and stability in the large. In their search for the feedback parameters the authors rely on the results of [4]. The algorithm described in [4] is available online at [5]. The ideas put forward in [4] were later summarized in [6], which is available online in Russian.

2 General Solutions for the Buck Converter

Let us analyze general solutions for the buck converter. Figure 1 shows its circuit. The mathematical model and parameters are the same as in [1]. The main difference is in the added capacity-current feedback, and the increased capacity up to 5 μF. Order of the differential equation system for our example 2.

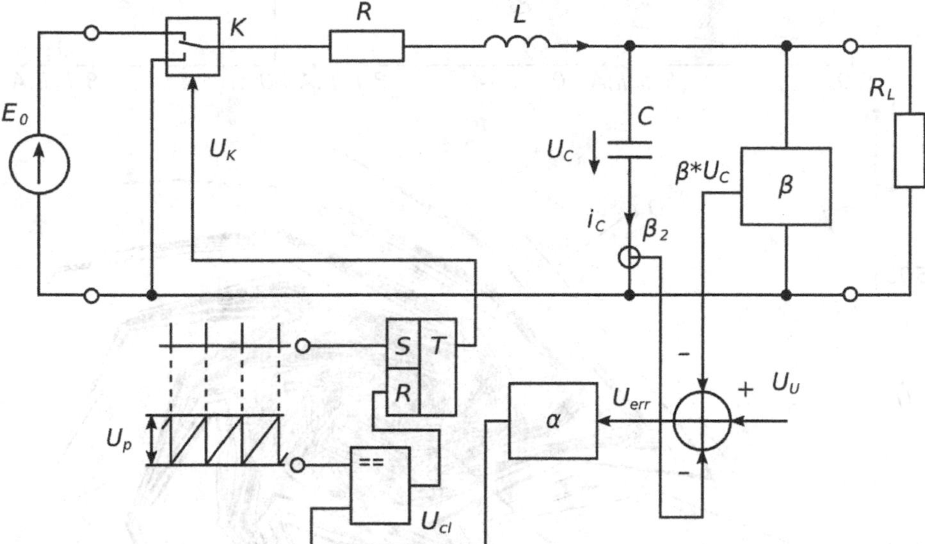

Fig. 1. Buck converter with feedback by capacitors voltage and capacitors current

In our analysis we must take into account the following:

1) D_0 is simply connected and limited. We shall designate its boundary – Γ_0. Transformation of $D(t)$ results in transformation of the boundary $\Gamma(t)$.

Since the converter state variables are almost always within the rectangle of initial conditions with the coordinates ($i_L=0, U_C=0$), ($i_L=i_{Ls}, U_C=U_{Cs}$), where i_{Ls}, U_{Cs} are the current of the reactor L and voltage of the capacitor C which are obtainable when the duty cycle is z=const=100%, there is no practical reason in taking D_0 from a wider range, e.g. from negative to positive infinity.

2) From the model in the continuous form let us move to the Poincaré map $X_k=X(t_k)=X(t=k*Tq)$ where $k=1,2,3\ldots$ and Tq is the saw period of the PWM. This is similar for the sets $D_k=D(t_k,D_0)$ and the boundary $\Gamma_k=\Gamma(t_k,\Gamma_0)$.

3) Quality of the dynamics can be assessed by observing the behavior of the general solution for transformation of its boundary Γ_k. Indeed, if the stabilizer is operating properly (i.e. is stable at large), the volume restricted by Γ_k will be compressing into a point surrounding the stationary state. Otherwise, if the stabilizer is not stable, it cannot happen. Therefore, instead of searching through all possible solutions within D_0, it is sufficient to analyze the behavior only of the boundary of the set, i.e. Γ_k.

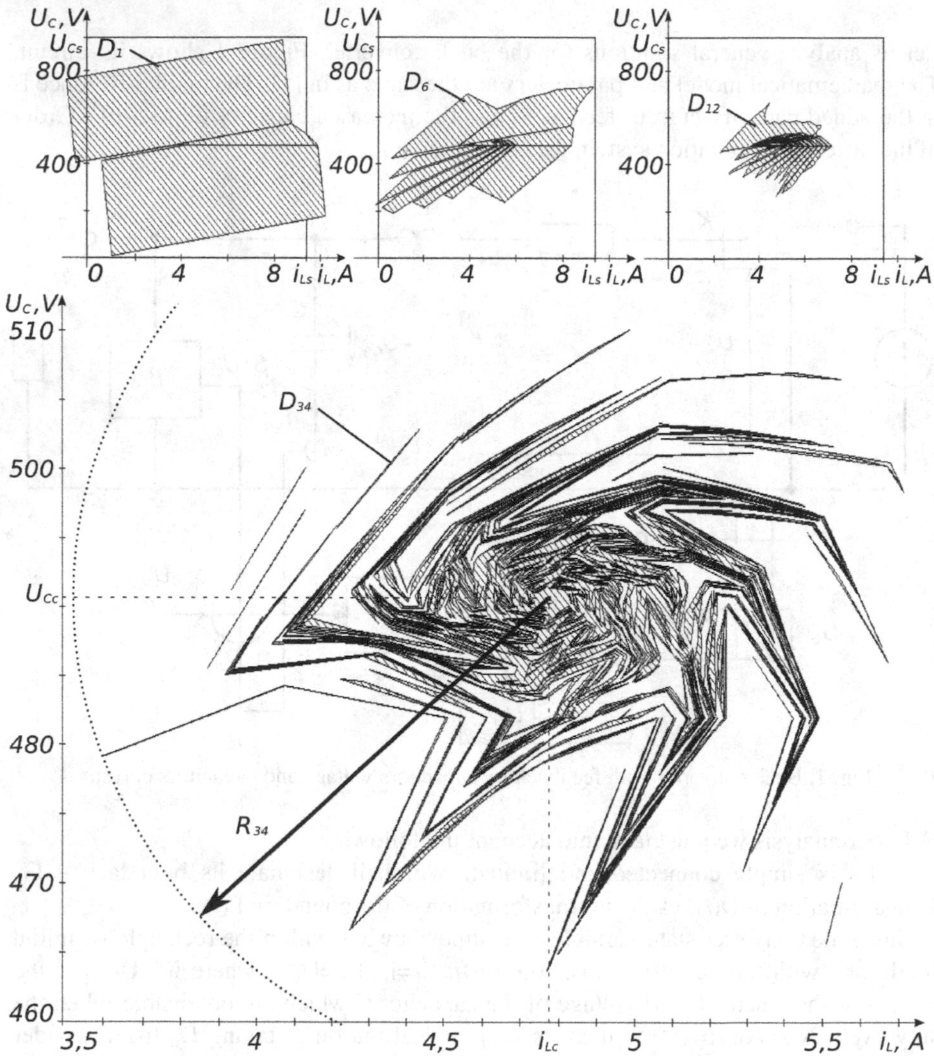

Fig. 2. Transformation of the initial set D_0 (which is a square bounded by the points (0,0) and (i_{Ls},U_{Cs})) with various values of k (k=1 is the upper left image, k=6 is the upper center image, k=12 is the upper right image, k=34 is the lower image, scaled up) resulting from feedback ratios (α=55.64, β_2=0)

Figure 2 shows transformation of the set D_0 with the growth of k. With the parameters chosen, the set is compressing and its structure becomes more complex. The boundaries of the sets $D_k - \Gamma_k$ are preserved and transformed with the use of polygons. The lower image in Figure 2 shows that, with the growth of k, D_k loses its simple connectedness and has an increasingly complex structure.

Given that each individual map D_k is complex, it is convenient to describe such maps using the radius R, which is the maximum distance from the steady basic operation mode X_C to the far point of the boundary.

$$R_k = \max \left\| N_r (\Gamma_k - X_c) \right\|_2 \cdot 100\% ,$$

where Γ_k specifies the set of points limiting D_k, N_r is the normalizing matrix

$$N_r = \begin{bmatrix} 1/i_{Ls} & 0 \\ 0 & 1/U_{Cs} \end{bmatrix}, \; X_c(t) \equiv Xc(t+Tq) \, , X_c = \begin{bmatrix} i_{Lc} \\ U_{Cc} \end{bmatrix}.$$

For example the lower image in Figure 2 shows R_{34}

It is clear that, if one is able to deliver solutions in general, it is possible to assess the rate of convergence in the large and specify the best parameters for fast response in the large. Figure 3 shows the overview of such analysis.

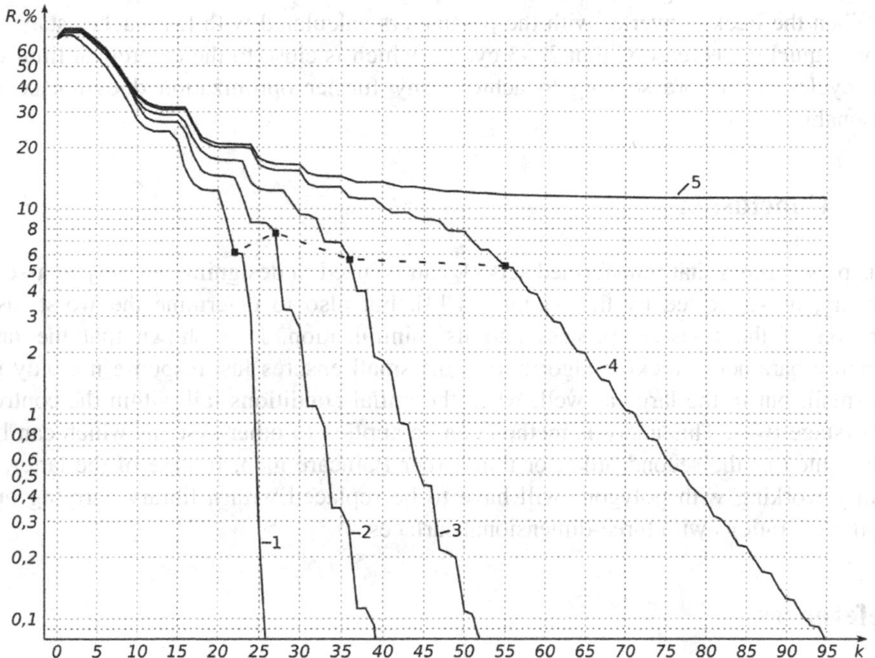

Fig. 3. Dependence of R on the step k with different feedback ratios. Curve 1 (α=148.3, β_2=0.1816), Curve 2 - (α=55.64, β_2=0.1), 3 - (α=70, β_2=0.05), 4 - (α=55.64, β_2=0), 5 - (α=100, β_2=0).

D_0 contains the worst case, where convergence is either nonexistent or has a maximum range. It is evident that the speed of radius reduction will be less than or equal to the speed of the corresponding worst case of the initial conditions.

The worst case corresponds to Curve 5 – the radius does not decrease and, therefore, movement is not stable in the large. Curve 4 corresponds to long transient processes and is similar to Curve 5, only it uses capacitor current feedback (β_2=0). Curve 4 follows the same parameters as figure 3. Introducing the capacitor current feedback makes it possible to reduce the duration of transient processes (Curves 3 and 2). Parameters of Curve 1 were determined using the method described in [4] and the online algorithm implementing the method [5].

We can observe two stages in compression of boundaries with the growth of k. At stage 1, a part of the sets D_k fall into the mode where PWM operates in a saturation mode during one or more periods, i.e. with z=1 or z=0.

At stage 2, D_k falls within the control area where $z \neq 0, z \neq 1$.

Within the saturated region, duration of transient processes is determined by the properties of R,L,C,R_L, since there is virtually no modulation present and its impact on the processes through feedback ratio is strongly limited.

Figure 3 shows Stages 1 above and 2 below the dotted line.

When the buck converter with the parameters calculated with [5] reaches stage 2, R_k will quickly decrease within 2—3 cycles, which is close to the theoretical limit of two cycles. Therefore seeking to achieve any further optimization does not seem reasonable.

3 Conclusion

This paper shows that transformed sets D_k can be used in designing not only to assess stability, as suggested by the authors in [3], but also to determine the worst-case duration of the transient process and its minimization. It is shown that the fast response parameter-tracking algorithm in the small ensures fast response not only in the small, but in the large as well, when the initial conditions fall within the control area (stage two). This analysis method can be applied to other systems which can be represented as the second order, or third order Poincare map. In case of the latter, a library working with polygons will have to be replaced with a library carrying out similar operations with three-dimensional surfaces.

References

1. Baushev, V.S., Zhusubaliev, Z.T., Mikhalchenko, S.G.: Stocastic Features in the Dynamic Characteristics of a Pulse-width Controlled Voltage Stabilizer. J. Electrical Techonology 1 (1996)
2. Banerjee, S., Ott, E., Yorke, J.A., Yuan, G.N.: Anomalous Bifurcation in DC-DC converters: borderline collisions in piecewise smooth maps. In: Proc. IEEE Power Electronics Specialists Conf., pp. 1337–1344 (1997)

3. Kobzev, G.A., Tanovitski, Y.N., Savin, D.A., Turan, V.V.: Method for Analysis of the Global Stability of Dynamic Systems with Pulse-Width Modulation that are Presented in the Form of Poincaré Mappings. Russian Physics Journal 54(6), 673–678 (2011)
4. Kobzev, G.A., Tanovitski, Y.N., Savin, D.A.: Control Algorithm for the Adaptive Buck Converter with Pulse-Width Modulation. Izvestiya Vuzov Rossii Radioelektronika 2007(3), 27–33 (2007) (in Russian)
5. Calculation of the buck converter parameters minimizing the duration of the transient process, `http://www.ie.tusur.ru/books/js/calc.htm`
6. Tanovitski, Y.N., Khalilyaev, T.F., Kobzev, G.A.: Adaptive Control Algorithm for Stabilized Buck Converters with Pulse-Width Modulation. In: Proceedings of Tomsk State University of Control Systems and Radioelectronics, vol. 1(21), part 2 (2010), `http://tusur.ru/filearchive/reports-magazine/2010-1-2/80-85.pdf`

Transient Response Analysis of Shunt Active Power Compensators under Asymmetric Voltage

Constantin-Vlad Suru, Cristina Alexandra Patrascu, and Mihaita Linca

University of Craiova, Faculty of Electrical Engineering
Craiova, Romania
{vsuru,apatrascu,mlinca}@em.ucv.ro

Abstract. This paper analyzes the dynamic response of a shunt active compensator, in three-phase, three-wire systems, based on the dSpace DS1103 control board and using the Conservative Power Theory for the compensating current computation, in non-sinusoidal and unbalanced conditions. The active filtering system was investigated both by simulation and experiment, the virtual filtering system being equivalent to the experimental one. The load under test was a balanced full-wave rectifier supplied from an unbalanced voltage source. The voltage unbalance was obtained by inserting a resistor between the grid and load on one phase, the absorbed current creating voltage unbalance, and also important voltage distortion.

Keywords: Dynamic response, Conservative Power Theory, harmonic distortion, active compensator.

1 Introduction

The widespread of static converters in the industrial environment produces a significant current harmonic distortion in the power grid with all its negative effects. The most versatile way to reduce the harmonic component of the grid current is the shunt static compensator which acts like a current generator, injecting harmonic currents in the PCC, thus, supplying the nonlinear loads with the necessary harmonic current [1], [2]. Consequently, the current drawn from the power grid by the nonlinear loads is only the active current (in fact, this depends on the compensation goal – unity power factor, sinusoidal grid current, etc [3]).

In the literature there are many compensating current computation methods, according to the desired compensation goals. These methods have good results with respect to the filtering performance [1],[2], [3]-[7], but it is interesting to investigate the dynamic response of the active filter regarding: the compensating current accuracy, the power grid phase and harmonic distortion, the compensating capacitor instantaneous voltage evolution.

After a brief introduction in the 1st section, the 2nd section presents the compensating current computation based on the Conservative Power Theory method. The 3rd section describes the Simulink model, which implements the complete active filtering system.

V.M. Mladenov and P.C. Ivanov (Eds.): NDES 2014, CCIS 438, pp. 148–155, 2014.

In this section, are also presented the results obtained by simulation, for a typical nonlinear load, and the power quality indicators.

The control block of the active power filter, used in the Simulink analysis, was also used in the experimental control model, discussed in the 4th section.

The 5th section treats the experimental determinations, in order to validate the virtual results, and the 6th section is for conclusions.

2 Current Terms Definition in Poly-Phase Networks According to the CPT

The definitions of the current components, used to calculate the compensating current, are defined taking into account the Conservative Power Theory.

Therefore, the compensating current will be computed using the load current which is defined as follows [4]-[6]:

$$\underline{i}_L = \underline{i}_a + \underline{i}_r + \underline{i}_v \,, \tag{1}$$

where the notation $\underline{i}$ means the column vector containing the components of i on each phase (i_R, i_S, i_T).

The active and reactive components are defined collectively with reference to an equivalent balanced load, as one can see [4]-[6]:

$$\underline{i}_a^b = \frac{\mathbf{P}}{\mathbf{U}^2}\underline{u} \,. \tag{2}$$

$$\underline{i}_r^b = \frac{W}{\widehat{\mathbf{U}}^2}\underline{\widehat{u}} = B^b\underline{\widehat{u}} \,. \tag{3}$$

where, **P** and W are the three-phase active power respective the reactive energy of the load.

3 Virtual Implementation of the Active Filtering System

The virtual filtering system was constructed so that is equivalent to the experimental active filtering system. This way, a comparison between the results obtained by simulation and the ones obtained by experiment can be relevant.

The virtual model of the system was built using distinct blocks for the corresponding system components, in order to be easy to follow (**Fig. 1**): the three-phase power grid, the nonlinear load, the three-phase power inverter, the T type interface filter, the DC-Link (compensating capacitor), the control algorithm of the active filter, various blocks for the data acquisition and monitoring.

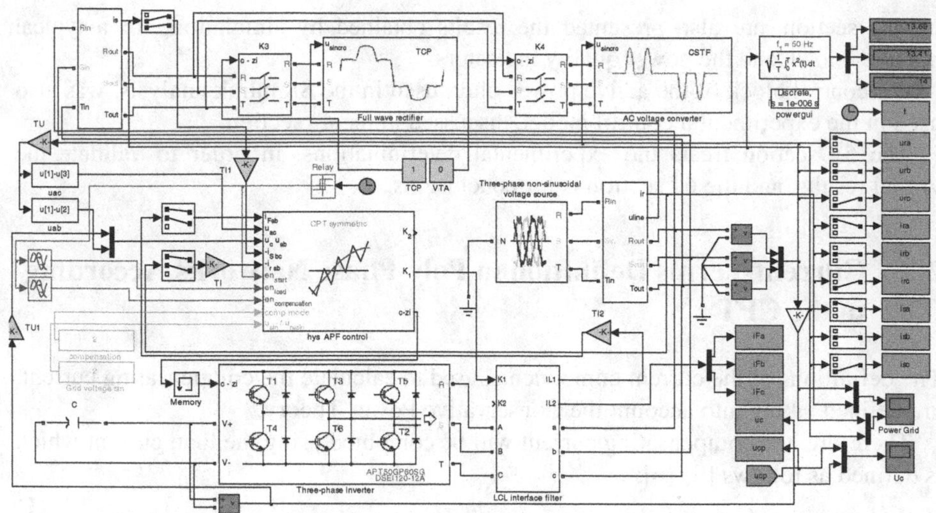

Fig. 1. The active filtering system Simulink model

The power grid voltage is non-sinusoidal, the distortion being obtained by inserting harmonic voltage sources on each phase, in series with the fundamental. The voltage unbalance is obtained by connecting a resistance between the voltage source and the PCC on phase b (the resistance value is 3.5 ohms). As a result, the voltage unbalance factor [8], and the THD on each phase will be influenced by the used nonlinear load.

A typical non-linear load was considered, i.e. a balanced three-phase full wave rectifier connected to the power grid through a Δ-Y transformer. The output line voltage of the transformer is 180 V, so the firing angle was adjusted so that the load absorbs from the grid a current of about 15 A on each phase.

The interface filter is a T passive filter having the inductances of 4 mH and 0.4 mH and the capacitor value of 10 nF. This type of interface filter will substantially reduce the distortion produced by the inverter switching [10], [11].

The control block of the active filter contains all the necessary components to control the APF, such as: the compensating current computation, the current and the voltage regulating loop, the auxiliary and validating blocks, etc. The advantage of using this block is the easy pass from the simulation study to the experimental study. This way the models corresponding to the power section are removed and the input and output ports of the acquisition platform are connected to the computation block [12].

For the study by simulation of the filtering system, the current absorbed from the grid by the nonlinear load before the compensation is illustrated in **Fig. 2**. Regarding the dynamic response of the active compensator, two situations were taken into consideration:

- the load is functional and the filter starts the compensation;
- the filter is fully functional (and compensating) and the load starts.

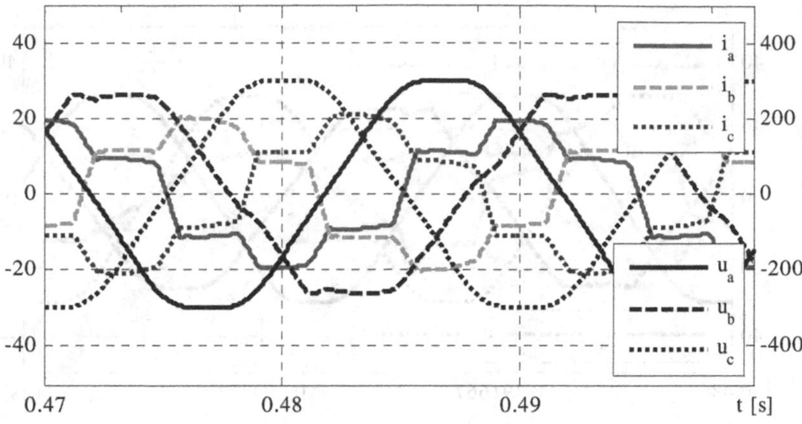

Fig. 2. The power grid currents and voltages at the PCC before the compensation

In the first case the dynamic response of the active filter is dependant only by the filter itself, while in the second case it depends also by the load startup sequence.

When the load is functional and the filter starts the compensation, the grid current waveform is illustrated in **Fig. 3**-a.

It can be seen that the compensation starts almost instantaneous, the delay between the compensation validation and the moment when the compensated current follows the desired shape being 985.92 μs. Also, it can be seen that the distortion and unbalance of the grid voltage are reduced. The numerical results (total harmonic distortion factors and the unbalance factors [8]) are synthesized in **Table 1**, showing the efficient reduction of these indicators for both grid voltage and current.

Table 1. Synthesized simulation results

	Balanced Rectifier					
	Before compensation			**After compensation**		
	a	**b**	**c**	**a**	**b**	**c**
THDi [%]	24.36	21.97	20.62	5.13	4.76	5.08
THDu [%]	3.26	7.10	3.17	3.25	3.26	3.18
CUF [%]	4.91			4.51		
VUF [%]	7.34			4.34		

Regarding the compensating current (**Fig. 3**-b), until the compensation starts, the filter generated current is only the sinusoidal active current necessary to maintain the DC-Link voltage to the imposed value (700 V). When the compensation is validated, the filter current starts to follow the reference current waveform, excepting the noticeable difference, which can be seen in the figure, but which extinguishes very shortly. This overshoot is due to the DC-Link voltage drop (seen in **Fig. 3**-c at a different time scale) and it appears because the voltage regulating loop restores the DC-Link voltage to the imposed value by absorbing an active current from the grid which is added to the compensating current. The overshoot time length is given by the necessary time for the voltage loop to reestablish the imposed voltage value.

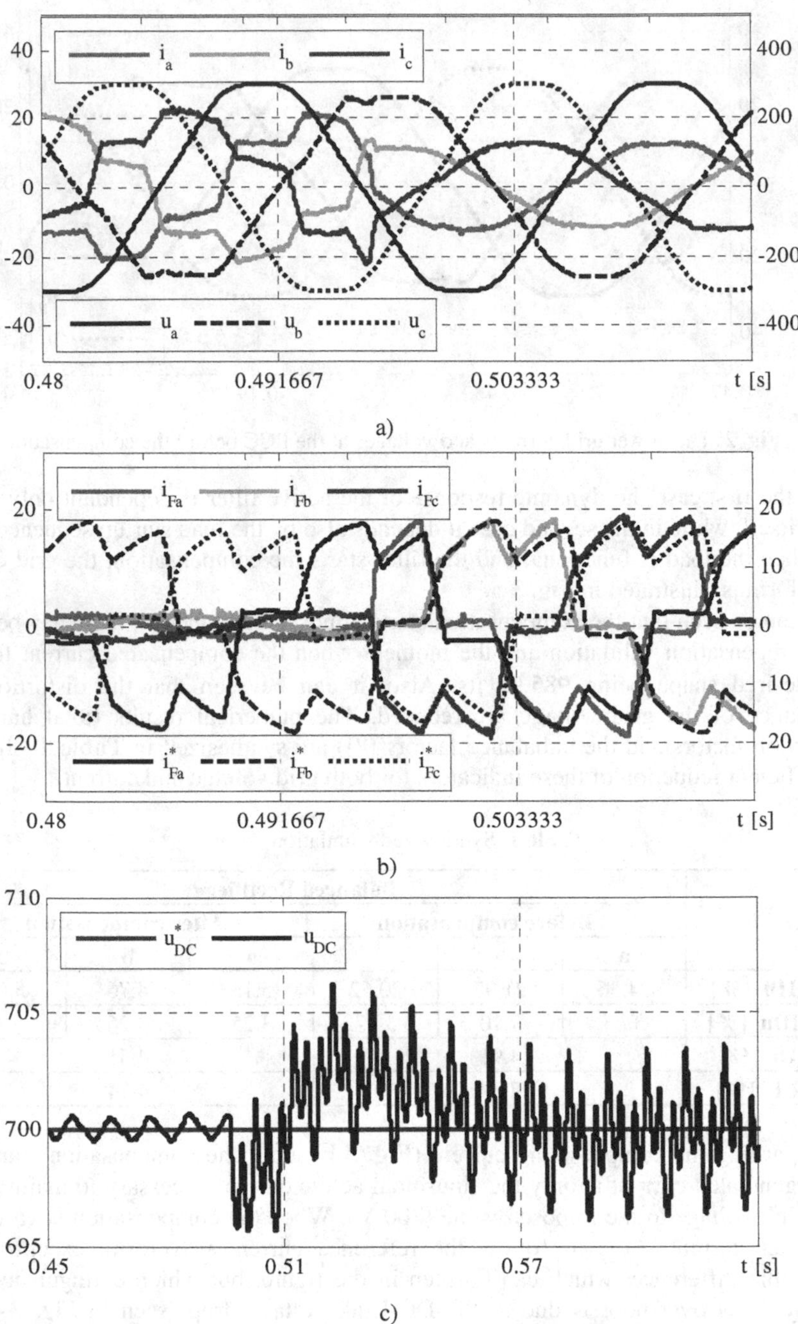

Fig. 3. The grid voltages and currents (a), the compensating currents (b) and the DC-Link voltage (c)

4 The Experimental Setup

The experimental active filtering system contains three major components [12]:

- *The software section* – all the computed components of the system (such as: the current and voltage regulators, the compensating current algorithm (CPT), the charging sequence control block, etc);
- *The hardware section* –three-phase inverter with 1200 V and 100 A IGBTs, 2200 μF compensating capacitor, LCL interface filter, current transducers and voltage transducers;
- *The nonlinear load*;

The command and control section of the experimental compensator consists in a dSpace DS1103 control board which uses a Matlab Simulink model compiled and loaded in the board memory and a control panel. This approach has the advantage of using a Simulink model as a control algorithm instead of programming a C code [12], [13]. The static compensator control panel is based on the DS1103 specific software (Control Desk), and is a virtual interface. This virtual panel (**Fig. 4**) contains all the necessary instruments for controlling the APF, such as check buttons, indicator lamps, etc, and also, the system monitoring devices (virtual panel meters, oscilloscopes, etc).

5 Experimental Results

Using the experimental platform described in the previous section, the load and the compensated current waveforms for phase a, captured with Metrix OX 7042 – M digital oscilloscope are illustrated in **Fig. 5**-a, for the case when the load is on and the compensation is validated. It can be seen that (like in the simulated results) the compensation starts almost instantaneous, with an almost negligible overshoot in the compensated current amplitude. This overshoot can be distinguished in **Fig. 5**-c, where the same signals are captured with the DS1103 board ADCs, and processed with ControlDesk and Matlab software.

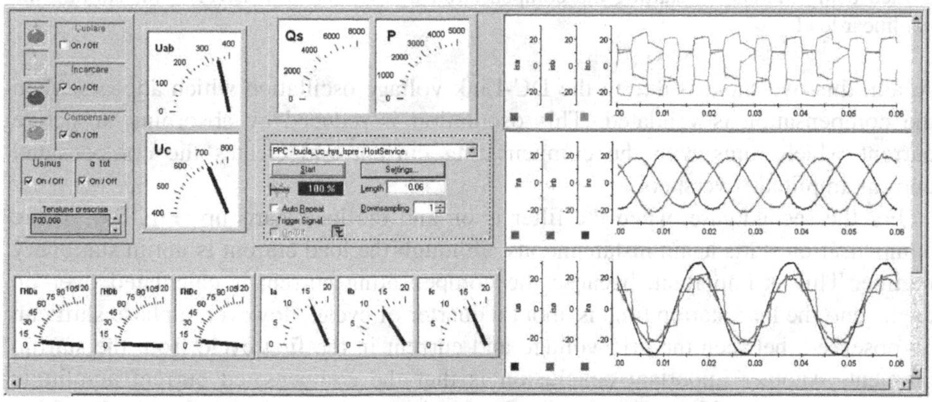

Fig. 4. The static compensator virtual control panel

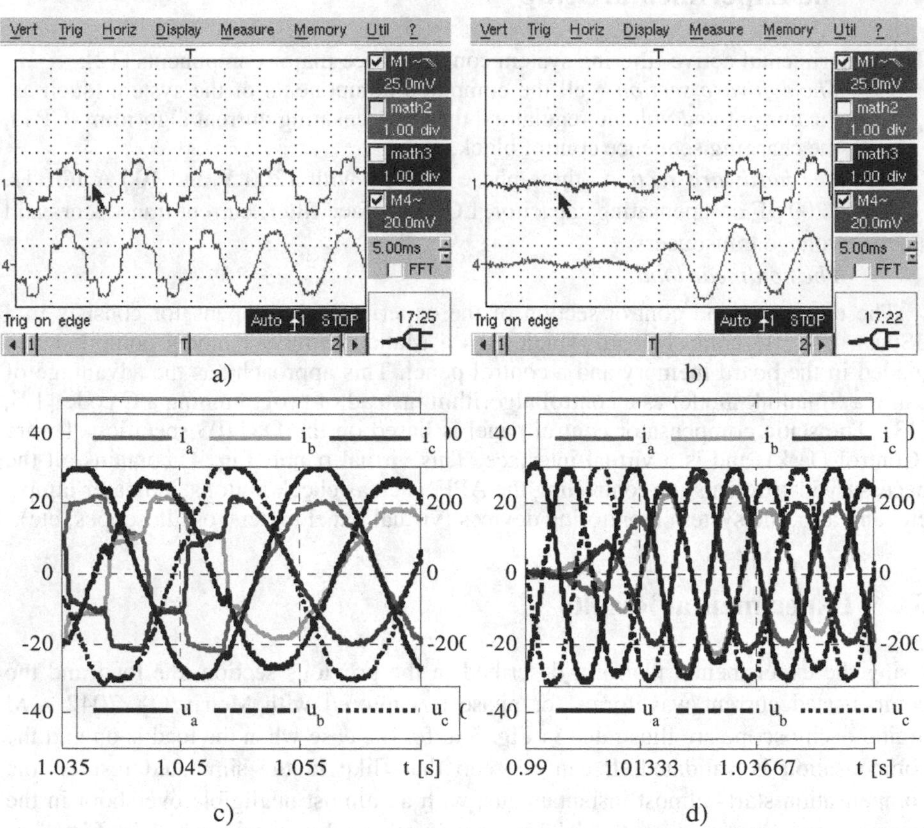

Fig. 5. The load and compensated current waveforms: a) captured with the Metrix oscilloscope for starting the compensation b) captured with the Metrix oscilloscope for starting the nonlinear load c) captured using the ControlDesk software, DS1103 acquisition board, and plotted using Matlab for starting the compensation d) captured with DS1103 for starting the nonlinear load

Again, this overshoot is due to the DC-Link voltage oscillation which appears when the compensation is validated. This oscillation is reduced by absorbing an active current which sums with the compensating current and creates the compensated current amplitude overshoot.

For the second case, when the filter is on and the load starts up (**Fig. 5**-b, d) the compensation starts again instantaneous, although the load current is not in stationary regime. This is important, because the compensating current is calculated over on cycle, and the load startup time is about a quarter of cycle. However, a phase shift can be observed, between the grid voltage and current in the first cycle from the startup moment. Another important conclusion is that the compensated current amplitude overshoot is noticeable higher than in the previous case, signifying that the DC-Link oscillation is significantly higher.

6 Conclusions

It can be concluded that the active compensator dynamic response is practically instantaneous, both for starting the compensation and for responding to a load change. The phase delay of the compensated current in the transitory regime is negligible (less than half of a cycle) even in the toughest conditions. Also, the compensated current overshoot due to the DC-Link voltage oscillation is reduced (11.95%) and quickly amortized (about four cycles). We can say that the equivalent results obtained in the experimental and simulation studies, in terms of quality as well as in terms of quantity, confirm the validity of the control model and of the virtual and experimental filtering systems, and also, the performances of these systems.

References

1. Asiminoaei, L., Blaabjerg, F., Hansen, S.: Detection is key. Harmonic detection methods for active power filter applications. IEEE Industry Applications Magazine, 22–33 (July/August 2007)
2. Asiminoaei, L., Blaajberg, F., Hansen, S.: Evaluation of Harmonic Detection Methods for Active Power Filter Applications
3. Popescu, M., Bitoleanu, A., Patrascu, C.A.: MATLAB/SIMULINK Library for Compensating Current Calculation in Three-Phase Shunt Active Filtering Systems, vol. 4/2012, pp. 1224–7928. Buletinul AGIR (2012) ISSN-L 1224-7928
4. Tenti, P.: Conservative Power Theory Seminar: A theoretical background to understand energy issues of electrical networks under non-sinusoidal conditions and to approach measurement, accountability and control problems in smart grids. UNICAMP – UNESP, Sorocaba (August 2012)
5. Tenti, P., Mattavelli, P.: A Time-Domain Approach to Power Term Definitions under Non-Sinusoidal Conditions. In: 6th International Workshop on Power Definitions and Measurements Under Non-Sinusoidal Conditions, Milano, October 13-15 (2003)
6. Tenti, P., Mattavelli, P., Tedeschi, E.: Compensation Techinques Based on Reactive Power Conservation. Electrical Power Quality and Utilisation Journal XIII(1) (2007)
7. Popescu, M., Bitoleanu, A., Suru, V.: Phase Coordinate System and p-q Theory Based Methods in Active Filtering Implementation. Advances in Electrical and Computer Engineering 13, 69–74 (2013) ISSN: 1582-7445, e-ISSN
8. Bollen, M.H.J.: Definitions of Voltage Unbalance. IEEE Power Engineering Review (November 2002) 0272-1724/02/$17.00©2002 IEEE
9. Suru, V., Pătrascu, A., Lincă, M.: Conservative Power Theory Implementation in Shunt Active Power Filtering. In: XI International School on Nonsinusoidal Currents and Compensation, ISNCC 2013, Zielona Gora, Poland (2013) Print ISBN: 978-1-4673-6312-9
10. Bitoleanu, A., Popescu, M., Marin, D., Dobriceanu, M.: LCL Interface Filter Design for Shunt Active Power Filters. Advances in Electrical and Computer Engineering 10(3), 55–60 (2010) ISSN: 1582-7445
11. Suru, V., Bitoleanu, A.: Design and Performances of Coupling Passive Filters in Three-Phase Shunt Active Power Filters. vol. 35, pp. 31–38. Analele Universitatii Din Craiova - Seria Inginerie Electrica (2011) ISSN 1842-4805
12. Suru, V., Popescu, M., Pătraşcu, A.: Using dSPACE Based System in Active Power Filter Control, vol. 37, pp. 94–99. Annals of The University of Craiova (2013) ISSN 1842-4805
13. Control Desk Experiment Guide for release 5.2. dSpace Gmbh (2006)

Unfolding the Threshold Switching Behavior of a Memristor

Stefan Slesazeck[1], Alon Ascoli[2], Hannes Mähne[1], Ronald Tetzlaff[2], and Thomas Mikolajick[1,3]

[1] NaMLab gGmbH
Nano-electronic Materials Laboratory, Dresden (Deutschland)
[2] Institut für Grundlagen der Elektrotechnik und Elektronik
Technische Universität Dresden, Dresden (Deutschland)
[3] Institut für Halbleiter- und Mikrosystemtechnik
Technische Universität Dresden, Dresden (Deutschland)
{stefan.slesazeck,hannes.maehne}@namlab.com,
{alon.ascoli,ronald.tetzlaff}@tu-dresden.de,
thomas.mikolajick@namlab.com

Abstract. Employing a mathematical model based upon Chua's unfolding theorem, some aspects of the nonlinear dynamics of a thermally-activated micro-scale NbO_x/Nb_2O_5 volatile memristor were modeled. Insights into the peculiar behavior of the device are gained through experiments and model-based simulations. Particularly, this enables us to reproduce its threshold switching behavior under quasi-static excitation, and to explain under which conditions the off-to-on switching is accompanied by the appearance of a negative differential resistance region on its current-voltage characteristic.

Keywords: Memristor, Unfolding Theorem, Nonlinear Dynamics, Threshold switching, Local activity.

1 Introduction

The recent introduction of a thermally-activated locally-active Niobium dioxide-based volatile memristor [1,2] is drawing a strong interest from both the industry and the research community since it may open up new opportunities in IC design, particularly for the hardware implementation of novel bio-inspired signal processing paradigms [3]. The most important tool for a circuit designer is the availability of a reliable [4] memristor model [5,6] capable to reproduce the nonlinear dynamics of the device [7] with sufficient accuracy and for a broad range of initial conditions and input signals [8]. At NaMLab we have direct access to the fabrication process and electrical characterization of a NbO_x/Nb_2O_5 [9] micro-scale memristor of the kind mentioned above [2]. Our aim is to develop a simple [10] yet accurate mathematical model [11] of our memristor so as to use it for design purposes [12]. After the electroforming process our two-terminal element acts as a large resistor under zero input. However, switching the power on, the

V.M. Mladenov and P.C. Ivanov (Eds.): NDES 2014, CCIS 438, pp. 156–164, 2014.

current flow through the device leads to an internal temperature increase due to Joule heating, and this phenomenon gradually lead to the *activation* of the device, which, when the current reaches a certain threshold value, exhibits a fast off-to-on switch. This switching behavior is at the origin of the recent proposal to use such two-terminal element as selector device [1,13] for non-volatile memristor crossbar arrays. Under suitable conditions, the switching phenomenon is accompanied by the appearance of a negative differential resistance (NDR) region in the current-voltage characteristic. Here during the turn-on process the quick increase in current induces an even larger drop in device resistance, and, consequently, the overall memristor voltage decreases. In the NDR portion of the current-voltage characteristic the memristor is locally-active [14] and may support the emergence of complex dynamical behaviors [1] including chaos [15,16]. Taking inspiration from Chua's unfolding theorem [17], in this work we first propose a novel mathematical model of our thermally-activated NbO_x/Nb_2O_5 volatile memory switch, and then use the model to capture some aspects of the nonlinear dynamics of the micro-scale device, particularly its threshold switching behavior, and the ability to exhibit a region of local activity in the current-voltage characteristic.

2 Polynomial Memristor Model Based on Chua's Unfolding Principle

The proposed mathematical model of our memristor falls into the class of memristive systems:

$$\frac{dx}{dt} = f(x, u), \tag{1}$$

$$y = g(x, u)u, \tag{2}$$

where x, u and y respectively denote state, input and output, $f(\cdot)$ stands for the state evolution function, and $g(\cdot)$ represents the memristance (memductance) function under current (voltage) excitation respectively. Adopting Chua's unfolding paradigm [17], $f(\cdot)$ and $g(\cdot)$ are represented as sums and/or products of state and input polynomials. In fact, according to the unfolding theorem [17], the mathematical framework of a memristive system[1], expressed in (1)-(2), may be expanded as

$$\frac{dx}{dt} = \sum_{k=-n_1}^{n_2} a_k x^k + \sum_{k=-m_1}^{m_2} b_k u^k + \sum_{k=-p_1}^{p_2} \sum_{l=-q_1}^{q_2} c_{kl} u^k x^l \tag{3}$$

$$y = \left(\sum_{k=-r_1}^{r_2} d_k x^k + \sum_{k=-s_1}^{s_2} e_k u^k + \sum_{k=-v_1}^{v_2} \sum_{l=-w_1}^{w_2} f_{kl} u^k x^l \right) u \tag{4}$$

[1] The ideal memristor [18] is an element from this class.

where $\{a_k, b_k, c_{kl}, d_k, e_k, f_{kl}\}$ is the set of real unfolding parameters. The polynomial coefficients, known as unfolding parameters, need to be fitted to the experimental data under disposal through some optimization algorithm. We employed a coefficient tuning procedure [4], which, using a certain number of sets of previously-collected experimental data relative to distinct operating conditions[2], leads to the selection of the simplest possible set of non-zero unfolding parameters capable to minimize the relative mean squared error between lab measurements and model predictions in each operating condition. The algorithm employs a bottom-up approach: only after ascertaining the failure of low-cardinality unfolding parameter sets to allow minimization of the chosen cost function in all case studies, are the number of parameters in the set increased and the process iterated.

This algorithm, recently employed [4,19] to model the switching kinetics of the Pickett's physical model [20] of the Hewlett Packard TiO_2-based [21] nonvolatile memristor [22], massages the time waveforms of the physical quantities of interest (e.g. the memristor current and voltage) into shapes matching as accurately as possible the laboratory measurements. Fitting equations (3)-(4) to the laboratory measurements at our disposal, the dynamics of our memristor were found to be properly uncovered by the following set of differential-algebraic equations (DAE)

$$\frac{dx}{dt} = a_0 + a_1 x + b_2 u^2 + c_{21} u^2 x + c_{22} u^2 x^2 + c_{23} u^2 x^3 + c_{24} u^2 x^4 + c_{25} u^2 x^5 \quad (5)$$

$$y = (d_0 + d_1 x + d_2 x^2 + d_3 x^3 + d_4 x^4 + d_5 x^5) u \quad (6)$$

where u denotes the voltage v_m across the device and y stands for the current i_m flowing through it (i.e. the device is a voltage-controlled memristor [23,24]), while, most importantly, the values for the non-zero unfolding parameters in equations (5) and (6) are respectively reported in Tables 1 and 2.

Table 1. Values for the optimal set of unfolding parameters in equation (5) according to our fitting procedure

a_0	a_1	b_2	c_{21}
$5.61 \cdot 10^9$	$-2.22 \cdot 10^7$	$1.37 \cdot 10^{10}$	$-1.33 \cdot 10^8$
c_{22}	c_{23}	c_{24}	c_{25}
$4.28 \cdot 10^5$	$-4.56 \cdot 10^2$	$2.38 \cdot 10^{-1}$	$-5.12 \cdot 10^{-5}$

[2] The distinct operating conditions included a scenario where the memristor was driven directly through a voltage source, one where a resistor-memristor series combination was excited via a voltage source, and one where an input voltage was applied across the series connection between a resistor and a memristor-capacitor parallel combination.

Table 2. Values of the optimal set of unfolding parameters in equation (6) resulting from fitting equation (4) to the experimental data.

d_0	d_1	d_2
$6.80 \cdot 10^{-3}$	$-6.66 \cdot 10^{-5}$	$2.14 \cdot 10^{-7}$
d_3	d_4	d_5
$-2.28 \cdot 10^{-10}$	$1.19 \cdot 10^{-13}$	$-2.56 \cdot 10^{-17}$

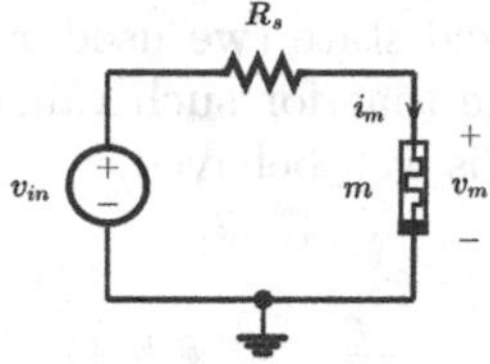

Fig. 1. Test circuit under study

3 Insights into the Threshold Switching Behavior

In this paper we model some aspects of the dynamical behavior of our memristor, which was explored using the circuit drawn in Fig. 1.

For the investigations described in this section the series resistance R_s is first replaced by a short circuit, and later set to a non-zero value. In the first study our memristor is simply driven by a voltage source v_{in}. The input waveform of a parameter analyzer Keithley 4200SCS (SCS stands for Semiconductor Characterization System) is a ramp first increasing from 0 V up to 1.5 V in 4.92 s, and then decreasing from its upper value back to its lower one over the same time interval. The signal is applied at time $t = 0.52$ s. This slowly-varying input is usually known as quasi-static excitation. Note that, besides generating signals, the Keithley 4200SCS does also possess the capability to measure currents and voltages through source measurement units (SMUs) and to further depict them on a display. Note that, in this experiment where no series resistor is employed, an oscilloscope would have been unable to measure the memristor current. In order to protect the device from an undesirable irreversible breakdown, the parameter analyzer delivers the specified input as long as the memristor current remains below a certain compliance level, here set to 5 mA. Should such level be ever attained, the input voltage would suddenly decrease to the value through which the memristor current-voltage characteristic is expected to pass for that particular current compliance level. Then the trajectory point would not move any further from the current location on the $i_m - v_m$ plane until the specified input signal was expected to decrease. Nevertheless the waveforms visualized on the Keithley 4200SCS refer to the specified driving conditions, therefore some interpretation of the experimental observations is required in case the current compliance level is ever hit for some time (see explanations below). Fig. 2 shows the ramping input signal (plot (a)), the current flowing through the memristor in response to the input (plot (b)), and the memristor quasi-static current-voltage characteristic (plot (c)). The red and blue curves respectively refer to laboratory measurements and model-based simulation results. In the model equations (5)-(6) we let $u = v_{in}$, where v_{in} was modeled using a triangular pulse with width 9.84 s and peak value 1.5 V at time $t = 5.44$ s, while $y = i_m$, and the state initial conditions was chosen as $x(0) = -a_0 a_1^{-1}$, which actually coincides with the equilibrium point of equation (5) for $u = 0$. The model is able to reproduce the typical threshold switching of our memristor. Fig. 2(d) depicts the model-based time evolution of the normalized state (we used $x(0)$ as normalization factor). From this plot one may realize that for such range of normalized state values the proposed unfolding model is well behaved.

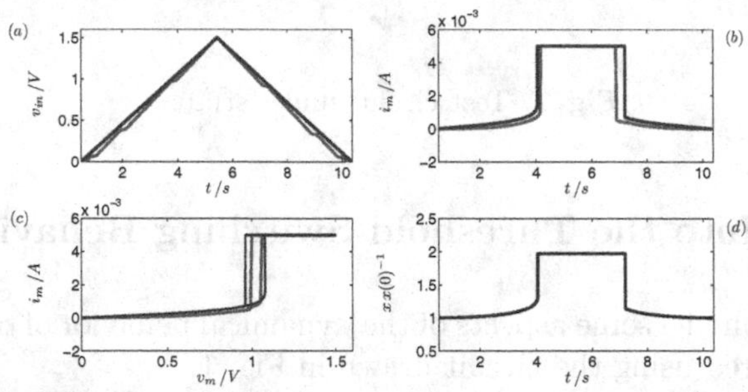

Fig. 2. Time waveforms of input voltage (plot(a)) and memristor current (plot (b)). Memristor quasi-static current-voltage characteristic (plot (c)). Experiment and model-based results are respectively depicted in red and blue. Plot (d): Normalized state over time resulting from the model simulation. Here $R_s = 0\Omega$.

During the threshold-activated turn-on switching process displayed in Fig. 2(c) a NDR region may not be observed for any value of v_{in}. In fact a local activity region is present on the $i_m - v_m$ plane, but is unstable for $R_s = 0\Omega$. In order to gain a deeper understanding of this phenomenon, we now include a finite series resistor $R_s = 330\Omega$ between the voltage source and the threshold switch (see Fig. 1). Here we use an oscilloscope Tektronix $TDS7154B$ to sense the memristor voltage and current (through an indirect measurement of the voltage drop over R_s) and to display the laboratory measurements, which in Fig. 3 are shown in red. As evident from the solid curve in plot (a), we set the input voltage to a triangular signal with amplitude 4 V, offset $2V$ (so that the signal is non-negative at all times), and frequency 10 Hz. The resulting memristor voltage and current over time are visualized in plots (b) and (c) respectively. It follows that a locally-active region appears on the quasi-static $i_m - v_m$ curve, as shown in plot (d). The blue curves in Fig. 3 refer to numerical simulations of equations (5)-(6) with $u = v_m$, where

$$v_m = v_{in} \frac{1}{1 + (d_0 + d_1 x + d_2 x^2 + d_3 x^3 + d_4 x^4 + d_5 x^5) R_s}, \tag{7}$$

v_{in} is the above specified periodic triangular signal, while $y = i_m$, and the state initial condition is kept unaltered in comparison to the simulation in Fig. 2. In Fig. 3(a) the normalized memristor state resulting from the model simulation is plotted over time in dashed line style.

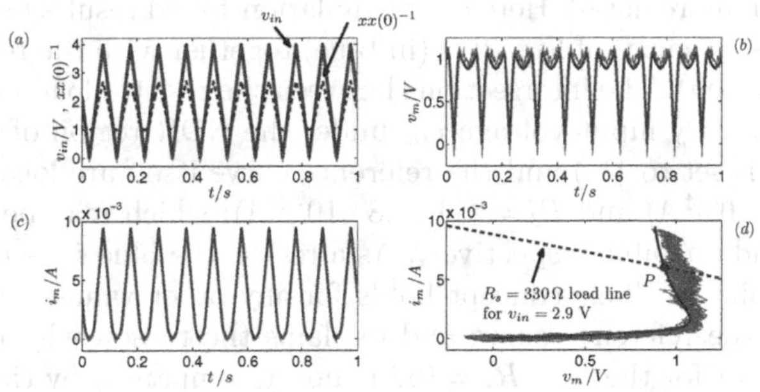

Fig. 3. Plot (a): Time evolution of the slowly-varying periodic triangular voltage source (solid curve) and of the normalized memristor state (dashed curve). Plot (b): Memristor voltage over time. Plot (c): Memristor current over time. Plot (d): $i_m(t)$ versus $v_m(t)$. Here $R_s = 330\Omega$.

From circuit theory it is well known that, under dc operation, for any value of the input voltage v_{in} the intersection between the resistor load line $v_m = v_{in} - R_s i_m$ and the $i_m - v_m$ relationship defined in equation (6) identifies the operating point of the threshold switch in the test circuit of Fig. 1. For example the operating point of the circuit with $R_s = 330\Omega$ for v_{in} set to the arbitrary value of 2.9V is located at $P = (V_m, I_m) = (0.94$ V, 6.00 mA) (dot blue marker in Fig. 3(d)). It may be demonstrated that, provided the absolute value of the slope of the load line, i.e. $G_s = R_s^{-1}$, is smaller than the absolute value of the minimum slope of the $i_m - v_m$ curve in the NDR region (let us call this inequality as condition C_1), by sweeping v_{in} it is possible to find locally-stable intersections over the entire extension of the locally-active part of the memristor current-voltage curve. This is the case for the scenario of Fig. 3(d), where the whole locally-active region of the memristor current-voltage characteristic is stable and thus appears during the input voltage sweep. Since the value for R_s chosen in the scenario of Fig. 3(d) allows the stabilization and thus the observation of all the points of the memristor quasi-static current-voltage characteristic (for the range of v_{in} values spanned by the input voltage signal in Fig. 3(a)), in the following the curve in Fig. 3(d) is taken as reference for the set of all possible (v_m, i_m) pairs through which the current-voltage characteristic of our memristor may get under DC driving conditions. If G_s is larger than the absolute value of the maximum slope of the $i_m - v_m$ curve in the NDR region, during the input voltage sweep, under any circumstance where the load line intersects the NDR region, it will intersect the $i_m - v_m$ curve in one further point (for the above chosen current compliance level)[3]. Regarding the stability of these intersection points, it may be proved that the one lying on the NDR region would be unstable, while the other one would be stable. Therefore the memristor would be unable to ever operate in some locally active-state. This is the case for the scenario depicted in Fig. 2(c).

Let us analyze the scenario with $R_s = 0\Omega$ (an example where condition C_1 does not hold) in more detail. Here only simulation-based results are shown. Fig. 4 shows the $i_m - v_m$ curve of Fig. 2(c) (in blue) together with the reference curve of Fig. 3(d) (in green). The intersections between the resistor load line (in dashed blue) for an arbitrary input voltage v_{in} under the NDR region of the reference curve (here v_{in} is set to $1V$) and the reference curve itself are located at points $P_1 = (1V, 8.16 \cdot 10^{-4}A)$ and $P_2 = (1V, 3.8 \cdot 10^{-3}A)$, which may be shown to be locally stable and unstable respectively. As a result, the blue $i_m - v_m$ curve goes only through point P_1. This concept holds for any other value of v_{in} under the NDR region of the reference curve and explains the reason why the threshold switching behavior for the case $R_s = 0\Omega$ is not accompanied by the appearance of a locally-active region.

Another aspect is worthy of mention, when condition C_1 is not met (we focus on the $R_s = 0\Omega$ case study once again). After the trajectory reaches point A (square blue marker in Fig. 4), the switch-on process sets in. The next stable point would lie somewhere above the current compliance level, but, due to the current bound, may not be reached. As briefly anticipated above, the blue trajectory on the $i_m - v_m$ plane is the curve which would be observed if the input voltage delivered by the parameter analyzer were the blue ramping signal in Fig. 2(a) at any time. But this is not possible when the current attains the compliance level, since taking the green reference curve of Fig. 4 into account, the memristor voltage may only assume one possible value at compliance. In fact, as i_m assumes the value of $5mA$, the applied input signal is suddenly decreased to the abscissa value of the only point through which the memristor current-voltage characteristic may get for the chosen current compliance level. Therefore the next stable trajectory point is B (square blue marker in Fig. 4). There the trajectory halts and remains until the specified input voltage is bound to decrease. At that time a switch off process takes place and the next stable point for the $i_m - v_m$ curve is C (square blue marker in Fig. 4).

With reference to Fig. 4, it is also interesting to observe that, as expected, the threshold switching points A and B on the blue $i_m - v_m$ curve for $R_s = 0\Omega$ lie on the green reference $i_m - v_m$ curve for $R_s = 330\Omega$.

4 Conclusions

Taking inspiration from Chua's unfolding principle, we expanded the state evolution function of the mathematical model of our NbO_x/Nb_2O_5 memristor into

[3] If the absolute value of the slope of the load line lies between the absolute value of the minimum and the absolute value of the maximum slope of the $i_m - v_m$ curve in the NDR region, by sweeping v_{in} it is possible to observe a portion of the locally-active region of the memristor current-voltage curve.

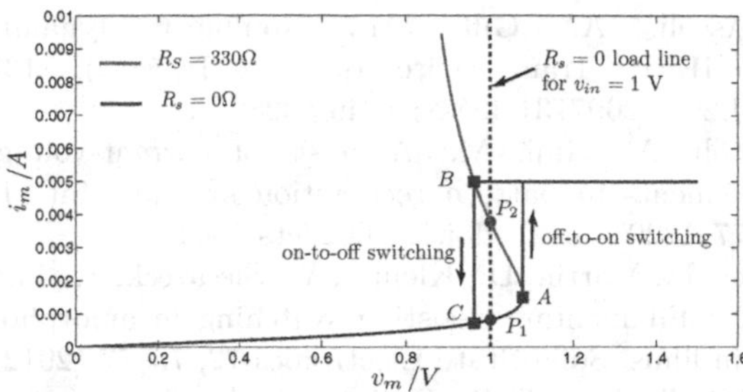

Fig. 4. Memristor current-voltage characteristics for $R_s = 330\Omega$ (reference curve in green, as from Fig. 3(d)) and for resistor $R_s = 0\Omega$ (blue curve, as from Fig. 2(c)), and $R_s = 0\Omega$ load line for an arbitrary input voltage under the NDR region of the reference $i_m - v_m$ curve (here v_{in} is set to 1 V)

a sum of a state series expansion, an input series expansion and a product of state and input polynomials. The memductance function was taken in the form of yet another state series expansion. After the collection of a large number of experimental data, we applied an optimization technique aimed at choosing the simplest possible set of non-zero unfolding parameters leading to the minimization of a certain cost function (error between experimental data and model predictions). Using this model we were able to reproduce some aspects of the static and dynamic behavior of our micro-scale device, particularly its threshold switching behavior, and the capability to exhibit a negative differential resistance region.

References

1. Pickett, M.D., Williams, R.S.: Sub-100 fJ and sub-nanosecond thermally driven threshold switching in niobium oxide crosspoint nanodevices. Nanotechnology 23(21), 215202(9pp) (2012)
2. Mähne, H., Wylezich, H., Slesazeck, S., Mikolajick, T., Vesely, J., Klemm, V., Rafaja, D., Room temperature fabricated NbO_x/Nb_2O_5 memory switching device with threshold switching effect. In: Proc. 5th IEEE Int. Memory Workshop (IMW), pp. 174–177 (2013)
3. Pickett, M.D., Medeiros-Ribeiro, G., Williams, R.S.: A scalable neuristor built with Mott memristors. Nature Materials 12(2), 114–117 (2012)
4. Ascoli, A., Corinto, F., Senger, V., Tetzlaff, R.: Memristor model comparison. IEEE Circuits and Systems Magazine 13(2), 89–105 (2013), doi:10.1109/MCAS.2013.2256272
5. Corinto, F., Ascoli, A.: A boundary condition-based approach to the modeling of memristor nanostructures. IEEE Trans. on Circuits and Systems–I 59(11), 2713–2726 (2012)
6. Corinto, F., Ascoli, A., Gilli, M.: Memristor models for pattern recognition systems. In: Kozma, R., Pino, R., Pazienza, G. (eds.) Advances in Neuromorphic Memristor Science and Applications. Springer Series in Cognitive and Neural Systems, vol. 4, part 3, pp. 245–268. Springer (2012)

7. Corinto, F., Ascoli, A., Gilli, M.: Nonlinear dynamics of Memristor Oscillators. IEEE Trans. Circuits Syst.–I 58(6), 1323–1336 (2011), doi:10.1109/TCSI.2010.2097731 ISSN: 1549-8328
8. Corinto, F., Ascoli, A., Gilli, M.: Analysis of current-voltage characteristics for memristive elements in pattern recognition systems. Int. J. Circuit Theory Appl. 40(12), 1277–1320 (2012), doi:10.1002/cta.1804
9. Mähne, H., Berger, L., Martin, D., Klemm, V., Slesazeck, S., Jakschik, S., Rafaja, D., Mikolajick, T.: Filamentary resistive switching in amorphous and polycrystalline Nb_2O_5 thin films. Solid-State Electronics 72, 73–77 (2012)
10. Ascoli, A., Schmidt, T., Tetzlaff, R., Corinto, F.: Application of the Volterra Series paradigm to memristive systems. In: Tetzlaff, R. (ed.) Memristors and Memristive Systems, ch. 5, pp. 163–191. Springer, New York (2014) ISBN: 978-1-4614-9067-8
11. Chua, L.O., Kang, S.-M.: Memristive devices and systems. Proc. IEEE 64(2), 209–223 (1976)
12. Pickett, M.D., Williams, R.S.: Phase transitions enable computational universality in neuristor-based cellular automata. Nanotechnology 24(38), 384002(7pp), (2013)
13. Kim, S., Lee, W., Hwang, H.: Selector devices for cross-point ReRAM. In: Proc. IEEE Int. Workshop on Cellular Nanoscale Networks and their Applications (CNNA), Turin, Italy (2012)
14. Chua, L.O.: Local activity is the origin of complexity. Int. J. on Bifurcation and Chaos 15(11), 3435–3456 (2005)
15. Chua, L.O.: CNN: A paradigm for complexity. Word Scientific Series on Nonlinear Science, Series Editor: L. O. Chua. Word Scientific Publishing Co. Pte. Ltd. (1998) ISBN: 981-02-3483-X
16. Ascoli, A., Corinto, F.: Memristor models in chaotic neural circuits. International Journal of Bifurcation and Chaos in Applied Sciences and Engineering, World Scientific 23(3), 1350052(28) (2013) ISSN: 0218-1274
17. Chua, L.O.: Resistance switching memories are memristors. Appl. Phys. A 102, 765–783 (2011)
18. Chua, L.O.: Memristor: The missing circuit element. IEEE Trans. on Circuit Theory 18(5), 507–519 (1971)
19. Ascoli, A., Senger, V., Tetzlaff, R., Corinto, F.: A novel memristor polynomial model. In: Proc. of Nonlinear Dynamics of Electronic Systems, Bari, Italy (2013)
20. Pickett, M.D., Strukov, D.B., Borghetti, J.L., Yang, J.J., Snider, G.S., Stewart, D.R., Williams, R.S.: Switching dynamics in titanium dioxide memristive devices. Journal of Applied Physics 106, 074508(1)–074508(6) (2009)
21. Mähne, H., Slesazeck, S., Jakschick, S., Dirnstorfer, I., Mikolajick, T.: The influence of crystallinity on the resistive switching behavior of TiO_2. Microelectronic Engineering 88, 1148–1151 (2011)
22. Strukov, D.B., Snider, G.S., Stewart, D.R., Williams, R.S.: The missing memristor found. Nature 453, 80–83 (2008)
23. Chua, L.O.: The Fourth Element. Proc. of the IEEE 100(6), 1920–1927 (2012)
24. Corinto, F., Ascoli, A.: Memristive diode bridge with LCR filter. IEEE IET Electronics Letters 48(14), 824–825 (2012), doi:10.1049/el.2012.1480
25. Strogatz, S.H.: Nonlinear dynamics and chaos: With applications to physics, biology, chemistry, and engineering. Westview Press, Perseus Books Publishing Group (1994) ISBN-13: 978-0-7382-0453-6

Spatiotemporal Dynamics of High-Intensity Ultrashort Laser Pulses in Strongly Nonlinear Regime

Maria Todorova[1], Todor Todorov[2], Michail Todorov[3], and Ivan Koprinkov[2,*]

[1] College of Energetic and Electronics, Technical University of Sofia, 1000 Sofia, Bulgaria
[2] Department of Applied Physics, Technical University of Sofia, 1000 Sofia, Bulgaria
[3] Faculty of Applied Mathematics and Informatics, Technical University of Sofia, 1000 Sofia, Bulgaria

Abstract. The spatiotemporal dynamics of high-intensity femtosecond laser pulses is studied in strongly nonlinear regime. The physical model is capable to describe ultrashort pulse propagation down to single-cycle regime at presence of ionization of the medium. The ionization contribution to the group velocity dispersion is introduced in the model. The pulse propagation is described by the nonlinear envelope equation. The propagation and material equations are solved self-consistently at realistic physical conditions. We have shown that, at typical laboratory scale distances, the linear processes, more particularly – the dispersion, play a secondary role, while the pulse propagation dynamics is ruled mainly by competitive nonlinear processes in neutrals and plasma.

Keywords: High-intensity ultrashort pulses, pulse propagation, self-compression, nonlinear dynamics.

1 Introduction

The high-intensity ultrashort laser pulse (HULP) dynamics is not yet completely investigated and understood due to a large number of simultaneously acting linear and nonlinear processes. Depending on the particular conditions, it may lead to variety of propagation regimes. The early experimental studies reveal broadening and splitting of the pulse at positive and large (as in the case of solid bulk material) value of the group velocity dispersion (GVD) [1-3]. At negative GVD, existence of spatiotemporal soliton has been predicted theoretically [4]. That understanding has been changed with the discovery of self-compression (SC) of femtosecond laser pulses at normal dispersion regime in a number of atomic and molecular gases and solid bulk material [5, 6]. The observation of SC in various optical media shows that it is a general phenomenon that may take place at given conditions. The overall behavior of the pulse at these studies can be summarized as [6]: self-focusing (SF) in the transversal direction leading to formation of a light filament, increasing of the peak intensity, SC in time, improvement of the spatio-temporal pulse shape, and stable propagation over given distance. The great potential of SC has been proved in a number of

*Corresponding author: igk@tu-sofia.bg

V.M. Mladenov and P.C. Ivanov (Eds.): NDES 2014, CCIS 438, pp. 165–172, 2014.

experimental and theoretical studies [5-12]. Thus, generation of few-cycle high-intensity pulses has been achieved experimentally [8]. The numerical simulations predict generation of even shorter pulses through the mechanism of SC - single-cycle [9] and even sub-cycle pulses [10]. The intensive studies in the field of propagation equations and physical models [11, 12] helps to deeper understand the HULP dynamics but the exact physical mechanism behind it is not yet completely established and further studies are required in that direction. Thus, for example, the earlier understanding of filamentation rests on the defocusing due to ionization of the medium so as to arrest the collapse of the pulse due to the SF. Recently, filamentation without ionization based on the interplay of various order of focusing and defocusing nonlinearities has been suggested and proved by numerically simulations as a possible mechanism of filamentation [13].

In this work, we investigate the pulse propagation dynamics of HULP emphasizing on the role of dispersion. We have found that the main pulse parameters, *i.e.*, the transversal width, the peak intensity, the time duration as well as the spatiotemporal pulse shape depend weakly on the dispersion over given typical laboratory scale distances, whereas the nonlinearities play a crucial role in the pulse dynamics. The decoupling of the dispersion from the nonlinear processes in the NEE is demonstrated for the first time, to our knowledge, at realistic physical conditions.

2 Physical Model and Numerical Method

Our approach is based on the physical model developed in [11]. The propagation equation is (3+1)D nonlinear envelope equation (NEE) for the field amplitude A (in standard notations [11])

$$\begin{aligned}\frac{\partial A}{\partial z} &= \frac{i}{2k_0}\hat{T}^{-1}\nabla_\perp^2 A + i\hat{D}A + i\frac{\omega_0}{c}n_2\hat{T}|A|^2 A - i\frac{\omega_0}{c}n_4\hat{T}|A|^4 A \\ &\quad - i\frac{k_0}{2n_0^2\rho_c}\hat{T}^{-1}\rho A - \frac{\sigma}{2}\rho A - \frac{\beta_{MPI}(A)}{2}A\end{aligned} \quad . \tag{1}$$

The non-instantaneous effects are neglected in the NEE due to the fast electronic response of the atomic medium (gaseous argon) considered here. The physical processes in the NEE can be referred to: *linear processes*, diffraction and dispersion; *nonlinear processes in neutrals*, cubic and quintic nonlinearity of neutral particles; *processes due to ionization*, ionization modification of the refractive index, collision ionization by inverse bremsstrahlung and multi-photon ionization. The electron number density ρ is described by the kinetic equation (in standard notations) [11]

$$\frac{\partial \rho}{\partial t} = W(I)(\rho_n - \rho) + \frac{\sigma(\omega_0)}{I_P}|A|^2 \rho - f(\rho) \tag{2}$$

where σ is inverse *bremsstrahlung* cross section, $f(\rho) = \alpha\rho^2$ is plasma recombination term. The ionization rate $W(I)$ is described by the multiphoton formula [11]

$$W = \sigma_k I^k \tag{3}$$

where σ_k is the k-photon ionization coefficient, I is the intensity and k is the number of photons for direct ionization of the atoms from their ground state.

In addition to the physical model of [11], we introduce the influence of ionization to the GVD, predicted in [14]. The total GVD of the medium $\beta^{(2)}$ can be presented as a sum of GVD of neutrals $\beta_0^{(2)}$ and GVD of plasma $\beta_i^{(2)}$, $\beta^{(2)} = \beta_0^{(2)} + \beta_i^{(2)}$, where the ionization contribution to the GVD is given by [14]

$$\beta_i^{(2)} = -\frac{e^2 \lambda^3 \rho}{2\pi^2 m_e c^4 \left(1 - \dfrac{e^2 \lambda^2 \rho}{\pi m_e c^2}\right)^{3/2}} \tag{4}$$

where λ is the wavelength, e, m_e, and c are the charge and the mass of electron and the velocity of light, respectively. The influence of ionization on the GVD accomplishes the physical model because its negative contribution is only the process that acts directly against the positive GVD of the neutrals. Thus, to each strong physical process in our model corresponds at least one other strong physical process acting in the opposite direction. That is why, the above model will be put into *a minimal sufficient model*.

The NEE is (3+1)-dimensional highly nonlinear equation. Although, seemingly, it is of cubic-quintic ($\chi^{(3)}$-$\chi^{(5)}$) type, the leading nonlinearity in the NEE becomes $\chi^{(21)}$, once the field intensity grows up and the multiphoton ionization term $0.5\beta_{MPI}(A)A$ becomes substantial. This is because the argon atom requires eleven photons absorption for ionization by electromagnetic field of 800 nm wavelength, considered here, which is ruled by $\chi^{(21)}$-nonlinearity.

To solve such highly nonlinear problem, we apply the following numerical method of solution. The original Eq. (1) is split twice. In the beginning, we split it into two z-evolutionary equations by physical processes - see, *e.g.*, [15]. On the next step, following [16], we apply a coordinate splitting to the resultant equations by using the Crank-Nicolson difference scheme direct in complex arithmetic combining it with inner iterations with respect to the nonlinear terms. We use 512 and 1024 grid nodes in the transversal direction and time, respectively, and the longitudinal step along the propagation direction z is 10^{-6} dimensionless units. We have found that such a small value of propagation step is crucial in order to achieve both independence of the numerical results on the spatial and temporal steps as well as the stability and fast convergence of the inner iterations with respect to the evolutionary z-step.

3 Results and Discussions

Equations (1)-(4) are solved self-consistently at realistic physical condition. The initial pulse, propagating in gaseous argon at 18atm pressure, has Gaussian shape in space and in time of 300μm beam diameter, 150fs pulse duration (full width at half maximum), 800nm central wavelength and 0.5mJ pulse energy. The following parameters of the neutral medium are used: the second and the third order dispersion $\beta_0^{(2)}=0.2Pfs^2/cm.atm$ and $\beta_0^{(3)}=0.9Pfs^3/cm.atm$, respectively, the nonlinear refractive indices $n_2=(1\pm0.09)\times10^{-19}Pcm^2/W.atm$ and $n_4=(-0.36\pm1.03)\times10^{-33}Pcm^4/W^2.atm$ [17], where P is pressure in atm, and the coefficient for eleven photon absorption is taken $\sigma_k=5\times10^{-140}cm^{22}/s.W^{11}$ [18]. The electron-neutral inverse bremsstrahlung cross section is calculated on the basis of Drude model, see, *e.g.*, [2].

To reveal the role of dispersion on the HULP dynamics, we solve Eqs.(1)-(4) twice. First, we solve the problem in its wholeness, where all terms in the Eqs.(1)-(4) are retained. Second, we solve the same equations, in which the dispersion operator D (in our case it includes the second and third order dispersion terms of neutral medium, $\beta_0^{(2)}$ and $\beta_0^{(3)}$) is totally excluded from Eq.(1). The ionization contribution to the GVD, Eq.(4), is also excluded in these studies. The results of both types of numerical simulations are shown in same graphs. The pulse propagation dynamics is presented in terms of evolution of main pulse parameters - the transversal width, the peak intensity, the time duration and the spatiotemporal pulse shape. The evolution of the transversal width of the pulse (full width at half maximum of the transversal intensity profile of the pulse), the peak intensity and the time duration (full width at half maximum of the intensity time profile of the pulse) at presence of dispersion, *full line*, and at all dispersion contributions excluded, *dotted line*, are shown in Figs. 1, 2, and 3, respectively.

The numerical simulations well reproduce the experimentally observed HULP dynamics [6]: SF and formation of stable light filament, Fig.1, increasing of the peak intensity, Fig.2, SC of the pulse, Fig.3, and stable propagation over given distance. The formation of stable light filament takes place between *14.1cm* and *17.4cm* from the beginning of the propagation distance, Fig.1. It results from a balance between SF from $\chi^{(3)}$ (and higher SF nonlinearities) and defocusing due to the common action of diffraction, $\chi^{(5)}$ (and higher defocusing nonlinearities), and ionization. Increasing and stabilization of the peak intensity of HULP results from the interplay between SF due to $\chi^{(3)}$ and dissipation of energy due to ionization ($\chi^{(21)}$-nonlinearity in our case), Fig.2. Stabilization of peak intensity around $4.2\times10^{13}W/cm^2$ takes place between *14.3cm* and *17.4cm* from the beginning of the propagation distance. In that range, the peak intensity varies not more than *2.5%* around the local mean value. Finally, the pulse also shows a partial stabilization in time. It takes place between *13.9cm* and *15.1cm* of the propagation distance. In that range, the pulse shape remains stable, while the pulse duration continue shortening but much slower than outside specified area, Fig.3. SC of the pulse from *150fs* (the initial pulse) to *25fs* (at the end of range

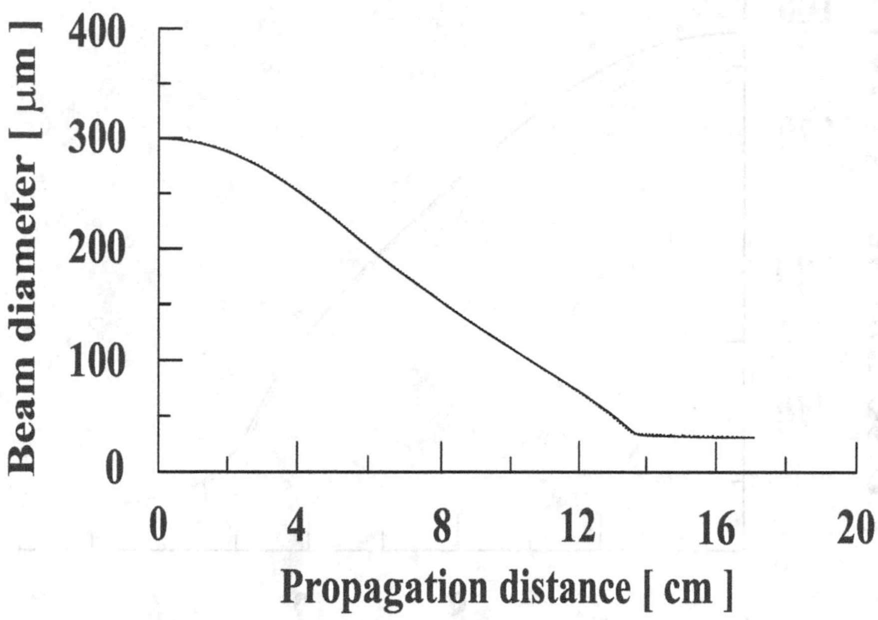

Fig. 1. Evolution of beam diameter versus propagation distance with (*full line*) and without (*dotted line*) dispersion contribution to the pulse propagation dynamics

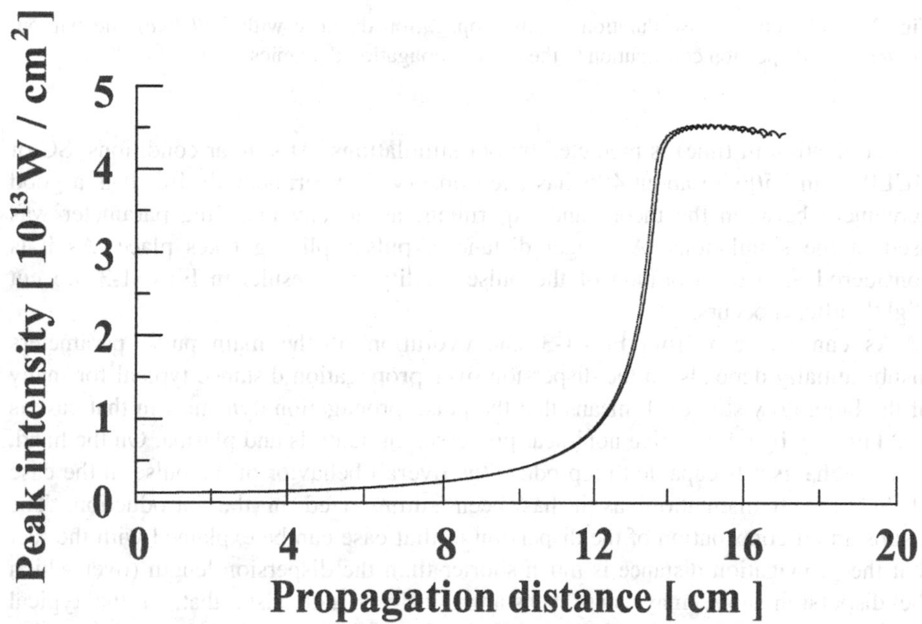

Fig. 2. Evolution of peak intensity versus propagation distance with (*full line*) and without (*dotted line*) dispersion contribution to the pulse propagation dynamics

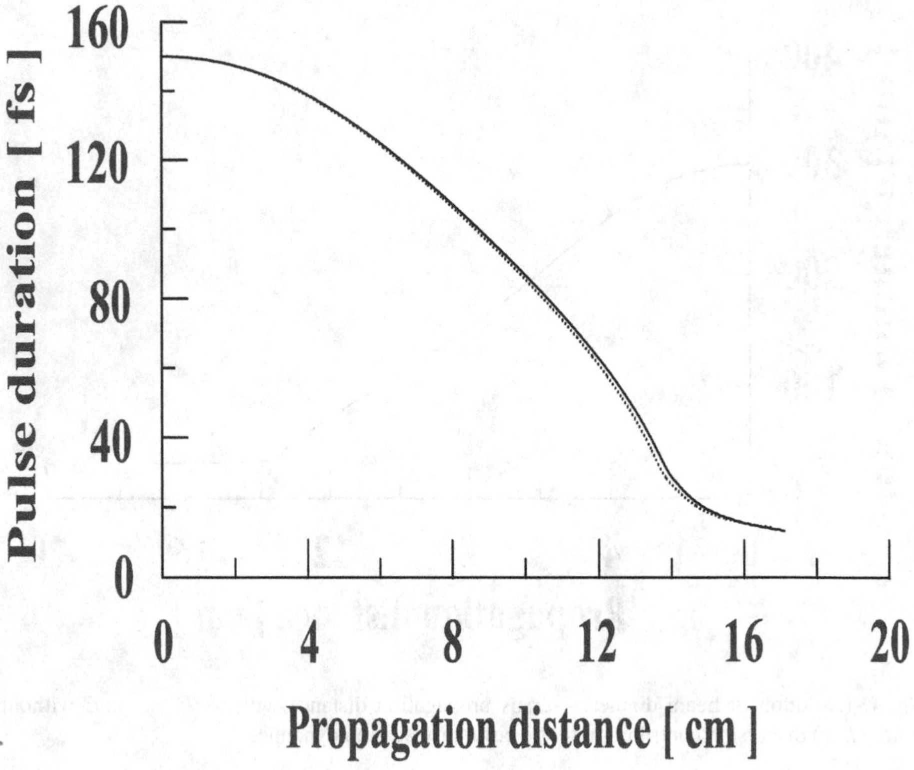

Fig. 3. Evolution of pulse duration versus propagation distance with (*full line*) and without (*dotted line*) dispersion contribution to the pulse propagation dynamics

of stabilization in time) is predicted by our simulations. At similar conditions, SC of HULP from *150fs* to about *40fs* has been observed experimentally [6]. It is a good agreement between the theory and experiment as no any matching parameter was used in the simulations. At longer distances, pulse splitting takes place. As it is considered as a deterioration of the pulse quality, the results in Figs. 1-3 are cut slightly after it occurs.

As can be seen from Figs.1-3, the evolution of the main pulse parameters insubstantially depends on the dispersion over propagation distance typical for many of the laboratory studies. It means that the pulse propagation dynamics in that case is ruled mainly by competitive nonlinear processes in neutrals and plasma. On the hand, such mechanism is capable to reproduce the overall behavior of the pulse in the case of SC and filamentation, as it has been summarized in the introduction. The insubstantial contribution of the dispersion in that case can be explained with the fact that the propagation distance is much shorter than the dispersion length (over which the dispersion alone may cause substantial effect) and, also, that, at the typical conditions of HULP, the nonlinear length L_{NL} is much shorter than the dispersion L_{DS} and the diffraction L_{DF} lengths, *i.e.*, $L_{NL} << L_{DF}$, L_{DS}. In our case, nonlinear, diffraction, and dispersion lengths are 0.16cm, 0.31cm, and 338cm, respectively, calculated by the pulse parameters in the region of stabilization. It means that the intensiveness of the nonlinearity strongly dominates that one of the dispersion and the diffraction, as well. The linear processes, dispersion and diffraction, are not among the leading processes in the SC of HULP and formation of stable pulse but play a secondary role, only. Instead, a number of nonlinear processes start playing the main role in that case.

4 Conclusions

High-intensity ultrashort laser pulse dynamics is investigated in regime of dominating nonlinear effects. The evolution of the main pulse parameters slightly depends on the dispersion within given laboratory scale distances. The pulse propagation dynamics in that case is ruled mainly by competitive nonlinear processes in neutrals and plasma. Such mechanism is capable to reproduce the overall behavior of the pulse in the case of self-compression and filamentation. The decoupling of the dispersion from the nonlinear pulse dynamics within the nonlinear envelop equation is found for the first time at realistic physical conditions.

References

1. Ranka, J.K., Schirmer, R.W., Gaeta, A.L.: Observation of Pulse Splitting in Nonlinear Dispersive Media. Phys. Rev. Lett. 77, 3783–3786 (1996)
2. Mlejnek, M., Wright, E.M., Moloney, J.V.: Femtosecond pulse propagation in argon: A pressure dependence study. Phys. Rev. E 58, 4903–4910 (1998)
3. Zozulya, A.A., Diddams, S.A., Van Engen, A.G., Clement, T.S.: Propagation Dynamics of Intense Femtosecond Pulses: Multiple Splitting, Coalescence, and Continuum Generation. Phys. Rev. Lett. 82, 1430–1433 (1999)
4. Silberberg, Y.: Collapse of Optical Pulses. Opt. Lett. 15, 1282–1284 (1990)
5. Koprinkov, I.G., Suda, A., Wang, P., Midorikawa, K.: Self-Shortening of Femtosecond Laser Pulses Propagating in Rare Gas Medium. Jap. J. Appl. Phys. 38, L978–L980 (1999)
6. Koprinkov, I.G., Suda, A., Wang, P., Midorikawa, K.: Self-Compression of High-Intensity Femtosecond Optical Pulses and Spatiotemporal Soliton Generation. Phys. Rev. Lett. 84, 3847–3850 (2000)
7. Tzortzakis, S., Lamouroux, B., Chiron, A., Moustaizis, S.D., Anglos, D., Franco, M., Prade, B., Mysyrowicz, A.: Femtosecond and Picosecond Ultraviolet Laser Filamentation in Air: Experiment and Simulations. Opt. Commun. 197, 13–143 (2001)
8. Hauri, C.P., Kornelis, W., Helbing, F.W., Heinrich, A., Couairon, A., Mysyrowicz, A., Biegert, J., Keller, U.: Generation of Intense, Carrier-Envelope Phase-Locked Few-Cycle Laser Pulses through Filamentation. Appl. Phys. B 79, 673–677 (2004)
9. Couairon, A., Francko, M., Mysyrowicz, A., Biegert, J., Keller, U.: Pulse Self-Compression to the Single-Cycle Limit by Filamentation in a Gas with a Pressure Gradient. Opt. Lett. 30, 2657–2659 (2005)
10. Gaarde, M.B., Couairon, A.: Intensity Spikes in Laser Filamentation: Diagnostics and Application. Phys. Rev. Lett. 103, 043901–043901 (2009)

11. Berge, L., Skupin, S., Nute, R., Casparian, J., Wolf, J.-P.: Ultrashort Filamentation of Light in Weakly Ionized, Optically Transparent Media. Rep. Prog. Phys. 70, 163–1713 (2007)
12. Couairon, A., Brambilla, E., Gorti, T., Majus, D., Ramires-Gongora, O., de, J., Kolesik, M.: Practitioner's Guide to Laser Pulse Propagation Models and Simulations. Eur. Phys. J. Special Topics 199, 5–76 (2011)
13. Béjot, P., Kasparian, J., Henin, S., Loriot, V., Vieillard, T., Hertz, E., Faucher, O., Lavorel, B., Wolf, J.-P.: Higher-Order Kerr Terms Allow Ionization-Free Filamentation in Gases. Phys. Rev. Lett. 104, 103903–103903 (2010)
14. Koprinkov, I.G.: Ionization Variation of the Group Velocity Dispersion by High-Intensity Optical Pulses. Appl. Phys. B - Lasers and Optics 79, 359–361 (2004)
15. Marchuk, G.I.: Mathematical Models in Environmental Problems (Studies in Mathematics and its Applications). North Holland, The Netherlands (1986)
16. Todorov, M.D., Christov, C.I.: Impact of the Large Cross-Modulation Parameter on the Collision Dynamics of Quasi-Particles Governed by Vector NLSE. Mathematics and Computers in Simulation 80, 46–55 (2009)
17. Loriot, V., Hertz, E., Faucher, O., Lavorel, B.: Measurement of High Order Kerr Refractive Index of Major Air Components: Erratum. Opt. Express 18, 3011–3012 (2010)
18. Zaïr, A., Guandalini, A., Schapper, F., Holler, M., Biegerti, J., Gallmann, L., Couairon, A., Franco, M., Mysyrowicz, A., Keller, U.: Spatio-Temporal Characterization of Few-Cycle Pulses Obtained by Filamentation. Opt. Express 15, 539–5405 (2007)

Laser Synapse

Alexander N. Pisarchik[1,2,⋆], Ricardo Sevilla-Escoboza[3], Rider Jaimes-Reátegui[3], Guillermo Huerta-Cuellar[3], and Victor B. Kazantsev[4]

[1] Centro de Investigaciones en Optica, Loma del Bosque 115, Lomas del Campestre, 37150 Leon, Guanajuato, Mexico
[2] Centre for Biomedical Technology, Technical University of Madrid, Campus Montegancedo, 28223 Pozuelo de Alarcon, Madrid, Spain
[3] Centro Universitario de Los Lagos, Universidad de Guadalajara, Enrique Díaz de León 1144, Paseo de la Montaña, Lagos de Moreno, Jalisco, Mexico
[4] Institute of Applied Physics of Russian Academy of Science, Uljanov Str. 46, 603950 Nizhny Novgorod, Russia

apisarch@cio.mx
http://www.cio.mx

Abstract. We implemented a laser synapse based on a diode-pumped erbium-doped fiber laser. The laser is driven by a presynaptic FitzHung-Nagumo electronic neuron and its intensity converted to the electric signal by a photodetector serves as an input signal for a postsynaptic FitzHung-Nagumo electronic neuron. The laser synapse provides very high flexibility for controlling the neuron dynamics to obtain different dynamical regimes, from silence to periodic oscillations, bursts, and chaos. The system complexity is demonstrated through the Shannon entropy. The proposed synapse can be beneficial for efficient biorobotics where behavioral flexibility and synaptic plasticity are challenges.

Keywords: Fiber laser, electronic neuron.

1 Introduction

Neurons usually interact via synapses which allow information transmission from one cell to the other. Being subserved by complex molecular mechanisms, the synapses are capable to change the efficiency of signal transmission between neurons by sensing current electric activity and chemical concentrations. It is believed that this phenomenon referred to as synaptic plasticity, underlies the implementation of computational and cognitive tasks in brain networks. While in living cells the synaptic plasticity is mediated by complex molecular transformation, in the artificial biosystem the synaptic transmission can be simply regulated by adjusting parameters of the artificial synapse.

Significant efforts are made to represent a synapse by a single device which mimics synaptic connection and behavioral flexibility. This challenging task for

⋆ Corresponding author.

V.M. Mladenov and P.C. Ivanov (Eds.): NDES 2014, CCIS 438, pp. 173–180, 2014.

biorobotics would allow a direct linkage between neuroscience and artificial intelligence. While constructing an artificial brain one must consider its complexity. Several attempts have already been made. For example, Sharp et al. [1] used an electronic circuit to couple two living stomatogastric ganglion neurons. Synaptic behavior has also been imitated by hardware-based neural networks such as hybrid complementary metal-oxide-semiconductor analogue circuits and other artificial neural devices [2,3,4] capable to mimic major features of the human memory; namely, the sensory, short-term, and long-term memories. A great progress in nanotechnology has provided significant advances in miniaturization of synthetic synapses, e.g., the fabrication of a carbon nanotube synaptic circuit [5]. Artificial synaptic devices based on ion migration have also been designed [6]; some of them [7,8] have demonstrated spike-timing-dependent plasticity [9], the important mechanism of the brain memory, related to the synaptic connection strength in biological circuits and synthetic devices. These devices require precise control of the signal timing to simulate the presynaptic and postsynaptic potentials in biological systems.

Another approach in constructing a synthetic brain is to make it completely artificial without linking to time and space scales of real neurons, in other words, to connect artificial neurons by artificial synapses. The main advantages of this approach is that such a device does not require very strong miniaturization, and that the time scale can be much shorter (i.e. much higher spike frequency) than in real neurons. The faster oscillations of electronic neurons in comparison with real neurons can be beneficial for an artificial biosystem because this would allow accelerating its activity. Several neuron models have been developed to mimic biological neuron dynamics. The simplified modification of the detailed Hodgkin-Huxley model [10], the FitzHung-Nagumo (FHN) model [11] has attracted much attention because of its easy implementation as an analog electronic circuit which simulates spike-timing neural activity [12,13,14].

In this paper we report on the experimental implementation of the laser synapse based on an erbium-doped fiber laser (EDFL) to establish a functional connection between FHN electronic neurons. Such a laser synapse allows controlling the spike transmission with a very high flexibility [15]. The distinguished features of our synapse from other artificial (electronic) synapses are (1) an optical carrier (optical radiation in the infrared spectral range) instead of an electric current, (2) optical fiber transmission instead of a metallic wire, and (3) very rich dynamics including fixed points, different periodic frequency-locking regimes, and chaos. We demonstrate high flexibility of the laser synapse in controlling signal transmission from a presynaptic to a postsynaptic neuron.

2 Experimental Setup

The experimental setup is shown in Fig. 1. The EDFL is pumped by a laser diode (wavelength 976 nm) through a wavelength-division multiplexing coupler and a polarization controller. The laser cavity of a 1.55-m length is formed by a piece of erbium-doped fiber of 70 cm in length and 2.7 μm in core diameter, and two fiber Bragg gratings with a 2-nm FWHM bandwidth and with 90.5% and 94% reflectivity at a 1550-nm wavelength. The diode pump laser is controlled by a laser diode controller (Thorlabs ITC510).

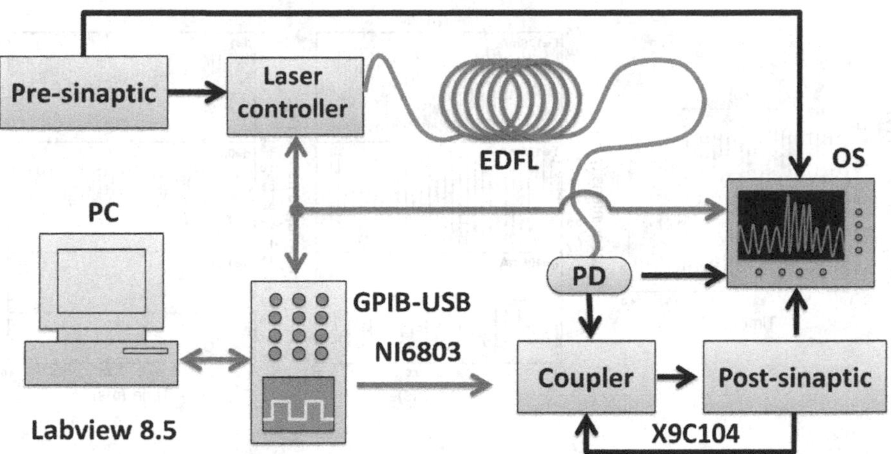

Fig. 1. Experimental setup. The presynaptic neuron drives the erbium-doped fiber laser (EDFL) via the laser diode controller. The signal from the photodetector (PD) after passing through the Coupler controls the postsynaptic neuron. The output signals from presynaptic neuron, laser, and postsynaptic neuron are recorded by the oscilloscope (OS) and stored in the computer (PC) using GPIB communication protocol and Labview 8.5. The square signal generated by the NI6803 card is applied to the digital potentiometer X9C104 to control the coupling strength.

The diode pump current of the EDFL is modulated by the presynaptic neuron. The optical output of the EDFL is converted to an electric signal by a photodetector and sent through the coupler to the postsynaptic neuron. The electronic schemes of the FHN circuit and coupler can be found in Refs. [13,14].

3 Results

3.1 Time Series

The presynaptic FHN electronic circuit generates periodic spikes shown in Fig. 2(a), which serve as an input signal to control the diode pump current with frequency of 30 kHz. The laser response to the pump modulation is shown in 2(b). Different pump currents I result in different laser waveforms, from periodic (upper and middle traces) to chaotic (lower trace). The response of the postsynaptic neuron to the laser input, i.e. the response to the signal from the presynaptic neuron transmitted through the laser synapse is shown in Fig. 2(c). To control the dynamics of the postsynaptic neuron, we can manipulate by the synaptic strength d between the laser and the postsynaptic neuron. The spike trains in Fig. 2(c) for different d are obtained for the laser input signals at the corresponding rows in Fig. 2(b).

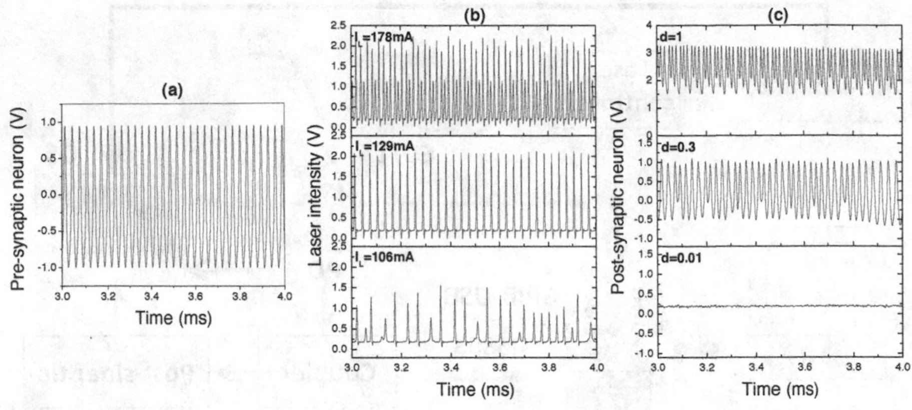

Fig. 2. Time series of (a) presynaptic neuron, (b) laser, and (c) postsynaptic neuron. The laser response in (b) displays the 2:3 (upper trace) and 1:1 (middle trace) frequency-locking, and chaotic (lower trace) regimes, while the postsynaptic neuron response in (c) exhibits the 2:3 (upper trace), unlocked (middle trace), and silent (lower trace) regimes.

3.2 Bifurcation Diagrams

The bifurcation diagrams of the laser output peak intensity versus the pump current and the peak potential of the postsynaptic neuron versus the coupling strength are shown respectively in Figs. 3(b) and 3(b). The EDFL modulated by the presynaptic neuron displays very rich dynamics. The complex dynamics, including chaos and multistability, in the EDFL under periodic pump modulation has been described in many experimental and theoretical papers [16,17,18,19]. For small pump currents, the laser generates chaotic spikes. When the pump current is increased from 115 mA to 140 mA, the laser oscillates periodically with the same frequency as the frequency of the presynaptic neuron spikes (1:1 frequency locking). For higher pump currents (from 140 mA to 160 mA), the laser again behaves chaotically, and for strong currents the laser oscillations are locked as 2:1 or 3:1.

We fix the pump current at 130 mA corresponding to 1:1 frequency locking between the laser and presynaptic neuron and measure the response of the postsynaptic neuron. As seen from the bifurcation diagram in Fig. 3(b), the

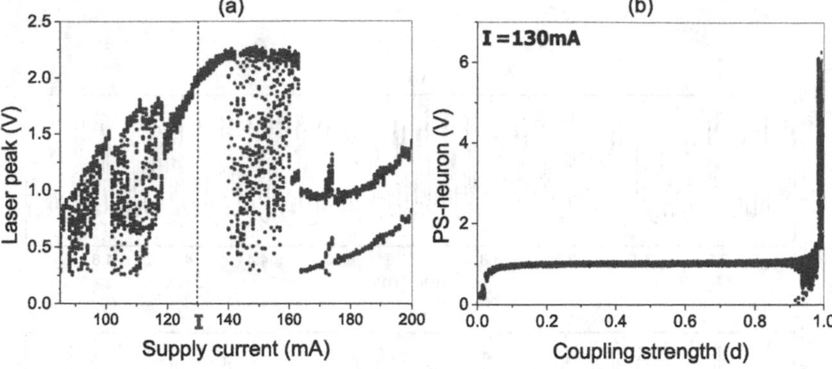

Fig. 3. Bifurcation diagrams of (a) laser peak intensity and (b) postsynaptic neuron peak potential

postsynaptic neuron peak potential increases for small coupling ($d < 0.1$) and remains almost the same for intermediate coupling strengths ($0.1 < d < 0.9$), whereas for strong coupling ($d > 0.9$) it highly increases. Although for medium couplings the peak potential of the postsynaptic neuron is almost constant and approximately equal to 1 V, the spike frequency is entrained to different rotation numbers $p : q$ (p and q being integers). One of such frequency-locking regimes with the 3:8 ratio is shown in Fig. 4. After 170 mA, the laser oscillations undergo period-doubling bifurcations (Fig. 4(a)) and the postsynaptic neuron spikes are locked to different ratios (Fig. 4(b)). For very strong coupling ($d > 0.92$), the postsynaptic neuron responds at high rotation numbers to the laser spikes.

3.3 System Complexity

The system complexity can be quantitatively characterized by the normalized Shannon entropy given as [20]

$$H[P] = S[P]/S_{max}, \tag{1}$$

where $S[P] = -\sum_{i=1}^{N} p_i \log p_i$ and $S_{max} = \log N$ with $N = D!$ being the total number of vectors over which the probability distribution P is computed, and D is the embedding dimension.

Figure 5 shows the Shannon entropy for the laser Fig. 5(a) and the postsynaptic neuron Fig. 5(b).

4 Conclusions

We constructed a laser synapse based on a diode-pumped erbium-doped fiber laser, whose pump current was controlled by a FinzHung-Nagumo electronic

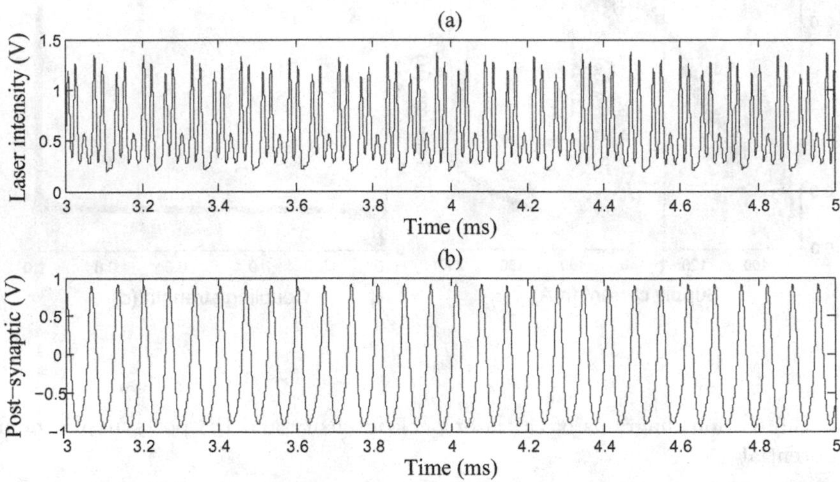

Fig. 4. Frequency locking of the postsynaptic neuron spikes by laser oscillations at 173 mA. (a) Laser oscillations, (b) postsynaptic neuron potential.

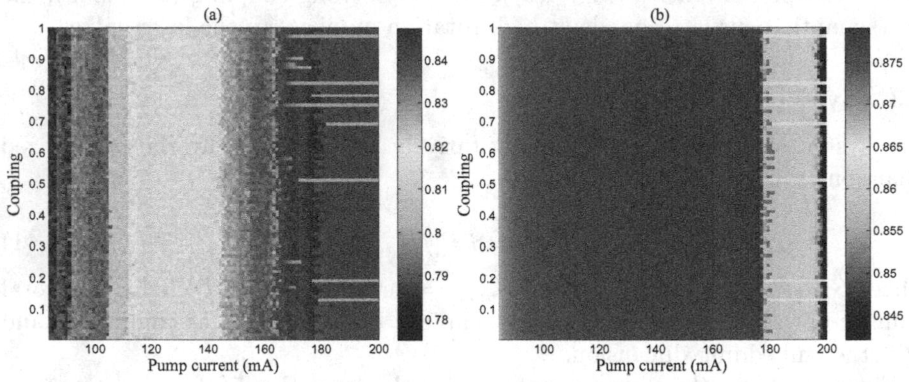

Fig. 5. Shannon entropy for (a) laser and (b) postsynaptic neuron for $D = 4$. The brown regions represent chaos.

circuit. We demonstrated high flexibility and easy functionality of this optical synapse to transmit information from a presynaptic to a postsynaptic neuron. Depending on both the diode pump current and the coupling strength between the laser and the postsynaptic neuron, the latter exhibited different dynamical regimes, from silence to periodic frequency locking and chaos. These experimental results are in good agreement with numerical simulations reported in Ref. [15]. This rich dynamics is very promising for biorobotic and neuroengineering applications due to high flexibility of the synapse, and can help in solving important problems in neuroengineering, such as reproducible neuron dynamics and enhanced functionality of neuromimmetic networks. The laser synapse can also be of interest for neuroscientists, because it allows efficient monitoring and manipulation of neuron functions in a optical way, in particular, it could be implemented as an optoelectronic interface between living and artificial neurons.

Acknowledgement. A.N.P. acknowledges support from CONACYT (Mexico).

References

1. Sharp, A.A., Abbott, L.F., Marder, E.: Artificial electrical synapses in oscillatory networks. J. Neurophysiology 67, 1691–1694 (1992)
2. Ramacher, U.: SYNAPSE - A neurocomputer that synthesizes neural algorithms on a parallel systolic engine. J. Parallel Distrib. Comput. 14, 306–318 (1992)
3. Diorio, C., Hasler, P., Minch, B.A., Mead, C.A.: A single-transistor silicon synapse. IEEE Trans. Electron Devices 43, 1972–1980 (1996)
4. Indiveri, G., Chicca, E., Douglas, R.: A VLSI array of low-power spiking neurons and bistable synapses with spike-timing dependent plasticity. IEEE Trans. Neural Netw. 17, 211–221 (2006)
5. Joshi, J., Zhang, J., Wang, C., Hsu, C.-C., Parker, A.C., Zhou, C., Ravishankar, U.: A biomimetic fabricated carbon nanotube synapse for prosthetic applications. In: Proc. 2011 IEEE/NIH Life Science Systems and Applications Workshop (LiSSA), pp. 139–142. IEEE Press, New York (2011)
6. Thakoor, S., Moopenn, A., Daud, T., Thakoor, A.P.: Solid-state thin-film memristor for electronic neural networks. J. Appl. Phys. 67, 3132–3135 (1990)
7. Jo, S.H., Chang, T., Ebong, I., Bhadviya, B.B., Mazumder, P., Lu, W.: Nanoscale memristor device as synapse in neuromorphic systems. Nano Lett. 10, 1297–1301 (2010)
8. Lai, Q., Zhang, L., Li, Z., Stickle, W.F., Williams, R.S., Chen, Y.: Ionic/electronic hybrid materials integrated in a synaptic transistor with signal processing and learning functions. Adv. Mater. 22, 2448–2453 (2010)
9. Bi, G.Q., Poo, M.M.: Synaptic modifications in cultured hippocampal neurons: Dependence on spike timing, synaptic strength, and postsynaptic cell type. J. Neurosci. 18, 10464–10472 (1998)
10. Hodgkin, L.A., Huley, A.F.: A quantitative description of membrane current and its application to conduction and excitation in nerve. J. Physiology 117, 500–544 (1952)
11. Fitzhugh, R.: Impulse and physiological states in models of nerve membrane. Biophys. J. 1, 445–466 (1961)

12. Kazantsev, V.B., Nekorkin, V.I., Binczak, S., Bilbault, J.M.: Spiking patterns emerging from wave instabilities in a one-dimensional neural lattice. Phys. Rev. E 68, 17201 (2003)
13. Kazantsev, V.B., Nekorkin, V.I., Binczak, S., Jacquir, S., Bilbault, J.M.: Spiking dynamics of interacting oscillatory neurons. Chaos 15, 023103 (2005)
14. Binczak, S., Jacquir, S., Bilbault, J.M., Kazantsev, V.B., Nekorkin, V.I.: Experimental study of electrical FitzHughNagumo neurons with modified excitability. Neural Networks 19, 684–693 (2006)
15. Pisarchik, A.N., Jaimes-Reátegui, R., Sevilla-Escoboza, R., García-Lopez, J.H., Kazantsev, V.B.: Optical fiber synaptic sensor. Optics and Lasers in Engineering 49, 736–742 (2011)
16. Saucedo-Solorio, J.M., Pisarchik, A.N., Kir'yanov, A.V., Aboites, V.: Generalized multistability in a fiber laser with modulated losses. J. Opt. Soc. Am. B 20, 490–496 (2003)
17. Pisarchik, A.N., Barmenkov, Y.O., Kir'yanov, A.V.: Experimental characterization of bifurcation structure in an erbium-doped fiber laser with pump modulation. IEEE J. Quantum. Electron. 39, 1567–1571 (2003)
18. Reategui, R.J., Kir'yanov, A.V., Pisarchik, A.N., Barmenkov, Y.O., Il'ichev, N.N.: Experimental study and modeling of coexisting attractors and bifurcations in an erbium-doped fiber laser with diode-pump modulation. Laser Physics 14, 1277–1281 (2004)
19. Pisarchik, A.N., Kir'yanov, A.V., Barmenkov, Y.O., Jaimes-Reátegui, R.: Dynamics of an erbium-doped fiber laser with pump modulation: Theory and experiment. J. Opt. Soc. Am. B 22, 2107–2114 (2005)
20. Bandt, C., Pompe, B.: Permutation entropy: a natural complexity measure for time series. Phys. Rev. Lett. 88, 174102 (2002)

EM Algorithm for Estimation of the Offspring Probabilities in Some Branching Models

Nina Daskalova

Sofia University "St.Kliment Ohridski",
Faculty of Mathematics and Informatics,
Sofia, Bulgaria
ninad@fmi.uni-sofia.bg

Abstract. Multitype branching processes (MTBP) model branching structures, where the nodes of the resulting tree are objects of different types. One field of application of such models in biology is in studies of cell proliferation. A sampling scheme that appears frequently is observing the cell count in several independent colonies at discrete time points (sometimes only one). Thus, the process is not observable in the sense of the whole tree, but only as the "generation" at given moment in time, which consist of the number of cells of every type. This requires an EM-type algorithm to obtain a maximum likelihood (ML) estimation of the parameters of the branching process. A computational approach for obtaining such estimation of the offspring distribution is presented in the class of Markov branching processes with terminal types.

Keywords: multitype branching processes, offspring probabilities, maximum likelihood estimation, EM algorithm.

1 Introduction

Branching processes (BP) are stochastic models widely used in population dynamics. They have traditionally served to model chemical reactions, biological growth and epidemics processes, but there are applications in fields as finance and engineering also. The theory of such processes is presented in [10], [1], and application in biology is discussed in [12], [13], [9]. Statistical inference in BP depends on the kind of observation available, whether the whole family tree has been observed, or only the generation sizes at given moments. Some estimators considering different sampling schemes could be found in [8] and [19]. The problems get more complicated for multitype branching processess (MTPB) where the objects are of different types ([16]). Computational approaches like simulation and Monte Carlo methods have been used in such models ([11], [6]).

When the entire tree was not observed, but only the objects existing at given moment, an Expectation Maximization (EM) algorithm could be used, regarding the tree as the hidden data. Guttorp [8] presents an EM algorithm for the single-type process knowing generation sizes and in [7] an EM algorithm is used for parametric estimation in a model of Y-linked gene in bisexual BP. Such algorithms exist for strictures, called Stochastic Context-free Grammars (SCFG).

V.M. Mladenov and P.C. Ivanov (Eds.): NDES 2014, CCIS 438, pp. 181–188, 2014.

There is a relation between MTBPs and SCFGs, presented in [18], [5]. This relation has been used in previous work [2] to propose a computational scheme for estimating the offspring distribution of MTBP using the Inside-Outside algorithm for SCFG ([14]). A new method, related to this, but constructed entirely for BP will be presented here.

The EM algorithm specifies a sequence that converges to the ML estimator under certain regularity conditions. The idea is to replace one difficult (sometimes impossible) maximization of the likelihood with a sequence of simpler maximization problems whose limit is the desired estimator. To define an EM algorithm two different likelihoods are considered – for the "incomplete-data problem" and for "complete-data problem". When the incomplete-data likelihood is difficult to work with, the complete-data could be used in order to solve the problem. More about the theory and applications of the EM algorithm could be found in [15].

In Section 2 the model of MTBP with terminal types is introduced. Section 3 shows the derivation of an EM algorithm for estimating the offspring probabilities in general, and then proposes a recurrence scheme that could be used to ease the computations. The results of a simulation study are shown in Section 4.

2 The Model

A MTBP could be represented as $\mathbf{Z}(t) = (Z_1(t), Z_2(t), \dots Z_d(t))$, where $Z_k(t)$ denotes the number of objects of type T_k at time t, $k = 1, 2, \dots d$. An individual of type T_k has offspring of different types according to a d-variate distribution $p_k(x_1, x_2, \dots, x_d)$, with $(x_1, x_2, \dots, x_d) \in \mathbb{N}_0^d$ and every object evolves independently. If $t = 0, 1, 2, \dots$, this is the Bienaymé-Galton-Watson (BGW) process. For the process with continuous time $t \in [0, \infty)$, define the *embedded generation process* ([1]) as follows. Let $\mathbf{Y}_n$ = *number of objects in the n-th generation of* $\mathbf{Z}(t)$. If we take the sample tree π and transform it to a tree π' having all its branches of unit length but otherwise identical to π, then $\mathbf{Y}_n(\pi) = \mathbf{Z}_n(\pi')$, where $\mathbf{Z}_n$ is a BGW process. We call $\mathbf{Y}_n$ the embedded generation process for $\mathbf{Z}(t)$. The trees associated either with BGW process, or the embedded generation BGW process will be used to estimate the offspring probabilities.

Now we consider MTBP where certain *terminal types* of objects, once created, neither die nor reproduce ([18]). If $T = \{T_1, T_2, \dots, T_m\}$ is the set of non-terminal types and $T^T = \{T_1^T, T_2^T, \dots, T_{d-m}^T\}$ is the set of terminal types, then an object of type T_i produces offspring of any type and an object of type T_j^T does not reproduce any more. Here each $T_j^T \in T^T$ constitutes a *final group* ([10]). This way we model a situation where "dead" objects do not disappear, but are registered and are included as "dead" in the succeeding generations.

We are interested in estimation of the offspring probabilities. If the whole tree π is observed the natural ML estimator for the offspring probabilities is

$$\hat{p}(T_v \to \mathcal{A}) = \frac{c(T_v \to \mathcal{A})}{c(T_v)}, \tag{1}$$

where $c(T_v)$ is the number of times a node of type T_v appears in the tree π and $c(T_v \to \mathcal{A})$ is the number of times a node of type T_v produces offspring $\mathcal{A}$ in π. It is not always possible to observe the whole tree though, often we have the following sampling scheme $\{\mathbf{Z}(0), \mathbf{Z}(t)\}$, for some $t > 0$. Let $\mathbf{Z}(0)$ consists of 1 object of some type. Suppose we are able to observe a number of independent instances of the process, meaning that they start with identical objects and reproduce according to the same offspring distribution. Such observational scheme is common in laboratory experiments in cell biology for example. If t is discrete $\mathbf{Z}(t)$ is the number of objects in the t-th generation. For continuous time it is a "generation" in the embedded generation BGW process. A simple illustration is presented in fig. 1 a)–c), where "alive" objects are grey, "dead" ones are white and numbers represent the different types. The bottom line repeats the observation that should be available at a particular moment t.

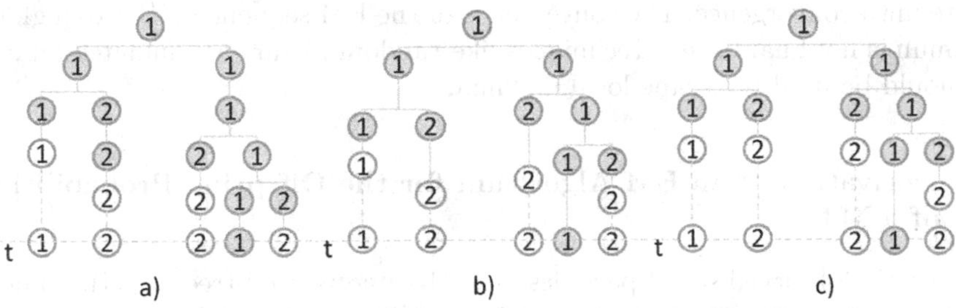

Fig. 1. A discrete time process a), continuous time process b) and its embedded process c)

Here the notion that "dead" objects are present and could be observed in succeeding generations as terminal types is crucial. If "dead" particles disappeared somewhere inside the "hidden" tree structure, estimation would be impossible (see fig. 2 for an example).

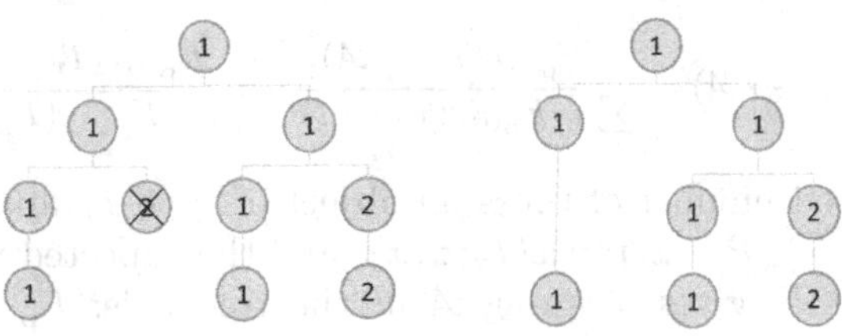

Fig. 2. Because of disappearing of a particle of type 2 in the first tree information about the reproduction has been lost and we have the same observation as in the second tree

3 The EM Proposal

3.1 The EM Algorithm

The EM algorithm was explained and given its name in a paper by Dempster, Laird, and Rubin in 1977 [4]. It is a method for finding ML estimates of parameters in statistical models, where the model depends on unobserved latent variables. Let a statistical model be determined by parameters θ, x is the observation and Y is some "hidden" data, which determines the probability distribution of x. Then the "complete" likelihood is denoted as $L(\theta|x, y)$ and the "incomplete" as $L(\theta|x)$. The aim is to maximize the log likelihood $\log L(\theta|x)$. In order to do this, define $Q(\theta|\theta') = E_{\theta'}[\log L(\theta|x, Y)]$. The *Expectation Maximization Algorithm* is usually stated formally like this:

E-step: Calculate function $Q(\theta|\theta^{(i)})$.

M-step: Maximize $Q(\theta|\theta^{(i)})$ with respect to θ.

Iterate until convergence. The convergence of the EM sequence $\{\hat{\theta}^{(i)}\}$ to a global maximun is not guarantied. Techniques like random restart or simulated annealing should be used to escape local maxima.

3.2 Derivation of an EM Algorithm for the Offspring Probabilities of a MTBP

Let x be the observed set of particles, π is the unobserved tree structure and $\mathbf{p}$ is the set of parameters – the offspring probabilities $p(T_v \to \mathcal{A})$ (the probability that an object of type T_v produces the set of objects $\mathcal{A}$). Then the likelihood of the "complete" observation is:

$$L(\mathbf{p}|\pi, x) = P(\pi, x|\mathbf{p}) = \prod_{v,\mathcal{A}:T_v \to \mathcal{A}} p(T_v \to \mathcal{A})^{c(T_v \to \mathcal{A};\pi,x)},$$

where c is a counting function – $c(T_v \to \mathcal{A}; \pi, x)$ is the number of times an object of type T_v produces the set of objects $\mathcal{A}$ in the tree π, observing x.

In [2] it has been proven that the re-estimating parameters are the normalized expected counts, which look like the ML estimators in the "complete" observation case (1), but the observed counts are replaced with their expectations.

$$p^{(i+1)}(T_v \to \mathcal{A}) = \frac{E_{\mathbf{p}^{(i)}} c(T_v \to \mathcal{A})}{\sum_{\mathcal{A}'} E_{\mathbf{p}^{(i)}} c(T_v \to \mathcal{A}')} = \frac{E_{\mathbf{p}^{(i)}} c(T_v \to \mathcal{A})}{E_{\mathbf{p}^{(i)}} c(T_v)}, \tag{2}$$

where the expected number of times an object of type T_v appears in the tree π is: $E_{\mathbf{p}^{(i)}} c(T_v) = \sum_\pi P(\pi|x, \mathbf{p}^{(i)}) c(T_v; \pi, x)$, and the expected number of times an object of type T_v gives offspring $\mathcal{A}$ in the tree π is: $E_{\mathbf{p}^{(i)}} c(T_v \to \mathcal{A}) = \sum_\pi P(\pi|x, \mathbf{p}^{(i)}) c(T_v \to \mathcal{A}; \pi, x)$.

It is a case of EM where the M-step is explicitly solved, so the computational effort will be on the E-step. The problem is that in general enumerating all possible trees π is of exponential complexity. The method proposed below is aimed to reduce complexity.

3.3 The Recurrence Scheme

We have shown that the E-step of the algorithm consists of determining the expected number of times a given type T_v or a given production $T_v \to \mathcal{A}$ appears in a tree π. A general method has been proposed in [3] for computing these counts. The algorithm consists of three parts – calculation of the inner probabilities, the outer probabilities and EM re-estimation, which are shown below.

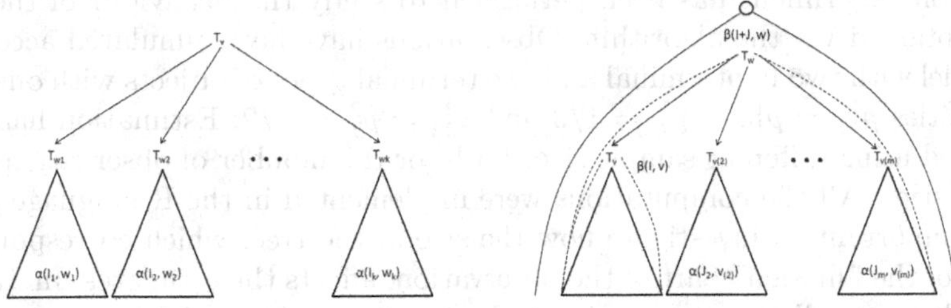

Fig. 3. The inner and outer probabilities recurrence

Let us define the **inner** probability $\alpha(\mathbf{I}, v)$ of a subtree rooted at a type T_v object to produce outcome $\mathbf{I} = (i_1, i_2, \ldots, i_d)$ where i_k is the number of objects of type T_k (fig. 3). From the basic branching property of the process we get the following recurrence:

$$\alpha(\mathbf{I}, v) = \sum_{\mathbf{w}} p(T_v \to \{T_{w_1}, \ldots, T_{w_k}\}) \sum_{\mathbf{I}_1 + \ldots + \mathbf{I}_k = \mathbf{I}} \alpha(\mathbf{I}_1, T_{w_1}) \ldots \alpha(\mathbf{I}_k, T_{w_k})$$

where $\mathbf{w} = \{T_{w_1}, \ldots, T_{w_k}\}$ are all possible sets of particles that T_v can produce. The **outer** probability $\beta(\mathbf{I}, v)$ is the probability of the entire tree excluded a subtree, rooted at a type T_v object and producing outcome $\mathbf{I} = (i_1, i_2, \ldots, i_d)$ (fig. 4). The recurrence here is:

$$\beta(\mathbf{I}, v) = \sum_{w} \sum_{\mathbf{v}} p(T_w \to \{T_v, T_{v_{(2)}}, \ldots, T_{v_{(m)}}\})$$
$$\times \sum_{\mathbf{J} \subset \mathbf{X} - \mathbf{I}} \beta(\mathbf{I} + \mathbf{J}, w) \sum_{\mathbf{J}_2 + \ldots + \mathbf{J}_m = \mathbf{J}} \alpha(\mathbf{J}_2, v_{(2)}) \ldots \alpha(\mathbf{J}_m, v_{(m)})$$

where $\{T_v, \mathbf{v}\} = \{T_v, T_{v_{(2)}}, \ldots, T_{v_{(m)}}\}$ are all possible sets of objects that T_w can produce and $\mathbf{X} = x_1, x_2, \ldots, x_d$ is the observation.

The **EM re-estimation** of the parameters is found to be:

$$p^{(i+1)}(T_v \to \{T_{w_1}, \ldots, T_{w_k}\})$$
$$= \frac{\sum_{\mathbf{I}} \sum_{\mathbf{I}_1 + \ldots + \mathbf{I}_k = \mathbf{I}} \beta(\mathbf{I}, v) \alpha(\mathbf{I}_1, w_1) \ldots \alpha(\mathbf{I}_k, w_k) p^{(i)}(T_v \to \{T_{w_1}, \ldots, T_{w_k}\})}{\sum_{\mathbf{I}} \alpha(\mathbf{I}, v) \beta(\mathbf{I}, v)}$$

For several observed sets of objects the expected numbers in the nominator and denominator are summed for all sets.

Such generally stated the algorithm still has high complexity. In practical applications though, often there are small number of types and possible productions in the offspring distributions. Thus, for a number of cases, a specific dynamic programming schemes based on the above recurrence could be proposed, which will be less complex.

4 Simulation Study

Simulation experiment has been performed to study the behaviour of the estimates obtained via the algorithm. Observations have been simulated according to a model with two nonterminal and two terminal types of objects with offspring probabilities $p^1_{11} = p^1_{12} = p^1_T = 1/3$ and $p^2_{22} = p^2_T = 1/2$. Estimation has been performed using different sample sizes both for the number of observations, and the tree sizes. All the computations were implemented in the R language [17].

It is important to investigate how the size of the tree, which corresponds to the size of the "hidden" part of the observation, affects the estimates. In Table 1 the result for small tree sizes and sample size 20 are shown. The most accurate estimates are obtained through averaging these results. It can be seen that there is great variation in the estimate of the individual distribution of type 2, thought the mean is close to the real values. For larger sample sizes the variance of the estimates is reduced, but there is some bias in the estimate for type 2 (Table 2). Larger sample trees also lead to biased estimates for the individual distribution for type 2 (Table 3). Using larger trees is also computationally more expensive.

The bias in the estimate for type 2 is due to the greater uncertainty in the process for type 2: these objects could be generated by an object of their own type, as well as, by an object of type 1. For example, production of one object of type 1 and two of type 2 could happen in two ways: once $T_1 \to \{T_1, T_2\}$ and then $T_2 \to \{T_2, T_2\}$, or twice $T_1 \to \{T_1, T_2\}$. So, in general, productions $T_1 \to \{T_1, T_2\}$ take part more often in the estimation than productions $T_2 \to \{T_2, T_2\}$. As the branching is hidden and all possible generations have to be taken in account, this results in underestimation of p^2_{22} and some overestimation of p^1_{12} when that hidden part gets larger.

5 Conclusion

A general EM algorithm has been proposed to find ML estimation of the offspring probabilities of MTBP with terminal types when only an observation of the generation at given moment is available. The algorithm is straightforward and convenient to apply for a particular model. Simulation study shows that better estimates are obtained using smaller samples. Such algorithms would be useful in biological models like cell proliferation, genetics, genomics, evolution, and wherever a model of MTBP with terminal types is suitable.

Acknowledgments. This work was supported by the European Social Fund through the Human Resource Development Operational Programme under contract BG051PO001-3.3.06-0052 (2012/2014).

Table 1. Estimation obtained using small tree samples of size 20

size 20	p_T^1	p_{11}^1	p_{12}^1	p_T^2	p_{22}^2
s.1	0.35	0.38	0.27	0.25	0.75
s.2	0.26	0.45	0.30	0.56	0.44
s.3	0.33	0.35	0.32	0.47	0.53
s.4	0.30	0.33	0.37	0.47	0.53
s.5	0.33	0.38	0.28	0.00	1.00
s.6	0.39	0.42	0.19	0.00	1.00
s.7	0.38	0.35	0.27	0.94	0.06
s.8	0.42	0.25	0.33	0.72	0.28
mean	0.34	0.36	0.29	0.43	0.57
st.dev.	0.05	0.06	0.05	0.33	0.33

size 20	p_T^1	p_{11}^1	p_{12}^1	p_T^2	p_{22}^2
s.9	0.34	0.39	0.27	0.00	1.00
s.10	0.36	0.29	0.36	0.77	0.23
s.11	0.31	0.29	0.40	0.46	0.54
s.12	0.33	0.41	0.26	0.59	0.41
s.13	0.48	0.17	0.36	0.92	0.08
s.14	0.34	0.25	0.41	0.64	0.36
s.15	0.20	0.33	0.47	0.93	0.07
s.16	0.18	0.25	0.57	0.66	0.34
mean	0.32	0.30	0.39	0.62	0.38
st.dev.	0.09	0.08	0.10	0.30	0.30

all samples	p_T^1	p_{11}^1	p_{12}^1	p_T^2	p_{22}^2
avg.	0.33	0.33	0.34	0.52	0.48
st.dev.	0.07	0.08	0.09	0.32	0.32

Table 2. Estimation obtained using small tree samples of size 50 and 100

size 50	p_T^1	p_{11}^1	p_{12}^1	p_T^2	p_{22}^2
s.1	0.29	0.40	0.31	0.39	0.61
s.2	0.34	0.35	0.31	0.45	0.55
s.3	0.36	0.36	0.27	0.82	0.18
s.4	0.39	0.32	0.30	0.52	0.48
s.5	0.31	0.35	0.34	0.55	0.45
s.6	0.27	0.43	0.30	0.81	0.19
mean	0.33	0.37	0.30	0.59	0.41
st.dev.	0.04	0.04	0.02	0.18	0.18

size 100	p_T^1	p_{11}^1	p_{12}^1	p_T^2	p_{22}^2
s.1	0.31	0.38	0.31	0.41	0.59
s.2	0.38	0.34	0.28	0.68	0.32
s.3	0.29	0.39	0.31	0.67	0.33
mean	0.33	0.37	0.30	0.59	0.41
st.dev.	0.04	0.03	0.02	0.16	0.16

Table 3. Estimation obtained using larger tree samples of size 50 and 100

size 50	p_T^1	p_{11}^1	p_{12}^1	p_T^2	p_{22}^2
s.1	0.33	0.32	0.35	0.64	0.36
s.2	0.33	0.30	0.36	0.72	0.28
s.3	0.33	0.28	0.39	0.80	0.20
s.4	0.34	0.33	0.33	0.68	0.32
mean	0.33	0.31	0.36	0.71	0.29
st.dev.	0.00	0.02	0.02	0.07	0.07

size 100	p_T^1	p_{11}^1	p_{12}^1	p_T^2	p_{22}^2
s.1	0.34	0.31	0.36	0.71	0.29
s.2	0.33	0.30	0.36	0.74	0.26
mean	0.34	0.31	0.36	0.73	0.27
st.dev.	0.00	0.00	0.00	0.02	0.02

References

1. Athreya, K.B., Ney, P.E.: Branching Processes. Springer, Berlin (1972)
2. Daskalova, N.: Using Inside-Outside Algorithm for Estimation of the Offspring Distribution in Multitype Branching Processes. Serdica Journal of Computing 4(4), 463–474 (2010)
3. Daskalova, N.: Maximum Likelihood Estimation in Multitype Branching Processes with Terminal Types. Comptes Rendus de lAcademie bulgare des Sciences 65(5), 575–580 (2012)
4. Dempster, A.P., Laird, N.M., Rubin, D.B.: Maximum likelihood from incomplete data via the EM algorithm. Journal of the Royal Statistical Society, B 39, 1–38 (1977)
5. Geman, S., Johnson, M.: Probability and statistics in computational linguistics, a brief review. In: Johnson, M., Khudanpur, S.P., Ostendorf, M., Rosenfeld, R. (eds.) Mathematical Foundations of Speech and Language Processing (2004)
6. González, M., Martín, J., Martínez, R., Mota, M.: Non-parametric Bayesian estimation for multitype branching processes through simulation-based methods. Computational Statistics & Data Analysis 52(3), 1281–1291 (2008)
7. González, M., Gutiérrez, C., Martínez, R.: Parametric inference for Y-linked gene branching model: Expectation-maximization method. In: Proceeding of Workshop on Branching Processes and their Applications. Lecture Notes in Statistics, vol. 197, pp. 191–204 (2010)
8. Guttorp, P.: Statistical inference for branching processes. Wiley, New York (1991)
9. Haccou, P., Jagers, P., Vatutin, V.A.: Branching Processes: Variation, Growth and Extinction of Populations. Cambridge University Press, Cambridge (2005)
10. Harris, T.E.: Branching Processes. Springer, New York (1963)
11. Hyrien, O.: Pseudo-likelihood estimation for discretely observed multitype Bellman-Harris branching processes. Journal of Statistical Planning and Inference 137(4), 1375–1388 (2007)
12. Jagers, P.: Branching Processes with Biological Applications. Wiley, London (1975)
13. Kimmel, M., Axelrod, D.E.: Branching Processes in Biology. Springer, New York (2002)
14. Lari, K., Young, S.J.: The Estimation of Stochastic Context-Free Grammars Using the Inside-Outside Algorithm. Computer Speech and Language 4, 35–36 (1990)
15. McLachlan, G.J., Krishnan, T.: The EM Algorithm and Extensions. Wiley (2008)
16. Mode, C.J.: Multitype Branching Processes: Theory and Applications. Elsevier, New York (1971)
17. R Development Core Team R: A language and environment for statistical computing. R Foundation for Statistical Computing, Vienna, Austria (2010), `http://www.R-project.org`
18. Sankoff, D.: Branching Processes with Terminal Types: Application to Context-Free Grammars. Journal of Applied Probability 8(2), 233–240 (1971)
19. Yanev, N.M.: Statistical Inference for Branching Processes. In: Ahsanullah, M., Yanev, G.P. (eds.) Records and Branching Processes, pp. 143–168. Nova Sci. Publishers, Inc. (2008)

An Introduction to Interdependent Networks

Michael M. Danziger[1], Amir Bashan[2], Yehiel Berezin[1], Louis M. Shekhtman[1], and Shlomo Havlin[1]

[1] Department of Physics, Bar Ilan University, Ramat Gan, Israel
[2] Channing Division of Network Medicine, Brigham Women's Hospital and Harvard Medical School, Boston, MA, USA

Abstract. Many real-world phenomena can be modelled using networks. Often, these networks interact with one another in non-trivial ways. Recently, a theory of interdependent networks has been developed which describes dependency between nodes across networks. Interdependent networks have a number of unique properties which are absent in single networks. In particular, systems of interdependent networks often undergo abrupt first-order percolation transitions induced by cascading failures. Here we present an overview of recent developments and significant findings regarding interdependent networks and networks of networks.

1 Background: From Single Networks to Networks of Networks

The field of network science emerged following the realization that in many complex systems, the way in which objects interact with each other could neither be reduced to their proximity (as in most traditional physical systems) nor was it random (as described in classical graph theory [1,2,3]). Rather, the topology of connections between objects in a wide range of real systems was shown to exhibit a broad spectrum of non-trivial structures: scale-free networks dominated by hubs [4,5], small-world networks which captured the familiar "six degrees of separation" idea [6,7] and countless other variations [8,9].

The late 1990s and early 2000s was an incredibly fruitful period in the study of complex networks and increasingly sophisticated models and measurements were developed. A strong catalyst for the growth of network science in this period was the rapid improvement of computers and new data from the world-wide web. The topological structure of networks was demonstrated to provide valuable insight into a wide range of real-world phenomena including percolation [10], epidemiology [11], marketing [12] and climate studies [13] amongst many others.

One of the most important properties of a network that was studied was its robustness following the failure of a subset of its nodes. Utilizing percolation theory, network robustness can be studied via the fraction occupied by its largest connected component P_∞ which is taken as a proxy for functionality of the network [14,15]. Consider, for example, a telephone network composed of

V.M. Mladenov and P.C. Ivanov (Eds.): NDES 2014, CCIS 438, pp. 189–202, 2014.

telephone lines and retransmitting stations. If $P_\infty \sim 1$ (the entire system), then information from one part of the network has a high probability of reaching any other part. If, however, $P_\infty \sim 0$, then information in one part cannot travel far and the network must be considered nonfunctional. Even if $P_\infty \sim 1$, some nodes may be detached from the largest connected component and those nodes are considered nonfunctional. We use the term *giant connected component* (GCC) to refer to P_∞ when it is of order 1. Percolation theory is concerned with determining $P_\infty(p)$ after a random fraction $1-p$ of nodes (or edges) are disabled in the network. Typically, $P_\infty(p)$ undergoes a second-order transition at a certain value p_c: for $p > p_c$, $P_\infty(p) > 0$ and it approaches zero as $p \to p_c$ but for $p < p_c$, $P_\infty(p) \equiv 0$. Thus there is a discontinuity in the derivative $P'_\infty(p)$ at p_c even though the function itself is continuous. It is in this sense that the phase transition is described as second-order [16,17]. It was shown, for example, that scale-free networks (SF)–which are extremely ubiquitous in nature–have $p_c = 0$ [10]. This is in marked contrast to Erdös-Rënyi (ER) networks ($p_c = 1/\langle k\rangle$) and 2D square lattices ($p_c = 0.5927$ [15]) and helps to explain the surprising robustness of many systems (e.g. the internet) with respect to random failures [18,10].

However, in reality, networks rarely appear in isolation. In epidemiology, diseases can spread within populations but can also transition to other populations, even to different species. In transportation networks, there are typically highway, bus, train and airplane networks covering the same areas but behaving differently [19]. Furthermore, the way in which one network affects another is not trivial and often specific nodes in one network interact with specific nodes in another network. This leads to the concept of interacting networks in which links exist between nodes within a single network as well as across networks. Just as ideal gases–which by definition are comprised of non-interacting particles–lack emergent critical phenomena such as phase transitions, we will see that the behavior of interacting networks has profound emergent properties which do not exist in single networks.

Since networks interact with one another selectively (and not generally all networks affecting all other networks), we can describe *networks of networks* with topologies between networks that are similar to the topology of nodes in a single network.

Multiplex networks are networks of networks in which the identity of the nodes is the same across different networks but the links are different [20,21,22]. Multiplex networks were first introduced to describe a person who participates in multiple social networks [23]. For instance, the networks of phone communication and email communication between individuals will have different topologies and different dynamics though the actors will be the same [24]. Also, each online social network shares the same individuals though the network topologies will be very different depending on the community which the social network represents.

One question that arises naturally in the discussion of networks of networks is, why describe this phenomenon as a "network of networks?" If we are describing a set of nodes and links then no matter how it is partitioned it is still a network.

Each description of interacting networks will answer this question differently but any attempt to describe a network of networks will be predicated on a claim that more is different—that by splitting the overall system into component networks, new phenomena can be uncovered and predicted. One way of describing the interaction between networks which yields qualitatively new phenomena is *interdependence*. This concept has been studied in the context of critical infrastucture and been formalized in several engineering models [25,26] (see Fig. 1a). However, as a theoretical property of interacting networks, interdependence was first introduced in a seminal study by Buldyrev et. al. in 2010 [27]. The remainder of this review will focus on the wealth of new phenomena which have been discovered using the ideas first described in that work.

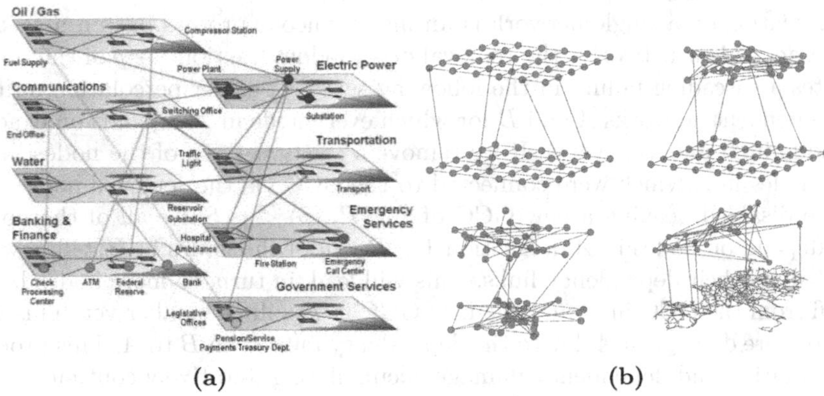

Fig. 1. An example of interdependent critical infrastructure systems and several modelled interdependent networks. (a) Schematic representation of interdependent critical infrastructure networks after [28]. (b) Illustration of interdependent networks composed of connectivity links (in blue, within the networks) and dependency links (in red, between the networks). Clockwise from upper-left: coupled lattices, a lattice coupled with a random regular (RR) network, two coupled RR networks and an RR network coupled to a real-world power grid. After [29]

2 Interdependence: Connectivity and Dependency Links

The fundamental property which characterizes interdependent networks is the existence of two qualitatively different kinds of links: *connectivity* links and *dependency* links [27,30,31] (see Fig. 1b). The connectivity links are the links which we are familiar with from single network theory and they connect nodes within the same network. They typically represent the ability of some quantity (information, electricity, traffic, disease etc) to flow from one node to another. From the perspective of percolation theory, if a node has multiple connectivity links leading to the GCC, it will only fail if all of those links cease to function. Dependency links, on the other hand, represent the idea that for a node to function, it requires support from another node which, in general, is in another network. In such a case, if the supporting node fails, the dependent node will also fail–even if it is still connected to the GCC in its network. If one network *depends on* and *supports* another network, we describe that pair of networks as interdependent. Interdependence is a common feature of critical infrastructure (see Fig. 1a) and many multiplex networks–often whatever causes a node to stop functioning in one layer will also disable it in other layers. Indeed, the percolation properties of interdependent networks describe the typical behavior in multiplex networks as well [27]. The properties of interdependence can affect a network's function in a variety of ways but here we focus on the response of a network of interdependent networks to the failure of a subset of its nodes using the tools of percolation theory [9]. We refer the reader to recent general reviews for other descriptions of interacting networks [23,32].

Percolation on a single network is an instantaneous process but on a system of interdependent networks, the removal of a random fraction $1-p$ of the nodes initiates a cascading failure in the following sense. Consider percolation on two interdependent networks A and B for which every node in A depends on exactly one node in B and vice versa. If we remove a fraction $1-p$ of the nodes in A, other nodes in A which were connected to the GCC via the removed nodes will also be disabled, leaving a new GCC of size $P_\infty(p) < p$. Since all of the nodes in B depend on nodes in A, a fraction $1-P_\infty(p)$ of the nodes in B will now be disabled via their dependency links. This will lead, in turn, to more nodes being cut off from the GCC in B and the new GCC in B will be smaller yet. This will lead to more damage in A due to the dependency links from B to A. This process of percolation and dependency damage accumulating iteratively continues until no more nodes are removed from iteration to iteration. This cascading failure is similar to the cascades described in flow and overload models on networks and the cascading failures in power grids which are linked to blackouts [33,34]. The cascade triggered by a single node removal has been called an "avalanche" [35] and the critical properties of this process have been studied extensively [35,36].

3 Interdependent Random Networks

This cascading failure was shown to lead to abrupt first-order transitions in systems of interdependent ER and SF networks that are qualitatively very different from the transitions in single networks (see Fig. 2b). Furthermore, p_c of a pair of ER networks was shown to increase from $1/\langle k\rangle$ to $2.4554/\langle k\rangle$. Surprisingly, it was found that scale-free networks, which are extremely robust to random failure on their own [18,10], become more vulnerable than equivalent ER networks when they are fully interdependent and for any $\lambda > 2$, $p_c > 0$. In general, a broader degree distribution leads to a higher p_c [27]. This is because the hubs in one network, which are the source of the stability of single scale-free networks, can be dependent on low degree nodes in the other network and are thus vulnerable to random damage via dependency links. These results were first demonstrated using the generating function formalism [37,27], though it has recently been shown that the same results can be obtained using the cavity method [38].

After the first results on interdependent networks were published in 2010 [27], the basic model described above was expanded to cover more diverse systems. One striking early result was that if less than an analytically calculable critical fraction q_c of the nodes in a system of two interdependent ER networks are interdependent, the phase transition reverts to the familiar second-order transition [30]. However, for scale-free networks, reducing the fraction of interdependent nodes leads to a hybrid transition, where a discontinuity in P_∞ is followed by a continuous decline to zero, as p decreases, (Fig. 2b)[39]. A similar transition was found when connectivity links between networks (which were first introduced in [40]) are combined with dependency links [41]. It has been shown that the same cascading failures emerge from systems with connectivity and dependency links within a single network [42,43,44].

In a series of articles, Gao et. al. extended the theory of pairs of interdependent networks to networks of interdependent networks (NoN) with general topologies [45,46,47,48]. Within this framework, analytic solutions for a number of key percolation quantities were presented including size of the GCC at each time-step t (see Fig. 2a), the size of the GCC at steady state (see Fig. 2b), p_c and other values.

The NoN topologies which were solved analytically include: a tree-like NoN of ER, SF or random regular (RR) networks ($q = 1$), a loop-like NoN of ER, SF or RR networks ($q \leq 1$), a star-like NoN of ER networks ($q \leq 1$) and a RR NoN of ER, SF or RR networks ($q \leq 1$). For tree-like NoNs, it was found [45,47] that the number of networks in the NoN (n) affects the overall robustness but the specific topology of the NoN does not. In contrast, for a RR NoN the number of networks n does not affect the robustness but the degree of each network within the NoN (m) does[46,48]. Because the topology of the loop-like and RR NoNs allows for chains of dependency links going throughout the system, there exists a quantity q_{max} above which the system will collapse with the removal of a single node, even if each network is highly connected ($p = 1$).

In light of these results, we can now see that single network percolation is simply a limiting case of NoN percolation theory. Furthermore, this framework can be extended to predict the percolation properties of NoNs of arbitrary networks, provided the percolation profile of the individual networks are known, even only numerically [49]. A more detailed summary of these results was recently published in [31].

The assumption that each node can depend on only one node was relaxed in [50] and it was shown that even if a node has many redundant dependency links, the first-order transition described above can still take place. If dependency links are assigned randomly, a situation can arise in which a chain of dependency links can be arbitrarily long and thus a single failure can propagate through the entire system. To avoid this scenario, most models for interdependent networks assume uniqueness or "no feedback" which limits the length of chains of dependency links [46,48] For a pair of fully interdependent networks, this reduces to the requirement that every dependency link is bidirectional. Under partial de-

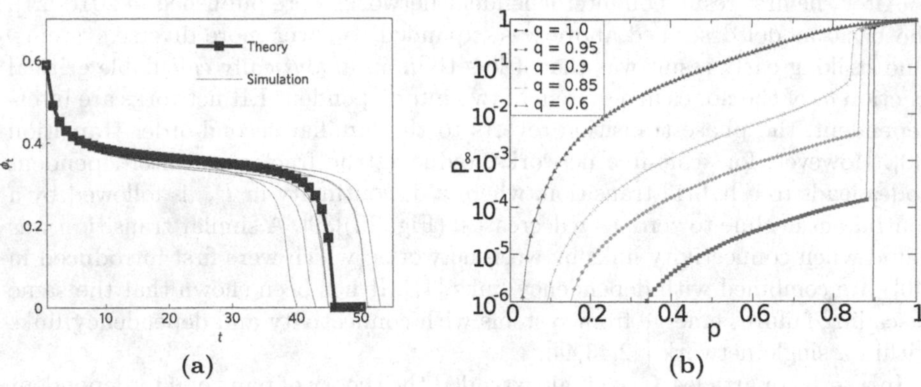

Fig. 2. Percolation of interdependent random networks. (a) The fraction of viable nodes at time t for a pair of partially interdependent ER networks. The gray lines represent individual realizations, the black squares are averages of all the realizations and the black line is calculated analytically. After [46]. (b) Percolation in a system of partially interdependent SF networks. We find three different kinds of phase transitions: first-order for $q = 1$ (black line), second-order for $q = 0.6$ and hybrid for $0.6 < q < 1$. This figure was generated for a system with $\lambda = 2.7$. After [39]. Cf. [41] for similar hybrid transitions.

pendency, this assumption is not necessary and the differences between systems with and without feedback have also been studied [48,29].

Though both the connectivity and dependency links were treated as random and uncorrelated in Refs. [27,30,45,46,47,48], the theory of interdependent networks has been expanded to more realistic cases. Assortativity of connectivity links was shown to decrease overall robustness [51]. Assortativity of dependency links was treated numerically [52] , analytically for the case of full degree-degree correlation [53] and analytically for the general case of degree-degree correlations with connectivity or dependency links using the cavity method [38]. Interestingly, if a fraction α of the highest degree nodes are made interdependent in each network, a three-phase system with a tricritical point emerges in the α–p plane [54]. If the system is a multiplex network, there may be overlapping links, i.e., two nodes which are linked in one layer may have a tendency to be linked in other layers [55,56,57]. In interdependent networks this phenomenon is referred to as intersimilarity [52,58]. Clustering, which has a negligible effect on the robustness of single networks [59], was shown to substantially reduce the robustness of interdependent networks [60,39].

4 Spatially Embedded Interdependent Networks

One of the most compelling motivations for developing a theory of interdependent networks is that many critical infrastructure networks depend on one another to function [25,26]. Essentially all critical infrastructure networks depend on electricity to function, which is why threats like electromagnetic pulses are taken so seriously (see Fig. 1a, Ref. [28]). The power grid itself, though, requires synchronization and control which it can only receive when the communication network is operational. One of the largest blackouts in recent history, the 2003 Italy blackout, was determined to have been caused by a cascading failure between electrical and communications networks [61].

In contrast to random networks, all infrastructure networks are embedded in space [19]. The nodes (e.g., power stations, communication lines, retransmitters etc.) occupy specific positions in a 2D plane and the fact that the cost of links increases with their length leads to a topology that is markedly different from random networks [62]. Thus infrastructure networks will tend to be approximately planar and the distribution of geographic link distances will be exponential with a characteristic length [63]. From universality principles, all such networks are expected to have the same general percolation behavior as standard 2D lattices [14,63]. As such, the first descriptions of spatially embedded interdependent networks were modelled with square lattices [64,29,65,66,49] and the results have been verified on synthetic and real-world power grids [29,65].

Analytic descriptions of percolation phenomena require that the network be "locally tree-like" and in the limit of large systems, this assumption is very accurate for random networks of arbitrary degree distribution [37]. However, lattices and other spatially embedded networks are not even remotely tree-like and analytic results on percolation properties are almost impossible to obtain [14,15]. Therefore most of the results on spatially embedded networks are based on numerical simulations.

One of the few major analytic results for spatially embedded systems is that for interdependent lattices, if there is no restriction on the length of the dependency links then any fraction of dependency leads to a first-order transition ($q_c = 0$). In [29], it was shown that the critical fraction q_c for which the system transitions from the first-order regime to the second order regime must fulfill:

$$1 = p_c^\star q_c P'_\infty(p_c) \tag{1}$$

in which $p_c^\star$ is the percolation threshold in the system of interdependent lattices, p_c is the percolation threshold in a single lattice and $P'_\infty(p)$ is the derivative of $P_\infty(p)$ for a single lattice. Since as $p \to p_c$, $P_\infty(p) = A(x - p_c)^\beta$ and for 2D lattices $\beta = 5/36$ [67], $P'_\infty(p)$ diverges as $p \to p_c$ and the only way to fulfill Eq. 1 is if $q_c = 0$. From universality arguments, all spatially embedded networks in $d < 6$ have $\beta < 1$ [14,15,63] and thus all systems composed of interdependent spatially embedded networks (in $d < 6$) with random dependency links will have $q_c = 0$. In Fig. 1b, all of the configurations shown except the RR-RR system have $q_c = 0$.

If the dependency links are of limited length, the percolation behavior is surprisingly complex and a new spreading failure emerges. Li et. al. [64] introduced the parameter r, called the "dependency length," to describe the fact that in most systems of interest the dependency links, too, will likely be costly to create and, like the connectivity links, will tend to be shorter than a certain characteristic length. In this model, dependency links between networks are selected at random but are always of length less than r (in lattice units). If $r = 0$, the system of interdependent lattices behaves identically to a single lattice. If $r = \infty$, the dependency links are unconstrained and purely random as in [29]. Li et. al. found that as long as r is below a critical length $r_c \approx 8$ the transition is second-order but for $r > r_c$ the transition is first order [64] (See Fig. 3b). The first-order transition for spatially embedded interdependent networks is unique in that it is characterized by a spreading process. Once damage of a certain size emerges at a given place on the lattice, it will begin to propagate outwards and destroy the entire system (See Fig. 3a).

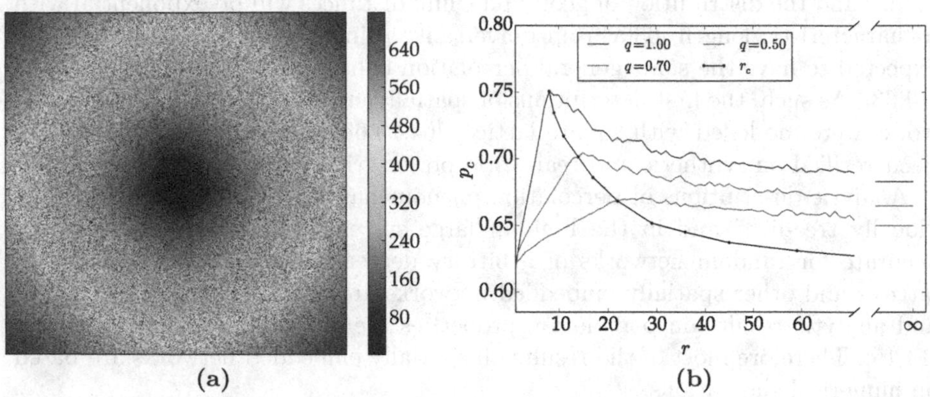

Fig. 3. Percolation of spatially embedded networks. (a) A snapshot of one lattice in a pair of interdependent lattices with nodes colored according to the time-step in which the node failed. The regularity of the color-change reflects the constant speed of the spreading failure in space (Generated for $q = 1, r = 11, L = 2900$). (b) The effect of r and q on p_c. As r increases, p_c increases until r reaches r_c. At that point the transition becomes first-order and p_c starts decreasing until it reaches its asymptotic value at $r = \infty$. Both after [66].

If the dependency is reduced from $q = 1$ to lower values, it is found that r_c increases and diverges at $q = 0$, consistent with the result from [29] that $q_c = 0$ for $r = \infty$ [66] (See Fig. 3b). Recently, the framework developed in [45,46,47,48] was extended to general networks of spatially embedded networks. Similar to the case of networks of random networks, the robustness of tree-like spatially embedded NoNs are affected by n but not by the topology while RR NoNs are affected by m but not by n [49].

5 Attack and Defense of Interdependent Networks

Due to their startling vulnerabilities with respect to random failures, it is of particular interest to understand how non-random attacks affect interdependent networks and how to improve the robustness of interdependent networks through topological changes. Huang et. al. [68] studied tunable degree-targeted attacks on interdependent networks. They found that even attacks which only affected low-degree nodes caused severe damage because high-degree nodes in one network can depend on low-degree nodes in another network. This framework was later expanded to general networks of networks [69].

Since high degree nodes in one network which depend on low degree nodes in another network can lead to extreme vulnerability, there have been several attempts mitigate this vulnerability by making small modifications to the inter-network topology. Schneider et. al. [70] demonstrated that selecting autonomous nodes by degree or betweenness can greatly reduce the chances of a catastrophic cascading failure. Valdez et. al. have also obtained promising results by selecting a small fraction of high-degree nodes and making them autonomous [71]. These mitigation strategies are methodologically related to the intersimilarity/overlap studies discussed above.

As we have seen, cascading failures are dynamic processes and the overall cascade lifetime can indeed be very long [36,66]. The slowness of the process opens the door for "healing" methods allowing the dynamic recovery of failed nodes in the midst of the cascade. Recently, a possible healing mechanism along these lines has been proposed and analyzed [72].

When considering infrastructure or other spatially embedded networks, not only is the network embedded in space but failures are also expected to be geographically localized. For instance, natural disasters can disable nodes across all networks in a given area while EMP or biological attacks can disable the power

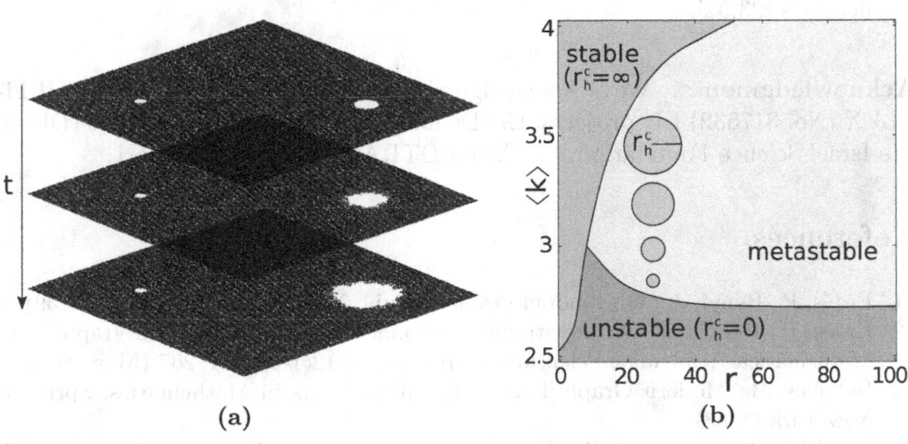

Fig. 4. Geographically localized attacks on interdependent networks. (a) The hole on the left is below r_h^c and stays in place while the hole on the right is larger than r_h^c and propagates through the system. (b) The phase space of localized attacks on interdependent networks. The increasing gray circles represent the dependence of r_h^c on $\langle k \rangle$. Both after [65].

grid or social network only in a given area. Geographically localized attacks of this sort have received attention in the context of single network percolation[73] and flow-based cascading failures [74]. However, the existence of dependency between networks leads to surprising new effects. Recently, Berezin et. al. [65] have shown that spatially embedded networks with dependencies can exist in three phases: stable, unstable and metastable (See Fig. 4b). In the metastable phase, the system is robust with respect to random attacks–even if finite fractions of the system are removed. However, if all of the nodes within a critical radius r_h^c fail, it causes a cascading failure which spreads throughout the system and destroys it (See Fig. 4a). Significantly, the value of r_h^c does not scale with system size and thus, in the limit of large systems, it constitutes a zero-fraction of the total system.

6 Applications of Networks of Networks

Many of the fields for which networks were seen as relevant models have been re-evaluated in light of the realization that interacting networks behave differently than single networks. Epidemics on interdependent and interconnected networks have received considerable attention [24,75,76,77,78]. Economic networks composed of individuals, firms and banks all interact with one another and are susceptible to large scale cascading failures [79,80] Interacting networks have also been found in physiological systems [81], ecology [82] and climate studies [83]. Multilevel transportation networks have also been studied from the perspective of interacting networks [84].

The breadth of applications of networks of networks is too great to address here and we refer the reader to recent reviews for more thorough treatment of applications [23,32].

Acknowledgments. We acknowledge the LINC (No. 289447) and MULTIPLEX (No. 317532) EU projects, the Deutsche Forschungsgemeinschaft (DFG), the Israel Science Foundation, ONR and DTRA for financial support.

References

1. Erdős, P., Rényi, A.: On random graphs i. Publ. Math. Debrecen 6, 290 (1959)
2. Erdős, P., Rényi, A.: On the strength of connectedness of a random graph. Acta Mathematica Academiae Scientiarum Hungaricae 12(1-2), 261–267 (Mar 1964)
3. Bollobás, B.: Modern Graph Theory. Graduate Texts in Mathematics. Springer, New York (1998)
4. Barabási, A.L., Albert, R.: Emergence of Scaling in Random Networks. Science 286(5439), 509–512 (1999)
5. Caldarelli, G.: Scale-free networks: complex webs in nature and technology. Oxford University Press (2007)
6. Watts, D.J., Strogatz, S.H.: Collective dynamics of 'small-world' networks. Nature 393(6684), 440–442 (1998)

7. Amaral, L.A.N., Scala, A., Barthélemy, M., Stanley, H.E.: Classes of small-world networks. Proceedings of the National Academy of Sciences 97(21), 11149–11152 (2000)
8. Newman, M.: Networks: An introduction. OUP, Oxford (2010)
9. Cohen, R., Havlin, S.: Complex Networks: Structure, Robustness and Function. Cambridge University Press (2010)
10. Cohen, R., Erez, K., ben Avraham, D., Havlin, S.: Resilience of the Internet to Random Breakdowns. Phys. Rev. Lett. 85, 4626–4628 (2000)
11. Pastor-Satorras, R., Vespignani, A.: Epidemic spreading in scale-free networks. Phys. Rev. Lett. 86, 3200–3203 (2001)
12. Goldenberg, J., Libai, B., Muller, E.: Talk of the network: A complex systems look at the underlying process of word-of-mouth. Marketing Letters 12(3), 211–223 (2001)
13. Yamasaki, K., Gozolchiani, A., Havlin, S.: Climate networks around the globe are significantly affected by el niño. Phys. Rev. Lett. 100, 228501 (2008)
14. Bunde, A., Havlin, S.: Fractals and disordered systems. Springer-Verlag New York, Inc. (1991)
15. Stauffer, D., Aharony, A.: Introduction To Percolation Theory. Taylor & Francis (1994)
16. Stanley, H.: Introduction to Phase Transitions and Critical Phenomena. International series of monographs on physics. Oxford University Press (1971)
17. Goldenfeld, N.: Lectures on Phase Transitions and the Renormalization Group. Frontiers in Physics. Addison-Wesley, Advanced Book Program (1992)
18. Albert, R., Jeong, H., Barabási, A.L.: Error and attack tolerance of complex networks. Nature 406(6794), 378–382 (2000)
19. Barthélemy, M.: Spatial networks. Physics Reports 499(1-3), 1–101 (2011)
20. Bianconi, G.: Statistical mechanics of multiplex networks: Entropy and overlap. Phys. Rev. E 87, 062806 (2013)
21. Nicosia, V., Bianconi, G., Latora, V., Barthelemy, M.: Growing multiplex networks. Phys. Rev. Lett. 111, 058701 (2013)
22. De Domenico, M., Solé-Ribalta, A., Cozzo, E., Kivelä, M., Moreno, Y., Porter, M.A., Gómez, S., Arenas, A.: Mathematical formulation of multilayer networks. Phys. Rev. X 3, 041022 (2013)
23. Kivelä, M., Arenas, A., Barthelemy, M., Gleeson, J.P., Moreno, Y., Porter, M.A.: Multilayer Networks. ArXiv e-prints (September 2013)
24. Goldenberg, J., Shavitt, Y., Shir, E., Solomon, S.: Distributive immunization of networks against viruses using the 'honey-pot' architecture Dimension of spatially embedded networks. Nature Physics 1(3), 184–188 (2005)
25. Rinaldi, S., Peerenboom, J., Kelly, T.: Identifying, understanding, and analyzing critical infrastructure interdependencies. IEEE Control Systems 21(6), 11–25 (2001)
26. Hokstad, P., Utne, I., Vatn, J.: Risk and Interdependencies in Critical Infrastructures: A Guideline for Analysis. Springer Series in Reliability Engineering. Springer (2012)
27. Buldyrev, S.V., Parshani, R., Paul, G., Stanley, H.E., Havlin, S.: Catastrophic cascade of failures in interdependent networks. Nature 464(7291), 1025–1028 (2010)
28. Foster Jr., J.S., Gjelde, E., Graham, W.R., Hermann, R.J., Kluepfel, H.M., Lawson, R.L., Soper, G.K., Wood, L.L., Woodard, J.B.: Report of the commission to assess the threat to the united states from electromagnetic pulse (emp) attack: Critical national infrastructures. Technical report, DTIC Document (2008)

29. Bashan, A., Berezin, Y., Buldyrev, S.V., Havlin, S.: The extreme vulnerability of interdependent spatially embedded networks. Nature Physics 9, 667–672 (2013)
30. Parshani, R., Buldyrev, S.V., Havlin, S.: Interdependent Networks: Reducing the Coupling Strength Leads to a Change from a First to Second Order Percolation Transition. Phys. Rev. Lett. 105, 048701 (2010)
31. Buldyrev, S.V., Paul, G., Stanley, H.E., Havlin, S.: Network of interdependent networks: Overview of theory and applications. In: D'Agostino, G., Scala, A. (eds.) Networks of Networks: The Last Frontier of Complexity. Understanding Complex Systems, pp. 3–36. Springer International Publishing (2014)
32. D'Agostino, G., Scala, A.: Networks of Networks: The Last Frontier of Complexity. Understanding Complex Systems. Springer International Publishing (2014)
33. Motter, A.E.: Cascade control and defense in complex networks. Phys. Rev. Lett. 93, 098701 (2004)
34. Dobson, I., Carreras, B.A., Lynch, V.E., Newman, D.E.: Complex systems analysis of series of blackouts: Cascading failure, critical points, and self-organization. Chaos: An Interdisciplinary Journal of Nonlinear Science 17(2), 026103 (2007)
35. Baxter, G.J., Dorogovtsev, S.N., Goltsev, A.V., Mendes, J.F.F.: Avalanche Collapse of Interdependent Networks. Phys. Rev. Lett. 109, 248701 (2012)
36. Zhou, D., Bashan, A., Berezin, Y., Cohen, R., Havlin, S.: On the Dynamics of Cascading Failures in Interdependent Networks. ArXiv e-prints (November 2012)
37. Newman, M.E.J., Strogatz, S.H., Watts, D.J.: Random graphs with arbitrary degree distributions and their applications. Phys. Rev. E 64, 026118 (2001)
38. Watanabe, S., Kabashima, Y.: Cavity-based robustness analysis of interdependent networks: Influences of intranetwork and internetwork degree-degree correlations. Phys. Rev. E 89, 012808 (2014)
39. Zhou, D., Gao, J., Stanley, H.E., Havlin, S.: Percolation of partially interdependent scale-free networks. Phys. Rev. E 87, 052812 (2013)
40. Leicht, E.A., D'Souza, R.M.: Percolation on interacting networks. ArXiv e-prints (July 2009)
41. Hu, Y., Ksherim, B., Cohen, R., Havlin, S.: Percolation in interdependent and interconnected networks: Abrupt change from second- to first-order transitions. Phys. Rev. E 84, 066116 (2011)
42. Parshani, R., Buldyrev, S.V., Havlin, S.: Critical effect of dependency groups on the function of networks. Proceedings of the National Academy of Sciences 108(3), 1007–1010 (2011)
43. Bashan, A., Parshani, R., Havlin, S.: Percolation in networks composed of connectivity and dependency links. Phys. Rev. E 83, 051127 (2011)
44. Zhao, J.H., Zhou, H.J., Liu, Y.Y.: Inducing effect on the percolation transition in complex networks. Nature Communications 4 (September 2013)
45. Gao, J., Buldyrev, S.V., Havlin, S., Stanley, H.E.: Robustness of a Network of Networks. Phys. Rev. Lett. 107, 195701 (2011)
46. Gao, J., Buldyrev, S.V., Stanley, H.E., Havlin, S.: Networks formed from interdependent networks. Nature Physics 8(1), 40–48 (2012)
47. Gao, J., Buldyrev, S.V., Havlin, S., Stanley, H.E.: Robustness of a network formed by n interdependent networks with a one-to-one correspondence of dependent nodes. Phys. Rev. E 85, 066134 (2012)
48. Gao, J., Buldyrev, S.V., Stanley, H.E., Xu, X., Havlin, S.: Percolation of a general network of networks. Phys. Rev. E 88, 062816 (2013)
49. Shekhtman, L.M., Berezin, Y., Danziger, M.M., Havlin, S.: Robustness of a Network Formed of Spatially Embedded Networks. ArXiv e-prints (February 2014)

50. Shao, J., Buldyrev, S.V., Havlin, S., Stanley, H.E.: Cascade of failures in coupled network systems with multiple support-dependence relations. Phys. Rev. E 83, 036116 (2011)
51. Zhou, D., Stanley, H.E., D'Agostino, G., Scala, A.: Assortativity decreases the robustness of interdependent networks. Phys. Rev. E 86, 066103 (2012)
52. Parshani, R., Rozenblat, C., Ietri, D., Ducruet, C., Havlin, S.: Inter-similarity between coupled networks. EPL (Europhysics Letters) 92(6), 68002 (2010)
53. Buldyrev, S.V., Shere, N.W., Cwilich, G.A.: Interdependent networks with identical degrees of mutually dependent nodes. Phys. Rev. E 83, 016112 (2011)
54. Valdez, L.D., Macri, P.A., Stanley, H.E., Braunstein, L.A.: Triple point in correlated interdependent networks. Phys. Rev. E 88, 050803 (2013)
55. Lee, K.M., Kim, J.Y., Cho, W.K., Goh, K.I., Kim, I.M.: Correlated multiplexity and connectivity of multiplex random networks. New Journal of Physics 14(3), 33027 (2012)
56. Cellai, D., López, E., Zhou, J., Gleeson, J.P., Bianconi, G.: Percolation in multiplex networks with overlap. Phys. Rev. E 88, 052811 (2013)
57. Li, M., Liu, R.R., Jia, C.X., Wang, B.H.: Critical effects of overlapping of connectivity and dependence links on percolation of networks. New Journal of Physics 15(9), 093013 (2013)
58. Hu, Y., Zhou, D., Zhang, R., Han, Z., Rozenblat, C., Havlin, S.: Percolation of interdependent networks with intersimilarity. Phys. Rev. E 88, 052805 (2013)
59. Newman, M.E.J.: Random graphs with clustering. Phys. Rev. Lett. 103, 058701 (2009)
60. Huang, X., Shao, S., Wang, H., Buldyrev, S.V., Eugene Stanley, H., Havlin, S.: The robustness of interdependent clustered networks. EPL 101(1), 18002 (2013)
61. Rosato, V., Issacharoff, L., Tiriticco, F., Meloni, S., Porcellinis, S.D., Setola, R.: Modelling interdependent infrastructures using interacting dynamical models. International Journal of Critical Infrastructures 4(1/2), 63 (2008)
62. Hines, P., Blumsack, S., Cotilla Sanchez, E., Barrows, C.: The Topological and Electrical Structure of Power Grids. In: 2010 43rd Hawaii International Conference on System Sciences (HICSS), pp. 1–10 (2010)
63. Li, D., Kosmidis, K., Bunde, A., Havlin, S.: Dimension of spatially embedded networks. Nature Physics 7(6), 481–484 (2011)
64. Li, W., Bashan, A., Buldyrev, S.V., Stanley, H.E., Havlin, S.: Cascading Failures in Interdependent Lattice Networks: The Critical Role of the Length of Dependency Links. Phys. Rev. Lett. 108, 228702 (2012)
65. Berezin, Y., Bashan, A., Danziger, M.M., Li, D., Havlin, S.: Spatially localized attacks on interdependent networks: The existence of a finite critical attack size. ArXiv e-prints (October 2013)
66. Danziger, M.M., Bashan, A., Berezin, Y., Havlin, S.: Interdependent spatially embedded networks: Dynamics at percolation threshold. In: 2013 International Conference on Signal-Image Technology Internet-Based Systems (SITIS), pp. 619–625 (December 2013)
67. Nienhuis, B.: Analytical calculation of two leading exponents of the dilute potts model. Journal of Physics A: Mathematical and General 15(1), 199 (1982)
68. Huang, X., Gao, J., Buldyrev, S.V., Havlin, S., Stanley, H.E.: Robustness of interdependent networks under targeted attack. Phys. Rev. E 83, 065101 (2011)
69. Dong, G., Gao, J., Du, R., Tian, L., Stanley, H.E., Havlin, S.: Robustness of network of networks under targeted attack. Phys. Rev. E 87, 052804 (2013)
70. Schneider, C.M., Yazdani, N., Araújo, N.A., Havlin, S., Herrmann, H.J.: Towards designing robust coupled networks. Scientific Reports 3 (2013)

71. Valdez, L.D., Macri, P.A., Braunstein, L.A.: A triple point induced by targeted autonomization on interdependent scale-free networks. Journal of Physics A: Mathematical and Theoretical 47(5), 055002 (2014)
72. Stippinger, M., Kertész, J.: Enhancing resilience of interdependent networks by healing. ArXiv e-prints (December 2013)
73. Agarwal, P.K., Efrat, A., Ganjugunte, S., Hay, D., Sankararaman, S., Zussman, G.: The resilience of WDM networks to probabilistic geographical failures. In: 2011 Proceedings of the IEEE, INFOCOM, pp. 1521–1529 (2011)
74. Bernstein, A., Bienstock, D., Hay, D., Uzunoglu, M., Zussman, G.: Sensitivity analysis of the power grid vulnerability to large-scale cascading failures. SIGMETRICS Perform. Eval. Rev. 40(3), 33–37 (2012)
75. Son, S.W., Bizhani, G., Christensen, C., Grassberger, P., Paczuski, M.: Percolation theory on interdependent networks based on epidemic spreading. EPL (Europhysics Letters) 97(1), 16006 (2012)
76. Saumell-Mendiola, A., Serrano, M.Á., Boguñá, M.: Epidemic spreading on interconnected networks. Phys. Rev. E 86, 026106 (2012)
77. Dickison, M., Havlin, S., Stanley, H.E.: Epidemics on interconnected networks. Phys. Rev. E 85, 066109 (2012)
78. Wang, H., Li, Q., D'Agostino, G., Havlin, S., Stanley, H.E., Van Mieghem, P.: Effect of the interconnected network structure on the epidemic threshold. Phys. Rev. E 88, 022801 (2013)
79. Erez, T., Hohnisch, M., Solomon, S.: Statistical economics on multi-variable layered networks. In: Salzano, M., Kirman, A. (eds.) Economics: Complex Windows. New Economic Windows, pp. 201–217. Springer, Milan (2005)
80. Huang, X., Vodenska, I., Havlin, S., Stanley, H.E.: Cascading failures in bi-partite graphs: Model for systemic risk propagation. Sci. Rep. 3 (February 2013)
81. Bashan, A., Bartsch, R.P., Kantelhardt, J.W., Havlin, S., Ivanov, P.C.: Network physiology reveals relations between network topology and physiological function. Nature Communications 3, 702 (2012)
82. Pocock, M.J.O., Evans, D.M., Memmott, J.: The robustness and restoration of a network of ecological networks. Science 335(6071), 973–977 (2012)
83. Donges, J., Schultz, H., Marwan, N., Zou, Y., Kurths, J.: Investigating the topology of interacting networks. The European Physical Journal B 84(4), 635–651 (2011)
84. Morris, R.G., Barthelemy, M.: Transport on Coupled Spatial Networks. Phys. Rev. Lett. 109, 128703 (2012)

Experimental Dynamics Observed in a Configurable Complex Network of Chaotic Oscillators

Carlo Petrarca, Soudeh Yaghouti, and Massimiliano de Magistris

Dipartimento di Ingegneria Elettrica e delle Tecnologie dell'Informazione
University of Naples FEDERICO II - Via Claudio 21, I-80125 Napoli, Italy
{carlo.petrarca,soudeh.yaghouti,m.demagistris}@unina.it

Abstract. New experimental results are presented, obtained with a recently developed set-up, designed and tailored to easily perform large scale experiments on networks of chaotic oscillators, and using Chua's circuits as nodes. A ring of 16 chaotic nodes, diffusively coupled with a large range of coupling strength values is considered, and accurately characterized. Complete synchronization, patterns and chaotic waves have been observed and described, so revealing the potential of the realized set-up for realizing accurate experiments in configurable complex networks.

Keywords: complex networks, chaotic oscillators, synchronization, patterns and waves.

1 Introduction

The paradigm of complex networks, intended as the proper interconnection of eventually non linear (oscillatory or chaotic) dynamical units, has received a plenty of scientific interest in recent years (for an extensive review refer to [1,2]). Their distinguishing characteristics from other modeling paradigms of real phenomena are: *i*) collective behavior ("emerging dynamics") differing substantially from individual "stand-alone" ones, *ii*) complexity arising from proper combination of non linearity of nodes and interconnection structure.

Worthless to say, a vast range of potential application domains is known, ranging from biological systems to social networks; in particular, in the specific area of circuit and systems, potential applications in the information processing are investigated [3]. The importance of studies on arrays of electronic systems, as prototypical models of different real systems, has been clearly recognized [4], motivated by the availability of well developed simulation tools and, in principle, the possibility of realizing prototypes as integrated structures.

Despite the vast range of theoretical and numerical results, the experimental description of high dimension networks with configurable interconnection structure is rarely faced in literature, mostly due to complexity in realization. For instance, in [5] a simple example with N=3 is analyzed, and in [6] a scalable system is considered, but referring to a prescribed CNN topology. We designed and realized a dedicate\d

V.M. Mladenov and P.C. Ivanov (Eds.): NDES 2014, CCIS 438, pp. 203–210, 2014.

experimental setup with the provision of scalable and settable complex networks of chaotic oscillators [7,8], based on robust implementation of Chua's circuits as nodes and programmable implementation of the interconnections. It is aimed to provide a complete and dynamic control on coupling topologies, link type, direction and strength and node's system parameters. As a result, flexible experiments can be carried out and the setup can be viewed as an "analog simulator" of a quite general complex network, with the double advantage of being more realistic than simulation and, at the same time, drastically reducing the time for getting results. In this work, as a sample of the experimental potential of the set-up, some characterization of the system is described in a ring configuration of 16 nodes with proportional (diffusive) links, and major findings in terms of synchronization, patterns and waves are reported. Canonical phenomena such as complete and lag-synchronization [9], travelling waves, clustering, appearance of patterns, etc. [9,11,12] have been observed. The real dynamical behavior exhibit more richness and complexity with respect to previously available accurate numerical results. This, if needed, provides a further push towards the experimental verification and implementation of complex networks.

2 The Experimental Setup

The structure and realization of the experimental setup has already been described in quite detail in previous papers [7,8], and will not be repeated here, where only its major features will be briefly summarized. The network is based on a modular set of Chua's circuits, taken as paradigmatic case of complex and chaotic dynamic nodes. Stand-alone dynamics of individual nodes can be individually settled onto periodic or chaotic trajectories, by properly setting the values for the linear resistor R, allowing different regions of operation. The network nodes are interconnected via a fully configurable link network, with adjustable topologies. Figure 1 depicts a schematic draw of the Chua's nodes and the interconnecting links along with symbols and parameters used within the text. Figure 2 shows the realty implemented setup.

In the design and simulation phase requirements in accuracies of realization have been determined, in order to ensure uniformity in the operating conditions between the circuits incorporated in the network. Components have been properly selected to fulfill such realization requirements, as described in [7].

Although the realized setup allows in principle arbitrary choice in determining the structure and state variables to be interconnected, we will consider in the following only the case of links via voltages v_{c1} across capacitor C_1, as indicated in figure 1. The nodes are connected through diffusive links via settable R_{link}; in particular the link resistances span in the range 0 to 6.4 kΩ in steps of 25 Ω.

A modular USB multi-channel acquisition system is used to measure and monitor the variables of interest (nodes' states) in real time. Up to 64 state variables can be synchronously acquired. The whole network is controlled via a USB interface from a PC running LabVIEW. Two executive modes are possible: "*Control mode*", which allows to set the parameters of the network and display the vaweforms of the signals;

"*Scan mode*", which performs a scan of some preset values for link network (at present the topology is fixed at beginning, link values are dynamically settled). Scan mode saves all state variables at each step of in records up to 2500 samples. A typical execution time for an accurate scan of 255 steps is of about 45 minutes, mainly limited by the setting time for the link network values. Evaluation of the synchronization level of the acquired waveforms is available by calculating the index:

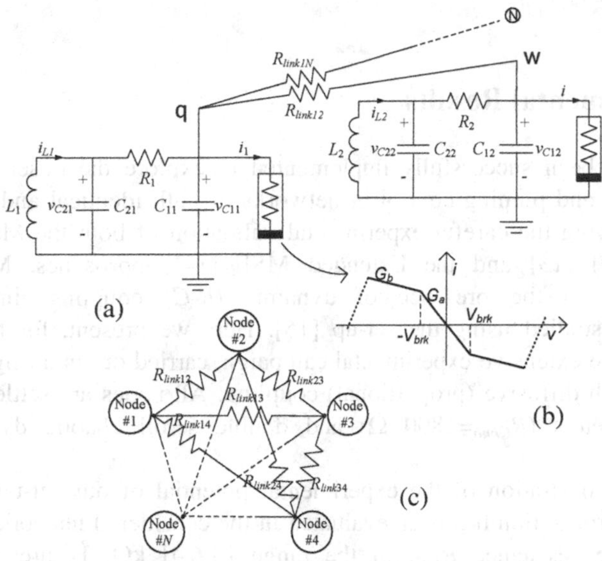

Fig. 1. a) Chua's circuit schematic, reference parameters, b) Chua's diode characteristic, c) general schematic of the link network

Fig. 2. The actual experimental setup, with 32 nodes and 32 links

$$I_s = 100\sqrt{\frac{1}{N_s N_s}\sum_{i=1}^{N_c}\sum_{k=1}^{N_s}\left(\frac{\left|v_{c1}^{i} - \langle v_{c1}(t)\rangle\right|}{\left|v_{c\max} - v_{c\min}\right|}\right)} \tag{1}$$

which expresses the (percent) *rms* distance of trajectories $v_{c1}^{i}(t)$ at *i-th* node from the "average" trajectory $\langle v_{c1}(t)\rangle$, where N_C is the number of node circuits, N_S the number of (measured) samples and $v_{c\max} - v_{c\min}$ is a normalizing factor for each set of measurements. Such synchronization index approaches zero when all traces are fully synchronized, and 100% when they are completely uncorrelated, and synchronization is definitely lost.

3 Experimental Results

The set-up has been successfully implemented to explore the onset of the global synchronization and pinning control in networks of both identical and non-identical nodes [7], allowing the careful experimental validation of both the Master Stability Function (MSF) [13] and the Extended MSF [14] approaches. Moreover, the synchronization in the presence of dynamic (*R-C*) coupling links has been experimentally studied using the set-up [15]. Here we present, for the first time, results on a quite extensive experimental campaign, carried out in a ring arrangement of 16 nodes with diffusive (proportional) coupling. All nodes are settled at identical nominal parameters (R_{Chua}=1800 Ω) and double scroll chaotic dynamics when uncoupled.

As first demonstration of the experimental potential of our set-up the onset of complete synchronization has been evaluated in the considered network as a function of the coupling resistance R_{link}, in the range [0-6.4] kΩ. Figure 3 depicts the synchronization index I_s defined as in eq. (1), evaluated for the state variable v_{c1}, as a function of the coupling resistance R_{link}; the red line corresponds to the theoretical value of the threshold (R_{th}=451 Ω) calculated with the MSF approach [9].

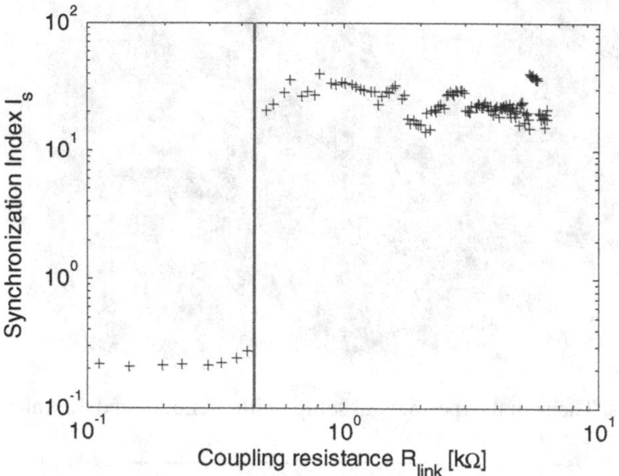

Fig. 3. 16 nodes ring topology: synchronization index vs. coupling resistance

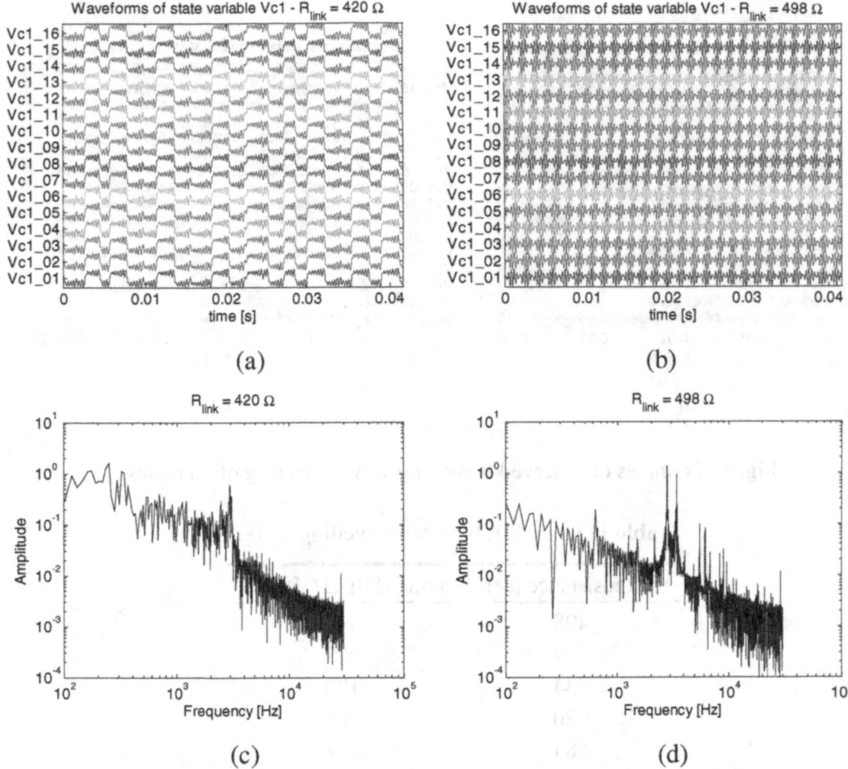

Fig. 4. vc1 waveforms and frequency spectra. Complete synchronization (a, c) and unsynchronized state (b, d).

A sharp transition between synchronized and unsynchronized state occurs at about R_{link}=420 Ω . v_{c1} waveforms significantly change around the transition values, as it can be seen by comparing Fig. 4a (R_{link}=420 Ω) and Fig. 4b (R_{link}=498 Ω). Also the amplitude frequency spectra change since, after the transition, the low frequency content sensibly decreases and two peaks occur at f_1=2.8 kHz and f_2=3.4 kHz. Complete synchronization is lost at R_{link}=498 Ω since the waveforms present a "lag synchronization", with a measured time shift Δt of about 90 μs (Fig. 4b).

As the second point, the characterization of chaotic travelling waves [16] in the chain structure has been carried out. Such dynamical behavior has been observed in the R_{link} range [498÷770] Ω, and examples of travelling waves are shown in Fig. 5a (R_{link}=684 Ω) and in Fig. 5b (R_{link}=770 Ω). It is noticeable as the coupling strength decreases, the time shift Δt between the waveforms in the ring sensibly increases, until the travelling waves disappear after the link among the nodes becomes too weak. Table I reports the measured time shifts for different values of the link resistance R_{link}.

As the coupling resistance increases, oscillatory patterns are observed, since node's dynamics are grouped in spatial structures, where the oscillation amplitude changes with well defined spatial periodicity. An example of this behavior is given in Fig. 6, where half of the ring (nodes 1 to 8) is behaving in the same way as the opposite half (nodes 9-16).

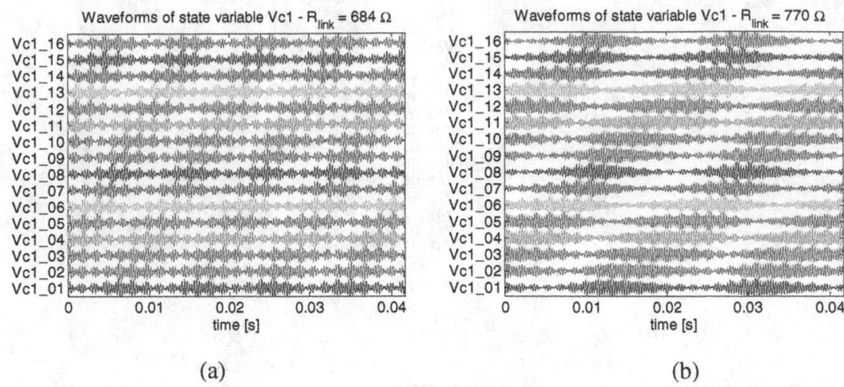

Fig. 5. Examples of observed travelling waves in a ring of 16 nodes

Table 1. Time shift between travelling waves

Link resistance [Ω]	Time shift Δt [μs]
498	80
534	85
584	100
620	700
684	1000
720	1200
770	2000

A usual way of visualizing spatial structures make use of Poincaré sections where the value of the state variable v_{c1} of each oscillator is reported while the correspondent variable v_{c2} crosses the zero level. For different values of the coupling parameter, different patterns are obtained. In Fig 6.b an example is given for R_{link}=1592 Ω. The patterns show the spatial periodicity of the waveforms, as confirmed by the dynamics exhibited by v_{c1} in each of Chua's circuits (Fig. 6a): the waveforms are replicated after 8 nodes of the ring. Such a behavior can be deduced also by phase plots (v_{c1}-v_{c2}) represented in Fig. 6c. The different colors used in the plots help the reader to verify the spatio-periodic structure if compared to the position of each node in the ring (Figure 6d).

At link resistance greater than 3000 Ω, all the waveforms become apparently uncorrelated, although settled on different dynamic behavior as they were isolated.

4 Conclusions

The accurate experimental characterization of an array of chaotic oscillators with diffusive dynamic links has been carried out, exploiting flexibility and automation level of the realized set-up which easily allow to afford complete scan of dynamics as function of parameters. Canonical phenomena such as complete and lag-synchronization, travelling waves, clustering, appearance of patterns, etc. have been evidenced and reported. The observed richness and complexity of dynamical behavior was even greater than what previously forecast by accurate simulations, thus providing a further push towards the experimental implementation of complex networks. Future work will be in the direction of investigating the network with increasing number of nodes and considering more complex interconnecting structure, as partly already available in the present set-up.

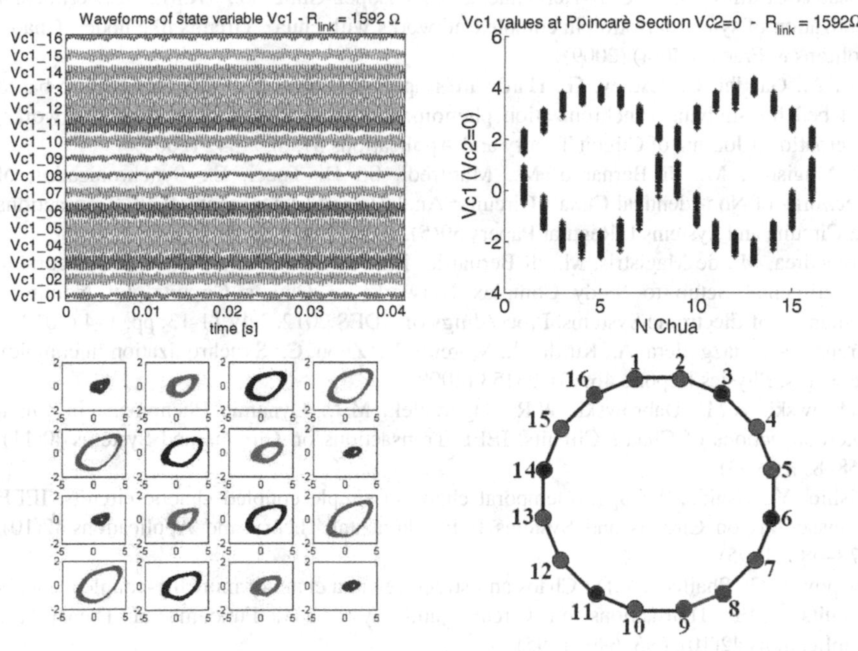

Fig. 6. Example of spatio-periodic structures (R_{link}=1592 Ω). Poincarè section (a), v_{c1} waveforms (b), phase plots (c), spatial structure (d)

Acknowledgments. The authors wish to acknowledge the valuable work of Dr. M. Colandrea for design and testing of electronic boards and Dr. M. Attanasio who supervised the activities at the Circuits Laboratory of the Electrical Engineering and Information Technology Department. This work was partially funded by the research program F.A.R.O. of the Science and Technology School of the University of Naples Federico II.

References

1. Strogatz, S.H.: Exploring complex networks. Nature 410, 268–276 (2001)
2. Boccaletti, S., Latora, V., Moreno, Y., Chavez, M., Hwang, D.-U.: Complex networks: Structure and dynamics. Physics Reports 424(4-5), 175–308 (2006)
3. Corinto, F., Biey, M., Gilli, M.: Non-linear coupled CNN models for multiscale image analysis. International Journal of Circuit Theory and Applications 34, 77–88 (2006)
4. Ogorzalek, M.J., Galias, Z., Dabrowski, A.M., Dabrowski, W.R.: Chaotic waves and spatio-temporal patterns in large arrays of doubly-coupled Chua's circuits. IEEE Transactions on Circuits and Systems I: Fundamental Theory and Applications 42(10), 706–714 (1995)
5. Posadas-Castillo, C., Cruz-Hernández, C., López-Gutiérrez, R.M.: Experimental realization of synchronization in complex networks with chua's circuits like nodes. Chaos, Solitons & Fractals 40(4) (2009)
6. Tar, A., Gandhi, G., Cserey, G.: Hardware implementation of a CNN architecture-based test bed for studying synchronization phenomenon in oscillatory and chaotic networks. International Journal of Circuit Theory and Applications 37, 529–542 (2009)
7. de Magistris, M., di Bernardo, M., Manfredi, S., Di Tucci, E.: Synchronization of Networks of Non-Identical Chua's Circuits: Analysis and Experiments. IEEE Transactions on Circuits and Systems I: Regular Papers 59(5), 1029–1041 (2012)
8. Colandrea, M., de Magistris, M., di Bernardo, M., Manfredi, S.: A Fully Reconfigurable Experimental Setup to Study Complex Networks of Chua's Circuits. In: Nonlinear Dynamics of Electronic Systems, Proceedings of NDES 2012, July 11-13, pp. 1–4 (2012)
9. Arenas, A., Diazguilera, A., Kurths, J., Moreno, Y., Zhou, C.: Synchronization in complex networks. Physics Reports 469(3), 93153 (2008)
10. Dabrowski, A.M., Dabrowski, W.R., Ogorzalek, M.J.: Dynamic Phenomena in Chain Interconnections of Chua's Circuits. IEEE Transactions on Circuits and Systems 40(11), 868–871 (1993)
11. Nishio, Y., Ushida, A.: Spatio-temporal chaos in simple coupled chaotic circuits. IEEE Transactions on Circuits and Systems I: Fundamental Theory and Applications 42(10), 678–686 (1995)
12. Osipov, V.G., Shalfeev, V.D.: Chaos and structures in a chain of mutually-coupled Chua's circuits. IEEE Transactions on Circuits and Systems I: Fundamental Theory and Applications 42(10), 693–699 (1995)
13. Pecora, L.M., Carroll, L.: Master Stability Functions for Synchronized Coupled Systems. Phys. Rev. Letters 80(10) (1998)
14. Sun, J., Bollt, E.M., Nishikawa, T.: Master stability functions for coupled nearly identical dynamical systems. Europhysics Letters 85 (2009)
15. de Magistris, M., di Bernardo, M., Petrarca, M.: Experiments on synchronization in networks of nonlinear oscillators with dynamic links. IEICE, Nonlinear Theory and its Applications 4(4), 462–472 (2013)
16. Shabunin, A., Astakhov, V., Anishchenko, V.: Developing chaos on base of traveling vawes in chain of coupled oscillators with period doubling synchronization and hierarchy of multistability formation. International Journal of Bifurcations and Chaos 12(8) (2002)

Basin Stability in Complex Oscillator Networks

Peng Ji[1,2] and Jürgen Kurths[1,2,3]

[1]Potsdam Institute for Climate Impact Research (PIK), 14473 Potsdam, Germany
[2]Department of Physics, Humboldt University, 12489 Berlin, Germany
[3]Institute for Complex Systems and Mathematical Biology, University of Aberdeen, Aberdeen AB24 3UE, United Kingdom

Abstract. How to quantify power system stability as the ability to regain an equilibrium state after being subjected to perturbations is of crucial interest. We employ basin stability, a nonlinear concept based on the volume of the basin of attraction, to investigate how stable one single node or the whole power grid is against large perturbations. Specifically, we start by illustrating basin stability using the simple mode with one generator and one load, and then extend it to IEEE Reliability Test System 30. A first-order transition in the plot of basin stability versus coupling strengths is observed in not only the simple model but also the whole network.

Keywords: Basins of attraction, Basin stability, Kuramoto model, Complex networks.

1 Introduction

The stability of an interconnected power system is the ability to regain an equilibrium state after being subjected to a disturbance [1, 2]. Often the stability of a particular node [3–5] or a group of nodes is also of interest. When suffered to a disturbance, the stability depends not only on the the initial operating condition but also on the nature of the disturbance [1, 2]. Local stability can be characterized in terms of Lyapunov exponents [6, 7]. When subjected a large disturbance, a nonlinear and nonlocal basin stability was proposed by Menck et al. [4] in order to quantify how stable a node or a group of nodes is.

In transmission grids, a mathematical derivation of the swing equation is simplified to the Kuramoto-like model in case of stability analysis [8–12] The simplified model was proposed by assuming that all generator properties, voltage magnitudes, line reactances to be all the same and voltage lossless. For the load buses with frequency dependent, the Kuramoto model was used [10]. The two models are known as the structure-preserving power network model [10].

In this paper, we seek to quantify the stability for each node in a grid using basin stability. More specifically, initially the whole network is in an equilibrium state, we perturb the node i and count the percentage of regaining the stationary state as basin stability BS_i. In particular, as an illustration, one generator and one load are used to deepen the understanding of basin stability in terms of the

V.M. Mladenov and P.C. Ivanov (Eds.): NDES 2014, CCIS 438, pp. 211–218, 2014.

basin of attraction and then we employ basin stability to investigate the stability of IEEE Reliability Test System 30 (RTS 30). RTS was developed to test methods for reliability analysis for power systems and RTS 30 represents a portion of the American Electric Power System [13]. A node with a poor basin stability tends to destroy the synchrony of the whole system even with a rather small perturbation. A first-order transition is observed in the plot of dependency of basin stability on coupling strengths not only in the case of two nodes but also in the network. Also of importance is the fact that the first-order phase transition between the strength of coupling and the network structure was recenltly observed also in netwroks of interactions between physiological systems [14].

2 Model and Basin Stability

A power grid consists of generators and loads linked to buses which are interconnected by transmission lines. If we take all lines to be lossless, then the dynamics of the generator at node i is modeled by the swing equation [3, 15]

$$M_i\ddot{\theta}_i + \alpha_i\dot{\theta}_i + \sum_{j\in C_i} V_i V_j b_{ij} \sin(\theta_i - \theta_j) = P_i, \tag{1}$$

where M_i is the inertia constant of the ith generator, α_i the damping constant, θ_i represents the rotational phase angle, $\dot{\theta}_i$ the time derivative of θ_i, $j \in C_i$ refers to the generators or loads connected to the generator i, V_i is a bus voltage, b_{ij} denotes the admittance of the line joining nodes i and j and P_i labels the net power input.

For simplicity, we assume all generator properties, line voltages and admittance to be the same and M_i to be unity for all i. Therefore α is the same at each node and $K \equiv V_i V_i b_{ij}$ is the same for all transmission lines. Thus a simple version of the dynamics of the generator is given by [3, 10, 15]

$$\ddot{\theta}_i = -\alpha\dot{\theta}_i + P_i + K\sum_{j=1}^{N} A_{ij}\sin(\theta_j - \theta_i). \tag{2}$$

The matrix A_{ij} reflects the network topology, with $A_{ij} = 1$ if nodes i and j are connected, and $A_{ij} = 0$ otherwise. $\sum_{j=1}^{N} a_{ij}\sin(\theta_j - \theta_i)$ is the restoring torque acting on the node i [10]. Here j could be either a generator or a load. The dynamics of the ith load bus is given by [10]

$$\dot{\theta}_i = P_i + K\sum_{j=1}^{N} A_{ij}\sin(\theta_j - \theta_i). \tag{3}$$

where θ_i represents the voltage phase angle of the ith load bus. P_i depicts the power consumption at the load i. In the network, a synchronous state is characterized by [10]

$$\dot{\theta}_1 = \dot{\theta}_2 = \cdots = \dot{\theta}_N. \tag{4}$$

In order to quantify how stable a synchronous state is even against large perturbations, basin stability was proposed by Menck et al. based on the volume of the basin of attraction B of fixed points as the nonlinear and non-local measure of a likehood of return to the synchronous state after suffering a large perturbation [3–5]. Specifically, basin stability BS_i at node i is defined as follows

$$BS_i = \int \chi_B(\theta_0, \dot{\theta}_0)\rho(\theta_0, \dot{\theta}_0)d\theta_0 d\dot{\theta}_0, \tag{5}$$

where $\rho(\theta_0, \dot{\theta}_0)$ denotes the distribution of perturbations θ_0 and $\dot{\theta}_0$, and $\chi_B(\theta_0, \dot{\theta}_0) = 1$ if $(\theta_0, \dot{\theta}_0)$ belong to B and $\chi_B(\theta_0, \dot{\theta}_0) = 0$ otherwise. The value of $BS_i \in [0, 1]$ quantifies how stable the node i is. If $BS_i = 0$, a rather small non-zero perturbation could desynchronize the whole system. Whereas if $BS_i = 1$, the whole system retains at the synchronous states even if the node i suffer a large perturbation. If $BS_i > BS_j$, we say the node i is more stable than the node j. In this paper, we select initial values randomly.

3 An Illustration of Basin Stability of a Simple Model

To illustrate the concept of basin stability, we first study a simple example: one generator $(\theta_g, \dot{\theta}_g)$ connected to one load θ_c. The dynamics of the two nodes are modeled by Eqs. (2) and (3) respectively. The energy generation $P_g > 0$ should be met by the consumption $P_c < 0$, and thus $P_g = -P_c = P$. The system has synchronized states at $\ddot{\theta}_g = \dot{\theta}_g = \dot{\theta}_c = 0$ and the phase difference $\theta_c - \theta_g = \arcsin(P/K)$ for $P < K$. As shown in fig. 1, the basin of attraction of the stable fixed point of the generator is colored in green and the white color indicates the basin of attraction of the stable limit cycle. The stable fixed point and the stable limit cycle coexist in fig. 1(a) and (b). The system will either converge to the stable fixed point or rotate periodically depending crucially on initial conditions of θ and $\dot{\theta}$. Fig. 1(d) shows the dependence of Basin stability BS on the coupling strengths K. For $K < 1$, only the stable limit cycle exists and $BS = 0$. The onset of a non-zero BS is located at $K = P$. BS increases smoothly from 0 to about 0.6, and then suddenly jumps to 1. To deepen an understanding of the dependence of basin stability on the parameter values α and P, we investigate the bifurcation diagram in the $P - \alpha$ parameter space. In the stability diagram of one pendulum [16], there are three types of bifurcations: homoclinic and infinite-period bifurcations periodic orbits, and a saddle-node bifurcation of fixed points. Therefore, the diagram is separated into three different areas corresponding to the stable fixed point, the stable limit cycle and bistability by these bifurcation lines [12, 16, 17].

As shown in fig. 2, there are also three different areas indicated according to BS. The deep red color denotes the area of the stable limit cycle with $BS = 0$. In the white area, the system will converge to the stable fixed point with $BS = 1$. Otherwise, the stationary solutions of the system depend crucially on the initial conditions.

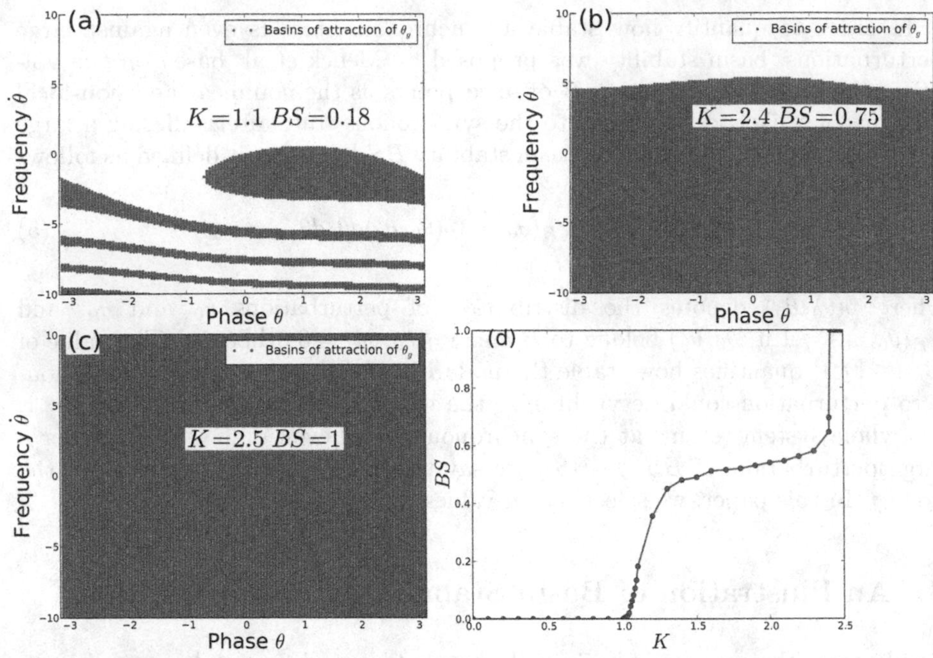

Fig. 1. (Color online) Basin stability of the generator. The green color indicates the basin of attraction of the fixed point for $K = 1.1$ (a), 2.4 (b) and 2.5 (c) at $\alpha = 0.1$ and $P_g = -P_c = P = 1$. Initially we set $\theta_c = 0$. (d) Basin stability against K at $\alpha = 0.1$ and $P = 1$. Here BS is calculated based on randomly selecting 500 initial values of θ_g, $\dot\theta_g$ and θ_c from $[-\pi, \pi] \times [10, 10] \times [-\pi, \pi]$.

4 Basin Stability in Complex Oscillator Networks

In the previous section, we have illustrated the basin stability using only two oscillators. What happens if the network have more oscillators? For this case we calculate the N single-node basin stabilities BS_i, $i \in 1, \cdots, N$ as follows. We find the synchronized states and create T initial value vectors with θ_i^0 and $\dot\theta_i^0$ drawn randomly, integrate the system, count the number R_i of initial conditions returning back to the synchronized states and estimate the basin stability $BS_i = R_i/T$, where $BS_i \in [0, 1]$. If $BS_i > BS_j$, we say node i is more stable than node j.

Next, we calculate the stability of the IEEE Reliability Test System 30 (RTS 30) illustrated in fig. 3. This grid has $N = 30$ nodes and 41 edges. Each node is randomly selected to be a generator or a load. Squares (resp. circles) denote loads with $P_i = -1$ (resp. generators with $P_i = +1$). With a high coupling strength, it is easy to find stationary synchronized states. Therefore, for the coupling strength $K = 4$, we find the synchronized solution and then vary the initial values of θ_i and $\dot\theta_i$ from $[-\pi, \pi] \times [-10, 10]$ at each node i. Once the synchronized solution is found, we increase/decrease the coupling strength and

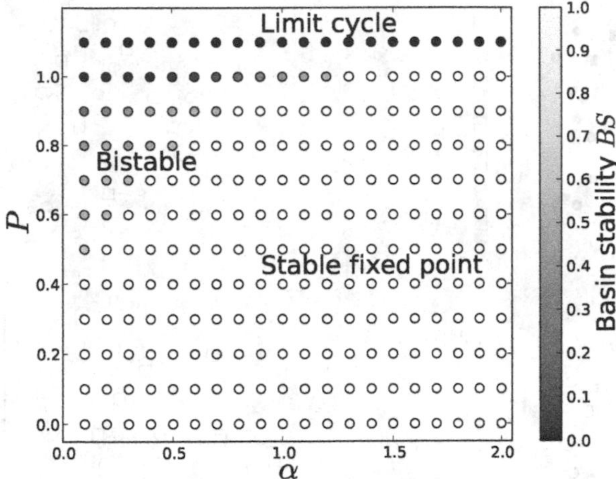

Fig. 2. (Color online) The bifurcation diagram over the $P - \alpha$ parameter space of the mode with one generator and one load for fixed the coupling strength $K = 1$. Color indicates the basin stability BS. We randomly draw 500 initial value vector $(\theta_g, \dot{\theta}_g, \theta_c)$ from $[-\pi, \pi] \times [-10, 10] \times [-\pi, \pi]$, integrate the system and calculate the percentage of initial conditions retaining synchronized states as BS

then check how the basin stability of each node change. In fig. 3, each node is colored according to its basin stability. From 3(a) to (d), basin stability BS_i increases with the coupling strength K. For large coupling strengths, in general, generators are less stable than loads and the results have no obvious correlation to the network's topology, which is opposite to the results of [5].

In order to evaluate the whole stability, we calculate the basin stability BS of the whole network as the average value of basin stability BS_i of each node

$$BS = \frac{1}{N} \sum_{i=1}^{N} BS_i. \tag{6}$$

In fig. 4, we observe a first-order transition at the onset of the basin stability (see from fig. 3(a) to (b)) and BS increases monotonically with increasing coupling strengths K as nodes (see the nodes with yellow color in fig. 3(b), (c) and (d)) becomes more stable. It is very interesting to see the different observations by comparing this plot with fig. 1 and [3, 5] ,in which BS becomes unity after the first-order transition.

5 Conclusions

We have introduced the Kuramoto model with and without inertia for a generator and a load respectively, and have illustrated the concept of basin stability

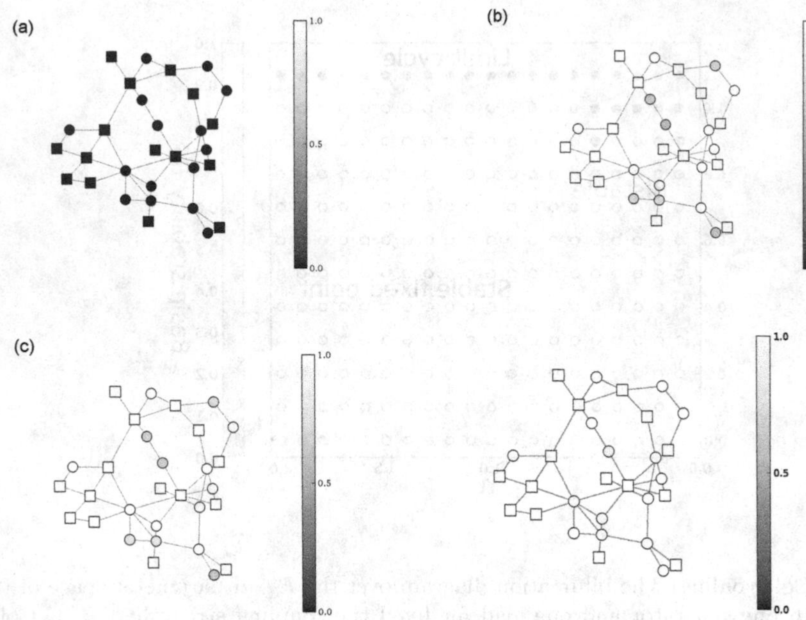

Fig. 3. (Color online) The dependence of BS_i on each node of RTS 30 and coupling strength with $K = 3.0$ (a), 3.5(b), 4(c) and 10(d)

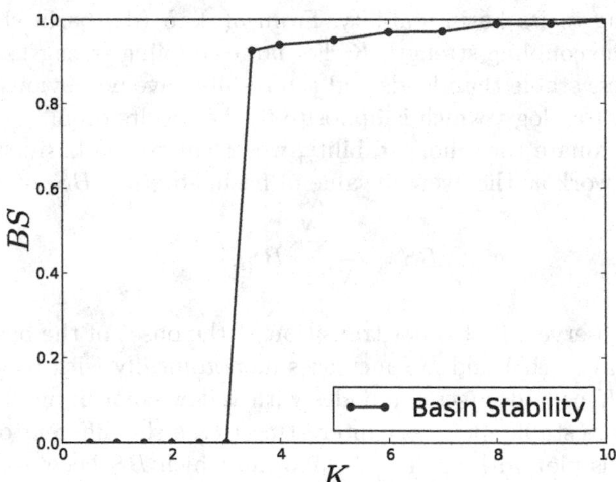

Fig. 4. (Color online) BS of the whole network against coupling strengths K

in terms of the basin of attraction of a simple model with one generator and one load. Using basin stability, we have determined the bifurcation diagram in the $P - \alpha$ (net energy input vs. damping constant) parameter space and have found that there are three different regions namely a stable limit cycle, bistability and a stable fixed point. The basin stability of each node against perturbations has been employed for the IEEE Reliability Test System 30 (RTS 30) in which generators have poor values of basin stability compared to loads. A first-order transition has been observed in the plot of basin stability of the synchronous state versus the coupling strength in not only the simplified model but also RTS 30.

Acknowledgments. P. Ji is funded by China Scholarschip Council (CSC) scholarship. J. Kurths would like to acknowledge IRTG 1740 (DFG and FAPESP) for the sponsorship provided.

References

1. Belykh, I.V., Belykh, V.N., Hasler, M.: Blinking model and synchronization in small-world networks with a time-varying coupling. Physica D 195, 188–206 (2004)
2. Machowski, J., Bialek, J., Bumby, J.: Power system dynamics: Stability and control. John Wiley & Sons (2011)
3. Menck, P.J., Kurths, J.: Topological identification of weak points in power grids. In: Nonlinear Dynamics of Electronic Systems, Proceedings of NDES 2012, pp. 1–4. VDE (2012)
4. Menck, P., Heitzig, J., Marwan, N., Kurths, J.: How basin stability complements the linear-stability paradigm. Nat. Phys. 9, 89 (2013)
5. Menck, P.J., et al: (submitted)
6. Pikovsky, A., Rosenblum, M., Kurths, J.: Synchronization: A universal concept in nonlinear sciences, vol. 12. Cambridge University Press (2003)
7. Arenas, A., Díaz-Guilera, A., Kurths, J., Moreno, Y., Zhou, C.: Synchronization in complex networks. Phys. Rep. 469(3), 93–153 (2008)
8. Filatrella, G., Nielsen, A.H., Pedersen, N.F.: Analysis of a power grid using a kuramoto-like model. The European Physical Journal B-Condensed Matter and Complex Systems 61(4), 485–491 (2008)
9. Rohden, M., Sorge, A., Timme, M., Witthaut, D.: Self-organized synchronization in decentralized power grids. Phys. Rev. Lett. 109, 064101 (2012)
10. Dorfler, F., Chertkov, M., Bullo, F.: Synchronization in complex oscillator networks and smart grids 110(6), 2005–2010 (2013)
11. Witthaut, D., Timme, M.: Braess's paradox in oscillator networks, desynchronization and power outage. New Journal of Physics 14(8), 083036 (2012)
12. Ji, P., Peron, T.K.D., Menck, P.J., Rodrigues, F.A., Kurths, J.: Cluster explosive synchronization in complex networks. Phys. Rev. Lett. 110, 218701 (2013)
13. PM Subcommittee. IEEE reliability test system. IEEE Transactions on Power Apparatus and Systems (6), 2047–2054 (1979)
14. Bashan, A., Bartsch, R.P., Kantelhardt, J.W., Havlin, S., Ivanov, P.C.: Network physiology reveals relations between network topology and physiological function. Nature Communications 3, 702 (2012)

15. Hill, D.J., Chen, G.: Power systems as dynamic networks. In: Proceedings of the 2006 IEEE International Symposium on Circuits and Systems, ISCAS 2006, 4 p. IEEE (2006)
16. Steven, H.S.: Nonlinear Dynamics and Chaos: With Applications to Physics, Biology, Chemistry, and Engineering. Addison-Wesley, Reading (1994)
17. Ji, P., Peron, T.K.D., Rodrigues, F.A., Kurths, J.: Analysis of cluster explosive synchronization in complex networks. ArXiv e-prints (February 2014)

Amplitude Death in Oscillators Network with a Fast Time-Varying Network Topology

Yohiski Sugitani[1,2], Keiji Konishi[1], and Naoyuki Hara[1]

[1] Department of Electrical and Information Systems, Osaka Prefecture University, 1-1 Gakuencho, Naka-ku, Sakai, Osaka 599-8531 Japan
mv104035@edu.osakafu-u.ac.jp
http://www.eis.osakafu-u.ac.jp/~ecs

[2] Research Fellow of Japan Society for the Promotion of Science

Abstract. This study investigates amplitude death in delayed coupled oscillators with a time-varying network topology. The local stability of amplitude death is governed by a linear time-varying system. This system is difficult to analyze; however, if the time-varying period is sufficiently short compared with the oscillation period, the linear time-varying system can be regarded as the linear time-invariant system. Thus, we can easily analyze the local stability of amplitude death in the fast time-varying network. Our analytical results are verified by some numerical simulations.

Keywords: Amplitude Death, Time-varying Network, Fast Switching.

1 Introduction

Various phenomena in coupled oscillators have been investigated in nonlinear science [1]. A stabilization of unstable steady state by a diffusive connection is called amplitude death [2]. It is proven that amplitude death never occurs in a static-coupled identical oscillators [3]. Reddy *et al.* found that a delayed connection can induce amplitude death even in coupled identical oscillators [4]. The delay induced death has been of great interest in nonlinear science [5] and observed in various network topologies: a pair of oscillators [4], an all-to-all network topology [6], a bi-directionally ring network topology [7], and one-way ring network topology [8]. Moreover, Atay showed that the bipartite network topologies are the hardest to induce amplitude death and the all-to-all topology with large number of oscillators is easiest to induce it [9].

These studies focused on amplitude death on the time-invariant network topologies. In real world, on the other hand, there are many systems with time-varying topology, such as mobile *ad hoc* network [10], opinion dynamics [11], epidemic system [12], and organ systems in the human body [13][1]. Thus, many researchers devoted to investigate the dynamics of coupled oscillators with time-varying network topologies [14]. However, they focused only on synchronization

[1] The network topology of organ systems greatly varies with transitions from one physiological state to another.

V.M. Mladenov and P.C. Ivanov (Eds.): NDES 2014, CCIS 438, pp. 219–226, 2014.

in coupled oscillators; unfortunately, there is a great lack of articles devoted to amplitude death in the time-varying network.

This study investigates amplitude death in delayed coupled oscillators with a fast time-varying network topology. The topology periodically changes with much short period. We show that, in such situation, the local stability of amplitude death can be considered as that of a linear time-invariant system. Thus, we can easily analyze the local stability of amplitude death in the fast time-varying network as same as that in a time-invariant network. Our analytical results are verified by some numerical simulations.

2 Coupled Oscillators with Time-varying Network Topology

Let us consider N coupled Landau-Stuart oscillators,

$$\dot{Z}_j(t) = \{\mu + i\omega - |Z_j(t)|^2\}Z_j(t) + U_j(t), \quad (j = 1, \ldots, N), \tag{1}$$

where $Z_j(t) \in \mathbb{C}$ is the state variable of oscillator j. The parameters $\mu > 0$ and $\omega > 0$ represent the instability of a fixed point and the natural frequency of oscillator, respectively. The coupling signal $U_j(t)$ is given by

$$U_j(t) = k\left[\sum_{l=1}^{N}\{\varepsilon_{jl}(t)Z_l(t-\tau)\} - d_j(t)Z_j(t)\right], \tag{2}$$

where k is the coupling strength and $Z_l(t-\tau)$ is a delayed signal. $\varepsilon_{jl}(t)$ governs the network topology as follows: if oscillator j is connected to oscillator l at time t, then $\varepsilon_{jl}(t) = \varepsilon_{lj}(t) = 1$, otherwise $\varepsilon_{jl}(t) = \varepsilon_{lj}(t) = 0$. The degree of oscillator j at time t is given by $d_j(t) := \Sigma_{l=1}^{N}\varepsilon_{jl}(t)$.

The network topology changes under the following rules: (a) The network topology changes at intervals of a period $\Delta t > 0$; (b) The network topology is restricted by a constraint matrix $\boldsymbol{H}$ as follows: if oscillator j is allowed to be connected with oscillator l, then $\{\boldsymbol{H}\}_{jl} = \{\boldsymbol{H}\}_{lj} = 1$, otherwise $\{\boldsymbol{H}\}_{jl} = \{\boldsymbol{H}\}_{lj} = 0$; (c) Under the constraint matrix $\boldsymbol{H}$, two oscillators are connected with a probability p. Figure 1 illustrates a concept of time-varying network topology. Assume that all the row-sum of the constraint matrix $\boldsymbol{H}$ are equivalent:

$$\sum_{l=1}^{N}\{\boldsymbol{H}\}_{jl} = D, \ \ \forall j \in \{1, \ldots, N\}, \tag{3}$$

where $D \in \mathbb{N}$ is a constant integer.

2.1 Stability Analysis

Oscillators (1) with connection (2) have the homogeneous steady state,

$$[Z_1^* \cdots Z_N^*]^{\mathrm{T}} = [0 \cdots 0]^{\mathrm{T}}. \tag{4}$$

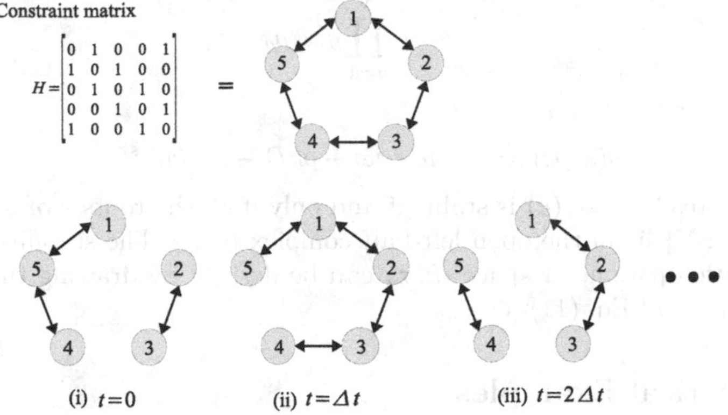

Fig. 1. Concept of a time-varying network topology

Coupled oscillators (1), (2) can be linearized around the steady state (4):

$$\dot{\boldsymbol{X}}(t) = (\mu + i\omega)\boldsymbol{X}(t) + k\left[\boldsymbol{E}(t)\boldsymbol{X}(t-\tau) - \mathrm{diag}\left\{d_1(t), \ldots, d_N(t)\right\}\boldsymbol{X}(t)\right], \quad (5)$$

where $\boldsymbol{X}(t) := [z_1(t) \cdots z_N(t)]^{\mathrm{T}}$ and $z_j(t) := Z_j(t) - Z_j^*$. The network topology at time t is governed by matrix $\boldsymbol{E}(t)$, whose elements are $\{\boldsymbol{E}(t)\}_{jl} = \varepsilon_{jl}(t)$ $(l \neq j)$ and $\{\boldsymbol{E}(t)\}_{jj} = 0$. Since linearized system (5) contains the time-varying parameters $\boldsymbol{E}(t)$ and $d_j(t)$, it is difficult to analyze its stability. However, they can be averaged by a period T_p, which is sufficiently long compared with Δt, as follows:

$$\frac{1}{T_p}\int_t^{t+T_p} \boldsymbol{E}(r)dr = p\boldsymbol{H}, \quad (6)$$

$$\frac{1}{T_p}\int_t^{t+T_p} \mathrm{diag}(d_1(r), \ldots, d_N(r))dr = pD\boldsymbol{I}_N, \quad (7)$$

for any t. Under the hypothesis of the fast time-varying topology [15, 16], there exists a period Δt^* such that, for $\Delta t^* > \Delta t$, linearized system (5) can be regarded as the following time-invariant system:

$$\dot{\boldsymbol{X}}(t) = (\mu + i\omega)\boldsymbol{X}(t) + pk\left\{\boldsymbol{H}\boldsymbol{X}(t-\tau) - D\boldsymbol{X}(t)\right\}. \quad (8)$$

It can be seen that Eq. (8) describes the local dynamics of coupled oscillators around the steady state with the connection matrix $\boldsymbol{H}$ and the coupling strength pk. The characteristic quasi-polynomial of system (8) is given by

$$G(s) = \det\left[(s - \mu - i\omega + pkD)\,\boldsymbol{I}_N - pke^{-s\tau}\boldsymbol{H}\right]. \quad (9)$$

The matrix $\boldsymbol{H}$ can be diagonalized into $\boldsymbol{T}^{-1}\boldsymbol{H}\boldsymbol{T} = \mathrm{diag}(\rho_1, \rho_2, \cdots, \rho_{\mathrm{N}})$ by a transformation matrix $\boldsymbol{T}$, where $\rho_1, \ldots, \rho_N$ are the eigenvalues of $\boldsymbol{H}$. Thus, the characteristic quasi-polynomial (9) can be rewritten by

$$G(s) = \prod_{q=1}^{N} g(s, \rho_q), \tag{10}$$

where

$$g(s, \rho_q) := s - \mu - i\omega + pkD - \rho_q pke^{-s\tau}. \tag{11}$$

Hence, the steady state (4) is stable if and only if all the roots s of $g(s, \rho_q) = 0$, $\forall q \in \{1, \ldots, N\}$ lie in the open left-half complex plane. The stability region on the connection parameter space (k, τ) can be derived by drawing the marginal stability curves of Eq. (11) [6].

3 Numerical Examples

This section provides some numerical examples for a pair of oscillators ($N = 2$) and oscillators network ($N > 2$). We will verify that our analytical results agree with the numerical results under the fast time-varying topology (i.e., the time-varying period Δt is sufficiently short compared with the oscillation period $T := 2\pi/\omega$). The parameters of oscillators are set to $\mu = 0.5$ and $\omega = \pi$ (i.e., $T = 2$) throughout this study.

3.1 A Pair of Oscillators

For $N = 2$, the constraint matrix $\boldsymbol{H}$ is set to

$$\boldsymbol{H} = \begin{bmatrix} 0 & 1 \\ 1 & 0 \end{bmatrix}. \tag{12}$$

The stability regions with the marginal stability curves of Eq. (11) for probabilities $p = 1/2$ and $p = 1$ are shown in Figs. 2(a) and 2(b), respectively. With increasing k, when the connection parameter set (k, τ) crosses the thin (bold) line, roots s of $g(s, \rho) = 0$ cross the imaginary axis from left to right (right to left). The shaded area shows the stability region where all the roots s of $g(s, \rho_q) = 0$, $\forall q \in \{1, \ldots, N\}$ lie in the open left-half complex plane. Comparing Fig. 2(a) with Fig. 2(b), the stability region of Fig. 2(a) extends along the direction of k. This arises from the following reasons: as stated in the previous section, Eq. (8) describes the local dynamics of coupled oscillators around the steady state with the coupling strength pk; thus, $p = 1/2$ and $p = 1$ represent the coupling strength $k/2$ and k, respectively.

Figures 3(a) and 3(b) show the time-series data of $\mathrm{Re}[Z_j(t)]$ at point A $(k, \tau) = (6.00, 1.50)$ and B $(k, \tau) = (6.00, 0.45)$ in Fig. 2(a), respectively. The two oscillators are independent (i.e., $\varepsilon_{12}(t) = \varepsilon_{21}(t) = 0$) until $t = 10$. At $t = 10$, they are connected with a short period $\Delta t = 0.0067 \ll T$. At point A (see Fig. 3(a)), $\mathrm{Re}[Z_j(t)]$ oscillates; in contrast, at point B (see Fig. 3(b)), $\mathrm{Re}[Z_j(t)]$ converges onto the fixed point.

Let us focus on point B in Fig. 2(a). At any frozen time t, the two oscillators have one of the following states: (i) they are not connected; (ii) they are

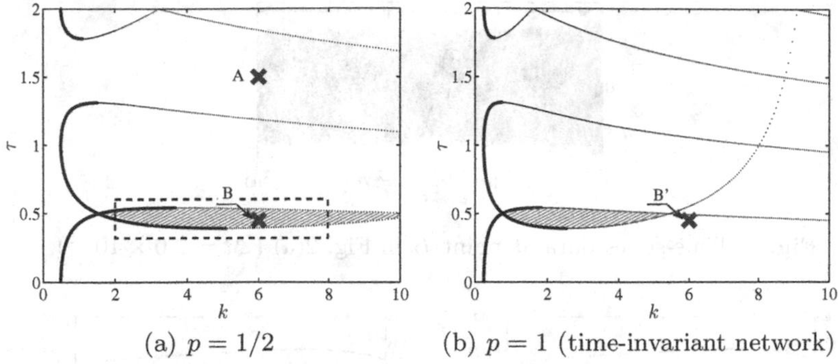

(a) $p = 1/2$ (b) $p = 1$ (time-invariant network)

Fig. 2. Stability regions for a pair of oscillators ($N = 2$)

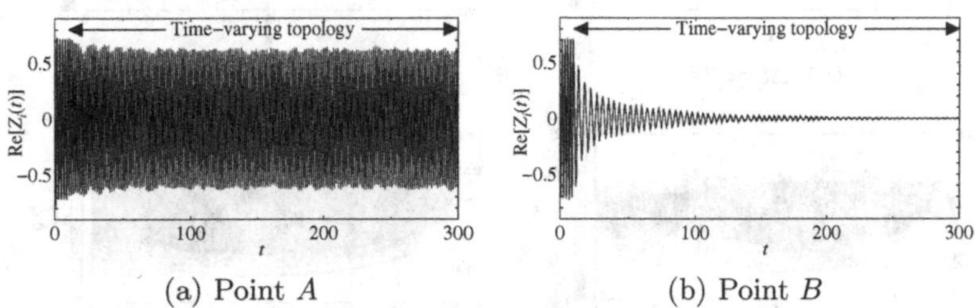

(a) Point A (b) Point B

Fig. 3. Time-series data at points A and B in Fig. 2(a) ($\Delta t = 6.7 \times 10^{-3}$)

connected. In state (i), the characteristic root has the positive real value (i.e., $s = \mu$). In state (ii), the conjugate pair of roots s lie in the open right-half complex plane (see Point B' in Fig. 2(b)). Thus, at any frozen time t, at least one root s lie in the open right-half complex plane at point B. Interestingly, the steady state becomes stable by switching states (i) and (ii) (see Fig. 3(b)).

Figure 4 shows the time-series data at point B in Fig. 2(a) for a long period $\Delta t = 0.2$. The oscillators do not converge onto the fixed point even though the connection parameter set (k, τ) are same as Fig. 3(b). Thus, we can see that our analytical results are valid when the period Δt is much shorter than T (i.e., the fast time-varying topology).

The influence of Δt on the stability of the steady state will be numerically investigated. The stability is evaluated by the error from the fixed point,

$$\delta := \frac{1}{N}\left\langle \sum_{j=1}^{N} |Z_j(t) - Z_j^*| \right\rangle, \tag{13}$$

where $< \cdot >$ represents the average over time interval $t \in [380, 400]$. We estimate the average error δ_{ave} by repeating the simulation 10 times at each connection parameter set (k, τ). The error δ_{ave} with $p = 1/2$ for the connection parameter space (rectangular bounded by dashed line in Fig. 2(a)) is shown in Fig. 5. The periods $\Delta t = 2.0, 2.0 \times 10^{-1}, 2.0 \times 10^{-2}$, and 6.7×10^{-3} are employed. The black

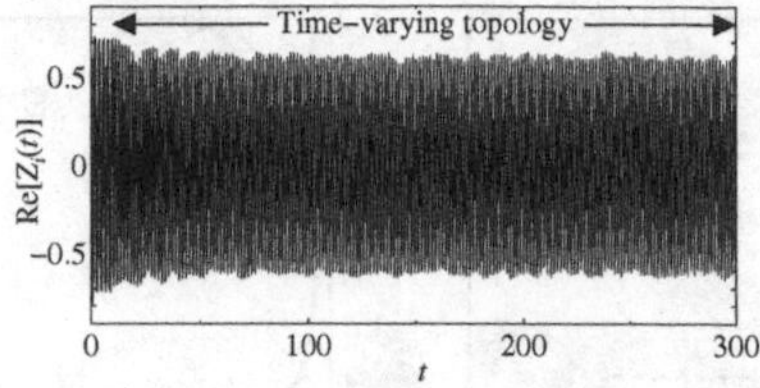

Fig. 4. Time-series data at point B in Fig. 2(a) ($\Delta t = 2.0 \times 10^{-1}$)

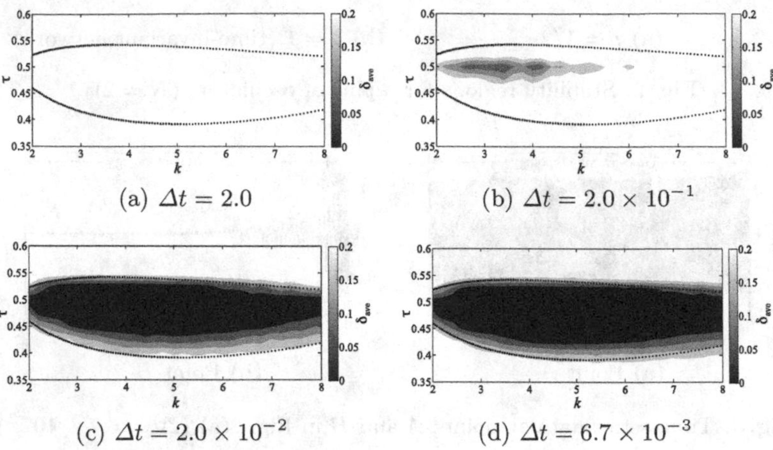

Fig. 5. Average error δ_{ave} in the connection space. (rectangular bounded dashed line in Fig. 2(a)).

region suggests that the average error δ_{ave} is small. A decrease in Δt enlarges the black region. For $\Delta t = 6.7 \times 10^{-3}$, the stability region almost agrees with the black region. In other words, our analytical results agree with the numerical results.

3.2 Oscillators Network

Let us consider the case of $N = 6$. There are some constraint matrices $\boldsymbol{H}$ which satisfy Eq. (3). Figure 6(a) illustrates examples of matrix $\boldsymbol{H}$ and its network topology. Here, we consider a ring topology with 4-nearest neighbors ($D = 4$) in Fig. 6(a) and draw the stability region for $p = 1/2$ as shown in Fig. 6(b). We estimate the average error δ_{ave} in the same way as the previous subsection. The estimated error δ_{ave} in the connection parameter space (rectangular bounded by dashed line in Fig. 6(b)) is shown in Fig. 7. The periods $\Delta t = 2.0$ and 2.0×10^{-1} are employed. For $\Delta t = 2.0$, the black region includes the stability region bounded by the white dashed line. On the other hand, for $\Delta t = 2.0 \times 10^{-1}$, the black region is almost equivalent to the stability region: the analytical result agrees with the numerical result. Comparing Fig. 5 with Fig. 7, we see that Δt^* would depend on the N and $\boldsymbol{H}$.

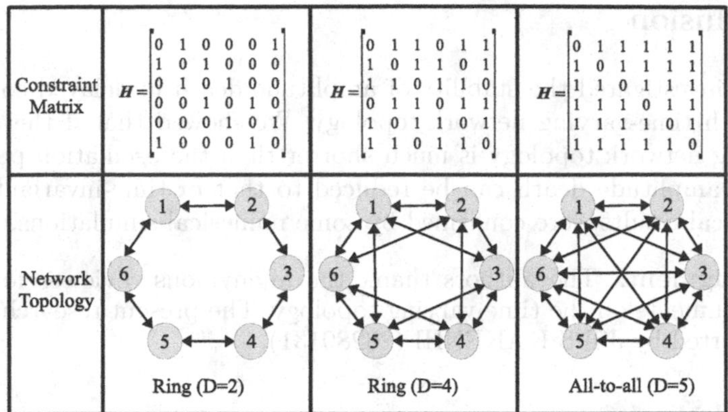

(a) examples of constraint network

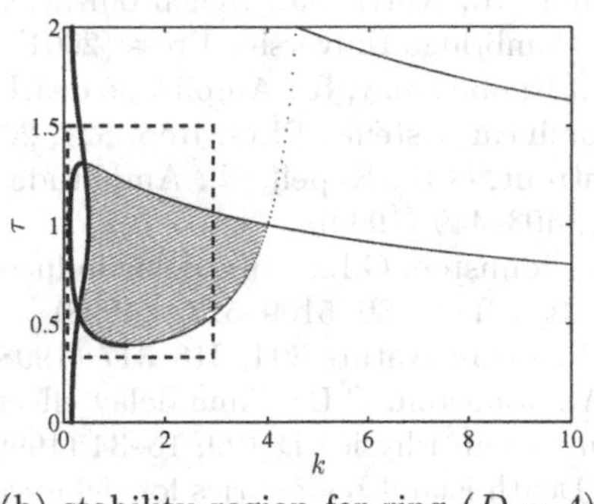

(b) stability region for ring ($D = 4$) with $p = 1/2$

Fig. 6. Oscillators network with time-varying topology ($N = 6$)

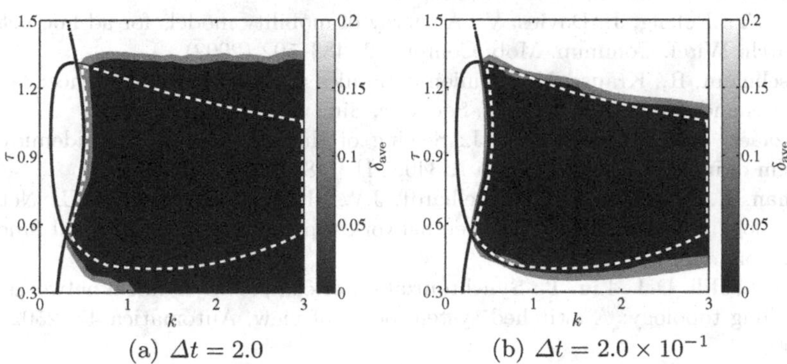

(a) $\Delta t = 2.0$ (b) $\Delta t = 2.0 \times 10^{-1}$

Fig. 7. Average error δ_{ave} in the connection space bounded dashed line in Fig. 6(b). The domain bounded by the white dashed line is the analytical stability region.

4 Conclusion

This study investigated the stability of amplitude death in delayed coupled oscillators with time-varying network topology. We showed that if the period of time-varying network topology is much shorter than the oscillation period, the stability of amplitude death can be reduced to that of time-invariant system. Our analytical results were confirmed by some numerical simulations.

Acknowledgment. The authors thank the anonymous reviewer for his/her valuable comments on the time-varying topology. The present research was partially supported by JSPS KAKENHI (26289131).

References

1. Pikovsky, A., Rosenblum, M., Kurths, J.: Synchronization: A Universal Concept in Nonlinear Sciences. Cambridge University Press (2001)
2. Saxena, G., Prasad, A., Ramaswamy, R.: Amplitude death: The emergence of stationarity in coupled nonlinear systems. Phys. Rep. 521, 205–228 (2012)
3. Aronson, D.G., Ermentrout, G.B., Kopell, N.: Amplitude response of coupled oscillators. Physica D 41, 403–449 (1990)
4. Reddy, D.V.R., Sen, A., Johnston, G.L.: Time delay induced death in coupled limit cycle oscillators. Phys. Rev. Lett. 80, 5109–5112 (1998)
5. Strogatz, S.H.: Death by delay. Nature 394, 316–317 (1998)
6. Reddy, D.V.R., Sen, A., Johnston, G.L.: Time delay effect on coupled limit cycle oscillators at Hopf bifurcation. Physica D 129, 15–34 (1999)
7. Mehta, M.P., Sen, A.: Death island boundaries for delay-coupled oscillator chains. Phys. Lett. A 355, 202–206 (2006)
8. Konishi, K.: Amplitude death in oscillators coupled by a one-way ring time-delay connection. Phys. Rev. E 70, 66201 (2004)
9. Atay, F.M.: Oscillator death in coupled functional differential equations near Hopf bifurcation. J. Diff. Equ. 221, 190–209 (2006)
10. Camp, T., Boleng, J., Davies, V.: A survey of mobility models for ad hoc network research. Wirel. Commun. Mob. Comput. 2, 483–502 (2002)
11. Hegselmann, R., Krause, U.: Opinion dynamics and bounded confidence: Models, analysis and simulation. J. Artif. Soc. Soc. Sim. 5, 1–32 (2002)
12. Gonzaáez, M.C., Herrmann, H.J.: Scaling of the propagation of epidemics in a system of mobile agents. Physica A 340, 741–748 (2004)
13. Bashan, A., Bartsch, R.P., Kantelhardt, J.W., Havlin, S., Ivanov, P.C.: Network physiology reveals relations between network topology and physiological function. Nat. Commun. 3, 702 (2012)
14. Zhao, J., Hill, D.J., Liu, T.: Synchronization of complex dynamical networks with switching topology: A switched system point of view. Automatica 45, 2502–2511 (2009)
15. Stilwell, D.J., Bollt, E.M., Roberson, D.G.: Sufficient conditions for fast switching synchronization in time-varying network topologies. SIAM J. Appl. Dyn. Syst. 5, 140–156 (2006)
16. Belykh, I.V., Belykh, V.N., Hasler, M.: Blinking model and synchronization in small-world networks with a time-varying coupling. Physica D 195, 188–206 (2004)

A Novel Concept Combining Neuro-computing and Cellular Neural Networks for Shortest Path Detection in Complex and Reconfigurable Graphs

Jean Chamberlain Chedjou and Kyandoghere Kyamakya

Universität Klagenfurt, Institute of Smart Systems Technologies,
Transportation Informatics Group (TIG), Lakeside B04.a,
L4.2.10, 9020, Klagenfurt, Austria
{kyandoghere.kyamakya,jean.chedjou}@aau.at

Abstract. This paper develops for the first time an analytical concept involving the Basic Differential Multiplier Method (BDMM) in a framework concept using Cellular neural networks (CNN) for finding shortest paths (SP) in reconfigurable graphs. The developed concept is modeled by coupled nonlinear ordinary differential equations (ODE). The resulting ODE parameters are the CNN templates that are, except from their dimension, independent of the different graph's elements. The main advantage of the CNN concept is that both the costs of arcs and the selection of the origin-destination (s-t) pair are insured by external commands which are inputs of the CNN-processor model. This allows a high flexibility and an easy re-configurability of the developed concept, thereby without any re-training need. Further, the concept can handle even graphs with negative arc's weights as well as graphs with nonlinear path's costs.

Keywords: Shortest path (SP) finding, basic differential multiplier method (BDMM), modeling based on ordinary differential equations (ODE), cellular neural network (CNN), computational efficiency, flexibility, re-configurability.

1 Introduction

The last decades have witnessed a tremendous interest in the development of new methods, concepts, algorithms and tools for finding SP in graphs [1-18]. The interest devoted to finding SP in graphs is explained by the multiple potential applications of SP problems in diverse fields of sciences and engineering. In intelligent transportation systems (ITS) for example, the development of efficient and flexible SP procedures is of necessary importance in order to fulfill some key requirements related to real time routing applications. These applications require fast computation as well as low memory consumption [1]. Some interesting applications of SP in ITS include vehicle routing, e.g. in-vehicle Route Guidance System (RGS) [2], Automated Vehicle Dispatching Systems (AVDS) [3], real-time traffic information sensing [4], etc. The information sensing for fleet management is an important step of the rerouting process within a traffic network in order to avoid areas of high traffic density [4].

V.M. Mladenov and P.C. Ivanov (Eds.): NDES 2014, CCIS 438, pp. 227–236, 2014.

In communication and transport networking, SP is used in various applications such as: traffic routing [5], optimization of the data transmission capacity [6], routing and scheduling over multi-hop wireless networks [7], network coding based multicasting [8], maximization of the life time of sensors networks [9], routing protocols in order to exchange topology information among routers [10], and path planning in robotic systems [11], just to name a few.

The SP problem has been studied extensively and several methods and concepts have been developed to handle or tackle this problem. Amongst these methods one can cite the linear programming methods (LPM) [12] which however lead to very poor performances (i.e. poor computational efficiency, low accuracy, low- precision and weak robustness) when dealing with the analysis of SP problems. The polynomial complexity algorithms (e.g. Bellman-Ford's [13] and Dijkstra's [14]) lead to a better performance than the LPM. However, these algorithms cannot analyze SP problems with negative costs of edges [15]. Further, the polynomial algorithms are very time consuming [16] and cannot efficiently handle or solve the resulting exploding complexity while solving SP problems for example in re-configurable graphs [15], [16] or in stochastic graphs [4]. Heuristic algorithms such as genetic algorithms (GA) [16] and particle swarm optimization (PSO) [15] can analyze SP problems in both deterministic and stochastic and even in reconfigurable graphs [16]. However, heuristic algorithms are less accurate, show less precision and are of weak robustness. They are also computationally expensive specifically when dealing with the analysis of complex and reconfigurable graphs [15], [16]. Regarding the methods based on Neural Networks (NN), various architectures of Neural Networks have been developed to analyze the SP problem. A first one, the NN-parallel architecture leads to poor computing performances and less accurate results [1]. Another one, the Hopfield Neural Network (HNN) architecture does however lead to a much better computing performance than the NN-parallel architecture. However, the reliability of the HNN remains an issue (i.e. not good). This reliability is expressed in terms of successful and valid convergence while searching for routing paths by the HNN [1]. Further, the concepts based on HNN lack of flexibility [16] and are not appropriate for SP determination in reconfigurable graphs (re-configurability being understood amongst others as the dynamic change of either source-destination pairs or that of the arc/edge weights or both) [1]. The Ali-Kamoun's method [17] proposes a new architecture of NN with a good feature related to the adaptability to variations of edges' weights. However, this method is only valid for single-source & single-destination problems. Further, this method also fails to always converge towards valid/correct solutions. Specifically, the convergence potential of the Ali's and Kamoun's method degrades drastically while increasing the number of nodes in the graph [1]. The Park-Choi's NN [18] was introduced as an extended version of the Ali-Kamoun's method. This method can efficiently analyze single-source multiple destination problems. However, the poor convergence potential/capability of this method is a serious limitation observed when the magnitude of the graph increases [18]. The Dependent Variable Hopfield Neural Network (DVHNN) was in the continuation introduced as an extended version of the HNN in order to solve some of the problems unsolved by the HNN, the Ali-Kamoun's and the Park-Choi's NN methods. In particular, this method is demonstrated (see [1]) as a best potential NN based candidate to improve the reliability of solutions when compared to the above mentioned methods based on NN. However, the DVHNN can solve only single-source & single-destination SP problems and consequently cannot handle or tackle SP problems in reconfigurable graphs.

Our core aim in this paper is to present and validate a comprehensive general and robust concept based on CNN that can be used to overcome all limitations of the methods described above for SP determination. Specifically, the proposed concept (CNN) is demonstrated being an efficient and universal concept for the solving of SP problems in complex and reconfigurable graphs. The key performance metrics for evaluating the CNN based novel concept developed are: computational efficiency, accuracy, precision, robustness (i.e. sure convergence independently of network size and magnitude), flexibility, re-configurability, and universality.

The rest of the paper is organized as follows. Section 2 describes the concept based on CNN for SP determination in complex and reconfigurable graphs. A synoptic representation of the concept is proposed and, a description of the key parameters of the complete system is presented. Section 3 concentrates on the mathematical modeling of the SP problem. The resulting equations of the mathematical model of the CNN-simulator for SP detection are then derived. Section 4 is then focused on providing a good proof of concept through a selected example of a directed and re-configurable graph of magnitude 14 and size 31. Several scenarios are envisaged (by choosing different combinations of source-destination pairs) and, the effectiveness of the concept developed in this paper is clearly demonstrated. Section 5 is devoted to concluding remarks. A summary of the core achievements of this work is presented. Further, selected interesting open research questions (under investigation in some of our ongoing works) are shortly listed in an outlook.

2 Concept for Finding SP in Complex and Reconfigurable Graphs

This section provides a description of the proposed concept for finding a SP in complex and reconfigurable networks. This concept does work on three basic inputs (see Fig. 1). The first input is a binary command α_i for the choice of the source node denoted by s ($\alpha_i = 1$ if the node s is assigned the attribute of source node and $\alpha_i = 0$ otherwise). The second input is a binary command β_i for the choice of the destination node denoted by t ($\beta_i = 1$ if the node t is assigned the attribute of destination node and $\beta_i = 0$ otherwise). The third input is a command $w_{i \to j}$ for assigning the weight-values to edges. The core of the full system is represented by the CNN-simulator (in Fig. 1). The CNN-simulator is made-up of five matrices $A(a_{k,l})$, $B(b_{k,l})$, $C(c_{k,l})$, $D(d_{k,l})$, and $E(e_{k,l})$ whose elements are denoted by $a_{k,l}$, $b_{k,l}$, $c_{k,l}$, $d_{k,l}$, and $e_{k,l}$. $A(a_{k,l})$ is a linear matrix, which is a characteristic matrix providing the signature of the graph topology under investigation. The matrices $B(b_{k,l})$ and $C(c_{k,l})$ reveal the degree of nonlinearity of the resulting mathematical ODEs (i.e. the equations modeling the CNN- simulator for SP findings). $D(d_{k,l})$, and $E(e_{k,l})$ are respectively the feedback- and forward- matrices. The output x_k of the CNN- simulator (see Fig. 1) is connected to a "sink" (e.g. a scope or a workspace in MATLAB) which displays the states of edges connectivity ($x_k = 1$ if the edge $i \longrightarrow j$ belongs to the shortest path and $x_k = 0$ otherwise). i and j denote the indexes of all nodes of the graph.

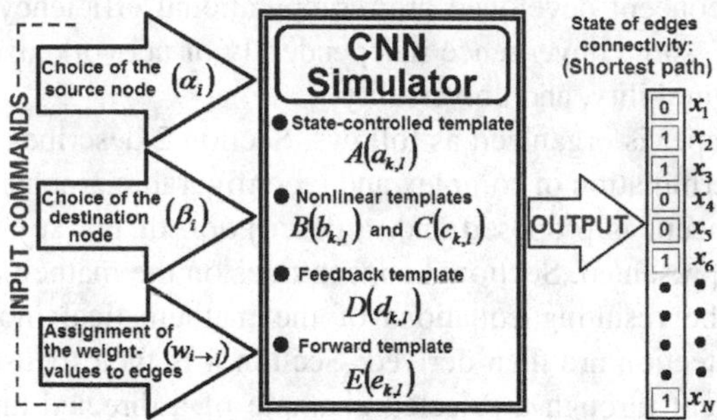

Fig. 1. Synoptic representation of the complete/full concept based on CNN for SP detection

Overall, α_i, β_i, and γ_i are binary commands used for assigning attributes to nodes in order to facilitate the re-configurability within a graph. Thus, using these commands, each node of the graph can be assigned the attribute of source-, destination- or intermediate- node (re-configurability). Nodes which are neither source nor destination are automatically assigned the attributes of intermediate- nodes. Regarding the modeling process, the first hypothesis is related to the assignment of an attribute to a given node and the second is concerned with the involvement/belonging of that node to the desired path (i.e. SP). These conditions will be derived analytically (from the constraints formulation) during the modeling phase.

3 Mathematical Modeling of the SP Problem and Resulting ODE Describing the CNN-simulator for SP Detection in Complex Graphs

3.1 Derivation of the Lagrange Function Corresponding to the SP Problem

This sub-section describes the key steps involved in the mathematical modeling of the shortest path problem as a multi-variable optimization problem. The first step is concerned with the derivation of the corresponding Lagrange function (see Eq. (1)) which is viewed as the total energy of the system. In this context, the optimization problem is represented by an objective function ($Min[\, f(\vec{x})\,]$) subject to constraints ($g_a(x)=0$).

For various discrete problems these constraints may make the problem NP-hard. In order to solve NP-hard problems the constraint optimization issues can be transformed into unconstraint optimization issues using relaxation methods. Several methods are proposed by the relevant state-of-the-art (see [19], [20]) to transform NP-hard problems into simple ones. Some of these methods are: direct substitution, constrained variation, and Lagrange multipliers (see [19], [20]), just to name a few.

The shortest path/walk problem can be considered as a multi-variable constrained optimization problem. This type of problem can be modeled mathematically by Eq. (1). In this equation, λ_a are multipliers variables and the decision variable $\vec{x} = [x_1, x_2, ..., x_n]$ is a binary vector. This vector reveals the state of all edges involved in the graph under consideration. $f(\vec{x})$ is a scalar value denoting the distance of the walk from a source node to a destination node. The function $g_a(x)$ stands for constraints. The parameter a corresponds to the index of constraints.

$$\tilde{L}(\vec{x}, \lambda_a) = f(\vec{x}) + \sum_a \lambda_a . [g_a(\vec{x})] \tag{1}$$

The second step exploits the concept of neuron dynamics as an optimization strategy that maps the optimization problem into the energy of a neural network (i.e., Hopfield network) in order to find the optimal solution [19], [20]. In this context, we represent this energy into the Lagrange form and the minimization of the Lagrange function leads to a stable state (convergence of the BDMM). Applying the BDMM to the Lagrange function in Eq. (1) leads to the derivation of Eq. (2).

$$\dot{x}_i = -\alpha \partial_{x_i} (\tilde{L}) \ \wedge \ \dot{\lambda}_a = +\beta \partial_{\lambda_a} (\tilde{L}) \tag{2}$$

The set of Eqs. (2) is the characteristic model of the BDMM [19], [20]. This model (from which the CNN-templates are derived) reveals the coupling between the dynamics of decision neurons (x_i) and multiplier neurons (λ_a).

3.2 Modeling of SP Problem: Problem Formulation and Resulting ODE

A convenient way of designing a general and robust concept for finding SP is to derive a universal mathematical model that can be used to handle (or tackle) SP problems regardless of the topologies of the graphs envisaged. This model is expressed in the form of nonlinear- coupled ODEs which are further used to design and implement the resulting SP CNN-simulator. This sub-section provides a full explanation of the systematic concept leading to the mathematical modeling of the SP problem as well as the design and implementation of the SP CNN- simulator. The SP problem is formulated by the objective function $\Im$ defined in Eq. (3), where $f(\vec{x})$ is the total "cost" corresponding to the full size of the graph under investigation.

$$\Im = Min\left[f(\vec{x}) = \sum_{i=1}^{M} \sum_{\substack{j=1 \\ i \neq j}}^{M} w_{i \to j} x_{i \to j} \right] \tag{3}$$

The limit of the summation M corresponds to the magnitude of the graph. The objective function in Eq. (3) is subject to constraints, which are formulated with the aim of fulfilling all key requirements related to the problem formulation. These key requirements are summarized by the following conditions: (a) Loops avoidance and connectivity of the intermediate nodes; (b) Connectivity of the source node; (c) Connectivity of the destination node. These conditions are expressed as follows:

$$\sum_{\substack{i=1 \\ i \neq k}}^{N} x_{k \to i} - \sum_{\substack{j=1 \\ j \neq k}}^{N} x_{j \to k} = \delta \quad \begin{cases} \delta = 0 \ (k \ is \ intermediate \ node) \\ \delta = +1 \ (k \ is \ source \ node) \\ \delta = -1 \ (k \ is \ destination \ node) \end{cases} \tag{4}$$

Eqs. (3) and (4) are combined (according to Eq. (1)) to obtain the function $\widetilde{L}$ as the total energy of the system. Substituting $\widetilde{L}$ into Eq. (2) leads to the derivation of Eq. (5) as the resulting mathematical model of the CNN- simulator (see [21], [22])

$$\frac{dx_k}{dt} = \left[-x_k + \sum_{l=1}^{M} (a_{k,l} x_l) + \sum_{l=1}^{M} (b_{k,l} x_k x_l) + \sum_{l=1}^{M} (c_{k,l} x_l^2) + \sum_{l=1}^{M} (d_{k,l} y_l) + \sum_{l=1}^{M} (e_{k,l} u_l) + I_k \right] \tag{5}$$

for SP finding. The coefficients of Eq. (5) are defined in section 2. I_k are elements of the threshold $\vec{I} = [I_1, \ldots, I_k]^T$. u_l are inputs of the elementary cells. y_l are the basic piecewise linear functions. k and l are indexes of rows and columns respectively.

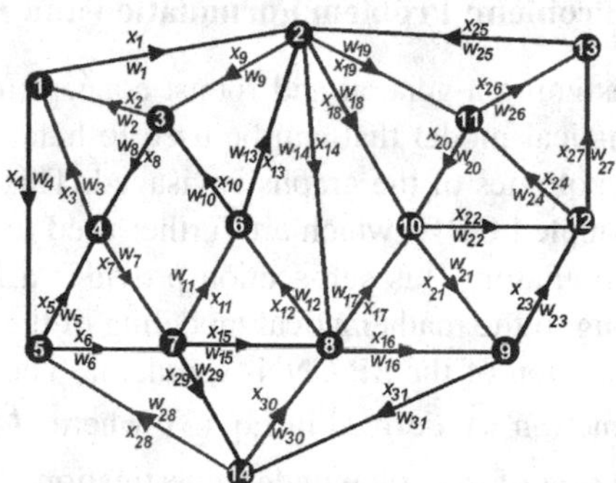

Fig. 2. Topology of a directed graph of magnitude 14 and size 31. The costs of edges are denoted by w_i The state of an edge connectivity is denoted by x_i .

4 Proof of Concepts: Application of the Concept Developed for SP Detection Based on CNN to a Concrete Graph (in Figure 2)

4.1 Derivation of the CNN-Templates Corresponding to the Graph in Figure 2

We now want to validate the concept developed by considering a directed graph of magnitude 14 and size 31 as shown in Fig. 2. The general expression in equations (5) is used to express the CNN- templates (corresponding to Fig. 2) as follow:

$$\begin{cases} b_{k,l} = -2 \ \text{if} \ 1 \le k \le 31 \ \text{and} \ l = k+31 \\ b_{k,l} = 0 \ \text{otherwise} \end{cases}; \qquad \begin{cases} d_{k,l} = +1 \ \text{if} \ k = l \\ d_{k,l} = 0 \ \text{otherwise} \end{cases}$$

$$\begin{cases} c_{k,l} = +1 \ \text{if} \ 32 \le k \le 62 \ \text{and} \ l = k-31 \\ c_{k,l} = 0 \ \text{otherwise} \end{cases}; \qquad \begin{cases} e_{k,l} = -1 \ \text{if} \ k = l \\ e_{k,l} = 0 \ \text{otherwise} \end{cases}$$

The corresponding values of the inputs u_i are expressed as follows:

Input-1: Commands for the weighting of edges *(see Fig. 1): $u_i=w_i$ (i=1,..,31).*

Input-2: Commands for assigning the attribute of source node *(see Fig. 1):*

$u_{63} = \alpha_1$, $u_{68} = \alpha_2$, $u_{73} = \alpha_3$, $u_{78} = \alpha_4$, $u_{83} = \alpha_5$, $u_{88} = \alpha_6$, $u_{93} = \alpha_7$, $u_{98} = \alpha_8$, $u_{103} = \alpha_9$, $u_{108} = \alpha_{10}$, $u_{113} = \alpha_{11}$, $u_{118} = \alpha_{12}$, $u_{123} = \alpha_{13}$, $u_{128} = \alpha_{14}$.

Input-3: Commands for assigning the attribute of destination node *(see Fig. 1):*

$u_{65} = \beta_1$, $u_{70} = \beta_2$, $u_{75} = \beta_3$, $u_{80} = \beta_4$, $u_{85} = \beta_5$, $u_{90} = \beta_6$, $u_{95} = \beta_7$, $u_{100} = \beta_8$, $u_{105} = \beta_9$, $u_{110} = \beta_{10}$, $u_{115} = \beta_{11}$, $u_{120} = \beta_{12}$, $u_{125} = \beta_{13}$, $u_{130} = \beta_{14}$.

Remark: The above mentioned input values are obtained using the Symbolic Toolbox in MATLAB when performing the mapping of Eqs. (2) with Eq.(5). Further, all non-specified indexes u_i are equal to zero.

The state-controlled template $A(a_{k,l})$ is a matrix of size 132x132. The components $a_{k,l}$ of this matrix are expressed only in terms of the binary input commands (α_i, β_i, and γ_i) of the CNN-processor in Fig. 1. The values of $a_{k,l}$ are obtained in the frame of the analytical mapping performed in the process of equating Eqs. (2) and Eq. (5). The *Symbolic toolbox* in MATLAB is used for calculating $a_{k,l}$ as follows: $a_{k,k+31} = -1 \ (1 \le k \le 31)$; $a_{k+31,k} = 1 \ (1 \le k \le 31)$; $a_{1,63} = \alpha_1$, $a_{1,69} = \alpha_2$, $a_{2,64} = \alpha_1$, $a_{2,73} = \alpha_3$, $a_{3,64} = \alpha_1$, $a_{3,78} = \alpha_4$, $a_{4,63} = \alpha_1$,

$a_{4,84} = \alpha_5$, $a_{5,79} = \alpha_4$. $a_{1,66} = \beta_1$, $a_{1,70} = \beta_2$, $a_{2,65} = \beta_1$, $a_{2,76} = \beta_3$, $a_{3,65} = \beta_1$, $a_{3,81} = \beta_4$, $a_{4,66} = \beta_1$, $a_{4,85} = \beta_5$, $a_{5,80} = \beta_4$, $a_{1,67} = \gamma_1$, $a_{1,72} = -\gamma_2$, $a_{2,67} = -\gamma_1$, $a_{2,77} = \gamma_3$, $a_{3,67} = -\gamma_1$, $a_{3,82} = \gamma_4$, $a_{4,67} = \gamma_1$, $a_{4,87} = -\gamma_5$, $a_{5,82} = -\gamma_4$. We have provided the $a_{k,l}$ corresponding to line 1 to 5. The others $a_{k,l}$ (line 6 to 132) can be deduced using the same process.

4.2 Simulation Results of the SP Finding in Fig.2 Using the CNN Model in Eqs.(5)

The CNN model in Eqs. (5) is implemented in MATLAB using the parameter settings $a_{k,l}$, $b_{k,l}$, $c_{k,l}$, $d_{k,l}$, $e_{k,l}$, and $u_{k,l}$ derived (in section 4.1). Further, several scenarios are envisaged in order to demonstrate the effectiveness of the concept developed in this work. The performance criteria used are the accuracy, precision, robustness, flexibility, re-configurability, and universality.

We consider the graph in Fig. 2 where the cost w_i of an edge i is defined by the following expressions: $w_i=i$. Table 1 presents sample results of the routing process (i.e. shortest path detection in Fig. 2) using the CNN-simulator implemented in MATLAB. The scenarios in Table 1 are considered such that nodes in Fig. 2 are assigned the attributes of sources-, destinations-, or intermediate- nodes. In Table 1, the states of edges connectivity are obtained as solutions of Eqs. (5). An illustrative example of the temporal evolution of these solutions is shown in Fig. 3. The solutions in Fig. 3 exhibit a very brief transient phase and finally a straightforward convergence to a binary state is observed. The solutions $x_{i\to j} = 1$ reveal the states belonging to the routing path and $x_{i\to j} = 0$ stand for the states that do not belong to the routing path. As it appears in Table 1 the CNN-simulator always detects the exact shortest path for the scenarios envisaged. This conclusion underscores the good accuracy of the CNN-simulator for shortest path finding/detection.

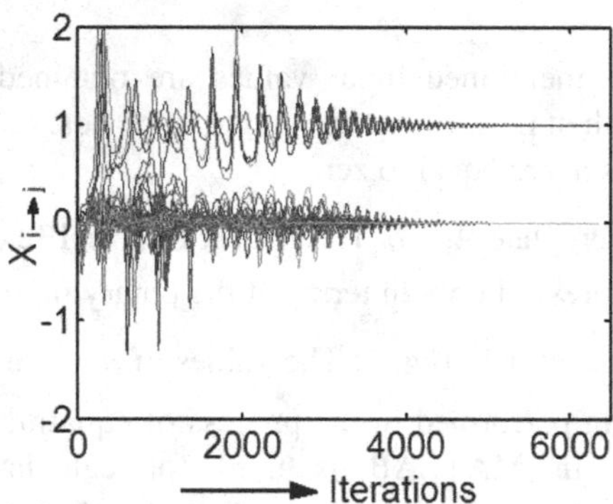

Fig. 3. Illustration of the temporal evolution of the costs of edges obtained as outputs of the CNN- simulator (i.e. solutions of equation (5))

Table 1. Results of the shortest path detection (in Fig. 2) using the CNN-simulator. The scenarios envisaged are considering large values of the costs of edges *(w_i=i)*.

From Source To Destination s-t	Shortest path results using CNN-simulator		CNN	
			Sim. Time (Tsim)	Convergence
	Edges in the shortest path	Costs of paths	CNN (ms)	CNN
14-1	x_3, x_5, x_{28}	36	3.453	Very Fast
13-5	x_2, x_4, x_9, x_{25}	40	5.400	Very Fast
12-3	x_9, x_{25}, x_{27}	61	18.320	Fast
11-4	x_8, x_9, x_{25}, x_{26}	68	16.935	Fast
10-5	x_{21}, x_{28}, x_{31}	80	15.754	Fast

5 Concluding Remarks

This work has developed and validated a general theoretical (i.e. analytical) concept based on the CNN- paradigm for the efficient computing of SP problems in complex and reconfigurable graphs. Since the analytical formula derived (i.e. the mathematical model of the CNN-simulator) are applicable to all graphs regardless the topology, the results in this work constitute a clear achievement which enriches the state-of-the-art as the concept developed is universally applicable. Another strong point of the concept developed in this work is that the state-controlled CNN-template corresponds to the s-t (origin-destination) matrix of the graph. The components of this matrix are expressed analytically in terms of the input commands (binary variables) for assigning attributes (source and destinations) to nodes. This underscores the offered flexibility through the CNN-concept to analyze SP problems regardless both the topology and routes of the graph.

A series of ongoing works under consideration does relate amongst others to: a) the extension of the concept where the cost of a path is a nonlinear function of the arc weights (for example for cases where the driving energy on a path must be minimized); and b) the extension of the concept developed in this work to the case of stochastic graphs. These graphs are characterized by arc weights that are stochastic processes. Thus, extending the concept developed in this work to both nonlinear and stochastic graphs models is possible. Regarding the later ones, this can be achieved by considering the key parameters i.e. the kth –moments (e.g. moment-1: average and moment-2: standard deviation, etc) of the distributions characterizing the stochastic dynamics of each edge of the graph. The extension of the concept developed in this work to stochastic graphs will lead to the mathematical modeling of various stochastic prone systems for example of sensor data with straightforward interesting applications in areas like intelligent transportation systems and many others.

References

1. Araújo, F., Ribeiro, B., Rodrigues, L.: A Neural Network for Shortest Path Computation. IEEE Transactions on Neural Networks 12, 1067–1073 (2001)
2. Fu, L.: An adaptive routing algorithm for in-vehicle route guidance systems with real-time information. Transportation Research Part B 35, 749–765 (2001)

3. Fu, L., Sun, D., Rilett, L.R.: Heuristic shortest path algorithms for transportation applications: State of the art. Computers & Operation Research 33, 3324–3343 (2006)
4. Kim, S., Lewis, M.E., White, C.C.: State Space Reduction for Nonstationary Stochastic Shortest Path Problems With Real-Time Traffic Information. IEEE Transactions on Intelligent Transportation Systems 6, 273–284 (2005)
5. Moy, J.: Open Shortest Path First Version 2. RFQ 1583. Internet Engineering Task Force (1994), http://www.ietf.org
6. Iraschko, R.R., MacGregor, M.H., Grover, W.D.: Optimal Capacity Placement for Path Restoration in STM or ATM Mesh-Survivable Networks. IEEE/ACM Transactions on Networking 6, 325–336 (1998)
7. Ying, L., Shakkottai, S., Reddy, A., Liu, S.: On Combining Shortest-Path and Back-Pressure Routing Over Multihop Wireless Networks. IEEE/ACM Transactions on Networking 19, 841–854 (2011)
8. Wu, Y., Kung, S.-Y.: Distributed Utility Maximization for Network Coding Based Multicasting: A Shortest Path Approach. IEEE Journal on Selected Areas in Communications 24, 1475–1488 (2006)
9. Chen, Y., Zhao, Q., Krishnamurthy, V., Djonin, D.: Transmission Scheduling for Optimizing Sensor Network Lifetime: A Stochastic Shortest Path Approach. IEEE Transactions on Signal Processing 55, 2294–2309 (2007)
10. Narváez, P., Siu, K.-Y., Tzeng, H.-Y.: New Dynamic Algorithms for Shortest Path Tree Computation. IEEE/ACM Transactions on Networking 8, 734–746 (2000)
11. Desaulniers, G., Soumis, F.: An Efficient Algorithm to Find a Shortest Path for a Car-like Robot. IEEE Trans. Robot. Automat. 11, 819–828 (1995)
12. Bertsekas, D.P. (ed.): Network Optimization: Continuous and Discrete Models. Belmont MA, Athena (1998)
13. Bellman, R.E. (ed.): Dynamic Programming. Princeton Univ. Press, Princeton (1957)
14. Dijkstra, E.: A note on two problems in connection with graphs. Numerical Math. 1, 269–271 (1959)
15. Mohemmed, A.W., Sahoo, N.C., Geok, T.K.: Solving Shortest Path Problem Using Particle Swarm Optimization. Applied Soft Computing 8, 1643–1653 (2008)
16. Ahn, C.W., Ramakrishna, R.S.: A Genetic Algorithm for Shortest Path Routing Problem and the Sizing of Populations. IEEE Transactions on Evolutionary Computation 6, 566–579 (2002)
17. Mustafa, K., Ali, M., Kamoun, F.: Neural Networks for Shortest Path Computation and Routing in Computer Networks. IEEE Transactions on Neural Networks 4, 941–953 (1993)
18. Park, D.C., Choi, S.E.: A Neural-Network-Based Multidestination Routing Algorithm for Communication Network. In: Proc. IEEE Int. Joint Conf. Neural Networks, vol. 2, pp. 1673–1678 (1998)
19. Li, S.Z.: Improving convergence and solution quality of Hopfield-type neural networks with augmented Lagrange multipliers. IEEE Transaction on Neural Networks 7, 1507–1523 (1996)
20. McFall, K., Mahan, J.: Artificial Neural Network Method for Solution of Boundary Value Problems With Exact Satisfaction of Arbitrary Boundary Conditions. IEEE Transactions on Neural Networks 20, 1221–1233 (2009)
21. Chua, L.O., Yang, L.: Cellular Neural Networks: Theory. IEEE Transactions on Circuits and Systems-I 35, 1257–1272 (1988)
22. Roska, T., Chua, L.O.: The CNN universal machine: 10 years later. Journal of Circuits Systems and Computers 12, 377–388 (2003)

Cellular Neural Networks Proposed for Image Predictive Coding

Tang Tang and Ronald Tetzlaff

Technische Universität Dresden, Faculty of Electrical Eng & Information Technology, Mommsenstraße 12, 01069 Dresden, Germany
{Tang.Tang,Ronald.Tetzlaff}@tu-dresden.de

Abstract. In this paper the feasibility of implementing Cellular Neural Networks (CNN) for image predictive coding is investigated. Various CNN structures as predictors are proposed. The performances are compared to the existing predictive coding methods. Thanks to their massive parallel nature, CNN have been proven well suitable for image predictive coding application.

Keywords: Cellular Neural Networks, predictive coding, parallel computing, image coding.

1 Introduction

Recently there is a tremendous interest in massive parallel computing. Due to the massive parallel nature, CNN are possible solutions for parallel computing.

Predictive image coding is one of the most important coding schemes. The key problem in predictive coding is to design a high performance predictor. Several publications have shown that the performance of predictors can be improved by introducing nonlinear dynamics [1], [2]. CNN serve not only as a parallel computing framework, but also as a kind of neural networks. In this paper, various CNN structures used as predictors are proposed. Their performances are compared to the existing image predictive coding methods.

2 CNN-Based Predictors

To keep a low complexity of system, here, only the pixels directly surrounding the pixel to be predicted are considered, and the neighboring structure is shown in Fig. 1 (assuming the scan order to be from the upper left to the bottom right in the image).

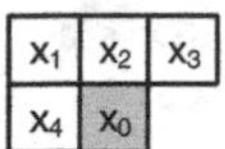

Fig. 1. Structure of neighborhood in predictor

V.M. Mladenov and P.C. Ivanov (Eds.): NDES 2014, CCIS 438, pp. 237–245, 2014.

2.1 Fundamentals of CNN

CNN have been introduced by Chua et al. CNN is any spatial arrangement of locally-coupled cells, where each cell is a dynamical system with an input, output and state evolving in accordance with some prescribed dynamical laws [3].

$$\dot{x}_{ij} = -x_{ij} + \sum_{kl \in S_{ij}(r)} a_{kl} y_{kl} + \sum_{kl \in S_{ij}(r)} b_{kl} u_{kl} + z, \tag{1a}$$

$$y_{ij} = f(x_{ij}) = \frac{1}{2}(|x_{ij} + 1| - |x_{ij} - 1|), \tag{1b}$$

$$i = 1,2,\dots, M, j = 1,2,\dots, N.$$

The mostly common CNN model is the standard CNN (1), where x_{ij}, u_{ij} and y_{ij} are state, input and output of a cell C_{ij}, respectively a_{kl} and b_{kl} are scalars called synaptic weights, and $S_{ij}(r)$ denotes the sphere of influence of radius r. With a standard translation invariant CNN, there is a uniform bias $z_{ij}=z$.

In this paper, for the purpose of prediction more complicated CNN models will be constructed and implemented.

2.2 Linear Predictor

The simplest predictor is a linear one, based on the assumption that the statistical model of image is an autoregressive model (AR). In an mth order causal AR process, the value of sample x(n) is dependent on the values of the *m* previous samples by

$$x(n) = \sum_{i=1}^{m} w_i x(n-i) + \varepsilon_n \tag{2}$$

where w_i are AR coefficients and ε_n is an independent and identically distributed random variable with zero-means. Under this assumption the predicted value is a linear combination of the neighboring pixels, i.e.

$$\hat{x}(n) = \sum_{i=1}^{m} w_i x(n-i). \tag{3}$$

The Equation (3) can be computed by a standard CNN. The template of CNN is defined as:

$$\mathbf{A} = \begin{pmatrix} 0 & 0 & 0 \\ 0 & 0 & 0 \\ 0 & 0 & 0 \end{pmatrix}, \mathbf{B} = \begin{pmatrix} w_1 & w_2 & w_3 \\ w_4 & 0 & 0 \\ 0 & 0 & 0 \end{pmatrix} z = 0. \tag{4}$$

The image to be predicted is fed into CNN as input **u** (the pixel values expressed as floating numbers from 0 to 1). In order to keep the predictive value in the range of [0, 1], the output function in (1b) is modified to

$$y = \begin{cases} 1 & x > 1 \\ x & 0 \le x \le 1 \\ 0 & x < 0 \end{cases}. \tag{5}$$

The linear predictor is well suitable for flat regions of an image, while shows big predictive errors in regions of texture and contour. Hence, a context-adapted predictor is needed. A possible way is introducing nonlinear dynamics to the predictor structure.

2.3 Polynomial CNN Predictor

Many contributions have proven that a nonlinear predictor could improve the prediction performance [1], [2]. Polynomial type CNN [4] is one of the possible structures. For the purpose of prediction, uncoupled polynomial CNN (all parameters in **A** template are 0) is proposed, which is defined by

$$\dot{x}_{ij} = -x_{ij} + \sum_{kl \in S_{ij}(r)} b_{kl}(u_{kl}) + z, \; b_{kl}(u_{kl}) = \sum_{d=1}^{D} b_{kl,d} \cdot (u_{kl})^{d}, \tag{6}$$
$$i = 1,2, \ldots, M, j = 1,2, \ldots, N,$$

where D denotes the polynomial order, z=0. The image to be predicted is taken as input **u**. The output function is defined in (5).

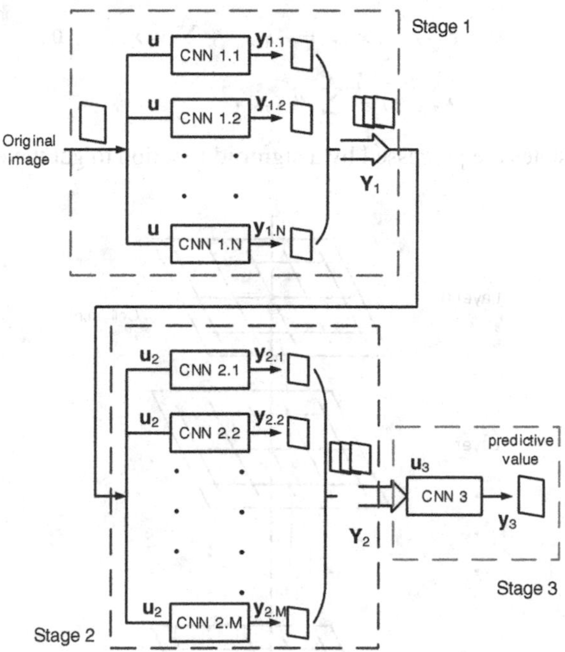

Fig. 2. Structure of multistage predictor

2.4 Multistage Predictor

This structure is inspired by the concept of multilayer perceptron in neural networks [5]. The structure is shown in Fig.2.

Here, the image to be predicted as input is fed into the first stage, where there are N parallel CNN channels. The CNN in each channel are denoted as *CNN 1.i* (i=1 to N), which are uncoupled standard CNN. The template structure is given in (4), while the concrete parameters of **B** template vary in different channels.

The outputs y1.i of each channel build a set of intermediate values denoted as $\mathbf{Y_1}$, sent to the second stage containing M channels of CNN. The CNN in each channel is labeled as *CNN 2.i* (i = 1 to M), which is a multilayer CNN structure (see Fig. 3). This structure contains N+1 layers (N is the number of the channels in the first stage). The outputs of the first CNN stage are loaded to Layer 1 to N as initial states respectively and keep unchanged during the processing. The cell (m,n) in Layer 0 is coupled to the cells in Layer 1 to N, which have the same coordinate (m,n). The state of cells in CNN 2.i is governed by equation:

$$\dot{x}_{mn\,i} = \begin{cases} -x_{mn,0} + \sum_{k=1}^{N} w_k x_{mn\,k} & i = 0 \\ 0 & i \neq 0 \end{cases} \tag{7}$$

Here, $x_{mn,i}$ denotes the state of cell (m,n) in Layer i, w_i is the coupling factor, w_0 is set to -1. Hence the final state of $x_{mn,0}$ is calculated to

$$\begin{aligned} &t \to \infty,\ \dot{x}_{mn0} = -x_{mn0} + \sum_{i=1}^{N} w_i x_{mni} = 0, \\ &x_{mn0}(\infty) = \sum_{i=1}^{N} w_i x_{mni}. \end{aligned} \tag{8}$$

The final cell states are processed by a sigmoid function to get the output values.

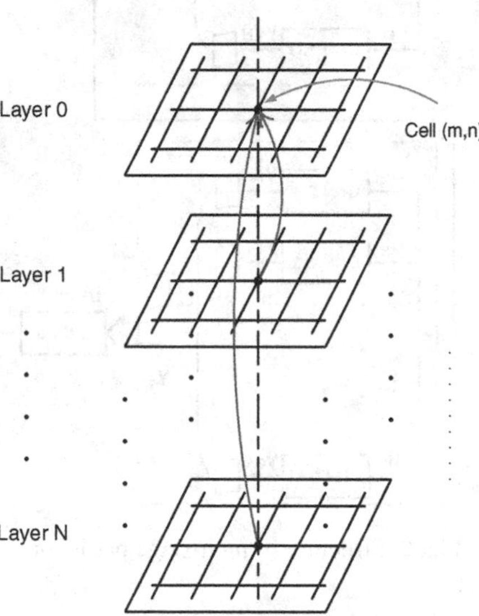

Fig. 3. Multilayer structure of CNN 2.i

Similar to the first stage, the output values y2.i of each channel in the second stage are combined as $\mathbf{Y}_2$ and sent to the third stage. The CNN in this stage labeled as CNN 3 reuse the structure of CNN 2.i. Here the number of layers is M+1 (M is the number of the channels in the second stage). The output values $\mathbf{Y}_2$ of the second stage are loaded as initial states to Layer 1 to M. Finally, a predictive value can be obtained as output of CNN 3.

The architecture described above has great scalability, except the first stage (as input stage) and the last stage (as output stage) must be kept, the number of stages between them can vary from 0 to a certain number (if this number is zero, the output values of CNN 1.i are directly send to CNN 3, in practical applications, this number should be kept small (below 3)).

With classical neural networks, above calculations are executed in a serial way, i.e. each time only four pixels are fed into the network to obtain one predictive value, then starting the next prediction. Thanks to the parallel computing nature of CNN, these calculations can be executed with CNN in parallel.

2.5 Higher Order Predictor

The higher order predictor is based on a generalized autoregressive model

$$\begin{aligned} x(n) &= \sum_i w_i x(n-i) + \sum_i \sum_j w_{ij}\, x(n-i)x(n-j) \\ &+ \sum_i \sum_j \sum_k w_{ijk}\, x(n-i)x(n-j)x(n-k) + \ldots + \varepsilon_n \end{aligned} \tag{9}$$

where ε_n is an independent and identically distributed random variable with zero-means. By introducing this model, the predictive values considering four neighboring pixels are calculated according to (definitions of x_0 to x_4 are shown in Fig.1)

$$\begin{aligned} \hat{x}_0 = {} & w_1x_1 + w_2x_2 + w_3x_3 + w_4x_4 && \cdots 1 \\ & + w_{12}x_1x_2 + w_{13}x_1x_3 + w_{14}x_1x_4 && \cdots 2 \\ & + w_{23}x_2x_3 + w_{24}x_2x_4 && \cdots 3 \\ & + w_{34}x_3x_4 && \cdots 4 \\ & + w_{123}x_1x_2x_3 + w_{124}x_1x_2x_4 && \cdots 5 \\ & + w_{134}x_1x_3x_4 && \cdots 6 \\ & + w_{234}x_2x_3x_4 && \cdots 7 \\ & + w_{1234}x_1x_2x_3x_4 && \cdots 8 \end{aligned} \tag{10}$$

Equation (10) can be calculated in CNN with a state equation in vector form:

$$\dot{\mathbf{x}} = -\mathbf{x} + (\mathbf{B}_1 + \mathbf{B}_2 + \mathbf{B}_3 + \mathbf{B}_4 + \mathbf{B}_5 + \mathbf{B}_6 + \mathbf{B}_7 + \mathbf{B}_8)\mathbf{u} \tag{11}$$

where the image to be predicted is used as input $\mathbf{u}$, $\mathbf{B}_1$ is template for the linear term 1 in (10), defined in (4), and the templates $\mathbf{B}_2$ to $\mathbf{B}_8$ for the cross terms 2 to 8 are determined in (12).

$$\text{terms 2}: B_2 = u_1\begin{pmatrix} 0 & w_{12} & w_{13} \\ w_{14} & 0 & 0 \\ 0 & 0 & 0 \end{pmatrix}, \text{terms 3}: B_3 = u_2\begin{pmatrix} 0 & 0 & w_{23} \\ w_{24} & 0 & 0 \\ 0 & 0 & 0 \end{pmatrix},$$

$$\text{terms 4}: B_4 = u_3\begin{pmatrix} 0 & 0 & 0 \\ w_{34} & 0 & 0 \\ 0 & 0 & 0 \end{pmatrix}, \text{terms 5}: B_5 = u_1u_2\begin{pmatrix} 0 & 0 & w_{123} \\ w_{124} & 0 & 0 \\ 0 & 0 & 0 \end{pmatrix},$$

$$\text{terms 6}: B_6 = u_1u_3\begin{pmatrix} 0 & 0 & 0 \\ w_{134} & 0 & 0 \\ 0 & 0 & 0 \end{pmatrix}, \text{terms 7}: B_7 = u_2u_3\begin{pmatrix} 0 & 0 & 0 \\ w_{234} & 0 & 0 \\ 0 & 0 & 0 \end{pmatrix}, \tag{12}$$

$$\text{terms 8}: B_8 = u_1u_2u_3\begin{pmatrix} 0 & 0 & 0 \\ w_{1234} & 0 & 0 \\ 0 & 0 & 0 \end{pmatrix}$$

$$\text{for cell}\,(i,j)$$

$$\mathbf{u} = \begin{pmatrix} u_1 & u_2 & u_3 \\ u_4 & u_0 & u_5 \\ u_6 & u_7 & u_8 \end{pmatrix} = \begin{pmatrix} u_{i-1,j-1} & u_{i-1,j} & u_{i-1,j+1} \\ u_{i,j-1} & u_{i,j} & u_{i,j+1} \\ u_{i+1,j-1} & u_{i+1,j} & u_{i+1,j+1} \end{pmatrix}.$$

The templates in (12) are position-dependent (translation variant), i.e. the coefficients in **B** are not only dependent on the position related to the center cell (i,j), but also dependent on the input values in the corresponding coordinates (except B_1).

2.6 Determination of CNN Parameters

In classical neural networks the parameters of feed-forward networks can be obtained by a backpropagation algorithm. Unfortunately this method is not easily to be implemented for the determination of CNN parameters, therefore, we use some global optimization methods. In this work, the simulated annealing method is used.

In order to determine the CNN parameters, firstly a training image set is chosen. This set contains images with different statistical characteristics.

During the optimization processing, the mean square error (MSE) is used as optimizing variable, which is defined as

$$MSE = \frac{1}{MN}\sum_{i=1}^{M}\sum_{j=1}^{N}(y_{ij} - \hat{y}_{ij})^2, \tag{13}$$

where y_{ij} are the pixel values of original image set and $\hat{y}_{ij}$ are the predictive values obtained from the CNN structure (y_{ij} and $\hat{y}_{ij}$ are expressed as integer number in the range of [0, 255]). According to this definition, the MSE value could be regarded as mean power of the predictive error in each pixel, thus should be as small as possible.

3 Test and Evaluation

In order to evaluate the performance of the CNN-based predictors proposed here, we have tested a wide variety of images with different statistical characteristics. Due to the restrictive size of this paper, only parts of results are listed here.

We take some prediction methods as references. The linear predictor evaluated as basic test is defined as (the definition of $x_1 \dots x_4$ are shown in Fig.1):

$$\hat{x}(n) = -0.5\, x_1 + 0.75 x_2 + 0.75 x_4. \tag{14}$$

We use two adaptive nonlinear predictors as references. One is LOCO-I algorithm used in JPEG-LS standard [6]. Another is an extension of LOCO-I by introducing more precise conditions to classify the context in the images [7].

Furthermore, we tested the images with the predictor in CALIC (a context-based, adaptive, lossless image codec) proposed in [8], which is considered to be one of the most efficient predictive coding schemes. In the CALIC algorithm, a larger neighborhood structure is used (7 pixels are involved) and a complicated gradient-adjusted prediction (GAP) is employed. In this paper, we take the results from the CALIC algorithm as high standard.

The performance comparison is given in Table 1. It is obvious, that in most of the cases, the CNN-based predictors have better performances (comparing the proposed methods to linear/LOCO-I) and the performance improvement is especially noticeable in the images "Mandrill" and "Barbara" which are texture/contour-dominant cases.

In many cases, the performances of proposed CNN predictors are better than LOCO-I and very close to the performance of the context-adapted predictor based on texture classification, which implements complicated conditions for classification of the context in the images to ensure a high performance.

Even compared to the performance of CALIC, which is taken as a high standard in this paper, the performances of the proposed CNN predictors are comparable to the performances of CALIC in most of the cases.

Among the proposed CNN-based predictors, the multilayer predictor with highest complexity (36 parameters) has the highest performance in the most cases. It is notable, even with lowest complexity, the polynomial type predictor has slightly weak performance compared to other two types.

Table 1. Performance Comparison (In MSE)

Images	Lena	Moon	Skull	Mandrill	Barbara
Linear	47.90	58.26	**10.40**	446.15	312.64
LOCI-I	51.17	63.67	12.47	431.94	270.56
Extended	42.90	55.20	11.03	**398.16**	250.24
CALIC	**41.05**	56.61	12.02	406.94	**235.93**
PPC	42.60	54.44	11.97	409.58	258.83
MPC	42.28	**54.42**	12.23	413.47	249.67
HPC	42.81	54.91	12.07	409.70	260.08

Linear: linear predictor, parameters see Fig. 4
LOCI-I: LOCI-I algorithm
Extend: a context-adapted predictor based on texture classification [7]
CALIC: CALIC algorithm
PPC: polynomial type predictor (2nd order, 8 non-zero parameters)
MPC: multilayer predictor (N=6, M=0, 30 parameters)
HPC: higher order predictor (15 parameters)

Fig. 3. Test images (from upper left to bottom rights are Lena, Moon, Skull, Mandrill and Barbara)

4 Conclusion and Outlook

In this paper, we proposed several kinds of nonlinear predictors by means of CNN. Experimental results have shown that the performances of the proposed CNN-based predictors are similar to the performances of some existing adaptive nonlinear predictors, which means the adaptability of predictors to images with different statistical characteristics can be realized by the nonlinear structure in CNN.

CNN supplies not only a parallel computing framework, but also a uniform calculation structure avoiding the decision operations with complicated conditions for image context classification, then leading to a decreasing system complexity. Due to the parallel computing nature, the proposed CNN-based predictors should have very high processing speed with the CNN-hardware support [9]. These predictors could be implemented in some time-critical applications. Considering the low complexity of systems and acceptable performance, the polynomial type CNN-based nonlinear predictor may be a good candidate for these application.

We propose to develop hardware-oriented solutions for predictive coding by means of CNN. With current CNN hardware platforms, a system with larger neighborhood structure (e.g. 5x5 cells) is difficult to realize. We restrict the neighborhood structure of CNN to 3x3 cells, thus in the predictive coding, only 4 previous pixels can be considered, and CALIC-like algorithms, including more pixels into prediction, cannot be realized by current CNN structures. In the further development of CNN hardware, and the computation of CNN with larger neighborhood would no more be a problem. It seems worth to develop some predictors applying CNN with bigger neighborhood structure to obtain better performance.

References

1. Li, J., Manikopoulos, C.N.: Nonlinear predictor in image coding with DPCM. Electron. Lett. 26, 1357–1359 (1990)
2. Haykin, S., Li, L.: Nonlinear adaptive prediction of non-stationary signals. To be published in IEEE Trans. Signal Processing (1995)
3. Chua, L.: CNN: A Paradigm for Complexity. World Scientific Publishing Co. Pte. Ltd. (1998)
4. Puffer, F., Tetzlaff, R.: Modeling nonlinear systems with cellular neural networks. ICASSP 96, 3513–3516 (1996)
5. Dianat, S.A., Nasrabadi, N.M., Venkataraman, S.: A non-linear predictor for differential pulse-code encoder (DPCM) using artificial neural networks. In: Proc. ICASSP, Toronto, pp. 2793–2796 (1991)
6. Weinberger, M.J., Seroussi, G., Sapiro, G.: The LOCO-I lossless image compression algorithm: Principles and standardization into JPEG-LS. IEEE Trans. Imge Processing 9(8), 1309–1324 (2000)
7. Xu, Y., Liu, B.: Predictive image coding scheme based on pixel texture classification. Computer Engineering and Applications 44(6), 62–64 (2008) (in Chinese)
8. Wu, X., Memon, N.: Context-based, adaptive, lossless image coding. IEEE Trans. on Communications 45(4), 437–444 (1997)
9. Mueller, J., Becker, R., Mueller, J., Tetzlaff, R.: CESAR: Emulating Cellular Networks on FPGA. CNNA, Turin (2012)

Phase Dynamics on Small Hexagonal Lattices with Repulsive Coupling

Petar Tomov and Michael Zaks

Institute of Mathematics, Humboldt University of Berlin
Rudower Chaussee 25, 12489 Berlin
tomov@mathematik.hu-berlin.edu

Abstract. We consider dynamics of identical phase oscillators on spatially periodic hexagonal lattices. Existence and stability properties of different clustering patterns are discussed and illustrated by numerical examples. In the case of the 4×4-lattice, the clustering pattern enables the existence of a constant of motion and, hence, of a continuous family of temporally periodic solutions.

Keywords: Active rotators, hexagonal lattice, clustering, stability.

1 Introduction

In diverse complex systems, from physics, chemistry and biology, to economy and sociology, synchronization manifests itself in identical behavior of all units in the coupled ensemble, or of certain groups of such units. Many of such systems can be viewed as coupled networks, where the level of synchrony depends on the network topology, internal dynamics of single elements, and the corresponding coupling functions. In a series of papers, summarized in [1], Golubitsky and Stewart develop a groupoid formalism enabling identification of synchronous states in networks with various coupling topologies. Within this general framework, the main accent is put on the existence of such states and not on their realizations as possible attractors under the specific choices of the functions governing the dynamics and of the system parameters.

Below, we consider an application of this formalism to finite hexagonal lattices. Dynamical systems on regular lattices with local coupling form a particular subset of coupled ensembles: constraints imposed by the lattice geometry can result in the occurrence of states and bifurcations which would look highly degenerate in the more generic context. Conditions for the existence of synchronized states on hexagonal latices have been stated by Wang, Golubitsky and coauthors in [2,3]. What we are interested in, is the realization of such states, their coexistence and stability. Reducing the complexity of the problem, we put into the lattice nodes identical phase oscillators and couple them to the nearest neighbors. We concentrate on the case of "negative" (repulsive) coupling: whilst positive coupling commonly favors the onset of globally synchronized states, repulsive coupling is known to result in less trivial effects like clustering and hysteresis [4].

V.M. Mladenov and P.C. Ivanov (Eds.): NDES 2014, CCIS 438, pp. 246–253, 2014.

In this respect, the commonly studied rectangular lattices are not especially rich dynamically; there, the checkerboard pattern of alternating antiphase states is the natural form of synchronized solution. In contrast, on hexagonal lattices with repulsive coupling not only completely synchronous steady solutions, but the two-state phase-antiphase patterns, as well, become potentially unstable due to inevitable presence of frustrated bonds between the nearest neighbors [5]. Hence, more complicated dynamics can be expected.

To start with, we focus on small hexagonal lattices with periodic boundary conditions. This choice is motivated by the fact that in a hexagonal lattice with nearest neighbor coupling, for a given number of possible clusters, there are finitely many patterns of synchrony which are always doubly-periodic in space [3]. Besides, the patterns of synchrony are uniquely determined on the whole (also infinitely large) lattice by their restriction to finite subsets that capture the whole nonperiodic motive of the pattern [3].

As a suitable model of on-site phase dynamics we take the model of sinusoidally coupled "active rotators", introduced by Kuramoto and Shinomoto [6], and widely used in studies of collective effects, mostly in the context of interaction between global coupling and noise (see e.g. [7,8,9]).

The equation in each node of the $M \times N$ hexagonal lattice reads

$$\dot{\phi}_i = \omega - b \sin \phi_i + \frac{\kappa}{6} \sum_j A_{ij} \sin(\phi_j - \phi_i), \quad i = 1, \ldots, MN \tag{1}$$

where A_{ij} is the adjacency matrix of the hexagonal lattice, with $A_{ii} = 0$, $A_{ij} = 1$ if there is a link between nodes i and j, and $A_{ij} = 0$ otherwise.

2 Robust Patterns of Synchrony and Their Stability

Here, without going into mathematical details, we shortly review some aspects of the groupoid formalism [1], emphasizing the necessary conditions for the existence of polysynchronous states in the system of interest (1). The fact that in (1) all elements have identical dynamics, simplifies the overall picture; hence, the conditions for the existence of such states, discussed below, are not the most general ones. We also present some observations on the conditions allowing to identify synchronous states with equivalent stability and bifurcation behavior.

In the context of coupled oscillators, it is usual to call the set of elements which at all times share instantaneous phase values, a *cluster*. If, on a lattice, we paint all the oscillators belonging to a given cluster in the same color and use different colors for different clusters, we end up with a colored pattern, called a pattern of synchrony. This pattern is *robust* if it is flow-invariant: if two oscillators belong to the same cluster at some moment of time, they stay in the same cluster forever. Within the framework of the groupoid formalism, it has been shown that a pattern of synchrony is robust if all elements bearing a given color receive input from the same set of colors [3]. This condition refers to the existence of possible patterns, but does not concern their stability and can therefore not predict which (stable) patterns will be observed in the simulations or realized in an experiment.

Although, without specifying the coupling function, general statements about stability are hardly possible, the language of the groupoid formalism allows us to identify patterns of synchrony that have the same dynamical evolution and in particular the same stability and bifurcation properties. The statements below refer to the system (1) but can be extended to generic coupled networks. Let a colored path on the network denote the sequence of colors that the elements along this path bear for the given pattern of synchrony. We deduce that two different, not necessarily robust, patterns of synchrony have the same dynamical evolution if for each colored element in one of the patterns there is a matching element in the other pattern with the same color, and for both elements there exist *equivalent* colored shortest paths connecting them to every other element ot a different color and *identical* colored shortest paths connecting them to every other element ot the same color. In particular, if the patterns of synchrony are robust, then they have the same stability and bifurcation properties. An application of this statement will be illustrated in Section 3.4 below.

3 Small Lattices: Temporary and Spatial Patterns

Here we describe a few exemplary small-size lattices and illustrate typical features of dynamics with the help of results of numerical integration.

3.1 General Features

We consider the plane hexagonal lattice with periodic boundary conditions; spatial periods are, respectively, M and N. On each lattice node the dynamics obeys the equation (1). Assuming $b \neq 0$, we turn b into 1 by rescaling the time units (and, if necessary, shifting all phases φ_i by π). At large values of $|\omega|$, the right-hand side of (1) never changes sign: the system performs monotonic rotations. At $|\omega| < 1$, in contrast, a decoupled rotator is locked at the equilibrium, hence time-dependence is a collective effect.

The phase space of the system is the $(2\pi)^{MN}$-torus. Invariance with respect to translations by integer number of nodes in either direction, as well as to rotations by $\pi n/3$ ($n = 1, \ldots, 5$) and to reflections with respect to the axes which pass through the nodes, ensures existence of numerous invariant subspaces and essentially influences the dynamics.

In absence of coupling, the system possesses 2^{MN} steady states, of which only the synchronous one, $\varphi_i = \varphi^{(0)} = \arcsin \omega$, is asymptotically stable. Invariant manifolds of the remaining steady states form in the phase space an intricate web, providing "building material" for heteroclinic connections. At $\kappa \geq 0$ the synchronous solution $\varphi^{(0)}$ is stable. For sufficiently small negative values of κ that state remains stable as well; at $\kappa_c = -\sqrt{1-\omega^2}/(1-\lambda/6)$ where λ is the *most negative* eigenvalue of the adjacency matrix, the synchronous state loses stability in the subcritical pitchfork bifurcation. For the hexagonal lattice the most negative eigenvalue is always degenerate; its multiplicity depends on the lattice size (see below). Hence, the subcritical pitchfork bifurcation is degenerate as well: several pairs of saddles simultaneously coalesce with the equilibrium, and the corresponding central manifold has a reasonably high (up to 9) dimension.

For our numerical studies, we took several different sizes of the lattice. For each of them, we fixed $\omega = 0.7$ and varied the coupling strength κ between -3 and κ_c (as well as slightly above κ_c, to check for a possible hysteresis with $\varphi^{(0)}$). For each value of κ, we started the trajectories from not less than 2×10^4 initial conditions, taken from the homogeneous random distribution in the $(2\pi)^{MN}$-torus. Below κ_c, we detected no time-independent attractors; observed regimes were either periodic or quasiperiodic or chaotic.

Like in other sets of coupled oscillators, equation (1) on an arbitrary lattice obeys variational dynamics [5], hence periodic motions in which every unit performs oscillations with the amplitude smaller than 2π, do not occur. Recalling mechanical distinction between *rotations* and *librations*, this means that pure librations are impossible: at least one of the units should perform full-scale rotations. This circumstance, in its turn, precludes the possibility of the Hopf bifurcations: in this context, periodic solutions can arise either from the saddle-node bifurcations or from trajectories, heteroclinic to the equilibria. Below, we discuss two qualitatively different classes of heteroclinic bifurcations.

3.2 Lattice 3×3

Within this smallest possible configuration of the hexagonal lattice with periodic boundary conditions, every unit is directly connected to nearly the whole rest of the ensemble: to 6 of its eight remaining members. Above κ_c all probing trajectories end up at the synchronous steady state. For this lattice, the most negative eigenvalue λ of the adjacency matrix equals -3 and has multiplicity 2.

The pitchfork bifurcation at κ_c creates a "heteroclinic channel" from the equilibrium at $\varphi^{(0)}$ to its replicas shifted by 2π. During the motion along the channel, the lattice consists of three clusters: each node is linked to all units from every other cluster, and to none of the units from its own cluster (Fig.1a).

When κ is decreased below κ_c, the attracting orbit detaches from the unstable fixed point at $\varphi^{(0)}$; temporal period (time of passage across the torus) becomes finite. Remarkably, asymptotics of the period is not the inverse square root law, typical for periodic orbits born from the homoclinics at a saddle-node bifurcation of equilibria: in the case of heteroclinics at degenerate pitchfork, the period gets inversely proportional to the distance from the critical parameter value:

$$T \sim (\kappa_c - \kappa)^{-1} \tag{2}$$

This limit cycle (along with its symmetric copies) appears to be the only attractor of the system in the whole studied range of the coupling strength κ.

3.3 Lattices 3×4, 3×5, 4×5

Bifurcation scenarios for small non-square lattices have much in common. This might be related to the fact that for all non-square lattices with hexagonal

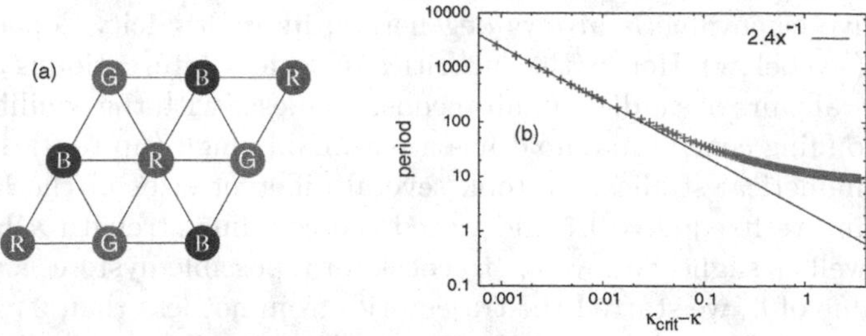

Fig. 1. (Color Online). Dynamics on the 3 × 3 lattice. (a) Clustering pattern for the limit cycle. (b) Growth of period near degenerate pitchfork. Crosses: numerical values. Solid curve: asymptotic dependence.

coupling, the multiplicity of the most negative eigenvalue of the adjacency matrix is the same: it equals 2. That eigenvalue itself is, respectively, $-1-\sqrt{3}$ for the 3×4 lattice, –2.82709... for the 3×5 lattice and –2.79360... for the 4×5 lattice, yielding critical values of κ between –0.485 and –0.491. In all these cases we found no other attractors at $\kappa > \kappa_c$, besides the synchronous $\varphi^{(0)}$. However, in all cases at a certain value κ_{het} at a small but *finite* distance *above* κ_c, the heteroclinic bifurcation occurs: a connection between one of the saddle points and one of its 2π-shifted replicas appears. [Needless to say, equivalent connections arise for all symmetric images of the point as well]. Subsequent breakup of this trajectory leaves in the toroidal phase space the closed orbit. The newborn periodic state is unstable: a few of its Floquet multipliers lie outside the unit circle. When κ is decreased across the critical value, instabilities get suppressed in several period-doubling and/or Neimark-Sacker bifurcations, until finally the periodic solution turns into the stable limit cycle. Since the original heteroclinicity concerns structurally stable saddle points (unlike heteroclinicity to the non-hyperbolic point from the previous section), the period of this solution displays the textbook logarithmic asymptotics (Fig. 2a). Lattice patterns of these periodic solutions show no clusters: all units assume different instantaneous values.

Further decrease of κ leads to multistability: new stable oscillatory states are born either via saddle-node bifurcations of periodic orbits or from heteroclinic contours. Among these periodic solutions, clustered patterns are present as well: the state with 6 clusters of 2 units each on the 3×4-lattice, and a state with 10 clusters of 2 units each on the 4×5-lattice.

In the interval between κ_c and stabilization of periodic trajectories, where neither steady nor periodic attractors are present, the lattice displays chaotic or quasiperiodic dynamics. The latter can be identified from one-dimensional curves along which the attracting trajectories intersect the suitably chosen Poincaré planes; an example is presented in Fig. 2b.

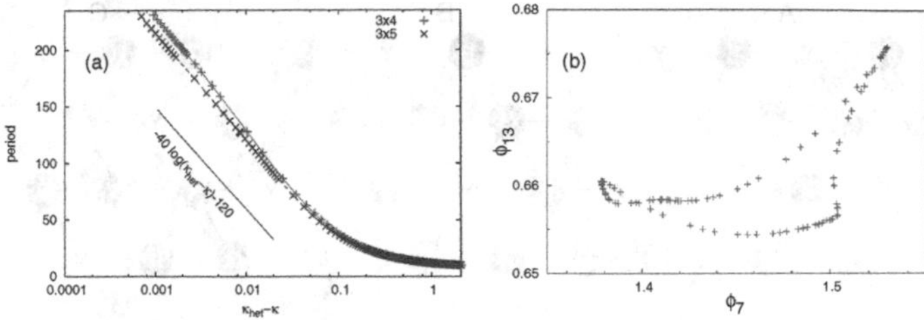

Fig. 2. (Color Online). Dynamics on lattices with unequal spatial periods. (a) Lengths of periods near heteroclinic bifurcations. Pluses: lattice 3×4 (κ_{het}=-0.47475). Crosses: lattice 3×5 (κ_{het}=-0.47731). Solid line: logarithmic dependence. (b) Quasiperiodic state on the 4 × 5-lattice at κ=–0.49. Crosses: intersections with the Poincarè plane $\varphi_1 = \pi$.

3.4 Lattice 4×4

For this lattice, the multiplicity of the most negative eigenvalue of the adjacency matrix is extraordinarily high: it equals 9. Accordingly, the central manifold which governs the system near the degenerate pitchfork bifurcation at $\kappa_c = -0.53560\ldots$, is 9-dimensional. It appears, however, that none of the states born in this bifurcation is attracting: among the lattices which we describe here, this is the only one where before the bifurcation the stable synchronous solution $\varphi^{(0)}$ coexists with another attractor, and immediately after the bifurcation the latter seems to overtake the whole basin of attraction of $\varphi^{(0)}$. This attractor is a periodic solution, born at $\kappa = -0.4817843$ in the saddle-node bifurcation. Remarkably, the lattice pattern in this state consists of 5 clusters: the big cluster is formed by 8 units which exhibit small-scale librations, whereas 4 small clusters, two units in each of them, perform full-scale rotations. The small clusters are phase-shifted with respect to each other by the quarter of period: effect, well understood in arrays of identical oscillators [10].

The 5-cluster periodic solution remains stable until $\kappa = -0.613$; it is replaced by quasiperiodic and later chaotic oscillations. At around $\kappa = -0.7$ a new continuous family of attractors appears in the phase space: these are the 4-cluster periodic solutions. Here, a remarkable situation takes place: a continuum of periodic states in the non-conservative system. This phenomenon is explained along the following lines. The coupling on the lattice is local, and each unit communicates only to the adjoining part of the ensemble. At a closer look we notice that in the arrangement of 4-cluster solutions (cf. Fig.3) each unit is linked to *two* units from *every other* cluster, and to none from its own cluster. Since all rotators are identical, contributions of the units from the same cluster coincide. Hence, in this state the 16 equations of the system (1) can be replaced by four equations – one for each cluster – which possess the same structure, save for the *doubled* coupling strength, and in which every unit interacts with every other unit. This is a spontaneous onset of global coupling in the setup with local coupling.

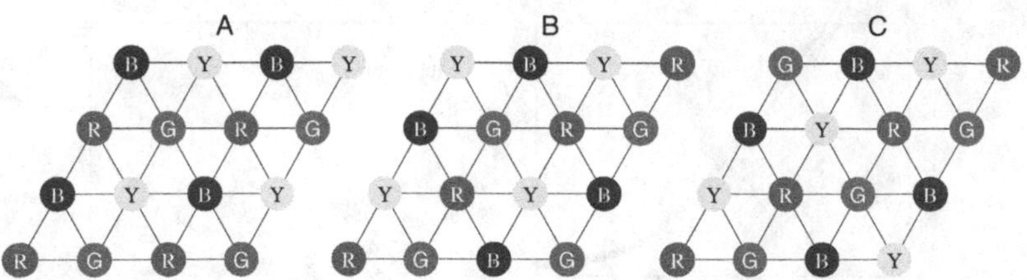

Fig. 3. (Color Online). Different clustering patterns of periodic states on the 4×4-lattice at $\kappa = -1.5$. Nodes belonging to the same cluster share the same color.

The coupling function in (1) includes only the first Fourier harmonics. Hence, the celebrated result of Watanabe and Strogatz holds: a system of N globally sinusoidally coupled phase oscillators possesses N–3 constants of motion [11]. In our case (N=4, the number of clusters), this means that there is just one constant of motion; varying it through the choice of appropriate initial conditions, we obtain, under fixed values of ω and κ, a continuous family of periodic solutions. In fact, this family is born from the heteroclinics to the degenerate saddle at κ_c, and the period of solutions obeys the asymptotics (2). However, near κ_c all these periodic solutions, stable in the 4-dimensional space of the reduced clustered system, are unstable with respect to the perturbations in the original 16-dimensional system which split the clusters, and stay invisible for direct numerical integration. At lower values of κ part of this continuum gets stabilized, and provides the bulk of the observed attractors: for $-2< \kappa <-1$ around 70% of initial conditions converge to the described 4-cluster solutions.

Considering the patterns of synchrony of the periodic solutions from Fig.3, we can apply the statement from Sect. 2 explaining why B and C have identical stability, different from that of A. Observe first that the patterns are *not* connected by symmetry transformations. Patterns B and C look similar, but differ in the way the units on the diagonal (from the top left to the bottom right) are ordered. These two configurations are related by an exchange of green and yellow units along this diagonal, which is not a symmetry transformation between the patterns. Nevertheless, both configurations have identical dynamical evolutions and therefore identical stability.

These three patterns of synchrony are robust: every unit of a given color receives input from exactly the same colors within the pattern. Therefore, all units of a given color within the pattern are equivalent, making also the sets of colored shortest paths starting from units with the same color equivalent. Consider, for example, the shortest paths of length 1 starting from a red unit. The sets of these paths are the same for the three patterns: $\{rg, rg, rb, rb, ry, ry\}$. The sets of shortest paths of length 2 are, however, different. In all three configurations each red unit is connected to three other red and respectively two blue, green and yellow ones. Table 1 shows the corresponding paths; paths towards the same unit are grouped in brackets. In all three patterns a way from a red unit to a blue one leads through either green or yellow. Those are *equivalent* paths but not

Table 1. Paths connecting a red element to its next nearest neighbors

Pattern		r	b	g	y
A	r	(bb) (gg) (yy)	(gy) (gy)	(by) (by)	(gb) (gb)
B	r	(bb) (gy) (gy)	(gg) (yy)	(by) (by)	(gb) (gb)
C	r	(bb) (gy) (gy)	(gy) (gy)	(by) (by)	(gb) (gb)

necessarily *identical*. In A and C a blue neighbor can be reached by going both through green *and* yellow (in those two cases the paths are *identical*), while in B it can be reached by going either through green *or* yellow, which are different but *equivalent* paths to those in A and C. The paths connecting red and green or yellow units are identical (thus also equivalent) for all three patherns. Hence, the first condition of the statement from section 2 is fullfiled for all three patterns. The paths connecting two red units, however, are identical only for patterns B and C. In the case of a 4×4 lattice, because of the periodic boundary conditions, there are no shortest paths of length greater than 2. In a similar manner we see that for all colors in B and C the conditions hold. Therefore, according to the statement, the patterns B and C have the same stability, different from that of A; this has been also confirmed by numerics.

Acknowledgment. Our research was supported by the IRTG 1740 of the DFG "Dynamical Phenomena in Complex Networks".

References

1. Golubitsky, M., Stewart, I.: Nonlinear dynamics of networks: The groupoid formalism. Bull. Amer. Math. Soc. 43(3), 305–364 (2006)
2. Wang, Y., Golubitsky, M.: Two-colour patterns of synchrony in lattice dynamical systems. Nonlinearity 18, 631–658 (2005)
3. Antoneli, F., Dias, A.P.S., Golubitsky, M., Wang, Y.: Patterns of synchrony in lattice dynamical systems. Nonlinearity 18, 2193–2210 (2005)
4. Tsimring, L.S., Rulkov, N.F., Larsen, M.L., Gabbay, M.: Repulsive synchronization in an array of phase oscillators. Phys. Rev. Lett. 95, 014101 (2005)
5. Ionita, F., Labavić, D., Zaks, M.A., Meyer-Ortmanns, H.: Order-by-disorder in classical oscillator systems. The Eur. Phys. Journal B 86, 511 (2013)
6. Shinomoto, S., Kuramoto, Y.: Phase transitions in active rotator systems. Progr. Theor. Phys. 75, 1105–1110 (1988)
7. Kurrer, C., Schulten, K.: Noise-induced synchronous neuronal oscillations. Phys. Rev. E 51, 6213–6216 (1995)
8. Zaks, M.A., Neiman, A.B., Feistel, S., Schimansky-Geier, L.: Noise-controlled oscillations and their bifurcations in coupled phase oscillators. Phys. Rev. E 68, 066206 (2003)
9. Tessone, C.J., Mirasso, C.R., Toral, R., Gunton, J.D.: Diversity-induced resonance. Phys. Rev. Lett. 97, 194101 (2006)
10. Aronson, D.G., Golubitsky, M., Mallet-Paret, J.: Ponies on a merry-go-round in arrays of Josephson junctions. Nonlinearity 4, 903–910 (1991)
11. Watanabe, S., Strogatz, S.H.: Constants of motion for superconducting Josephson arrays. Physica D 74, 197–253 (1994)

Novel Forecasting Techniques Using Big Data, Network Science and Economics

Irena Vodenska[1,3,⋆], Andreas Joseph[2,3], Eugene Stanley[3], and Guanrong Chen[2]

[1] Administrative Sciences Department, Metropolitan College, Boston University, Boston, MA 02215 USA
[2] Center for Chaos and Complex Networks, Department of Electronic Engineering, City University of Hong Kong, Hong Kong S.A.R., China
[3] Center for Polymer Studies and Department of Physics, Boston University, Boston, MA 02215, USA
vodenska@bu.edu

Abstract. The combination of theoretical network approach with recently available abundant economic data leads to the development of novel analytic and computational tools for modeling and forecasting key economic indicators. The main idea of this study is to introduce a topological component into economic analysis, consistently taking into account higher-order interactions in the economic network. We present a multiple linear regression optimization algorithm to generate a relational network between individual components of national balance of payment accounts. Our model describes well annual country statistics using the explanatory power and best fits of related global financial and trade indicators. The proposed algorithm delivers good forecasts with high accuracy for the majority of the indicators.

Keywords: Complex Systems, Econometrics, Interdependent Networks, Finance and Economics.

1 Introduction

Since the latest global financial crisis of 2008 and the resulting European Sovereign Debt Crisis of 2011, policy makers, academics and the public have shown increased awareness of the strong and important interconnectedness and interdependence of the global financial and economic architecture [1,2,3,4]. Additionally, standard techniques in macroeconomics partly failed to describe or foresee these major downturns, as pointed out by Jean-Claude Trichet in his opening address at the ECB Central Banking Conference (Frankfurt, 18 November 2010): "Macro models failed to predict the crisis and seemed incapable of explaining what was happening to the economy in a convincing manner. As a policy-maker during the crisis, I found the available models of limited help. In fact, I would go further: in the face of the crisis, we felt abandoned by conventional tools." In recent years we

⋆ Corresponding author.

V.M. Mladenov and P.C. Ivanov (Eds.): NDES 2014, CCIS 438, pp. 254–261, 2014.

have seen the rise of *Big Data*, i.e. the availability of large amounts of high-quality digitalized data, as well as the development of *network science*, which investigates the properties of systems composed of a large number of connected components. The combination of Big Data and network science offers potential applications for the design of data-driven analysis and regulatory tools, covering many aspects of our society [5]. More specifically, *econometrics* is a field where merging current analysis techniques with data- and network science is expected to provide large gains to modern economics.

The main idea behind network science is to generate a *network gain* by consistently considering multiple, as well as higher-order interactions between the individual components of a large system, such as users in a social network, banks in the financial system, or the interaction of entire economies on a global scale. Here we consider network dynamics and topological structure constructed by the flow of goods or cross-border investments between certain agents in a marketplace.

2 The MLR-Fit Network of Global Balance of Payment Accounts

Recently, the understanding of economic contagion has attracted growing interest, aiming to explain and measure the spreading of economic downturns between countries and across asset classes [6,7,3,8]. The modeling of global economic interactions between the different macro-components originating from multiple countries faces substantial difficulties due to the large number of possible interaction channels and due to the underlying computational complexity of the problem.

In this section, we present how a standard technique from economic analysis, namely multiple regression analysis (MLR), can be combined with Big Data and network science to tackle the above-stated problem by delivering an accurate phenomenological description of network relations and exhibiting good predictive power for macroeconomic indicators. The methodology that we apply here is not confined to any particular field, but has a large number of potential applications, whenever the criterion of sufficient data availability is fulfilled.

2.1 Balance of Payments Network Analysis

To study global trade and investment flows, we construct a relational network for 60 countries and eight financial and trade indicators for eleven consecutive years (2002-2012). These indicators constitute major parts of a country's balance of payments and represent nodes in our network. The eight indicators are the trade of goods (exports and imports) [9], inbound and outbound foreign direct investment (in- and out-FDI) [10] and inbound and outbound cross-border portfolio investment (in- and out-CPI) of equity and debt securities [11] on the nationally aggregated level, where we differentiate between in- and outbound relations. We show in Fig. 1 that the aggregated global trade flow within one year tracks the corresponding start-of-the-year investment positions, CPI and FDI. This is the reason why we will focus on a country's total exports and imports, as well as in- and outbound investment positions, in this analysis. Example indicators, which are included in this analysis, are the total exports of China (China: Exports) and the aggregated foreign holdings of US debt securities (USA: Debt (in)). All indicators have been adjusted for yearly changes in GDP, using the global GDP-deflator [12] (constant year-2012 values).

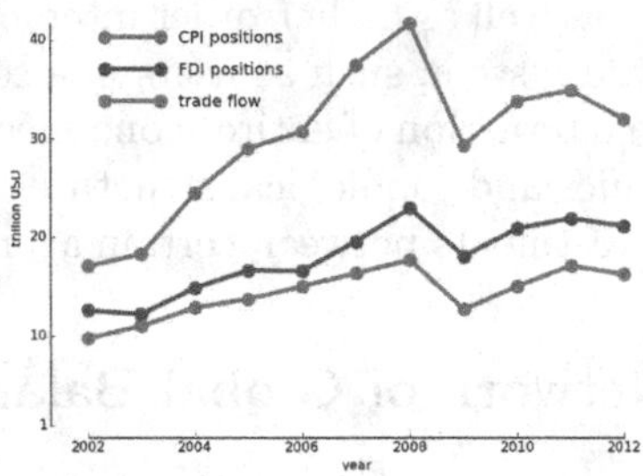

Fig. 1. Comparison of the **magnitudes of global trade flows and investment positions** (CPI and FDI) between 2002-2012. Because the total trade flow within one year tracks well the start-of-the-year positions of both types of investment (debt and equity), we study a combination of trade flows and investment positions.

We generate the links (relationships) between the trading and financial indicators (nodes) using a multiple linear regression (MLR) algorithm, connecting every indicator to one or several other indicators. We call the resulting relational network *Global Balance of Payments Network* (GBoPN). The network itself represents an evolution framework which describes and even forecasts most of the macroeconomic indicators with high accuracy. In algebraic terms, a network is represented by its *adjacency matrix* $\mathcal{A}$, where the element $a_{ij} \neq 0$ if there is an edge between node i and node j. One distinguishes between different classes of networks, depending on the possible values that the matrix elements a_{ij} are allowed to take. If $a_{ij} \in \{0,1\}$, $a_{ij} = a_{ji}$ and $a_{ii} = 0$, the resulting network is called a simple graph, which captures the topological structure of the network in which connections between nodes are symmetric. If the only requirement is that $a_{ij} \in \{0,1\}$, one obtains a directed graph, where connections between nodes are not required to be reciprocal. In the most general network, $a_{ij} \in \mathbb{R}$, meaning that one can have asymmetric connections between any two nodes accounting for possibly negative feedback. The underlying idea of constructing this network is that global financial and trade statistics are related to each other, such that it is possible to find a set of indicators (regressors) which best describes *another indicator* (regressand), using MLR.

2.2 MLR-fit Model for Forecasting Global Macroeconomic Indicators

Let $\boldsymbol{I}(t)$ denote the set of all indicators of all countries at time t, e.g. in 2007. We are interested in finding the best-fit network coefficient matrix $\boldsymbol{\beta} = \beta_{ij}$ which describes $\boldsymbol{I}(t+1)$. This can be written in terms of a linear matrix equation

$$I_i\,(t+1) \overset{\text{MLR}}{=} \sum_{j=1,j\neq i}^{N} \beta_{ij}\, I_j\,(t) + c_i\,, \tag{1}$$

where N is the total number of indicators from all countries and c_i is the intercept of indicator I_i. A link between indicators I_i and I_j is established if $\beta_{ij} \neq 0$. Note that the model (1) is highly appealing from a mathematical point of view because of its simplicity, as well as its built-in predictive power, since the coefficient matrix $\boldsymbol{\beta}$ can be interpreted as an *evolution operator*, which takes the indicator vector $\boldsymbol{I}$ from $t \to t+1$.

However, finding the optimal $\boldsymbol{\beta}$ according to fixed statistical criteria on the MLR-fit is a computationally hard problem because the number of possible solutions is growing super-exponentially with the number of indicators. Finding a good solution to this problem is much more likely with the availability of a large amount of data, while an efficient search method will considerably shorten the time to find a solution. We use an iterative least-square algorithm, which is based on the assumption that a regressor which individually describes a regressand well (simple regression) is likely to be contained in a group of regressors (multiple regression). The optimization algorithm that we use is as follows: On each $I_i\,(t+1)$ time series we perform a simple linear regression (SLR) with each *other* time-lagged time series $I_j\,(t)$ (Step 1). We pick the regressor I_j which generates the smallest error (residuum) for the starting model $f^1\left(I_i^{t+1}\right)$ (Step 2). We then add additional regressors to the model,according to the ordering of residua from Step 1 (smallest to largest), and test the resulting model for error reduction, statistical significance of all regressors (t-test) and collinearity (variance inflation factor (VIF) and condition number of the normalized design matrix, Steps 3 and 4). Based on the test results, we update the model (or not) and go back to Step 3. We repeat this step for $N-2$ times (Step 5). Finally, we test the statistical significance (F-test) and error of the final model $f^n\left(I_i^{t+1}\right)$, where n is the final number of regressors, then accept or reject the model (Step 6).

This procedure yields one row of the coefficient matrix $\boldsymbol{\beta}$ for each indicator $I_i\,(t+1)$. The final result crucially depends on the chosen requirements regarding the maximally allowable error, the statistical significance of each coefficient, and on collinearity bounds where there is a general trade off between the maximal error on one side and the statistical significance and collinearity between regressors on the other. Depending on the achievable balance between these quantities, the MLR-fit model may be accepted or rejected.

We test our model on a set of 60 countries, selected according to a 95%-criterion on the cumulative amount of the total monetary value (in USD) of all eight indicators taken together. Theoretically, we obtain a total of $60 \times 8 = 480$

indicators. Unfortunately the data are partly incomplete, so that we are left with a total of 405 indicators, or about 84% of the expected number. This is still a considerable number and enough data to achieve good fits for the great majority of indicators. Rather strict statistical criteria had been set in order to accept or reject each single fit at each step. Namely, a significance level for the t- and F- tests of $\alpha = 2.5\%$, a maximal time-averaged final fit error of each I_i of 10% and a maximal condition number and VIF on the design matrix of ten and five, respectively.

In order to test the forecasting capability of the MLR-network model (1), we remove the year-2012 data before performing the fit to use it for an out-of-sample test later. Taking the one-year time shift between regressor and regressand variables into account, we are left with a series of nine data points to do the fitting, which will turn out to be sufficient to do proper forecasting.

The resulting GBoPN is generated from the coefficient matrix $\boldsymbol{\beta}$, where an edge from indicator i to indicator j is drawn if $\beta_{ij} \neq 0$. This methodology gives us a non-trivial insight into selected macroeconomic indicators from network perspective, since it tells us that any of the initial 405 indicators is significantly coupled to at least one other indicator and that the resulting network covers all indicators for each of the 60 countries. This analysis does not show a separation into geographically localized clusters, which could have been a valid expected outcome, but rather it depicts a picture of a globally interacting multi-layered economic system. A feature which strongly underlines the importance of cross-border economic relations is that basically all best-fit edges (more than 98.4%) are connecting indicators from different countries and about three quarter of the links connects indicators from different classes, such as trade and FDI.

The GBoPN turns out to be extremely sparse with an edge density[1] of less than 0.5%. This means that a very small number of relations between all possible pairs of indicators is enough to describe nearly all indicators with high accuracy. A great majority of 298 indicators are *tracked* by two indicators or even just one. On the other side of the spectrum, there is a smaller number of indicators which *track* a large number of other indicators. This, in the language of graph theory, means that these indicators have a large out-degree and might prove potentially useful for the purpose of macroeconomic monitoring and forecasting because their observation is expected to provide information about other indicators and their host-countries.

2.3 Tracking Centrality of Macroeconomic Indicators

An additional criterion for identifying nodes of interest is the monetary value of indicators to which they point, because the analyzed indicators vary by several orders of magnitude. To account for the number and size (in terms of their time-averaged values in USD) of a node's regressands, we define its *tracking centrality* as

[1] The number of maximally possible edges $N(N-1)$ dived by the number of actual edges, which, in our case, is the number of non-zero coefficient β_{ij}.

$$T_i \equiv \sum_{j=1, j \neq i}^{N} \sqrt{R_{ij}^2}\, S_j\,, \tag{2}$$

were R_{ij}^2 is the *coefficient of determination* between indicators i and j, and S_j is the time-averaged monetary value of indicator i. By definition, $R_{ij}^2 \neq 0$, whenever $\beta_{ij} \neq 0$. $\sqrt{R_{ij}^2}$ equals the absolute value of the Pearson product correlation coefficient between indicators i and j and measures the fraction of variation which is mutually described.

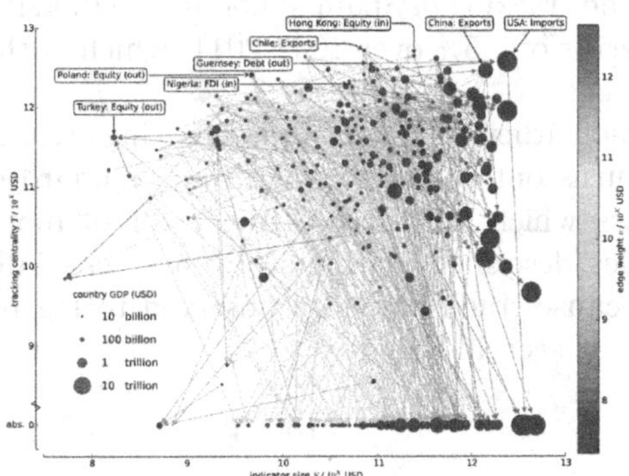

Fig. 2. Visualization of the **Global Balance of Payments Network** (GBoPN), constructed with the MLR-fit algorithm described in section 2. Indicators (nodes) are represented by dots with sizes corresponding to countries' GDP. The x- and y-positions of each node are set according to the indicator's time-averaged size S and tracking centrality T, respectively. Directed edges, pointing from a regressor i to its regressand(s) j, are color-coded according to the tracked value v_{ij}. A key-result is the high tracking capabilities of many indicators of small sizes or from small economies, which might be used as proxies or "thermometers" for larger economic changes. Highlighted examples include the outbound equity securities positions of Poland or the outbound debt securities positions of Guernsey. Indicators which have not been identified as statistically significant regressors, but are tracked by others, are positioned on the x-axis.

A comprehensive visualization of the entire GBoPN is shown in Fig. 2. Each indicator is represented by a dot of the size of its host-country's GDP. The x-position of each node is its size S and the y-position is its tracking centrality T, where a higher value means that a larger amount of economic value is quantitatively described by this indicator. The edges of the GBoPN are given through color-coded arrows pointing from a regressor to its regressand(s). The color is set according to the relation $v_{ij} = \sqrt{R_{ij}^2}\, S_j$, quantifying the value of this relationship. As such, the tracking centrality of an indicator is given by the values of its outbound connections. A key-result from Fig. 2 is the high tracking capabilities of many indicators of small sizes or from small economies, which might, in turn, be used as proxies or "thermometers" for larger economic changes. Such nodes can be easily spotted as having strong connection spanning a large horizontal distance (several orders of magnitude) from left to right. Highlighted examples include the outbound equity securities positions of Poland or the outbound debt securities positions of Guernsey. We note that the so-called off-shore financial centers [13] have high tracking centralities on average, due to their inherently strong coupling to the financial system.

Given the small average fit error, it is possible to make explicit use of the network structure, taking higher-order interactions into account, and track indicators over short paths, where the averaged errors are small. The maximally expected error is then the sum over all contributing errors along this path. A simple but instructive example is the tracking of Spain's inbound FDI, using Greece's imports with an aggregated error of 4.5% over 2002-2011, which is the only regressor in this case.

Besides having some outliers, the median forecasting error for the 372 macroeconomic indicators turns out to be 8.5% with an standard deviation of 15.5% excluding 3 indicators which have forecasting error of more than 100%. The forecasting capabilities described in this study are especially powerful for international trade, because trade flows are observed to lag behind cross-border investment positions, as seen in Fig. 1.

3 Conclusion

International trade is one of the main catalysts of globalization and social and economic development. The volume of global trade grew by more than a factor of five, in nominal terms, over the past 20 years[9], thereby closely linking up many of the involved countries and forming an international web of trade. In this sense, the investigation of the relation of global trade and economic growth is highly suited for the application of techniques from network science. In this study we use trade and financial (investment) flows between individual economies (nodes) to construct MLR-fit based edges of a *Global Balance of Payments Network* (GBoPN).

We introduce an MLR-fit algorithm to model multiple links between individual components of balance of payment accounts of a group of countries, which encompasses the majority of global macroeconomic activity. The derived network model delivers quite accurate description of most indicators, while the built-in time shift between regressors and regressands lead to good indicator forecasts. We have introduced the concept of an indicator's tracking centrality, which allowed for the identification of "macroeconomic thermometers", i.e. small indicators which track multiple large indicators with a relatively high precision. This novel methodology, in combination with other economic models, could be used as a monitoring tool of complex processes of economic cross-border interactions.

The methodology that we have developed here is not meant to stand alone, but to be merged with established economic analytics tools to generate a network-based approach in understanding global economic interactions. Similarly to [14], where the human organism is described as integrated network of complex physiological systems characterized by distinct network structure and topology, here we study the economic "organism" as dynamic complex network. By investigating the global balance of payment system as complex network, we introduce novel aspect of econometric analyses that may contribute to emergence of new research filed of *network economics.*

Acknowledgments. We would like to thank the European Commission FET Open Project FOC 255987 and FOC-INCO 297149 for financial support.

References

1. Catanzaro, M., Buchanan, M.: Network opportunity. Nature 9(3), 121–123 (2013)
2. Crotty, J.R.: Structural causes of the global financial crisis: A critical assessment of the 'new financial architecture'. Camb. J. Econ. 33(4), 563–580 (2009)
3. Preis, T., Kenett, D.Y., Stanley, H.E., Helbing, D., Ben-Jacob, E.: Quantifying the behavior of stock correlations under market stress. Sci. Rep. 2, 752 (2012), doi:10.1038/srep00752
4. Stulz, R.M.: Credit default swaps and the credit crisis. J. Econ. Perspect. 24(1), 73–92 (2010)
5. FuturICT: Global Computing for Our Complex World, `http://www.futurict.eu/`
6. Constancio, V.: Contagion and the european debt crisis. Financial Stability Review (16), 109–121 (2012)
7. Hartmann, P., Straetmans, S., de Vries, C.G.: Asset market linkages in crisis periods. The Review of Economics and Statistics 86(1), 313–326 (2004)
8. Kolanovic, M., et al.: Rise of cross-asset correlations. Global equity derivatives & delta one strategy report, J.P. Morgan Securities LLC (2011), `http://www.cboe.com/Institutional/JPMCrossAssetCorrelations.pdf` (accessed: November 15, 2013)
9. United Nations Statistics Division: `http://comtrade.un.org`: United Nations Commodity Trade Statistics Database (UN comtrade) (accessed: November 15, 2013)
10. United Nations Conference on Trade and Development Statistics (UNCTAD STAT): Inward and outward foreign direct investment stock, annual (1980-2012), `http://unctadstat.unctad.org/ReportFolders/reportFolders.aspx?sRF_ActivePath=P,5,27&sRF_Expanded=,P,5,27` (accessed: November 15, 2013)
11. International Monetary Fund: Coordinated Portfolio Investment Survey (Table 8): `http://cpis.imf.org` (acessed: April 1, 2013)
12. The World Bank, Data: `http://data.worldbank.org/indicator`: GDP (current USD), Stocks traded, total value (% of GDP), GDP deflator (annual %) (accessed: April-November 2013)
13. Zoromé, A.: Concept of offshore financial centers: In search of an operational definition. IMF Working Papers 07/87, International Monetary Fund (IMF) (2007): `http://www.imf.org/external/pubs/ft/wp/2007/wp0787.pdf` (accessed: November 15, 2013)
14. Bashan, A., Bartsch, R.P., Kantelhardt, J.W., Havlin, S., Ivanov, P.C.: Network physiology reveals relations between network topology and physiological function. Nature Communications 3, 702 (2012)

Complex Networks of Harmonic Structure in Classical Music

Florian Gomez, Tom Lorimer, and Ruedi Stoop

Institute of Neuroinformatics,
University of Zurich and ETH Zurich,
Winterthurerstrasse 190, 8057 Zurich, Switzerland
{fgomez,lorimert,ruedi}@ini.phys.ethz.ch

Abstract. Music is a ubiquitous, complex and defining phenomenon of human culture. We create and analyze complex networks representing harmonic transitions in eight selected compositions of Johann Sebastian Bach's Well-Tempered Clavier. While all resulting networks exhibit the typical 'small-world'-characteristics, they clearly differ in their degree distributions. Some of the degree distributions are well fit by a power-law, others by an exponential, and some by neither. This seems to preclude the necessity of a scale-free degree distribution for music to be appealing. To obtain a quality measure for the network representation, we design a simple algorithm that generates artificial polyphonic music, which also exhibits the different styles of composition underlying the various pieces.

Keywords: Complex Networks, Music, Harmony, Artificial Music.

1 Introduction

In recent years, the network-based approach to complex biological, social or technical systems has become increasingly popular. Especially since the terms 'small-world' and 'scale-free' networks were coined [1,2], the structural properties of systems as diverse as power grids, co-authorship networks, protein- interaction networks, or the brain have been analyzed for these properties (see [3] for a review). More recently, pitch transitions/fluctuations in classical and popular music were studied from a similar complex network perspective [4,5,6]. This approach, however, poses one relevant and fundamental question: it is *a priori* not clear how to meaningfully define *nodes* and *links* for a network representation of a (complex) musical piece. In this article, we apply a simple but musically justified procedure to create networks from classical music compositions by restricting ourselves to a representation solely based on *harmony* (co-occurring notes). In contrast to the popular corpus-based studies (e.g. [5,7]), we adhere to the view that "a single piece is normally the largest unit of artistic significance" [8,9] and therefore treat the musical pieces individually. As a measure for the quality of the representation, we use the appeal of polyphonic music generated from a simple artificial music generating algorithm. This allows us to assess the suitability of the general approach for the different compositions.

V.M. Mladenov and P.C. Ivanov (Eds.): NDES 2014, CCIS 438, pp. 262–269, 2014.

Music is often regarded as a truly universal language, yet it can provoke strikingly different emotions among different people. Trying to understand the beauty (and often the associated complexity) of music has thus been a subject of interest for thousands of years, and continues to be a topic across different fields. From a physical or mathematical point of view, different aspects of music may be addressed, such as the physical properties of sound itself (acoustics), the notion of salient perception-related characteristics such as 'pitch', or the mathematical rules underlying consonance and dissonance. In the past decades, some of the attention has shifted towards the investigation of the statistical properties of music. In 1978, Voss and Clarke reported that the spectral density of loudness fluctuations in music and speech approximately follows a 1/f-law [10] (which is related to the notion of self-organized criticality [11]). A later and more detailed study then showed that distinct power-law scaling exponents of these fluctuations characterize different genres of music [12]. Extending the time series based approach with the help of the tools from dynamical systems theory, entropy and complexity measures have been derived and applied to a variety of compositions dating from three centuries [9,13]. Paralleling work on text corpora, comparisons between occurrence distributions and Zipf's law have been made [14,15]. In the complex network field, different approaches were taken to create networks from musical pieces. Liu, Tse and Small [4] based their study on individual notes (including duration), pooled over different collections of compositions and used an algorithm to create artificial (however single-voiced) music. More closely related to our study, a thorough investigation by Serrà *et al.* [5] characterized the evolution of popular music in terms of harmony-network properties, where the notes within one beat determined the network nodes. While both of these studies addressed overall characteristics of music, in the following we focus on individual compositions, the characteristics they share, and the differences that separate them.

2 Methods

2.1 Network Creation

We created networks from 4 preludes and 4 fugues from Johann Sebastian Bach's Well-Tempered Clavier, volume I. The music notes were automatically read-in from freely available MIDI-files, but, with the help of the original score, carefully corrected by hand (a necessary step due to the often imperfect MIDI-files; moreover, trills and mordents were removed). From the score, the networks were generated as depicted in Fig. 1.

Nodes are defined by co-occurring notes, and a link a_{ij} is drawn if node j follows node i. Since we only focus on the harmonic structure, absolute pitch values, note duration, multiple occurrences of notes and possible permutations of notes among the voices are ignored. For example, in a piece with four voices, the co-occurrence of the notes $(C4, G4, C5, E5)$ is described by the node (C, E, G) (i.e., C Major). In this way, nodes correspond to single notes (e.g. (C)), double notes (e.g. (C, E)), triads (e.g. (C, E, G)), four notes (e.g. (C, E, G, A)), etc.

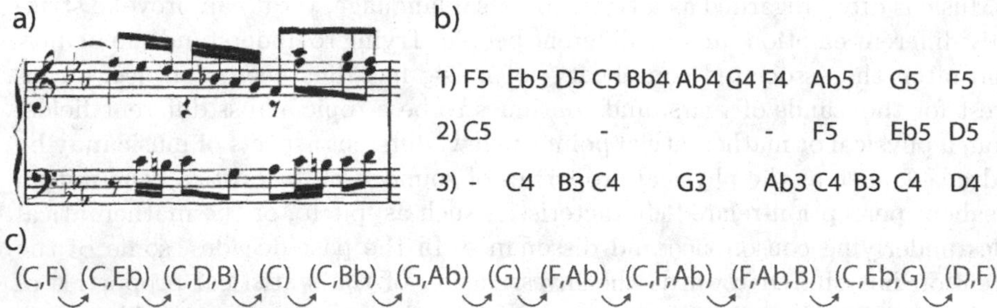

Fig. 1. Network creation. a) Original score (Fugue c-minor, bar 7), b) transcribed score, c) network nodes and links.

While this is similar to the approach of Ref. [5], here we apply a criterion of strict note co-occurrence, which seems (from the point of view of harmonic structure) more justified for classical music than a beat-based procedure. (We do, however, avoid more music-theoretically elaborate grouping methods, see e.g. [16].) To put weights on the links, we simply increase the weight w_{ij} by 1 whenever the transition a_{ij} occurs (see also [4]). For simplicity and without much loss of generality (see below), we concentrate on undirected networks, where in the case of the weighted networks, the weights w_{ij} and w_{ji} are added.

2.2 Generation of Artificial Music

The mapping from the musical score to the network is a projection-like operation, with considerable associated losses of musical information. To check the validity of the approach, a simple algorithm to generate artificial polyphonic music is designed. Given a network, a simple random walk on the network creates a sequence of nodes $\{s_k\}$. In the case of the weighted network, the next node is chosen with a probability corresponding to the respective link weight (similar to Ref. [4]). To map this sequence back to a truly polyphonic piece, the notes have to be assigned to the different voices in some way (a problem known as 'voice leading', see also [17]). Here, we adopt a simple procedure which is inspired by the old style of *basso continuo* playing. In accord with the first node s_1 in the artificial sequence (e.g. $s_1 = (C, G, Bb)$), we assign a starting point (in the case of three voices, e.g. $(C3, G3, Bb3)$). To obtain the next state, of all possible note permutations and absolute pitches consistent with the next node s_2, the configuration is chosen that minimizes the total variation of the voices (measuring the variation in units of semitones). For the example above, if $s_2 = (C, F, A)$, the next state would be $(C3, F3, A3)$: the lower is held constant, the middle voice moves 2 semitones, the upper voice 1 semitone down. In the case of multiple optimal configurations, the first solution found is chosen. In a final step, consecutive identical notes within one voice are connected (otherwise all notes have the same length, typically 0.25s).

3 Results

Using the procedure described above, 8 (weighted and unweighted) networks were created and analyzed, see Table 1 and Fig. 2.

Table 1. Overview over the compositions and network measures for undirected networks. The length is in units of 1/16-notes (1/8-notes for the fugue in Bb minor), C denotes the average local clustering coefficient, and L the average shortest path length. Values in brackets correspond to random networks with the same number of nodes and links.

Composition	Voices	Length	Nodes/Links	C	L
Prelude C Major	3	560	59/120	0.50 (0.07)	3.30 (2.91)
Prelude C# Major	2	624	59/232	0.31 (0.13)	2.29 (1.98)
Prelude F minor	3-4	352	101/228	0.23 (0.04)	3.16 (3.06)
Prelude Ab Major	2-4	528	56/261	0.42 (0.17)	2.16 (1.80)
Fugue C minor	3	496	103/259	0.20 (0.05)	2.90 (2.87)
Fugue E minor	2	504	67/279	0.39 (0.12)	2.22 (1.98)
Fugue A minor	4(-5)	1392	223/826	0.25 (0.03)	2.87 (2.70)
Fugue Bb minor	5	600	128/282	0.16 (0.03)	3.42 (3.27)

Comparing to random networks of the same number of nodes and links, all networks exhibit a high clustering coefficient and a comparably low average shortest path length thus fulfilling the 'small-world' property [1]. Looking at the individual networks in more detail, one finds a few well-connected nodes, which typically correspond to the double notes and triads closely related to the composition's key (often nodes pertaining to the tonic or the dominant). This reveals that despite the *contrapuntal* technique used (for the fugues), the composition harmonically still has its weight on the classic tonal chords.

The individuality of the compositions is reflected in the degree distributions as depicted in Fig. 3. We fitted the the degree distributions of both the weighted and the unweighted networks (undirected in both cases) with truncated power-laws and truncated exponentials, using maximum likelihood estimation. To statistically corroborate the observations, we used an approach similar to Refs. [18,19] and compared Kolmogorov-Smirnov statistics between the original degree data and corresponding surrogate data. The value p then denotes the proportion of surrogate datasets that have a worse fit (larger Kolmogorov-Smirnov distance) than the original data. The results suggest that some of the compositions' degree distributions are consistent with a (truncated) power-law, but that others clearly tend towards an exponential distribution. Fig. 3 shows two such examples (the Prelude Ab Major exhibiting a power-law-like distribution, and the Fugue E minor a more exponential distribution). Other degree distributions (such as the one corresponding to the largest network, the Fugue A minor) were neither consistent with an power-law nor with an exponential (e.g. $p < 0.1$ in both cases).

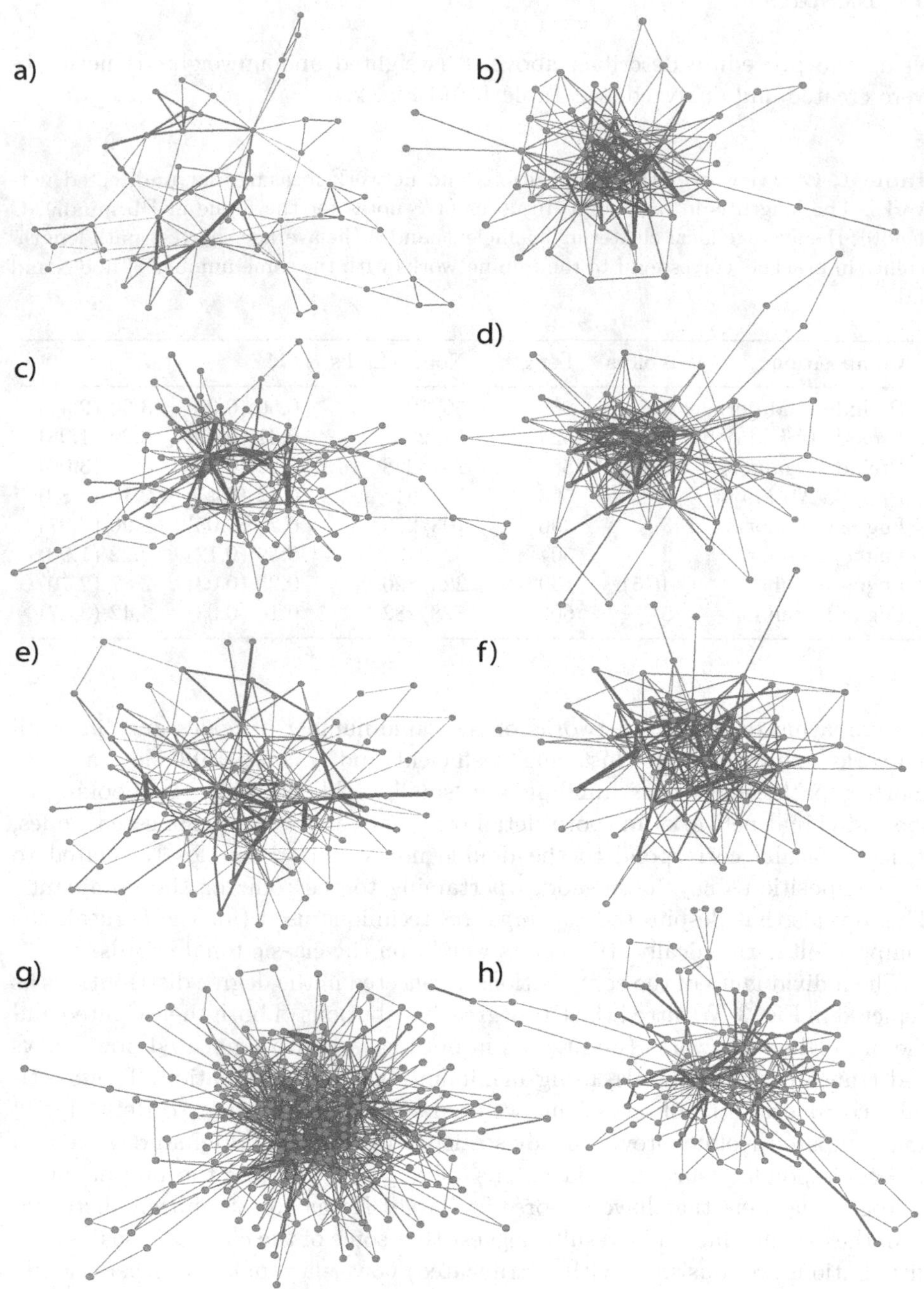

Fig. 2. Networks corresponding to the compositions of Table 1. a) Prelude C Major, b) Prelude C# Major, c) Prelude F minor, d) Prelude Ab Major, e) Fugue C minor, f) Fugue E minor, g) Fugue A minor, and h) Fugue Bb minor.

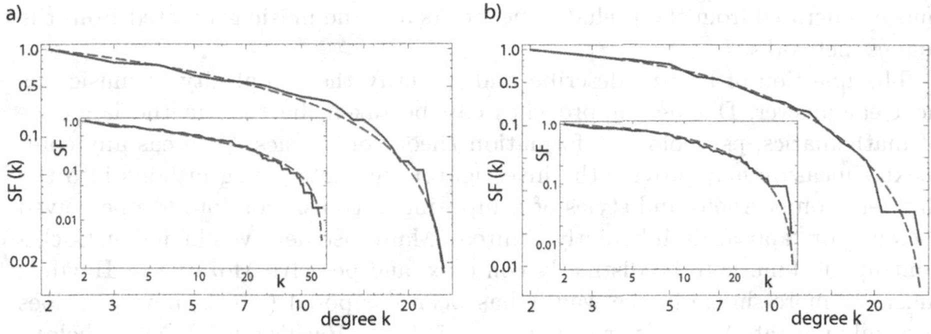

Fig. 3. Survival functions ($SF(k) = 1 - P(k)$, with the cumulative distribution $P(k)$) for a) the unweighted Prelude Ab Major network, and b) the unweighted Fugue E minor network. Fit (*dashed line*) corresponds in a) to a power-law fit (on the interval $(2, k_{max})$; exponent 0.88), and in b) to an exponential fit (same interval; decay constant 0.12). Insets: Same for weighted networks. Corresponding p-values (power-law/exponential): a) 0.54/0.02 (weighted: 0.30/0.01), and b) 0.08/0.79 (weighted: 0.18/0.40).

Using the algorithm described in the methods section, we generated artificial polyphonic music for 1) the unweighted and undirected, 2) the weighted and undirected, and 3) the weighted and directed networks. Typical examples are available online [20]. As general observation, the music from the weighted networks (2) sounds much more appealing than the music from the unweighted networks (1). The step from the undirected (2) to the directed (3) case, however, did usually not result in any further increase of appeal (rather the opposite). Comparing the underlying compositions of the artificial music, a trend emerges: whereas the music generated from most preludes' networks sounds for many listeners astonishingly pleasant and complex (see [20] for 'good' examples), the fugues usually give rise to rather unstructured and unpleasant music. A direct relation to the degree distribution, however, seems not to be evident.

4 Discussion and Conclusion

The approach we have taken here considers the immediate harmonic transitions of a musical composition and represents them as a complex network. In this way, salient characteristics of the evolution of a piece are captured (e.g. a dominant-tonic-transition), but longer time-scales (higher-order transitions, higher-level structures) are lost. Such higher-order transitions, which often exist due to intermediate notes of a voice (or multiple voices), are especially present in contrapuntal pieces, the most well-known of which are the fugues. While our investigations have demonstrated that there still exists a musically meaningful 'harmonic backbone' in these fugues (the nodes with highest degree), the independence of the voices in this truly polyphonic music renders any representation of harmony difficult. This probably explains the difference between the artificial music generated from the preludes' networks and the music generated from the fugues' networks.

The question of how to describe and quantify the complexity of music has no clear answer. Different approaches can be taken, be they in the language of mathematics, psychology, information theory or physics. Whereas any complexity measure may provide the investigator with interesting insights into the different compositions and styles of composing, a translation into the perceived 'beauty' or 'appeal' is inherently limited. Many listeners would judge Bach's contrapuntal music to be distinctly complex, and perceive Mozart's or Haydn's music as more simple in the way it has been composed (which, however, does certainly not rate Mozart's or Haydn's music below Bach's music). Nevertheless, and in spite of the subjectivity of 'beauty', there is considerable agreement on which composers were 'great' and one may therefore still ask whether there are any universal features that make music 'beautiful'.

In physics, the natural objects to look for when talking about universality are power-laws. A scale-free behavior of any attribute of musical compositions (be it a rank-frequency plot or the degree distribution of a network) seems, in this sense, obviously promising. Scale-free degree distributions of related (weighted) networks were found both in classical and popular music ([4], and statistically well-founded in the Supplementary Information of [5]). Whereas these studies pooled over entire collections of musical pieces, we analyzed the degree distributions of both the weighted and unweighted networks for single compositions. As a result, we find that some distributions are indeed well-fit by a power-law, but others clearly deviate and are much more consistent with an exponential. We conclude that a scale-free distribution of harmonic transitions seems not to be a prerequisite for 'appealing' musical compositions, but only a possibility.

Acknowledgments. This work was supported by the Swiss National Science Foundation.

References

1. Watts, D.J., Strogatz, S.H.: Collective dynamics of 'small-world' networks. Nature 393, 440–442 (1998)
2. Barabási, A., Albert, R.: Emergence of scaling in random networks. Science 286, 509–512 (1999)
3. Boccaletti, S., Latora, V., Moreno, Y., Chavez, M., Hwang, D.-U.: Complex networks: Structure and dynamics. Phys. Rep. 424, 175–308 (2006)
4. Liu, X.F., Tse, C.K., Small, M.: Complex network structure of musical compositions: Algorithmic generation of appealing music. Physica A 389, 126–132 (2010)
5. Serrà, J., Corral, A., Boguñá, M., Haro, M., Arcos, J.L.: Measuring the Evolution of Contemporary Western Popular Music. Sci. Rep. 2, 00521 (2012)
6. Liu, L., Wei, J., Zhang, H., Xin, J., Huang, J.: A Statistical Physics View of Pitch Fluctuations in the Classical Music from Bach to Chopin: Evidence for Scaling. PLoS ONE 8, e58710 (2013)

7. Zivic, P.H.R., Shifres, F., Cecchi, G.A.: Perceptual basis of evolving Western musical styles. Proc. Natl. Acad. Sci. U.S.A. 110, 10034–10038 (2013)
8. Nettheim, N.: On the Spectral Analysis of Melody. Interface 21, 135–148 (1992)
9. Boon, J.P., Decroly, O.: Dynamical Systems theory for music dynamics. Chaos 5, 501–508 (1995)
10. Voss, R.F., Clarke, C.: "1/f noise" in music: Music from 1/f noise. J. Acoust. Soc. Am. 63, 258–263 (1978)
11. Bak, P., Tang, C., Wiesenfeld, K.: Self-organized criticality: an explanation of 1/f noise. Phys. Rev. Lett. 59, 381–384 (1987)
12. Jennings, H.D., Ivanov, P.C., Martins, A.M., da Silva, P.C., Viswanathan, G.M.: Variance fluctuations in nonstationary time series: A comparative study of music genres. Physica A 336, 585–594 (2004)
13. Boon, J.P., Noullez, A., Mommen, C.: Complex Dynamics and Musical Structure. Interface 19, 3–14 (1990)
14. Zanette, D.H.: Zipf's law and the creation of musical context. Music Sci. 10, 3–18 (2006)
15. Beltrán del Río, M., Cocho, G., Naumis, G.G.: Universality in the tail of musical note rank distribution. Physica A 387, 5552–5560 (2008)
16. Lerdahl, F.: Tonal Pitch Space. Oxford University Press (2001)
17. Tymoczko, D.: The Geometry of Musical Chords. Science 313, 72–74 (2006)
18. Clauset, A., Shalizi, C.R., Newman, M.E.J.: Power-Law Distributions in Empirical Data. SIAM Rev. 51, 661–703 (2009)
19. Deluca, A., Corral, Á.: Fitting and goodness-of-fit test of non-truncated and truncated power-law distributions. Acta Geophys. 61, 1351–1394 (2013)
20. Artificially generated music is available on `http://stoop.ini.uzh.ch/artmus`

Coexisting Forms of Coupling and Phase-Transitions in Physiological Networks

Ronny P. Bartsch[1,2,3] and Plamen Ch. Ivanov[1,3,4,⋆]

[1] Harvard Medical School and Division of Sleep Medicine, Brigham and Women's Hospital, Boston, MA 02115, USA
[2] Department of Physics, Bar-Ilan University, Ramat Gan, 52900, Israel
[3] Department of Physics, Boston University, Boston, MA 02215, USA
[4] Institute of Solid State Physics, Bulgarian Academy of Sciences, Sofia 1784, Bulgaria
plamen@buphy.bu.edu

Abstract. Utilizing methods from nonlinear dynamics and a network approach we investigate the interactions between physiologic organ systems. We demonstrate that these systems can exhibit multiple forms of coupling that are independent from each other and act on different time scales. We also find that physiologic systems interaction is of transient nature with intermittent "on" and "off" periods, and that different forms of coupling can simultaneously coexist representing different aspects of physiologic regulation. We investigate the network of physiologic interactions between the brain, cardiac and respiratory systems across different sleep stages, well-defined physiologic states with distinct neuroautonomic regulation, and we uncover a strong relationship between network connectivity, patterns in network links strength and physiologic function. We show that physiologic networks exhibit pronounced phase transitions associated with reorganization in network structure and links strength in response to transitions across physiologic states.

Keywords: Cardio-respiratory coupling, phase synchronization, sleep, networks.

1 Introduction

Physiologic organ systems exhibit complex nonlinear dynamics characterized by nonstationary, intermittent, scale-invariant and multifractal behaviors [1–6]. These dynamics result from underlying feedback mechanisms of neural regulation acting over a range of time scales [7–9] and exhibit pronounced phase transitions associated with changes across physiologic states and pathologic conditions [10–12]. Further, nonlinear coupling and interactions between organ systems influence their output dynamics and coordinate their functions, leading to another level of complexity. It is an open problem to adequately determine interactions

⋆ Correspondence author.

V.M. Mladenov and P.C. Ivanov (Eds.): NDES 2014, CCIS 438, pp. 270–287, 2014.

between complex systems where their coupling is not known a priori, and where the only available information is contained in the output signals of the systems. For integrated physiological systems, this is further complicated by transient non-linear characteristics and continuous fluctuations in their dynamics. In recent years, research in nonlinear dynamics has focused on developing an analytic framework and novel measures to detect and quantify interactions between physiologic systems and to elucidate the nature of their coupling [13–15].

Here we hypothesize that integrated organ systems can communicate through several independent mechanisms of interaction which operate at different time scales, and that different forms of coupling can simultaneously coexist. Because the neuroautonomic regulation of each organ system changes with transition from one physiologic state to another leading to transitions in scaling and non-linear features of the output dynamics [12, 16–19], we further hypothesize that physiologic coupling also undergoes phase transitions in order to facilitate and optimize organ interactions during different physiologic states. Specifically, we hypothesize that at any given moment pairs of systems interact through complementary forms of coupling that may exhibit different strength and different stratification patterns across physiologic states.

2 Data and Methods

Data: To test our hypotheses, we analyze physiologic data recordings from a group of 189 healthy subjects (99 female and 90 male, ages ranging from 20-95 years) during night-time sleep (average record duration is 7.8 h), tracking changes in physiologic coupling across different sleep stages. We focus on physiological dynamics during sleep as sleep stages are well-defined physiological states, and external influences due to physical activity or sensory inputs are reduced during sleep. Sleep stages are scored in 30 s epochs by sleep lab technicians based on standard Rechtschaffen & Kales criteria. We consider EEG, ECG and respiratory signals. In particular, we focus on cardio-respiratory coupling and on the network of interactions between different brain areas and the cardiac and respiratory system.

Methods: Cardiac, respiratory and brain dynamics exhibit transient changes associated with different physiologic states and conditions. How their coupling responds to these changes in relation to the underlying mechanisms of physiologic control remains not understood. Moreover, whether the systems interact via different functional forms of coupling and whether these forms of coupling are influenced differently by the same physiologic state, is not known.

Employing ideas from the theory of non-equilibrium systems [22], synchronization of coupled nonlinear systems [14], time delay stability [23] and complex networks [24], we investigate aspects of physiologic coupling and networks of interaction between physiological systems, and how they change with transitions from one physiologic state to another.

Specifically, we investigate three forms of cardio-respiratory interaction and how they change during different sleep stages. Utilizing multichannel physiologic

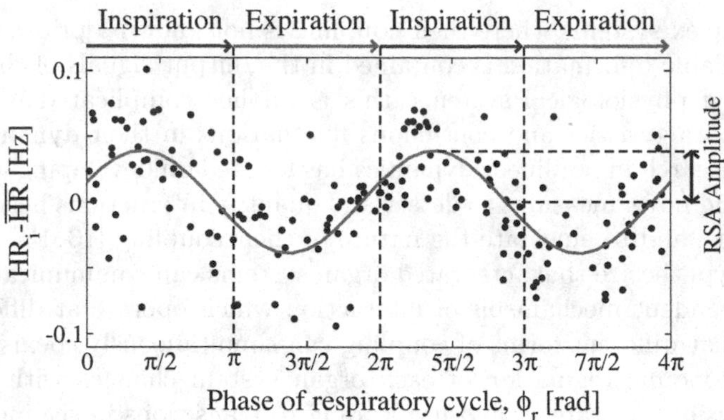

Fig. 1. Cardio-respiratory coupling: Respiratory sinus arrhythmia (RSA) This form of coupling is characterized by a periodic variation of the heart rate (HR) within each breathing cycle: increase of HR with inspiration and decrease of HR with expiration for normal breathing frequencies (≈ 7-12 breaths per min). The strength of the coupling is defined by the amplitude of the HR variation measured relative to the mean heart rate $\overline{\text{HR}}$, called RSA amplitude. RSA is most pronounced at low breathing frequencies and nonlinearly decreases with increasing breathing frequency [20, 21]. Data points represent instantaneous heart rate (inverse heartbeat intervals), normalized to the mean heart rate $\overline{\text{HR}}$ within each breathing cycle, for a period of 200 sec over pairs of consecutive breathing cycles. RSA is highlighted by a sinusoidal least-square-fit line to the data points. Data are recorded from a healthy subject during sleep.

data during sleep as an example, we demonstrate that a network approach to physiological interactions is necessary to understand how modulations in the regulatory mechanism of individual systems translate into reorganization of physiological interactions across the human organism.

Respiratory sinus arrhythmia (RSA): In physiological studies, the interaction between the cardiac and the respiratory system, is traditionally identified through the respiratory sinus arrhythmia (RSA), which accounts for the periodic variation of the heart rate within a breathing cycle [20, 25]. Typically, for normal breathing rates in the range 7-12 breaths per minute, RSA is characterized by the increase of heart rate during inspiration and decrease during expiration, and is quantified by the amplitude of this heart rate modulation. To obtain the RSA amplitude, we first estimate the instantaneous heart rate (inverse heartbeat RR interval) associated with each heartbeat within a breathing cycle, and from each instantaneous heart rate we subtract the average heart rate for this breathing cycle. We consider artifact-free segments of consecutive normal heartbeats with duration ≥300 sec. We plot the instantaneous heart rates (normalized to the mean) for each pair of consecutive breathing cycles within each artifact-free data segment (Fig. 1), and we define the RSA amplitude as one half of the peak-through difference in the least-square-fit sinusoid line to the data points. This fit line represents the respiratory modulation of the heart rate (Fig. 1).

For our RSA analysis, heartbeat data (R-peaks) were extracted from the ECGs utilizing a semi-automatic peak detector (*Raschlab*, `www.librasch.org`). RR time intervals were calculated between each pair of consecutive R-peaks. A RR interval was labeled as artifact and excluded from the analysis if (i) the interval was shorter than 300 ms or longer than 2000 ms, or (ii) the interval was more than 30% shorter or more than 60% longer than the preceding RR interval. These exclusion criteria effectively eliminate ectopic heartbeats and artifacts. Respiration was measured by the oronasal airflow through a thermistor and by belts around the chest and abdomen. These three respiratory signals were resampled to 4Hz to eliminate high frequency fluctuations and to assure that the signal is narrow-banded, and thus its Hilbert transform can be used to calculate the respiratory phase.

Cardio-respiratory phase synchronization (CRPS): Phase-synchronization analysis was recently developed to identify interrelations between the output signals of coupled nonlinear oscillatory systems even when these output signals are not cross-correlated [14, 26, 27]. Thus, phase-synchronization analysis can quantify the degree of coupling between nonlinear systems when other conventional methods can not.

Cardio-respiratory synchronization can systematically be studied in an automated way by utilizing an algorithm that evaluates cardio-respiratory synchrograms (Fig. 2(d)). The synchrogram is a method in which the phase of a continuous signal (e.g., respiration $r(t)$) is plotted at incidents t_k of a second signal described by a point process (e.g., the occurrence of R-peaks in the ECG at times t_k). The instantaneous respiratory phase $\phi_r(t)$ can be calculated by the analytic signal approach [26] and, in the complex plane, $\phi_r(t)$ represents the angle between the respiratory signal $r(t)$ and its Hilbert transform $r_H(t)$ which is the imaginary part of the respiratory signal (Fig. 2(c)). The plot of $\phi_r(t_k)$ over t_k defines the cardio-respiratory synchrogram (Fig. 2(d)). Cardio-respiratory phase synchronization exists when n parallel horizontal lines are observed, where n is the number of heartbeats per breathing cycle ($n = 3$ in Fig. 2(d)). In our automated synchrogram algorithm, the times t_k of the occurrence of heartbeats are mapped on the cumulative respiratory phase $\Phi_r(t)$, and $\phi_r^m(t_k) = \Phi_r(t)$ mod $2\pi m$ is plotted versus t_k. For each m respiratory cycles where n heartbeats occur at times t_i^c (t_i^c corresponds to the times of $i = 1, \ldots, n$ heartbeats within m respiratory cycles, denoted by c), we replace the phase points $\phi_r^m(t_i^c)$ by averages $\langle\phi_r^m(t_i^c)\rangle$ and standard deviations $\sigma(t_i^c)$ calculated over all phase points in the time window $\mathfrak{T}_i^c = [t_i^c - \tau/2, t_i^c + \tau/2]$ and in the phase interval $\mathfrak{R}_i = [\phi_r^m(t_i^c) - \frac{2\pi m}{n}, \phi_r^m(t_i^c) + \frac{2\pi m}{n})$ (i.e., we average the phase points along the horizontal lines in Fig. 2(d) over the time window $\mathfrak{T}_i^c$). Thus, the phase average and standard deviation are defined by $\langle\phi_r^m(t_i^c)\rangle = \frac{1}{N_i}\sum_{t_k\in\mathfrak{T}_i^c}\phi_r^m(t_k)$ and $\sigma(t_i^c) = \sqrt{\frac{1}{N_i}\sum_{t_k\in\mathfrak{T}_i^c}\left(\phi_r^m(t_k) - \langle\phi_r^m(t_i^c)\rangle\right)^2}$, where N_i is the number of phase points (heartbeats) in the time window $\mathfrak{T}_i^c$ and the phase interval $\mathfrak{R}_i$, and where m is the number of respiratory cycles in which n heartbeats occur. Next, for each breathing cycle we average the standard deviation $\sigma(t_i^c)$ for all $i = 1, \ldots, n$

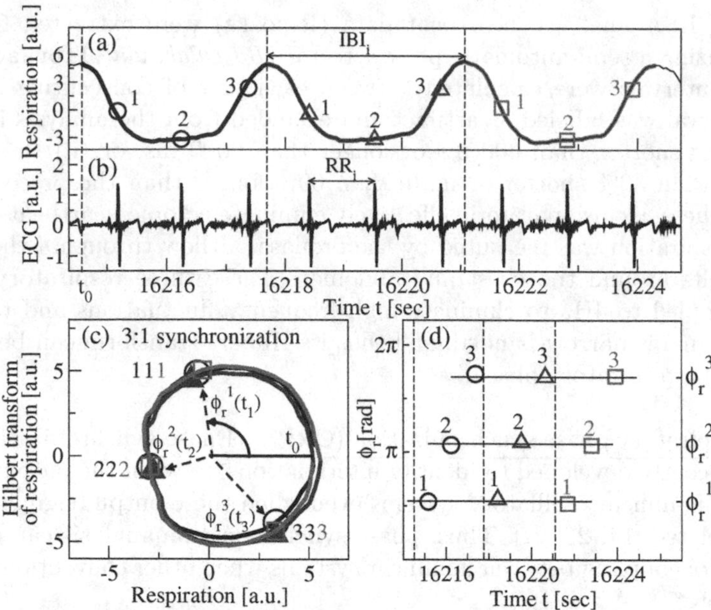

Fig. 2. Cardio-respiratory coupling: Phase synchronization (CRPS) Three consecutive breathing cycles (in black, blue and red) are shown in (a) and a simultaneously recorded ECG signal in (b). The interbreath interval (IBI) is approximately 3 times longer than the beat-to-beat interval (RR). Horizontal arrows indicate an interbreath interval (IBI) and a RR beat-to-beat interval. (c) Demonstration of 3 : 1 phase synchronization between the heartbeats and respiratory cycles shown in (a) and (b). For each breathing cycle the 1st heartbeat occurs at the same respiratory phase $\phi_r^1(t)$, and the 2nd and 3rd heartbeats within each cycle occur at $\phi_r^2(t)$ and $\phi_r^3(t)$ (symbols collapse), indicating robust CRPS. (d) Three horizontal parallel lines formed respectively by the 1st, 2nd and 3rd heartbeats in the three consecutive breathing cycles indicate 3 : 1 phase synchronization. Different symbols represent heartbeats in different breathing cycles as in (a) and (c), and vertical dashed lines show the beginning of each breathing cycle.

phase points $\langle \phi_r^m(t_i^c) \rangle$ to obtain $\langle \sigma \rangle_n$. Only breathing cycles with $\langle \sigma \rangle_n \leq \frac{2\pi m}{n\Delta}$ are considered, and synchronization segments are identified only when such consecutive breathing cycles span over time intervals $\geq T$. Finally, we relate the phase synchronization segments to the time intervals of the different sleep stages throughout the night, and for each sleep stage we calculate the % synchronization as the ratio between the time duration of the sum of all synchronization segments and the total time duration of the sleep stage during the night. Segments of data artifacts in both cardiac and respiratory signals are disregarded in these calculations. Since we have three respiratory signals from oronasal airflow, chest and abdomen belts, for each consecutive episode in a given sleep stage we consider the pair of cardiac and respiratory signals that yields the highest % synchronization (thus, optimally reducing the influence of breathing artifacts).

In our analyses we use $\Delta = 5$, and $T = \tau = 30$ sec corresponding to standard 30 sec sleep-stage scoring epochs.

For the phase-synchronization analysis we used the same signal pre-processing procedure for the heart rate and respiratory signals as described for the RSA analysis.

Time delay stability analysis (TDS): Integrated physiologic systems are coupled by feedback and/or feed forward loops with a broad range of time delays. To probe physiologic coupling we propose an approach based on the concept of time delay stability [23]: in the presence of strong stable interactions between two systems, transient modulations (e.g., bursts) in the output signal of one system lead to corresponding changes that occur with a stable time lag in the output signal of another coupled system. Long periods of constant time delay indicate strong physiologic coupling. We demonstrate the TDS method on the example of cardio-respiratory interaction, considering the time delay interrelation between bursts in the heart rate and the respiratory rate (Fig. 3(a,b)).

The TDS method consists of the following steps:
(1.) To probe the interaction between two physiological systems X and Y, we consider their output signals $\{x\}$ and $\{y\}$ each of length N and sampled at 1 Hz. We divide both signals $\{x\}$ and $\{y\}$ into N_L overlapping time windows ν of equal length $L = 60$ sec. We choose an overlap of $L/2 = 30$ sec which corresponds to the time resolution of the conventional sleep-stage scoring epochs, and thus $N_L = [2N/L] - 1$. Prior to the analysis, the signal in each time window ν is normalized separately to zero mean and unit standard deviation, in order to remove constant trends in the data and to obtain dimensionless signals. This normalization procedure assures that the estimated coupling between the signals $\{x\}$ and $\{y\}$ is not affected by the difference in their amplitudes.
(2.) Next, within each time window $\nu = 1, \ldots, N_L$, we calculate the cross-correlation function $C_{xy}^{\nu}(\tau) = \frac{1}{L}\sum_{i=1}^{L} x_{i+(\nu-1)\frac{L}{2}}^{\nu} \, y_{i+(\nu-1)\frac{L}{2}+\tau}^{\nu}$, by applying periodic boundary conditions (Fig. 3(c)). For each time window ν we define the time delay τ_0^{ν} to correspond to the maximum in the absolute value of the cross-correlation function $C_{xy}^{\nu}(\tau)$ in this time window $\tau_0^{\nu} = \tau|_{|C_{xy}^{\nu}(\tau)| \geq |C_{xy}^{\nu}(\tau')| \ \forall \tau'}$. Time periods of stable interrelation between two signals are represented by segments of approximately constant τ_0 in the newly defined series of time delays, $\{\tau_0^{\nu}\}_{\nu=1,\ldots,N_L}$ — e.g., the flat region in Fig. 3(d) corresponding to a period of stable coupling between heart rate and respiratory rate during light sleep. In contrast, absence of stable coupling between the signals is characterized by large fluctuations in τ_0, as shown in the gray-shaded region in Fig. 3(d) corresponding to deep sleep.
(3.) We identify two systems as linked if their corresponding signals exhibit a time delay that does not change by more than ± 1 sec for several consecutive time windows ν. We track the values of τ_0 along the series $\{\tau_0^{\nu}\}$: when for at least four out of five consecutive time windows ν (corresponding to a period of 5×30 sec) the time delay remains in the interval $[\tau_0 - 1, \tau_0 + 1]$ these segments are labeled as stable. This procedure is repeated for a sliding time window with

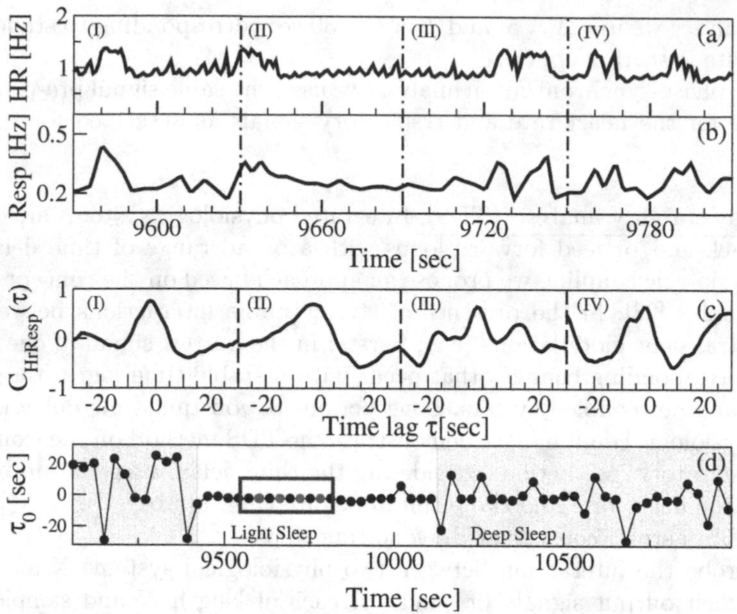

Fig. 3. Cardio-respiratory coupling: Time delay stability (TDS). Segments of (a) heart rate (HR) and (b) respiratory rate (Resp) normalized to zero mean and unit standard deviation in 60 sec time windows (I), (II), (III) and (IV), in order to remove trends in data and to obtain dimensionless signals. Synchronous bursts in the HR and Resp signal leading to pronounced cross-correlation within each time window as shown in (c), and stable time delay characterized by segments of constant τ_0 as shown in (d) — four red dots highlighted by a blue box in panel (d) represents the time delay for the 4 time windows. Note the transition from strongly fluctuating behavior in τ_0 to a stable time delay regime at the transition from deep sleep to light sleep at around 9400 sec (gray shaded area) in panel (d). The TDS analysis is performed on overlapping moving windows with a step of 30 sec represented by red and black dots in the blue box in panel (d). Long periods of constant τ_0 indicate strong physiological coupling.

a step size one along the entire series $\{\tau_0^\nu\}$. The % TDS is finally calculated as the fraction of stable points in the time series $\{\tau_0^\nu\}$.

Longer periods of TDS between the output signals of two systems reflect more stable interaction/coupling between these systems. Thus, the strength of coupling is determined by the percentage of time when TDS is observed: higher percentage of TDS corresponds to stronger coupling. To identify physiologically relevant interactions, we determine a significance threshold level for the TDS based on comparison with surrogate data derived from uncoupled systems, e.g., heart rate from subject A and respiratory rate from subject B. Only interactions characterized by TDS values above the significance threshold are considered.

The TDS method is general, and can be applied to diverse systems to identify and quantify links in networks of physiologic interactions that are not a priori known (Fig. 7). It is more reliable in identifying physiologic coupling compared to traditional cross-correlation and cross-coherence analyses which are not suitable for heterogeneous and nonstationary signals, and are affected by the degree of auto-correlations embedded in these signals [28]. TDS is also suitable to identify coupling between systems that are not characterized by oscillatory output dynamics, where the phase-synchronization approach can not be applied.

In order to study interrelations and coupling between the brain and the cardiac and respiratory system, we extract the following time series from their output signals: the spectral power of seven frequency bands of the EEG in moving windows of 2 sec with a 1 sec overlap: δ waves (0.5-3.5 Hz), θ waves (4-7.5 Hz), α waves (8-11.5 Hz), σ waves (12-15.5 Hz), β waves (16-19.5 Hz) and $\gamma 1$ (20-33.5 Hz) and $\gamma 2$ (34-100 Hz) waves; heartbeat RR intervals and interbreath intervals are both re-sampled to 1 Hz (1 sec bins) after which values are inverted to obtain heart rate and respiratory rate. Thus, all time series have the same time resolution of 1 sec before the TDS analysis is applied. EEG data were recorded from 6 scalp locations: frontal left (Fp1), frontal right (Fp2), central left (C3), central right (C4), occipital left (O1) and occipital right (O2).

Applying the TDS method to these data we investigate the network of interactions between the brain, cardiac and respiratory system, and how this network changes with transition from one physiologic state to another (Figs. 7 and 8).

3 Results

3.1 Coexisting Forms of Coupling and Phase Transitions across Physiological States

To probe for coexistence of different forms of cardio-respiratory coupling we apply three different methods: (i) analysis of respiratory sinus arrhythmia (RSA), (ii) phase-synchronization analysis, and (iii) time delay stability analysis (TDS). These three methods quantify independent characteristics of the interaction between the cardiac and respiratory systems at different time scales: (i) amplitude of heart rate modulation during a breathing cycle (Fig. 1), (ii) phase-synchronized activity between the heart rate and the respiratory rate so that within a breathing cycle heartbeats occur at specific respiratory phases and this behavior is consistent for several consecutive breathing cycles (Fig. 2), and (iii) stable time delay in bursting activity of the heart and respiratory rate over long periods of time from several minutes to hours (Fig. 3).

Empirical studies have reported a strong variation in linear and nonlinear characteristics in both cardiac and respiratory dynamics with the sleep-wake cycle [16], across circadian phases [19, 29] and sleep-stage transitions [10–12, 30, 31], indicating significant changes in the regulatory mechanisms with transitions across physiologic states. Thus, we test whether cardio-respiratory coupling may also undergo phase transitions with transitions across physiologic states. Since sleep stages are well-defined physiologic states and are associated with distinct mechanisms of autonomic control [11, 12], we also investigate the strength of different forms of cardio-respiratory coupling during different sleep stages.

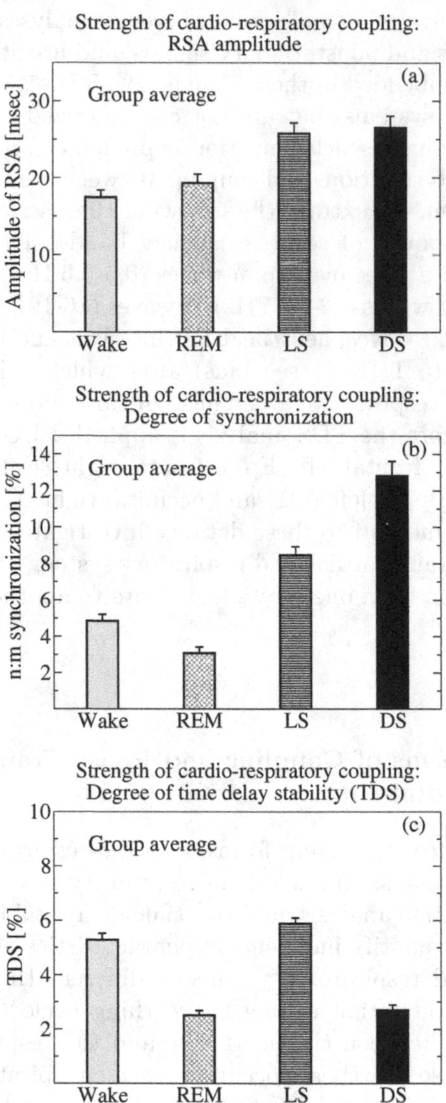

Fig. 4. Phase transitions in physiologic coupling across physiological states. Stratification patterns in different measures of cardio-respiratory coupling with transitions across sleep stages indicate complex re-organization in physiological interaction during different physiologic states. Three independent forms of coupling — (a) Respiratory sinus arrhythmia (RSA), (b) Phase synchronization, and (c) Time delay stability (TDS) — show significantly different coupling strength during different sleep stages. Columns in each panel represent the average over a group of 189 healthy subjects; error bars correspond to the standard error. Phase synchronization and TDS coupling strength in (b) and (c) is estimated as the percent of data segments in the entire recording for a given sleep stage where heart rate and respiratory rate are phase-synchronized or exhibit TDS.

We find that the amplitude of RSA exhibits a pronounced and statistically significant stratification pattern across different sleep stages. The amplitude of RSA is lowest during wake periods and gradually increases during REM, light sleep and deep sleep. In particular, periods of non-REM sleep (light sleep and deep sleep) are characterized by $\approx 40\%$ higher RSA amplitude compared to REM and wake (Fig. 4(a)), indicating a strong dependence of this form of coupling on the underlying neuroautonomic control. With transition from wake to REM, light and deep sleep, the sympathetic tone of autonomic control decreases while the level of parasympathetic activity remains unchanged [11]. This yields a relative dominance of parasympathetic tone in cardiac and respiratory regulation, which in turn leads to lower respiratory rates in non-REM sleep associated with higher RSA amplitude. Indeed, experimental studies in healthy subjects show a decrease in the RSA amplitude with increasing respiratory frequency under paced respiration [20, 25]. However, we note that while the respiratory frequency decreases just with $\approx$ 6-8% during transitions from wake to light and deep sleep, we find a highly nonlinear response in the RSA amplitude which increases with $\approx$40% during these transitions.

Our analyses show that in addition to RSA cardio-respiratory interaction is characterized by two other forms of coupling — phase synchronization and time delay stability — which also exhibit a pronounced stratification in the strength of coupling across different sleep stages (Fig. 4(b,c)). However, in contrast to RSA cardio-respiratory phase synchronization is weaker during REM as compared to wake, while during light and deep sleep there is a dramatic increase in the degree of synchronization of $\approx 400\%$ compared to REM (Fig. 4(b)). Notably, this form of cardio-respiratory coupling exhibits a factor of 10 stronger response to sleep-stage transitions compared to RSA.

While the sleep-stage stratification patterns for both forms of coupling, RSA and synchronization, show a relative separation between non-REM and REM sleep, we find a very different stratification pattern for the strength of TDS: (i) much higher degree of TDS during wake and light sleep compared to REM and deep sleep, and (ii) comparable strength of TDS during wake and light sleep and during REM and deep sleep (Fig. 4(c)). Our empirical observation of significant difference in the degree of cardio-respiratory TDS coupling during light sleep compared to deep sleep is surprising, given the similarity in spectral, scale-invariant and nonlinear properties of cardiac and respiratory dynamics during light sleep and deep sleep [12, 30, 32] (light and deep sleep are traditionally classified as non-REM), and indicate that previously unrecognized aspects of cardiac and respiratory neuroautonomic regulation during sleep are captured by the TDS analysis.

Our investigations demonstrate the existence of three distinct forms of cardio-respiratory interaction that represent the dynamics at different time scales. Our findings of different stratification patterns indicate that these forms of coupling are independent and represent different aspects of physiologic interaction that are affected in a different way by changes in neuroautonomic control across physiologic states. Moreover, we find that these three forms of cardio-respiratory coupling are

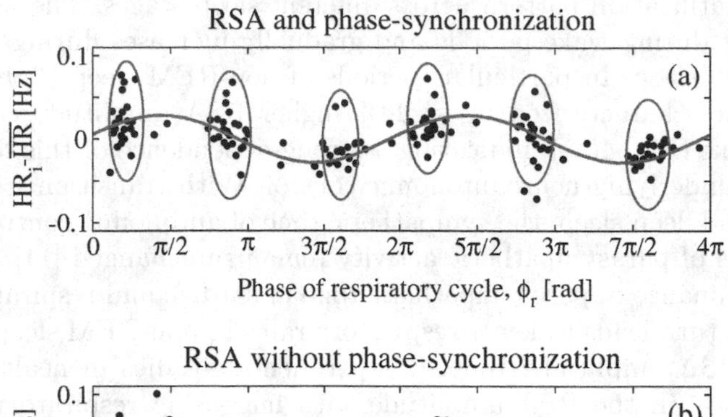

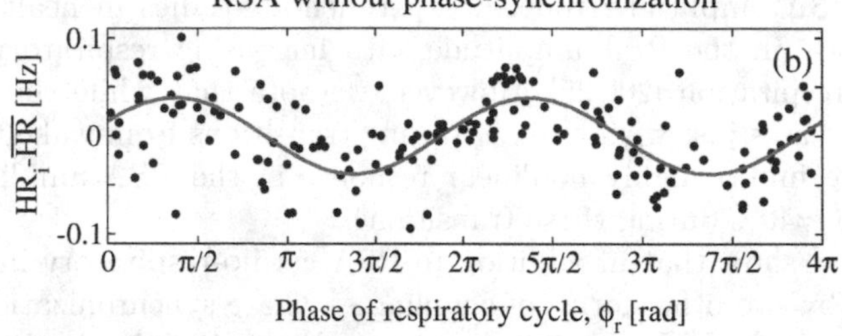

Fig. 5. Coexisting forms of cardio-respiratory coupling. Phase synchronization (CRPS) and respiratory sinus arrhythmia (RSA) represent different and independent forms of cardio-respiratory coupling. (a) While RSA leads to periodic modulation of the heart rate within each breathing cycle (green sinusoidal line) and is quantified by the amplitude of the heart rate modulation, CRPS leads to clustering of heartbeats at certain phases ϕ_r of the breathing cycle (red ovals). Shown are consecutive heartbeats over a period of 200 sec. The x-axis indicates the phases ϕ_r of the breathing cycle where heartbeats occur, and the y-axis indicates the deviation of the heart rate (inversed heartbeat intervals) from the mean heart rate $\overline{HR}$ within a given breathing cycle. Instantaneous heart rates are plotted over pairs of consecutive breathing cycles, $\phi_r \in [0, 4\pi]$, to better visualize rhythmicity. Data are selected from a subject during deep sleep. (b) For the same subject as in (a), instantaneous heart rates from another period of 200 sec also during a deep sleep episode are plotted over pairs of consecutive breathing cycles. Data show well-pronounced RSA with a similar amplitude as in (a), however, heartbeats are homogeneously distributed across all phases of the respiratory cycles, indicating absence of CRPS.

not continuously present and are not of constant strength but are rather transient and intermittent with “on” and “off” periods. Indeed, cardio-respiratory phase synchronization is usually observed in relatively short epochs rarely exceeding 100 sec in duration (Fig. 6). Even for the same subject within the same sleep stage, we find time periods when synchronization is not present (Fig. 5), indicating that “on” and “off” switching of this form of cardio-respiratory interaction is not always triggered by transitions across physiological states. Further, the total time when cardio-respiratory phase synchronization is observed even under controlled conditions in resting subjects is a relatively small fraction of the entire duration of the recordings [33]. Moreover, some subjects may even not exhibit cardio-respiratory phase synchronization [33]. Our cardio-respiratory phase synchronization analysis shows that most of the synchronization epochs are of 20–45 sec duration (Fig. 6). Further, we find that for each subject the total sum of the synchronization epochs does not exceed 20–25% of the entire sleep duration or of the duration of the individual sleep stages, indicating that the cardio-respiratory coupling, as manifested by phase synchronization, is not stable in time.

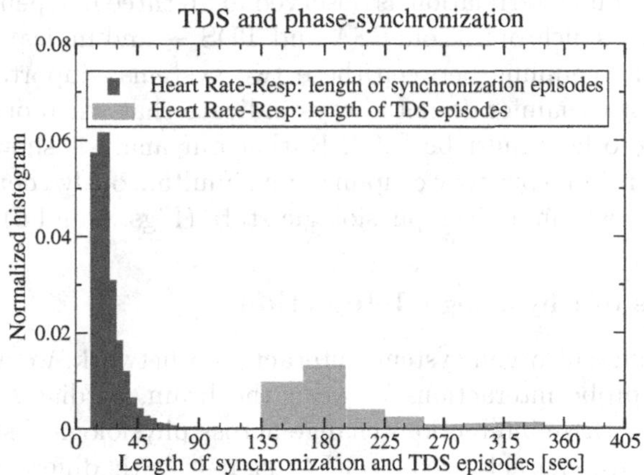

Fig. 6. Coexisting forms of cardio-respiratory coupling. Phase synchronization (CRPS) and time delay stability (TDS) probe the coupling at different time scales. Probability distributions of the length of cardio-respiratory phase synchronization epochs (red color) and of cardio-respiratory TDS epochs (light-blue color) indicating significant difference in the average epoch length. Thus, different aspects of cardio-respiratory coupling are captured by these two measures. However, both measures reflect transient changes in cardio-respiratory coupling with "on" and "off" periods even within a single sleep stage, where the "on" periods form only a fraction of the entire heart rate and respiratory recordings. Results are obtained from 189 healthy subjects, taking into account all sleep stages during a one-night sleep period.

The amplitude of RSA, another measure of cardio-respiratory coupling also continuously changes with breathing frequency and sympatho-vagal balance [20, 25], and under certain conditions RSA is even absent. Thus, the strength of RSA as a representative measure of cardio-respiratory coupling is also not stable in time. Finally, we find such intermittent behavior also in the TDS measure which probes cardio-respiratory coupling at larger time scales above 2 min. We note that TDS is a more stringent condition to quantify cardio-respiratory coupling, since TDS epochs longer than 2 min are more rare than the epochs of ≈ 20–45 sec typical for phase synchronization (Fig. 6). Indeed, our analyses show that most of the TDS epochs for the cardio-respiratory interaction are of 150–220 sec duration (Fig. 6). Since, epochs of TDS shorter than 2 min are neglected, we have a lower cardio-respiratory TDS (not exceeding $\approx 7\%$ for the entire recording) compared to what we find for phase synchronization (Fig. 4(b,c)). We note that our choice of a 2 min window over which TDS is determined is not arbitrary because this is the minimal size window over which results obtained from the TDS method applied to real data are significantly different (t-test: $p < 10^{-3}$) from the results obtained based on surrogate data, where signals from one subject are paired with signals from different subjects.

The presence of "on" and "off" periods in the interaction between heart rate and respiration, where relatively short episodes of effective interrelation are separated by periods of no interrelation, is observed in all three independent measures — degree of phase synchronization, RSA and TDS — and indicates an intermittent nature of the coupling between these two systems. Importantly however, these intermittent dynamics do not always indicate that when one form of coupling is "on" the others must be "off". Rather our analysis shows that two or more forms of cardio-respiratory coupling can simultaneously coexist during the same time period within a given physiologic state (Figs. 5 and 6).

3.2 Networks of Physiologic Interaction

To study how different organ systems interact as a network, we apply the TDS method [23] to probe interactions between the brain, cardiac and respiratory systems, and how these interactions change across physiological states. Because brain dynamics are characterized by EEG signals with different spectral frequencies dominant at different scalp locations and during different physiological states, the TDS method allows us to study how bursts in EEG frequency bands from certain brain areas are coupled with corresponding bursts in the heart and respiratory rate.

We consider six scalp locations represented by EEG channels: frontal left (Fp1), frontal right (Fp2), central left (C3), central right (C4), occipital left (O1) and occipital right (O2). For each EEG channel, we estimate the coupling with the respiratory and cardiac systems by the degree of TDS for the following physiologically relevant EEG frequency bands: δ (0.5-3.5 Hz), θ (4-7.5 Hz), α (8-11.5 Hz), σ (12-15.5 Hz), β (16-19.5 Hz), $\gamma1$ (20-33.5 Hz) and $\gamma2$ (34-100 Hz). We represent each brain-respiration and brain-heart link in the network by the average of the TDS values obtained from all 7 frequency bands.

We find that the network of brain, respiratory and cardiac interactions exhibits different topology during different sleep stages. Specifically, we find that the physiologic network is characterized by high connectivity during wake and light sleep, lower connectivity during REM and is "disconnected" (below threshold TDS link strength) during deep sleep (Fig. 7). Such transitions in network structure indicate strong relation between network topology and physiologic function. Traditionally, differences between sleep stages are attributed to modulation in the sympatho-vagal balance with dominant sympathetic tone during wake and REM [32]: spectral, scale-invariant and nonlinear characteristics of the dynamics of individual physiologic systems indicate higher degree of temporal correlations

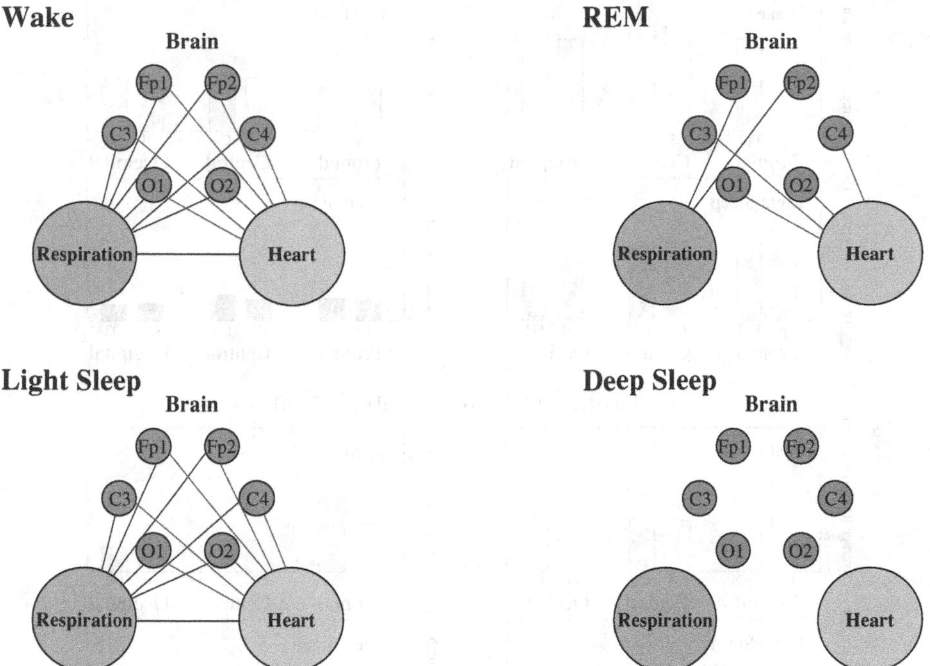

Fig. 7. Phase transitions in network of interactions between physiological organ systems. Networks of cardiac, respiratory and brain interactions during different sleep stages. Different brain areas are represented by frontal (Fp1 and Fp2), central (C3 and C4) and occipital (O1 and O2) EEG channels. The strength of the links between pairs of systems is determined based on the TDS measure shown in Fig. 3. Plotted are only links the strength of which exceeds 50% of the maximum TDS value for all sleep stages: 7.5% TDS threshold for brain-heart links, 2% for brain-respiration links, and 3% for heart-respiration links. There is a pronounced network re-organization with transition from one sleep stage to another. Remarkably network connectivity is very different for light sleep and deep sleep although both sleep stages correspond to non-REM sleep. A similar contrast is observed between wake and REM. Networks are obtained from 8-hour recordings during sleep by averaging over a group of 36 healthy subjects.

and nonlinearity during wake and REM compared to non-REM (light and deep sleep) where physiologic dynamics exhibit weaker correlations and loss of nonlinearity [12, 17, 30]. In contrast, the network of physiologic interactions shows a completely different picture: the network characteristics during light sleep are much closer to those during wake and very different from deep sleep (Fig. 7). Further, we find that not only network connectivity but also the overall strength of physiologic interactions is significantly higher during wake and light sleep, intermediate during REM and much lower during deep sleep (Fig. 8). Thus,

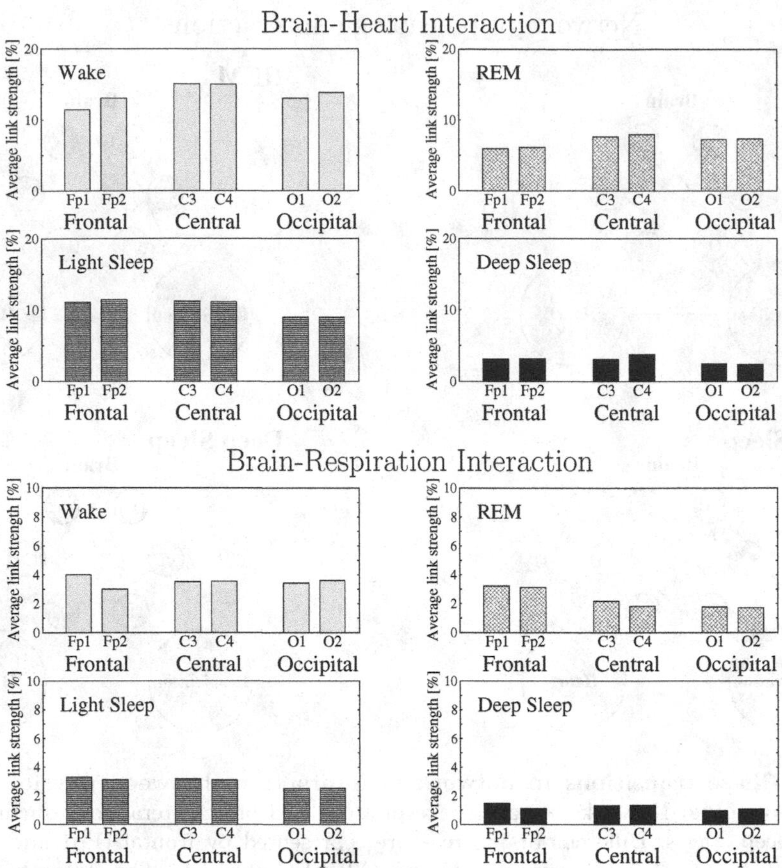

Fig. 8. Sleep-stage stratification pattern in the strength of network links between organ systems. Histograms show group averaged TDS links strength during different sleep stages for brain-heart (top four panels) and brain-respiration interaction (bottom four panels). Cardio-respiratory links strength for different sleep stages is shown in Fig. 4(c). A consistent stratification pattern with higher links strength during wake and light sleep compared to REM and deep sleep is observed for all pairs of coupling between brain, heart and respiration. Note that for all sleep stages, the brain-heart coupling strength as measured by TDS is consistently more than twice higher compared to the brain-respiration coupling. Histograms are obtained from 8-hour recordings during sleep by averaging over the same group of 36 healthy subjects as in Fig. 7.

our empirical observations indicate that while sleep-stage related modulation in sympatho-vagal balance plays a key role in regulating individual physiologic systems, it does not account for the physiologic network topology and dynamics across sleep stages, showing that the proposed TDS network approach captures principally new information.

We note that TDS quantifies the stability of coupling over large time scales. Absence of a link in the physiologic networks shown in Fig. 7 does not mean absence of coupling but rather that this coupling is present only for a small fraction of the data recordings. Specifically, networks links are not show when TDS is below the threshold value of 7.5% for brain-heart links, 2% for brain-respiration links, and 3% for heart-respiration links. Further, our investigations of the links strength in the physiologic networks show that brain-heart links as measured by the TDS are more than twice stronger than the brain-respiration links (Fig. 8), while the heart-respiration link is of intermediate strength (Fig. 4(c)). Notably this order in network links strength between the brain, heart and respiratory systems is preserved for all sleep stages. Remarkably, the sleep-stage stratification pattern we find for network connectivity, with high number of links during wake and light sleep, lower connectivity during REM and lowest during deep sleep (Fig. 7), is also present in the links strength — we find strongest links during wake and light sleep, weaker during REM and weakest links during deep sleep (Figs. 8 and 4(c)). Finally, our network analysis demonstrates a certain degree of heterogeneity in the network links strength between different brain areas and the heart and respiratory systems: frontal EEG channels show stronger coupling with respiration while central and occipital channels exhibit stronger coupling with the cardiac system. This heterogeneity in links strength is tentatively present at all sleep stages but most pronounced during REM (Fig. 8).

4 Conclusion

Utilizing concepts and methods from nonlinear dynamics, we demonstrate that the communication between pairs of physiologic organ systems can be facilitated by different forms of coupling. Considering the cardiac and respiratory systems, we show that there are three forms of coupling — respiratory sinus arrhythmia, phase synchronization and time delay stability — which are independent from each other and act on different time scales. Further, the strength of these forms of cardio-respiratory coupling changes with transitions across physiologic states such as sleep stages and exhibit markedly different sleep-stage stratification patterns. We find that all these forms of physiologic interaction are of transient nature with intermittent "on" and "off" periods, and can simultaneously coexist while at the same time representing different aspects of physiologic coupling and neuroautonomic regulation of the cardiac and respiratory systems. We extend these investigations to a network of physiologic interactions between the brain, cardiac and respiratory systems, and we uncover a strong relationship between network connectivity, patterns in network links strength and physiologic function. Studying the evolution of the physiologic network across distinct physiologic states such as wake, light sleep, deep sleep and REM, we show that physiologic networks exhibit pronounced phase transitions associated with reorganization in network structure and links strength in response to changes in the underlying neuroautonomic regulation. Our investigations are first steps towards understanding the network of communications between organ systems and developing a new field, Network Physiology.

Acknowledgments. We thank Kang Liu and Qianli Ma for helpful discussions. We acknowledge support from National Institutes of Health (NIH Grant 1R01-HL098437), the US-Israel Binational Science Foundation (BSF Grant 2012219) and the Office of Naval Research (ONR Grant 000141010078).

References

1. Ivanov, P.C., Rosenblum, M.G., Peng, C.K., Mietus, J., Havlin, S., Stanley, H.E., Goldberger, A.L.: Scaling behaviour of heartbeat intervals obtained by wavelet-based time-series analysis. Nature 383, 323–327 (1996)
2. Ivanov, P.C., Amaral, L.A.N., Goldberger, A.L., Havlin, S., Rosenblum, M.G., Stanley, H.E., Struzik, Z.: From 1/f Noise to Multifractal Cascades in Heartbeat Dynamics. Chaos 11, 641–652 (2001)
3. Ivanov, P.C., Chen, Z., Hu, K., Stanley, H.E.: Multiscale aspects of cardiac control. Physica A 344, 685–704 (2004)
4. Bartsch, R., Hennig, T., Heinen, A., Heinrichs, S., Maass, P.: Statistical analysis of fluctuations in the ECG morphology. Physica A 354, 415–431 (2005)
5. Bartsch, R., Plotnik, M., Kantelhardt, J.W., Havlin, S., Giladi, N., Hausdorff, J.M.: Fluctuation and synchronization of gait intervals and gait force profiles distinguish stages of Parkinsons disease. Physica A 383, 455–465 (2007)
6. Ivanov, P.C., Ma, Q.D.Y., Bartsch, R.P., Hausdorff, J.M., Amaral, L.A.N., Schulte-Frohlinde, V., Stanley, H.E., Yonseyama, M.: Levels of complexity in scaleinvariant neural signals. Phys. Rev. E 79, 041920 (2009)
7. Ivanov, P.C., Amaral, L.A.N., Goldberger, A.L., Stanley, H.E.: Stochastic feedback and the regulation of biological rhythms. Europhys. Lett. 43, 363–368 (1998)
8. Ashkenazy, Y., Hausdorff, J., Ivanov, P.C., Stanley, H.E.: A stochastic model of human gait dynamics. Physica A 316, 662–670 (2002)
9. Lo, C.C., Amaral, L.A.N., Havlin, S., Ivanov, P.C., Penzel, T., Peter, J.H., Stanley, H.E.: Dynamics of sleep-wake transitions during sleep. Europhys. Lett. 57, 625–631 (2002)
10. Ivanov, P.C.: Scale-invariant aspects of cardiac dynamics - Observing sleep stages and circadian phases. IEEE Eng. Med. Biol. 26, 33–37 (2007)
11. Schmitt, D.T., Stein, P.K., Ivanov, P.C.: Stratification pattern of static and scaleinvariant dynamic measures of heartbeat fluctuations across sleep stages in young and elderly. IEEE Trans. Biomed. Eng. 56, 1564–1573 (2009)
12. Schumann, A.Y., Bartsch, R.P., Penzel, T., Ivanov, P.C., Kantelhardt, J.W.: Aging effects on cardiac and respiratory dynamics in healthy subjects across sleep stages. Sleep 33, 943–955 (2010)
13. Tass, P., Rosenblum, M., Weule, J., Kurths, J., Pikovsky, A., Volkmann, J., Schnitzler, A., Freund, H.: Detection of n: m phase locking from noisy data: Application to magnetoencephalography. Phys. Rev. Lett. 81, 3291–3294 (1998)
14. Pikovsky, A.S., Rosenblum, M.G., Kurths, J.: Synchronization: A Universal Concept in Nonlinear Sciences. Cambridge University Press, Cambridge (2001)
15. Schindler, K.A., Bialonski, S., Horstmann, M.T., Elger, C.E., Lehnertz, K.: Evolving functional network properties and synchronizability during human epileptic seizures. Chaos 18, 033119 (2008)
16. Ivanov, P.C., Bunde, A., Amaral, L.A.N., Havlin, S., Fritsch-Yelle, J., Baevsky, R.M., Stanley, H.E., Goldberger, A.L.: Sleep-wake differences in scaling behavior of the human heartbeat: analysis of terrestrial and long-term space flight data. Europhys. Lett. 48, 594–600 (1999)

17. Kantelhardt, J.W., Ashkenazy, Y., Ivanov, P.C., Bunde, A., Havlin, S., Penzel, T., Peter, J.H., Stanley, H.E.: Characterization of sleep stages by correlations in the magnitude and sign of heartbeat increments. Phys. Rev. E 65, 051908 (2002)
18. Karasik, R., Sapir, N., Ashkenazy, Y., Ivanov, P.C., Dvir, I., Lavie, P., Havlin, S.: Correlation differences in heartbeat fluctuations during rest and exercise. Phys. Rev. E 66, 062902 (2002)
19. Ivanov, P.C., Hu, K., Hilton, M.F., Shea, S.A., Stanley, H.E.: Endogenous circadian rhythm in human motor activity uncoupled from circadian influences on cardiac dynamics. Proc. Natl. Acad. Sci. USA 104, 20702–20707 (2007)
20. Angelone, A., Coulter, N.A.: Respiratory sinus arrhythmia: A frequency dependent phenomenon. J. Appl. Physiol. 19, 479–482 (1964)
21. Bartsch, R.P., Schumann, A.Y., Kantelhardt, J.W., Penzel, T., Ivanov, P.C.: Phase transitions in physiologic coupling. Proc. Natl. Acad. Sci. USA 109, 10181–10186 (2012)
22. Sornette, D.: Critical phenomena in natural sciences. Chaos, fractals, selforganization, and disorder — Concepts and tools, 2nd edn. Springer, Berlin (2004)
23. Bashan, A., Bartsch, R.P., Kantelhardt, J.W., Havlin, S., Ivanov, P.C.: Network physiology reveals relations between network topology and physiological function. Nat. Commun. 3, 702 (2012)
24. Dorogovtsev, S.N., Mendes, J.F.F.: Evolution of networks. Advances in Physics 51, 1079–1187 (2002)
25. Song, H.S., Lehrer, P.M.: The effects of specific respiratory rates on heart rate and heart rate variability. Appl. Psychophysiol. Biofeedback 28, 13–23 (2003)
26. Rosenblum, M.G., Pikovsky, A.S., Kurths, J.: Phase synchronization of chaotic oscillators. Phys. Rev. Lett. 76, 1804–1807 (1996)
27. Xu, L., Chen, Z., Hu, K., Stanley, H.E., Ivanov, P.C.: Spurious detection of phase synchronization in coupled nonlinear oscillators. Phys. Rev. E 73, 065201 (2006)
28. Podobnik, B., Fu, D.F., Stanley, H.E., Ivanov, P.C.: Power-law autocorrelated stochastic processes with long-range cross-correlations. Eur. Phys. J. B 56, 47–52 (2007)
29. Hu, K., Ivanov, P.C., Hilton, M.F., Chen, Z., Ayers, R.T., Stanley, H.E., Shea, S.A.: Endogenous circadian rhythm in an index of cardiac vulnerability independent of changes in behavior. Proc. Natl. Acad. Sci. USA 101, 18223–18227 (2004)
30. Bunde, A., Havlin, S., Kantelhardt, J.W., Penzel, T., Peter, J.H., Voigt, K.: Correlated and Uncorrelated Regions in Heart-Rate Fluctuations during Sleep. Phys. Rev. Lett. 85, 3736–3739 (2000)
31. Kantelhardt, J.W., Penzel, T., Rostig, S., Becker, H.F., Havlin, S., Bunde, A.: Breathing during REM and non-REM sleep: correlated versus uncorrelated behaviour. Physica A 319, 447–457 (2003)
32. Otzenberger, H., Gronfier, C., Simon, C., Charloux, A., Ehrhart, J., Piquard, F., Brandenberger, G.: Dynamic heart rate variability: A tool for exploring sympathovagal balance continuously during sleep in men. Am. J. Physiol. 275, H946–H950 (1998)
33. Schäfer, C., Rosenblum, M.G., Kurths, J., Abel, H.H.: Heartbeat synchronized with ventilation. Nature 392, 239–240 (1998)

A Wavelet-Based Method for Multifractal Analysis of Medical Signals: Application to Dynamic Infrared Thermograms of Breast Cancer

Evgeniya Gerasimova[1], Benjamin Audit[2], Stephane-G. Roux[2], André Khalil[3], Olga Gileva[4], Françoise Argoul[2], Oleg Naimark[1], and Alain Arneodo[2,5,⋆]

[1] Laboratory of Physical Foundation of Strength, Institute of Continuous Media Mechanics UB RAS, Perm, Russia
[2] Université de Lyon and Laboratoire de Physique, ENS de Lyon, CNRS, UMR 5672, Lyon, France
[3] University of Maine, Department of Mathematics and Statistics, Orono, Maine, USA
[4] Perm State Academy of Medicine, Department of Therapeutic and Propedeutic Dentistry, Perm, Russia
[5] Laboratoire de Physique, CNRS, ENS Lyon, 46 Alle d'Italie, Lyon, F69007, France
alain.arneodo@ens-lyon.fr

Abstract. We use the wavelet transform modulus maxima (WTMM) method to perform multifractal analysis of the temporal fluctuations of breast skin temperature recorded using infrared (IR) thermography. When investigating thermograms collected from a panel of patients with breast cancer and some female volunteers with healthy breasts, we show that the multifractal complexity of temperature fluctuations observed on intact breast is lost in mammary glands with malignant tumors. These results highlight dynamics IR imaging as a very valuable non-invasive technique for preliminary screening in asymptomatic women to identify those with risk of breast cancer. Besides potential clinical impact, they also shed a new light on physiological changes that may precede anatomical alterations in breast cancer development.

Keywords: Breast cancer, infrared thermography, multifractal analysis, wavelet transform modulus maxima method.

1 Introduction

Many biomedical signals are extremely inhomogeneous and nonstationary, fluctuating very irregularly and likely displaying scale invariance properties [1–5]. This manifests as a power-law behavior of the power spectrum that extends over

⋆ Correspondence author.

V.M. Mladenov and P.C. Ivanov (Eds.): NDES 2014, CCIS 438, pp. 288–300, 2014.

a wide range of spatial or/and temporal frequencies. But power spectrum analysis intrinsically fails to fully characterize the multifractal complexity of intermittent signals [6, 7]. Multifractal methods were developed either empirically like the detrended fluctuation analysis [8, 9] or based on mathematically grounded approaches like the wavelet-based methods built on either the WTMM skeleton [6, 7, 10–12] or the wavelet leaders [13–15]. The WTMM method was originally developed to remedy for the limitations of the structure functions method to perform multifractal analysis of one-dimensional (1D) velocity signal in fully developed turbulence [6, 7, 10, 11]. It has proved very efficient to estimate scaling exponents and multifractal spectra [6, 16, 17]. This method has been generalized in 2D for the multifractal analysis of rough surfaces [18] and then for the analysis of 3D scalar and vector fields [19–21]. Successful applications of the WTMM methodology include areas such as fully-developed turbulence [10, 16, 22], econophysics [23, 24], astrophysics [25], geophysics [18, 26, 27], surface science [28], genomics [5, 17] and medical and biological image analysis [18, 27, 29–32]. In the context of the present study, the 1D WTMM method has proved very efficient at discriminating between healthy and sick heart beat dynamics [33–35], whereas the 2D WTMM method can be used to detect microcalcifications and has great potential to assist in cancer diagnosis from digitized mammograms [17, 36, 37]. In this work, we use the 1D WTMM method to characterize the multifractal properties of 1D temperature time-series recorded by dynamic IR thermography of normal and cancerous breasts [38, 39].

Breast cancer is the most common type of cancer among women and despite recent advances in the medical field, there are still some inherent limitations in the currently used screening techniques. The radiological interpretation of X-ray mammograms is a difficult task since the mammographic appearance of normal tissue is highly variable. It often lead to over-diagnoses and, as a consequence, to unnecessary traumatic and painful biopsies [40–43]. Since the key to breast cancer survival lies upon its earliest possible detection, more accurate methods and techniques need to be developed to improve screening procedures while minimizing side effects. As sensitive to abnormal increase of metabolic activity and vascular circulation in breast tissue [44–46], IR thermography has been considered as a promising non-invasive screening method of breast cancer [47, 48]. For many years the suitability of static IR imaging for routine screening has been severely questioned [49–52]. Renewed interest in dynamic IR imaging to identify breast cancer comes from the rapid development in IR thermal imaging technology with new digital IR thermography cameras with higher temperature resolution (0.08 °C or better) and faster frame rate (70 Hz) [53], combined with increasing knowledge of angiogenesis and nitric oxide production of the cancer tissue causing local disturbances in vasomotor (autonomic nervous control of smooth muscles forcing blood through capillaries) and cardiogenic phenomena as compared to normal tissues [54, 55]. Dynamic IR imaging is currently used to detect variations in temperature rhythms generated by the cardiogenic and vasomotor frequencies [56, 57]. Human cardiogenic frequency levels lie between 1 and 1.5 Hz and vasomotor frequencies between 0.1 and 0.2 Hz, with higher harmonic frequency rhythms. As discussed below, beyond intensity differences in these rhythms between normal and tumor tissues, there is much more to learn from dynamical IR imaging of breast cancer. Using a wavelet-based multiscale analysis of the temperature intensity fluctuations about these perfusion oscillations, we propose to characterize the multifractal properties of these temperature fluctuations as a novel computer-aided method that can be easily implemented in early screening procedures to identify women with high risk of breast cancer.

2 The Wavelet Transform Modulus Maxim Method

The wavelet transform (WT) is a mathematical microscope [6, 7, 58] that is well suited for the analysis of complex non-stationary time-series such as those found in physiological systems [3, 34, 35], thanks to its ability to filter out low-frequency trends in the analyzed signal. The WT is a space (or time in our study)-scale analysis which consists in expanding signals in terms of wavelets which are constructed from a single function, the "analyzing wavelet" ψ, by means of translations and dilations. The WT of a real-valued function Σ is defined as [58]:

$$W_\psi[\Sigma](t_0, a) = \frac{1}{a}\int_{-\infty}^{+\infty} \Sigma(t)\psi(\frac{t-t_0}{a})dt \;, \tag{1}$$

where t_0 and a (> 0) are the time and scale parameters. The main advantage of using the WT for analyzing the fluctuations in regularity of the function Σ lies in its ability to eliminate order-n polynomial behavior (*i.e.* filtering fluctuations from low-frequency components) by simply choosing a wavelet ψ whose $n+1$ first moments are zero $[\int t^m \psi(t)\,dt = 0, 0 \leq m \leq n]$ [10]. Here, we use the compactly supported version $\psi_{(3)}^{(2)}$ of the Mexican hat introduced in Ref. [59] and which has two vanishing moments. The WTMM method [10–12] consists in investigating the scaling behavior of some partition function defined in terms of wavelet coefficients:

$$Z(q,a) = \sum_{l \in \mathcal{L}(a)} \left[\sup_{\substack{(t,a') \in l \\ a' \leq a}} |W_\psi(t,a')| \right]^q \sim a^{\tau(q)}, \tag{2}$$

where $q \in \mathbb{R}$. The sum is taken over the WT skeleton $\mathcal{L}(a)$ defined, at each fixed scale a, by the local maxima of $|W_\psi(t,a)|$. These WTMM are disposed on curves connected across scales called maxima lines $l_t(a)$, along which the WTMM behave as $a^{h(t)}$, where $h(t)$ is the Hölder exponent characterizing the singularity of Σ located at time t where $l_t(a)$ points to, in the limit $a \to 0^+$. Indeed, the Legendre transform of the scaling function $\tau(q)$ in Eq. (2), is the singularity spectrum $D(h) = \min_q[qh - \tau(q)]$ defined as the Hausdorf dimension of the set of points t where the Hölder exponent value is h [10–12]:

$$D(h) = \min_q [qh - \tau(q)]. \tag{3}$$

As originally pointed out in Ref. [6, 7], one can avoid some practical difficulties that occur when directly performing the Legendre transform of $\tau(q)$, by computing the following mean quantities $h(q,a)$ and $D(q,a)$:

$$h(q,a) = \sum_{l\in\mathcal{L}(a)} \ln|W_\psi(t,a)|\mathcal{W}_\psi(q,\mathcal{L},a) \sim h(q)\ln a, \tag{4}$$

$$D(q,a) = \sum_{l\in\mathcal{L}(a)} \mathcal{W}_\psi(q,\mathcal{L},a)\ln[\mathcal{W}_\psi(q,\mathcal{L},a)] \sim D(q)\ln a, \tag{5}$$

where $\mathcal{W}_\psi(q,\mathcal{L},a) = |W_\psi(t,a)|^q/Z(q,a)$ is a Boltzmann weight computed from the WT skeleton. From the scaling behavior of these quantities, in the limit $a \to 0^+$, one extracts $h(q)$ and $D(h)$ and therefore the $D(h)$ spectrum as a curve parametrized by q [6].

Homogeneous *monofractal* functions, *i.e.* functions with singularities of unique Hölder exponent H, are characterized by a linear $\tau(q)$ curve of slope H. Monofractal scaling implies that the shape of the probability distribution function (pdf) of (rescaled) wavelet coefficients does not depend on the scale a. A nonlinear $\tau(q)$ is the signature of nonhomogeneous *multifractal* functions, meaning that the Hölder exponent $h(t)$ is a fluctuating quantity [6, 7, 10–12] that depends on t. In this study, we fit the $\tau(q)$ data by the log-normal quadratic approximation $\tau(q) = -c_0 + c_1 q - c_2 q^2/2$, where the coefficients $c_n > 0$ [16]. The corresponding singularity spectrum has a characteristic single-humped shape $D(h) = c_0 - (h - c_1)^2/2c_2$, where $c_0 = -\tau(0)$ is the fractal dimension of the support of singularities of Σ, c_1 is the value of h that maximizes $D(h)$ and c_2, the so-called *intermittency coefficient* [6, 7, 10, 16], characterizes the width of the $D(h)$ spectrum, *i.e.* the signature of a change in WT coefficient statistics across scales [6, 7, 10, 27].

3 Results

3.1 Experimental Protocol

Subjects were recruited for the present study from Perm Region Cancer Hospital using procedures approved by the Human Studies Committee [39, 60]. They all gave Informed Consent to participate in this study via the recording of the IR thermograms of both mammary glands, the cancerous one and the opposite undiagnosed one with no visible sign of pathology. Our database includes 33 females with ages between 37 and 83 (average 57 years) who all went through surgery to remove the histologically confirmed malignant tumor (invasive ductal and/or lobular cancer) a few weeks after thermograms were recorded. The tumors were found at different depths from 1 cm down to 12 cm with a size varying from 1.2 cm up to 6.5 cm. As a control, we also investigated 14 women with intact mammary glands and of ages between 23 and 79 (average 49.6 years). Both breasts of healthy volunteers and patients with breast cancer were imaged with a InSb photovoltaic (PV) detector camera [53]. Imaging was performed with the patient in sitting position with arms down to avoid too much discomfort during

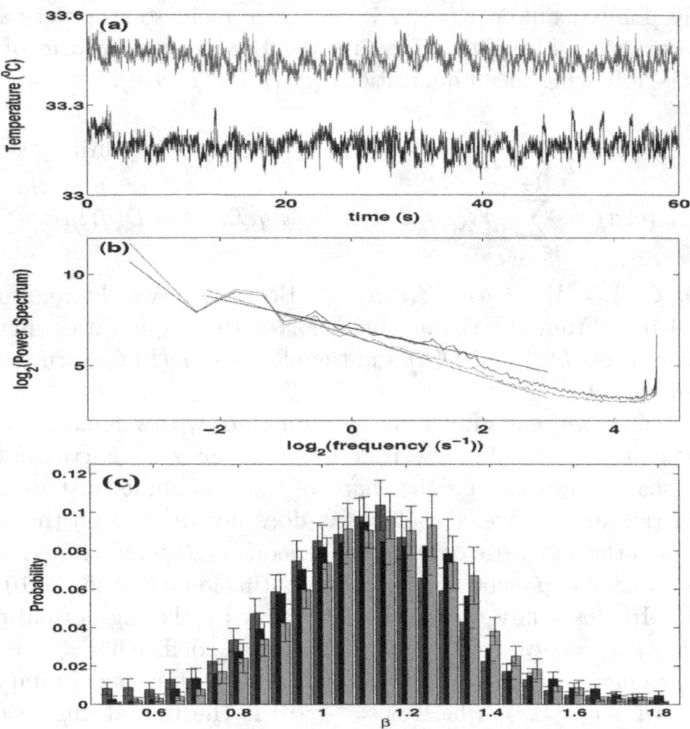

Fig. 1. Comparative power-spectrum analysis of temperature time-series recorded on both breasts of 33 patients with breast cancer and 14 healthy volunteers [38, 39]. (a) Pixel temperature time-series recorded in the cancerous right (red, top) and opposite left (black, bottom) breasts of a 54 years old patient. (b) Averaged temperature power spectra for the cancerous (red) and opposite (black) breasts. The straight lines correspond to power-law scaling $1/f^{\beta}$ with exponent $\beta = \tau(2) = 1.13$ (malignant) and 0.79 (normal), as estimated with the WTMM method. (c) Overall normalized histograms of power-law scaling exponent $\beta = \tau(2)$ values obtained in the breasts with malignant tumor (red: $N = 4032$ 8×8 pixel^2 squares), the opposite breasts (black: $N = 3606$) and the healthy breasts (green: $N = 3185$).

imaging. Frontal images were taken at a distance $\sim$ 1 m of the patient in an environmental room temperature of 20–22 °C. The image frame rate was set to 50 Hz. Each image set comprised 30 000 image frames during the 10 min immobile imaging phase. To eliminate low frequency patient movements, skin surface markers were successfully used as reference points for motion correction in the analysis. Pixel based and windowed regional power spectra and wavelet-based multifractal analysis of normal and cancer breasts were tested to define the best procedure to minimize the effect of the camera noise and to ensure statistical convergence in the multifractal spectra estimation. The results reported correspond to averaged power spectra, partition functions and singularity spectra over 64 temperature time-series in these 8×8 subareas spanning 10×10 mm^2 and covering the entire breast [38, 39].

3.2 Power Spectrum Analysis

Figure 1 summarizes the results obtained for a 54 years cancer patient. Fig. 1(a) shows two 1-pixel temperature time-series taken from a 8×8-pixel^2 window inside the tumor region in the malignant right breast and in a symmetrically positioned window in the normal left breast, respectively. As expected, the former fluctuates at a higher temperature than the latter. The corresponding average power spectra over the 64 pixels of the considered squares are shown in Fig. 1(b) in a logarithmic representation. The solid lines represent the rather convincing $1/f^{\beta}$ power-law scaling observed at frequencies higher than the characteristic human respiratory frequency ($\gtrsim 0.3$ Hz) and smaller than the cross-over frequency ($\lesssim 4$ Hz) towards (instrumental) white noise. Their respective slopes $\beta = \tau(q = 2) = 1.13$ (± 0.14) and 0.79 (± 0.21) are obtained from the WTMM analysis in Sect. 3.3. This difference looks quite significant and very promising in a discriminatory perspective. Unfortunately, as shown in Fig. 1(c), the histograms of β values obtained for all 8×8-pixel^2 squares covering 33 cancerous breasts, the 32 opposite breasts (one patient had mastectomy of the opposite breast [39]) and the 28 volunteer healthy breasts are quite similar with mean values $\bar{\beta} = 1.09 \pm 0.01$ (cancer),

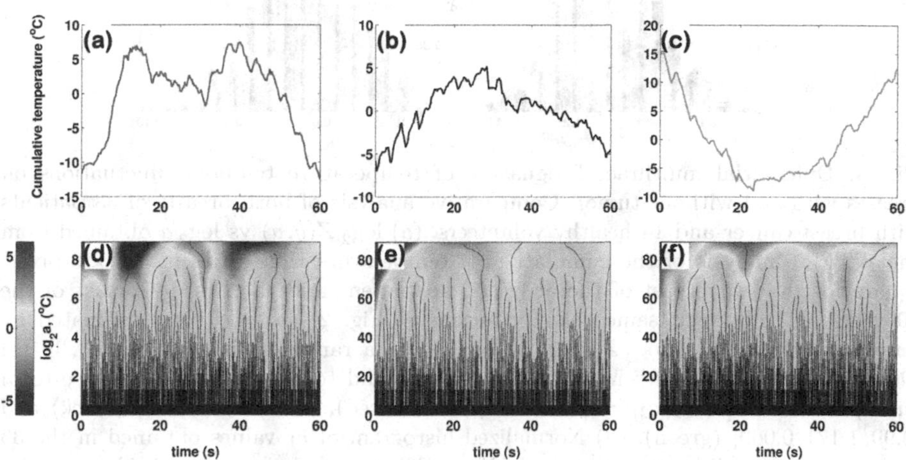

Fig. 2. Comparative wavelet analysis of the cancerous right breast (red) and the opposite left breast (black) of a 56 years old patient and of the healthy right breast (green) of a 60 years old volunteer. (a-c) 1 min portion of cumulative pixel temperature time-series (after removing the overall linear trend). (d-f) WT of the cumulative time-series as coded, from black (min $|W(\cdot,a)|$) to red (max $W(\cdot,a)$); solid black lines are the WTMM lines that define the WT skeleton. The analyzing wavelet is the second-order compactly supported analyzing wavelet $\psi_{(3)}^{(2)}$ [59].

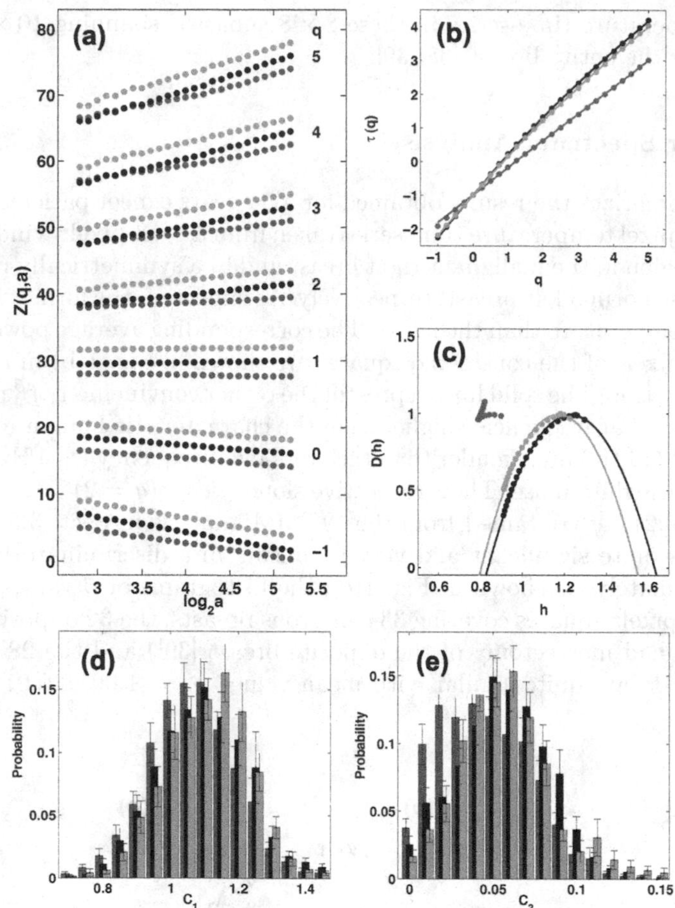

Fig. 3. Differential multifractal signature of temperature temporal fluctuations on breasts with and without tumor. Comparative analysis of both breasts of 33 patients with breast cancer and 14 healthy volunteers. (a) $\log_2 Z(q, a)$ vs $\log_2 a$ obtained from the WTMM analysis of the cumulative IR temperature time-series in a 8×8-pixel^2 square for the two breast of the 56 years old patient and the healthy breast of the 60 years old volunteer; same color coding as in Fig. 2. (b) $\tau(q)$ vs q estimated by linear regression fit of $\log_2 Z(q, a)$ vs $\log_2 a$ over a range of time scales [0.3, 3]. (c) $D(h)$ vs h. The solid lines in (b) and (c) correspond to quadratic spectra (see text) with parameters $[c_0, c_1, c_2] = [0.99, 0.81, 0.0044]$ (red), $[0.99, 1.23, 0.080]$ (black) and $[0.99, 1.171, 0.069]$ (green). (d) Normalized histogram of c_1 values obtained in the 33 breasts with a malignant tumor (red: $N = 4032$ 8×8 pixel^2 squares), the opposite breasts (black: $N = 3606$) and the healthy breasts (green: $N = 3185$). (e) Normalized histograms of c_2 values.

1.14 ± 0.01 (opposite) and 1.14 ± 0.01 (healthy). Indeed these histograms extend over a rather wide range of β values: $0.5 \lesssim \beta \lesssim 1.9$.

3.3 WWTM Multifractal Analysis

As previously experienced when applying the 1D WTMM method to rainfall time-series [26], the analysis was performed on the cumulative (or integral) of the temperature time-series. Hence the singularities with possible negative Hölder exponent $-1 < h < 0$, become singularities $0 < h_c = h + 1 < 1$ in the cumulative, thus posing no estimation problem when taking the sup along maxima lines in Eq. (2). As illustrated in Figs 2 and 3(a) for the two breasts of a 56 years old patient and the healthy breast of a 60 years old volunteer, when applying the WTMM method (Sect. 2) to the cumulative of the corresponding time-series (Fig. 2(a-c)), we confirm that the partition functions $Z(q,a)$ (Eq. (2)), $h(q,a)$ (Eq. (4)) and $D(q,a)$ (Eq. (5)) computed on the corresponding WT skeletons (Fig. 2(d-f)), display rather convincing scaling properties. As illustrated in Fig. 3(a) for $Z(q,a)$, power-law scaling is observed for $q = -1$ to 5, over a range of time-scales that we strictly limited to [0.43 s, 2.30 s] for linear regression fit estimates in a logarithmic representation. The $\tau(q)$ so-obtained are well approximated by quadratic spectra (Fig. 3(b)). For the malignant breast of patient with cancer, $\tau(q)$ is nearly linear as quantified by a very small value of the intermittency coefficient $c_2 = (4.4 \pm 0.6) \cdot 10^{-3}$. This signature of monofractality is confirmed, when respectively plotting $h(q,a)$ and $D(q,a)$ vs $\log_2 a$ (data not shown), where the slopes $h(q) = c_1 = 0.81 \pm 0.01$ and $D(q) = 0.99 \pm 0.03$, do not significantly depend on q, meaning that the $D(h)$ singularity spectrum nearly reduces to a single point $D(h = c_1 = 0.81) = 1$ (Fig. 3(c)). This monofractal diagnosis is confirmed when comparing the WT pdfs obtained at different time-scales; when rescaling the WT coefficients by a^H with $H = c_1$, they all collapse on a single curve as an indication of scale-invariant statistics [27, 38, 39].

In contrast, the $\tau(q)$ spectrum obtained for the opposite breast of this patient, is definitely nonlinear with a no longer negligible (one order of magnitude larger) quadratic term $c_2 = 0.080 \pm 0.001$ (Fig. 3(b)), the hallmark of multifractal scaling. The slopes $h(q)$ and $D(q)$ of $h(q,a)$ and $D(q,a)$ vs $\log_2 a$ now depend on q (data not shown). From the estimate of $h(q)$ and $D(q)$, we get the single-humped $D(h)$ spectrum shown in Fig. 3(c), which is well approximated by a quadratic spectrum with parameters $c_0 = 0.99 \pm 0.05, c_1 = 1.23 \pm 0.01$ and $c_2 = 0.080 \pm 0.001$. Because there is no longer a unique scaling exponent c_1, the shape of the WT coefficient pdf now evolves across scales, with fatter tails appearing at small scales (data not shown). Interestingly, the $\tau(q)$ (Fig. 3(b)) and $D(h)$ (Fig. 3(c)) spectra obtained for the healthy breast of the 60 years old volunteer are quite similar quadratic spectra with parameter values $c_0 = 0.99 \pm 0.03, c_1 = 1.17 \pm 0.01$ and $c_2 = 0.069 \pm 0.002$. Again this multifractal diagnosis is strengthened by the observation that the WT coefficient pdf has a shape that evolves across time-scales [38, 39].

The results of our comparative wavelet-based multifractal analysis of (cumulative) temperature fluctuations over the 33 cancerous breasts, the 32 opposite breasts (no right breast for patient 6) and the 28 volunteer healthy breasts are reported in Fig. 3(d,f) [39]. In Fig. 3(d), the corresponding histograms of c_1 values extend over a rather wide range $0.6 \lesssim c_1 \lesssim 1.8$ but turn out to be quite similar with mean values $\bar{c}_1 = 1.066 \pm 0.002$ (cancer), 1.104 ± 0.002 (opposite) and 1.103 ± 0.002 (healthy). This reminds that similar histograms were also obtained for the power-spectrum exponent β (Fig. 1(c)). In contrast, the intermittency parameter c_2 has a definite discriminatory power. The histogram for cancerous breasts ($\bar{c}_2 = 0.045 \pm 0.001$) is definitely shifted towards smaller values relative to the ones for opposite ($\bar{c}_2 = 0.056 \pm 0.001$) and healthy breasts ($\bar{c}_2 = 0.058 \pm 0.001$) (Fig. 3(e)). The small-value left-side of the c_2 histogram is much more populated in cancerous than in healthy breasts, confirming that cancerous breasts are enriched in squares where temperature fluctuations display significantly reduced multifractal properties with $c_2 \leq 0.03$ [38, 39]

4 Conclusion

To summarize, the wavelet-based multifractal analysis of dynamic IR thermograms is able to discriminate between cancerous breasts with monofractal temperature fluctuations characterized by a unique singularity exponent ($h_c = c_1$), and healthy breasts with multifractal temperature fluctuations requiring a wide range of singularity exponents as quantified by the intermittency coefficient $c_2 \gg 0$. This is analogous to the results of a similar wavelet-based multifractal analysis of human heart beat dynamics in Refs [34, 35], where the multifractal character and nonlinear properties of the healthy heart rate were shown to be lost in pathological condition, congestive heart failure. Fundamentally, our results indicate that skin temperature fluctuations of healthy breasts are more complex (multifractal) than previously suspected, posing a challenge to ongoing efforts to develop 3D breast models based on the physiology of the breast and that account for the skin surface temperature distribution in the presence, or absence, of an internal tumor [61–64]. The observed drastic simplification from multifractal to monofractal skin temperature dynamics may result from some increase in blood flow and cellular activity associated with the presence of a tumor [54–57]. More likely it can be the signature of some architectural change in the microenvironment of breast tumor [65] that may deeply affect heat transfer and related thermomechanics in breast tissue [63, 66]. Identifying the regulation mechanisms that originate in a loss of multifractal temperature dynamics will be an important step towards understanding breast cancer development. Dynamic IR thermography is a non-invasive screening method that is inexpensive, quick and painless for the patient. These promising findings could lead to future use of wavelet-based multifractal processing of dynamic IR thermography, to help identify women with high risk of breast cancer, prior to more traumatic and painful examination such as mammography and biopsy.

Acknowledgments. We are very grateful to INSERM, ITMO Cancer for its financial support under contract PC201201-084862 “Physiques, mathématiques ou sciences de l’ingénieur appliqués au Cancer”, to Perm Regional Government (Russia) for the contract “Multiscale approaches in mechanobiology for early cancer diagnosis” and to the Maine Cancer Foundation. We are thankful to

Prof. O. Orlov for his support in IR thermal imaging at Perm Region Cancer Hospital and to I. Panteleev, Y. Bayandin, O. Plekhov and S. Uvarov for fruitful discussions.

References

1. Schlesinger, M., West, B.: Random Fluctuations and Pattern Growth: Experiments and Models. Kluwer, Boston (1988)
2. Bassingthwaighte, J., Liebovitch, L., West, B.: Fractal Physiology. Oxford University Press, New York (1994)
3. Goldberger, A.L., Amaral, L.A.N., Hausdorff, J.M., Ivanov, P.C., Peng, C.K., Stanley, H.E.: Fractal dynamics in physiology: alterations with disease and aging. Proc. Natl. Acad. Sci. USA 99, 2466–2472 (2002)
4. Bunde, A., Kropp, J., Schellnhuber, H.: The Science of Disasters: Climate Disruptions, Heart Attacks, and Market Crashes. Springer, Berlin (2002)
5. Arneodo, A., Vaillant, C., Audit, B., Argoul, F., d'Aubenton-Carafa, Y., Thermes, C.: Multi-scale coding of genomic information: From DNA sequence to genome structure and function. Phys. Rep. 498, 45–188 (2011)
6. Muzy, J.F., Bacry, E., Arneodo, A.: The multifractal formalism revisited with wavelets. International Journal of Bifurcation and Chaos 4, 245–302 (1994)
7. Arneodo, A., Bacry, E., Muzy, J.F.: The thermodynamics of fractals revisited with wavelets. Physica A 213(1-2), 232–275 (1995)
8. Kantelhardt, J.W., Zschiegner, S.A., Koscielny-Bunde, E., Havlin, S., Bunde, A., Stanley, H.E.: Multifractal detrended fluctuation analysis of nonstationary time series. Physica A 316, 87–114 (2002)
9. Ihlen, E.A.F.: Introduction to multifractal detrended fluctuation analysis in Matlab. Front. Physiol. 3(141), 1–18 (2012)
10. Muzy, J.F., Bacry, E., Arneodo, A.: Wavelets and multifractal formalism for singular signals: Application to turbulence data. Phys. Rev. Lett. 67(25), 3515–3518 (1991)
11. Muzy, J.F., Bacry, E., Arneodo, A.: Multifractal formalism for fractal signals: The structure-function approach versus the wavelet-transform modulus-maxima methods. Phys. Rev. E 47(2), 875–884 (1993)
12. Bacry, E., Muzy, J.F., Arneodo, A.: Singularity spectrum of fractal signals from wavelet analysis: Exact results. Journal of Statistical Physics 70(3-4), 635–674 (1993)
13. Jaffard, S., Lashermes, B., Abry, P.: Wavelet leaders in multifractal analysis. In: Wavelet Analysis and Applications, pp. 219–264. Birkhuser (2006)
14. Wendt, H., Abry, P., Jaffard, S.: Bootstrap for empirical multifractal analysis. IEEE Signal Proc. Mag. 24, 38–48 (2007)
15. Lashermes, B., Roux, S.G., Abry, P., Jaffard, S.: Comprehensive multifractal analysis of turbulent velocity using wavelet leaders. Eur. Phys. J. B 61, 201–215 (2008)
16. Delour, J., Muzy, J.F., Arneodo, A.: Intermittency of 1D velocity spatial profiles in turbulence: A magnitude cumulant analysis. Eur. Phys. J. B 23(2), 243–248 (2001)
17. Audit, B., Bacry, E., Muzy, J.F., Arneodo, A.: Wavelet-based estimators of scaling behavior. IEEE Trans. Info. Theory 48, 2938–2954 (2002)
18. Arneodo, A., Decoster, N., Kestener, P., Roux, S.G.: A wavelet-based method for multifractal image analysis: From theoretical concepts to experimental applications. Adv. Imaging Electr. Phys. 126, 1–92 (2003)

19. Kestener, P., Arneodo, A.: Three-dimensional wavelet-based multifractal method: The need for revisiting the multifractal description of turbulence dissipation data. Phys. Rev. Lett. 91(19), 194501 (2003)
20. Kestener, P., Arneodo, A.: Generalizing the wavelet-based multifractal formalism to random vector fields: Application to three-dimensional turbulence velocity and vorticity data. Phys. Rev. Lett. 93(4), 044501 (2004)
21. Arneodo, A., Audit, B., Kestener, P., Roux, S.G.: Wavelet-based multifractal analysis. Scholarpedia 3, 4103 (2008)
22. Arneodo, A., Manneville, S., Muzy, J.F.: Towards log-normal statistics in high Reynolds number turbulence. Eur. Phys. J. B 1(1), 129–140 (1998)
23. Arneodo, A., Muzy, J.F., Sornette, D.: "Direct" causal cascade in the stock market. Eur. Phys. J. B 2(2), 277–282 (1998)
24. Muzy, J.F., Sornette, D., Delour, J., Arneodo, A.: Multifractal returns and hierarchical portfolio theory. Quant. Finance 1, 131–148 (2001)
25. Khalil, A., Joncas, G., Nekka, F., Kestener, P., Arneodo, A.: Morphological analysis of H_I features. II. Wavelet-based multifractal formalism. Astrophys. J. Suppl. Ser. 165(2), 512–550 (2006)
26. Venugopal, V., Roux, S.G., Foufoula-Georgiou, E., Arneodo, A.: Revisiting multifractality of high-resolution temporal rainfall using a wavelet-based formalism. Water Resour. Res. 42(6), W06D14 (2006)
27. Arneodo, A., Audit, B., Decoster, N., Muzy, J.F., Vaillant, C.: The Science of Disasters: Climate Disruptions, Heart Attacks, and Market Crashes, pp. 26–102. Springer, Berlin (2002)
28. Roland, T., Khalil, A., Tanenbaum, A., Berguiga, L., Delichère, P., Bonneviot, L., Elezgaray, J., Arneodo, A., Argoul, F.: Revisiting the physical processes of vapodeposited thin gold films on chemically modified glass by atomic force and surface plasmon microscopies. Surf. Sci. 603(22), 3307–3320 (2009)
29. Khalil, A., Grant, J.L., Caddle, L.B., Atzema, E., Mills, K.D., Arneodo, A.: Chromosome territories have a highly nonspherical morphology and nonrandom positioning. Chromosome Res. 15(7), 899–916 (2007)
30. Grant, J., Verrill, C., Coustham, V., Arneodo, A., Palladino, F., Monier, K., Khalil, A.: Perinuclear distribution of heterochromatin in developing C. elegans embryos. Chromosome Res. 18(8), 873–885 (2010)
31. Snow, C.J., Goody, M., Kelly, M.W., Oster, E.C., Jones, R., Khalil, A., Henry, C.A.: Time-lapse analysis and mathematical characterization elucidate novel mechanisms underlying muscle morphogenesis. PLoS Genet. 4(10), e1000219 (2008)
32. Khalil, A., Aponte, C., Zhang, R., Davisson, T., Dickey, I., Engelman, D., Hawkins, M., Mason, M.: Image analysis of soft-tissue in-growth and attachment into highly porous alumina ceramic foam metals. Med. Eng. Phys. 31(7), 775–783 (2009)
33. Ivanov, P.C., Rosenblum, M.G., Peng, C.K., Mietus, J., Havlin, S., Stanley, H.E., Goldberger, A.L.: Scaling behaviour of heartbeat intervals obtained by wavelet-based time-series analysis. Nature 383(6598), 323–327 (1996)
34. Ivanov, P.C., Amaral, L.A., Goldberger, A.L., Havlin, S., Rosenblum, M.G., Struzik, Z.R., Stanley, H.E.: Multifractality in Human Heartbeat Dynamics. Nature 399(6735), 461–465 (1999)
35. Ivanov, P., Goldberger, A., Stanley, H.: Fractal and multifractal approaches in physiology. In: Bunde, A., Kropp, J., Schellnhuber, H. (eds.) The Science of Disasters: Climate Disruptions, Heart Attacks, and Market Crashes, pp. 219–258. Springer (2002)

36. Kestener, P., Lina, J.M., Saint-Jean, P., Arneodo, A.: Wavelet-based multifractal formalism to assist in diagnosis in digitized mammograms. Image Anal. Stereol. 20, 169–174 (2001)
37. Batchelder, K.A., Tanenbaum, A.B., Albert, J., Guimond, L., Arneodo, A., Kestener, P., Khalil, A.: Wavelet-based 3D reconstruction of microcalcification clusters from two mammographic views: Fractal tumors are malignant and Euclidean tumors are benign. PLoS ONE (submitted, 2014)
38. Gerasimova, E., Audit, B., Roux, S.G., Khalil, A., Argoul, F., Naimark, O., Arneodo, A.: Multifractal analysis of dynamic infrared imaging of breast cancer. Europhys. Lett. 104, 68001 (2013)
39. Gerasimova, E., Audit, B., Roux, S.G., Khalil, A., Gileva, O., Argoul, F., Naimark, O., Arneodo, A.: Wavelet-based multifractal analysis of dynamic infrared thermograms to assist in early breast cancer diagnosis. Front. Physiol. 5, Article 176, 1–11 (2014)
40. Nass, S.J., Henderson, I.C., Lashof, J.C.: Mammography and Beyond: Developing Technologies for the Early Detection of Breast Cancer. National Academy Press, Washington (2000)
41. Bronzino, J.D.: Biomedical Engeneering Handbook. CRC Press, Boca Raton (2006)
42. Tamini, R.M., Colditz, G.A., Hankinson, S.E.: Circulating carotenoids, mammographic density, and subsequent risk of breast cancer. Cancer Res. 69, 9323–9329 (2009)
43. Vinitha Sree, S., Ng, E.Y.K., Kaw, G., Acharya, U.R., Chong, B.K.: The use of skin surface electropotentials for breast cancer detection–preliminary clinical trial results obtained using the biofield diagnostic system. J. Med. Syst. 35, 79–86 (2009)
44. Sterns, E.E., Zee, B., SenGupta, S., Saunders, F.W.: Thermography: Its relation to pathologic characteristics, vascularity, proliferation rate, and survival of patients with invasive ductal carcinoma of the breast. Cancer 77, 1324–1328 (1996)
45. Yahara, T., Koga, T., Yoshida, S., Nakagawa, S., Deguchi, H., Shirouzou, K.: Relationship between micro vessel density and thermographic hot areas in breast cancer. Surg. Today 33, 243–248 (2003)
46. Keyserlingk, J.R., Ahlgren, P.D., Yu, E., Belliveau, N., Yassa, M.: Biomedical Engineering Handbook, pp. 97–158. CRC Press, Boca Raton (2006)
47. Head, J.F., Wang, F., Lipari, C., Elliott, R.L.: The important role of infrared imaging in breast cancer. IEEE Eng. Med. Biol. Mag. 19(3), 52–57 (2000)
48. Amalu, W.C., Hobbins, W.B., Head, J.F., Elliott, R.L.: Biomedical Engineering Handbook, pp. 1–21. CRC Press, Boca Raton (2006)
49. Itoh, T., Kato, T., Igarashi, Y., Ishihara, S., Sasamon, H., Sekiya, T.: Contact-thermography in breast cancer mass screening (in Japanese with English abstract). Biomed. Thermol. 10, 49–51 (1990)
50. Iwase, T., Yoshimoto, M., Watanabe, S., Fujio, K., Nishi, M., Ohashi, Y.: Relation between hot spot and tumor location in the thermogram of breast cancer (in Japanese with English abstract). Biomed. Thermol. 10, 60 (1990)
51. Yokoe, T.: The relationship between thermographic positive pattern and pathological factors of breast cancer (in Japanese with English abstract). Biomed. Thermol. 12, 60 (1990)
52. Ng, E.Y.K.: A review of thermography as promising non-invasive detection modality for breast tumor. Int. J. Therm. Sci. 49, 849–859 (2009)
53. Joro, R., Lääperi, A.L., Dastidar, P., Soimakallio, S., Kuukasjäri, T., Toivonen, T., Saaristo, R., Järvenpää, R.: Imaging of breast cancer with mid- and long-wave infrared camera. J. Med. Eng. Technol. 32, 189–197 (2008)

54. Thomsen, L.L., Miles, D.W.: Role of nitric oxide in tumour progression: Lessons from human tumours. Cancer Metast. Rev. 17(1), 107–118 (1998)
55. Anbar, M., Milescu, L., Naumov, A., Brown, C., Button, T., Carty, C., Dulaimy, K.: Detection of cancerous breasts by dynamic area telethermometry. IEEE Eng. Med. Biol. Mag. 20(5), 80–91 (2001)
56. Button, T.M., Li, H., Fisher, P., Rosenblatt, R., Dulaimy, K., Li, S., O'Hea, B., Salvitti, M., Geronimo, V., Jambawalikar, S., Weiss, R.: Dynamic infrared imaging for the detection of malignancy. Phys. Med. Biol. 49(14), 3105–3116 (2004)
57. Joro, R., Lääperi, A.L., Soimakallio, S., Järvenpää, R., Kuukasjäri, T., Toivonen, T., Saaristo, R., Dastida, P.: Dynamic infrared imaging in identification of breast cancer tissue with combined image processing and frequency analysis. J. Med. Eng. Technol. 32, 325–335 (2008)
58. Mallat, S.: A Wavelet Tour of Signal Processing. Academic Press, New York (1998)
59. Roux, S., Muzy, J.F., Arneodo, A.: Detecting vorticity filaments using wavelet analysis: About the statistical contribution of vorticity filaments to intermittency in swirling turbulent flows. Eur. Phys. J. B 8(2), 301–322 (1999)
60. Gileva, O.S., Freynd, G.G., Orlov, O.A., Libik, T.V., Gerasimova, E.I., Plekhov, O.A., Bayandin, Y.V., Panteleev, I.A.: Interdisciplinary approaches to early diagnosis and screening of tumors and precancerous diseases (for example, breast cancer). Vestnik RFFI (2-3), 93–99 (2012)
61. Ng, E.Y.K., Sudharsan, N.M.: Computer simulation in conjunction with medical thermography as an adjunct tool for early detection of breast cancer. BMC Cancer 4, 17 (2004)
62. Panteleev, I.A., Plekhov, O.A., Naimark, O.: Mechanibiology study of structural homeostasis in tumor using infrared thermography data. Phys. Mesomech. 15(3), 105–113 (2007)
63. Xu, F., Lu, T.J., Steffen, K.A.: Biothermomechanical behavior of skin tissue. Acta Mech. Sin. 24, 1–23 (2008)
64. Lin, Q.Y., Yang, H.Q., Xie, S.S., Wang, Y.H., Ye, Z., Chen, S.Q.: Detecting early breast tumour by finite element thermal analysis. J. Med. Eng. Technol. 33, 274–280 (2009)
65. Bissel, M.G., Hines, W.C.: Why don't we get more cancer? a proposed role of the microenvironment in restraining cancer progression. Nat. Med. 17(3), 320–329 (2011)
66. Quail, D.F., Joyce, J.A.: Microenvironmental regulation of tumor progression and metastasis. Nat. Med. 19(11), 1423–1437 (2013)

Information Flow in Ising Models on Brain Networks

Sebastiano Stramaglia[1,2,3], Jesus M. Cortes[2,3], Leonardo Angelini[1], Mario Pellicoro[1], and Daniele Marinazzo[4]

[1] Dipartimento di Fisica, Universitá degli Studi di Bari and INFN, 70126 Bari, Italy
[2] Biocruces Health Research Institute. Hospital Universitario de Cruces. E-48903, Barakaldo, Spain
[3] Ikerbasque, The Basque Foundation for Science, E-48011, Bilbao, Spain
[4] Faculty of Psychology and Educational Sciences, Department of Data Analysis, Ghent University, B-9000 Ghent, Belgium

Abstract. We analyze the information flow in the Ising model on two real networks, describing the brain at the mesoscale, with Glauber dynamics. We find that the critical state is characterized by the maximal amount of information flow in the system, and that this does not happen when the Ising model is implemented on the two-dimensional regular grid. At criticality the system shows signatures of the law of diminishing marginal returns, some nodes showing disparity between incoming and outgoing information. We also implement the Ising model with conserved dynamics and show that there are regions of the systems exhibiting anticorrelation, in spite of the fact that all couplings are positive; this phenomenon may be connected with some evidences in real brains (the default mode network is characterized by anticorrelated components).

Keywords: Ising model, criticality, brain, Transfer entropy, Granger causality.

1 Introduction

The prototypical example of system exhibiting *criticality* is the Ising model, and criticality has been proposed to characterize brain signals [1]. Recently, the occurrence of a nonequilibrium critical dynamics in brain activity during sleep has been has been reported in [2]. In this contribution we implement the Ising model, with Glauber dynamics, on two real brain networks and analyze its critical behavior in terms of information flow, and find signatures of the law of diminishing marginal returns. We also compare with the results of the Ising model on a regular two-dimensional lattice, and show that, differently from this network, the critical state for the brain networks is characterized by the maximal amount of circulating information. We measure the information flow by transfer entropy [3], a technique that has been introduced in the context of the inference of the underlying network structure of complex systems from these time series. Recently [4] it has been shown that transfer entropy is strongly related

V.M. Mladenov and P.C. Ivanov (Eds.): NDES 2014, CCIS 438, pp. 301–308, 2014.

to Granger causality, a major tool to reveal drive-response relationships among variables, and it is connected to the inverse ising problem [5]. Initially developed for econometric applications, Granger causality has gained popularity also among physicists (see, e.g., [6,7,8,9]) and eventually became one of the methods of choice to study brain connectivity in neuroscience [10,11]. In a recent paper it has been shown that the pattern of information flow among variables of a complex system is the result of the interplay between the topology of the underlying network and the capacity of nodes to handle the incoming information, and that, under suitable conditions, it may exhibit the law of diminishing marginal returns, a fundamental principle of economics which states that when the amount of a variable resource is increased, while other resources are kept fixed, the resulting change in the output will eventually diminish (see [12] and references therein). The origin of such behavior resides in the structural constraint related to the fact that each node of the network may handle a limited amount of information. In [12] the information flow pattern of several dynamical models on hierarchical networks has been considered and found to be characterized by exponential distribution of the incoming information and a fat-tailed distribution of the outgoing information, a clear signature of the law of diminishing marginal returns.

In this work we also implement the Ising model with conserved dynamics (Kawasaki [13]) on a brain network and show that, although all couplings are positive, different regions of the brain exhibit anticorrelation, a feature that is observed in real brain (the default mode network is characterized by anticorrelated components [14]).

2 Bivariate Transfer Entropy of Ising Systems

In this section we describe how the transfer entropy is evaluated. Let us consider the configurations $\{\sigma_i(t)\}_{i=1,\dots,n}$ of an Ising system of n spins living on network, obtained by Montecarlo with Glauber or Kawasaki dynamics. The lagged spin vectors are denoted $\Sigma_i(t) = \sigma_i(t-1)$.

For each pair of spins (i, j) connected by a link in the underlying network, the bivariate transfer entropy c, measuring the information flow $i \to j$ is evaluated as follows:

$$c_{ij} = \sum_{\sigma_j=\pm1} \sum_{\Sigma_j=\pm} \sum_{\Sigma_i=\pm1} p(\sigma_j, \Sigma_j, \Sigma_i) \log \frac{p(\sigma_j, \Sigma_j)\, p(\Sigma_j, \Sigma_i)}{p(\sigma_j, \Sigma_j, \Sigma_i)\, p(\Sigma_j)}, \tag{1}$$

where $p(\Sigma_j, \Sigma_i)$ is the fraction of times that the configuration (Σ_j, Σ_i) is observed in the data set, and similar definitions hold for the other probabilities. c_{ij} is zero if spins σ_i and σ_j are not connected by a link in the network. The total information flow is then given by $C = \sum_{i=1}^{n} \sum_{j=1}^{n} c_{ij}$.

3 Two Dimensional Ising Model and Brain Networks

In this section we report the results obtained on the two-dimensional Ising model (on a regular grid), and implemented on real structural networks. Considering the 2D Ising model, we find that the total transfer entropy C has a peak in the paramagnetic phase, see figure (1): the critical point is not the one with the maximal circulation of information. Moreover we stress that the amount of information flow depends on the updating scheme, but the maximum is attained in correspondence of the same temperature, see e.g. figure (2) where Metropolis and Glauber dynamics are compared for the 2D Ising model.

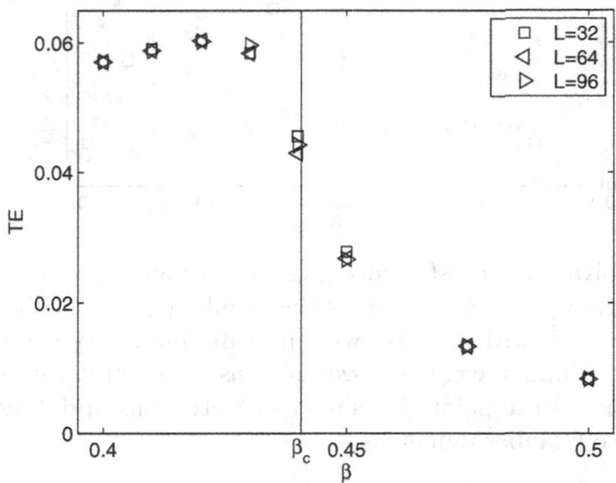

Fig. 1. The sum of bivariate transfer entropies for all pairs of spins, connected by a non-vanishing interaction, is depicted versus the coupling β for the 2D Ising model on a square lattice of size L^2, with $L = 32, 64$, and 96, with periodic boundary conditions. Transfer entropies have been evaluated averaging over 20 runs of 10000 iterations. The vertical line corresponds to the critical point. Since the maximum of the heat capacity matches quite well the critical point, this suggests that the system of linear size 96 is a reasonably good approximation to the thermodynamic limit, and that the transfer entropy has its maximum in the paramagnetic phase.

Turning to consider the Ising model with Glauber dynamics on a 74 nodes brain network obtained from diffusion tractography and contained in The Virtual Brain platform, a simulator of primate brain network dynamics [15], with couplings $J = \beta A$ (A is the connectivity matrix), in figure (3) we depict four quantities as a function of the inverse temperature β: R, the ratio between the standard deviation of c_{out} and those of c_{in}, where c is the information flow estimated by transfer entropy; the total causality C, i.e. the sum of all information flows; the ratio S between the intra-communities information flow and the inter-communities information flow, measuring the *segregation* of the network; the susceptibility χ. C, S and χ attain their maximum value in correspondance of the same temperature, whilst R is maximal at a lower temperature.

As another example, we consider a network which describes at low resolution the anatomical connectivity in brain (66 nodes), obtained via diffusion spectrum imaging (DSI) and white matter tractography [16]. In figure (4) we depict R, C, S and χ for the Ising model with Glauber dynamics where couplings are $J = \beta A$,

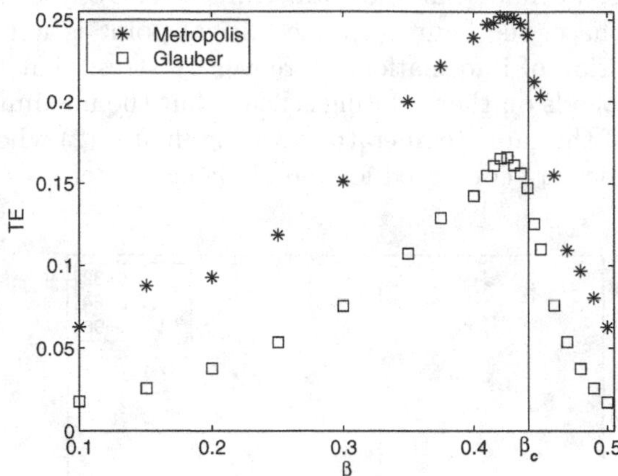

Fig. 2. The sum of bivariate transfer entropies for all pairs of spins, connected by a non-vanishing interaction, is depicted versus the coupling β for the 2D Ising model on a square lattice of size L^2, with $L = 16$, with periodic boundary conditions. Transfer entropies have been evaluated averaging over 20 runs of 10000 iterations. The vertical line corresponds to the critical point. Stars refer to Metropolis updating scheme, whilst empty squares refer to Glauber dynamics.

where A is the brain connectivity. In this case R, C and χ are unimodal and they peak nearly at the same temperature, so the critical state in this case is also the one with maximal circulation of information and the most affected by the law of diminishing marginal returns. The segregation of the system increases as the temperature is lowered. In both the brain networks considered here, the critical state is also the one maximizing the information flow.

It is interesting to report the analysis of equal-time correlations for the Ising model on the brain network. For each temperature we evaluate the correlational pattern obtained by Glauber dynamics and we find the corresponding modular decomposition by modularity maximization. Before the transition, the system has a stable decomposition in four moduli, after the transition by a single module for all nodes. In figure (5) the modularity of the best decomposition is plotted versus the temperature.

In order to compare the spin correlations by Glauber dynamics with those obtained by implementing the Ising model with conserved dynamics on the same brain network, using the Kawasaki update rule [13]. At the critical point the spin correlations for the Kawasaki dynamics can be both positive and negative: negative correlations emerge due to the conservation constraint. This is clear if we plot the Kawasaki correlations versus the Glauber correlations for the same pair of spins, at the critical point , see figure (6). We note that the pattern of correlations drastically change due to the conservation of the magnetization

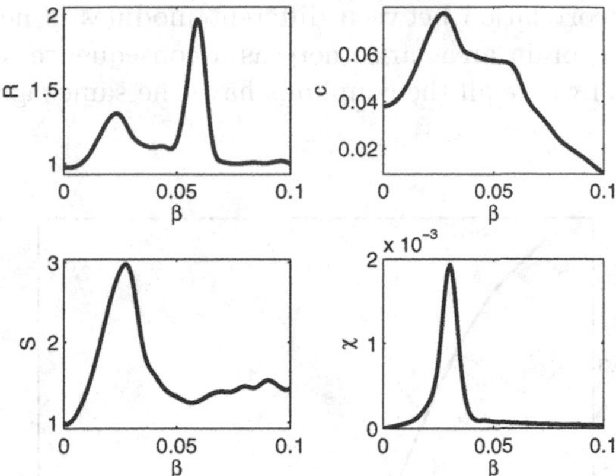

Fig. 3. Concerning the Ising model on the 74 nodes brain network, the following quantities are depicted versus the inverse temperature β: R, the ratio between the standard deviations of outgoing and incoming information flows (Top Left); the total causality C, i.e. the sum of all information flows in the network (Top Right); the ratio S between the intra-hemispheres and the inter-hemispheres information flows, measuring the segregation of the network (Bottom Left); the susceptibility χ (Bottom Right)

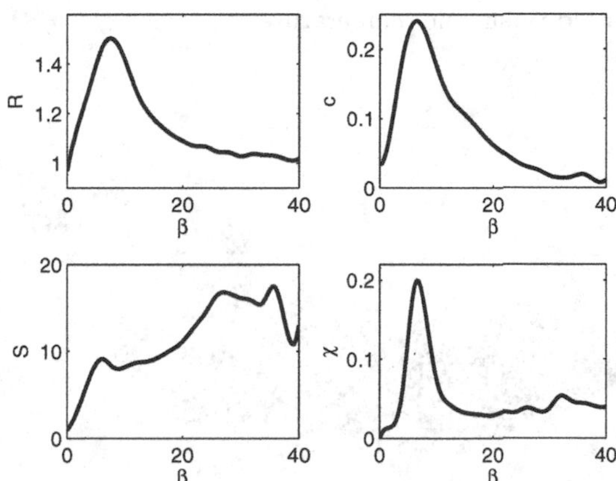

Fig. 4. Concerning the Ising model on the 66 nodes brain network, the following quantities are depicted versus the inverse temperature β: R, the ratio between the standard deviations of outgoing and incoming information flows (Top Left); the total causality C, i.e. the sum of all information flows in the network (Top Right); the ratio S between the intra-hemispheres and the inter-hemispheres information flows, measuring the segregation of the network (Bottom Left); the susceptibility χ (Bottom Right)

(in the spirit of the neural interpretation of the Ising model, we may speak of conservation of the average neural activation in the system). The modular decomposition of the Kawasaki correlational pattern corresponds to four modules:

in figure (7) we depict the average intra and inter correlations among moduli, and note that the correlation between different modules is negative: negative correlations between brain areas arise here as a consequence of a conservation law, even in a model where all the couplings have the same sign.

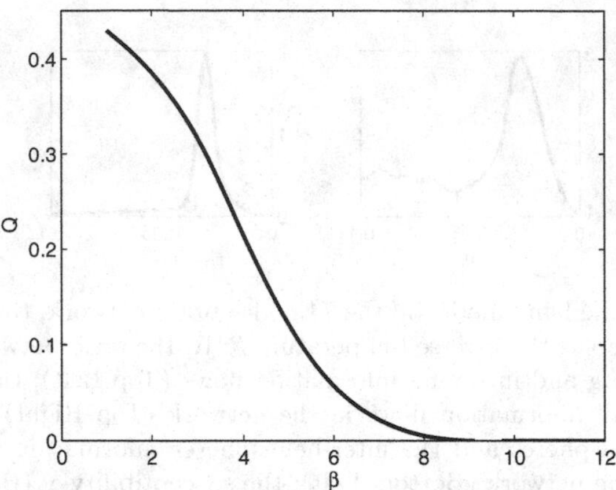

Fig. 5. The modularity of the best partition of the correlational network, for the 66 nodes network, is plotted versus the temperature

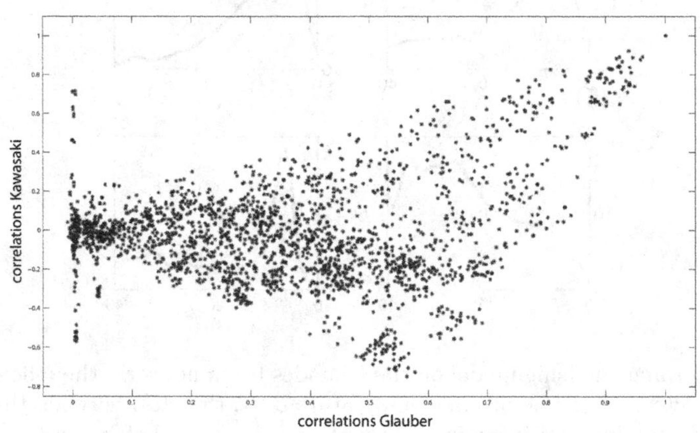

Fig. 6. For the brain network, the spin correlations by Kawasaki dynamics is plotted versus the spin correlations by Glauber dynamics, at criticality for the 66 nodes brain network

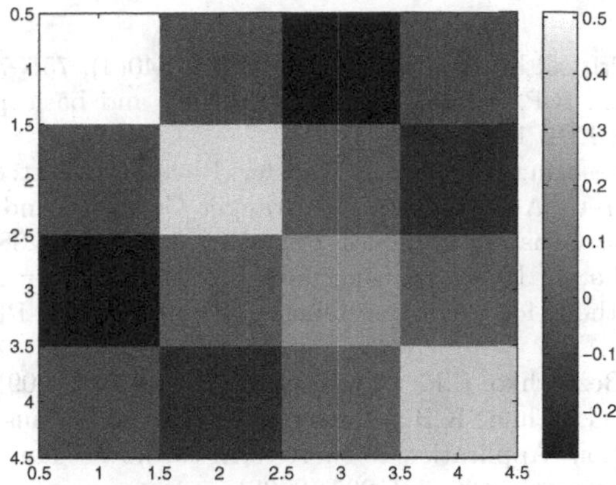

Fig. 7. The pattern of correlations, at criticality, by Kawasaki dynamics leads to four moduli in the 66 nodes brain network, obtained by maximization of the modularity. In the figure, along the diagonal the average intra correlation in the moduli is depicted, the off diagonal squares correspond to the average inter-correlations among the four moduli.

4 Conclusions

We have shown that the critical state of the Ising model on a brain network is characterized by the maximal amount of information transfer among units and brain effective connectivity networks may also be considered in the light of the law of diminishing marginal returns. This property does not hold for a generic network: we have shown that on the regular 2D grid the maximal circulation of information is not attained at criticality. It is a matter for further research the search for the largest class of networks in which the peaks of C and χ coincide.

We also reported that, in Ising models of brain, there is no need to introduce negative couplings to get anticorrelation between regions, indeed the implementation of a conserved dynamics leads to negative correlations. Negative correlations are observed, at the mesoscopic brain scale, for example w.r.t. the default mode network: whilst at the microscopic level the presence of both inhibitory and excitatory synapses justifies positive and negative couplings, at the mesoscopic scale the fibers connecting regions are similar hence all the couplings should all have the same sign. If the conservation law of the Kawasaki dynamics admits or not a physiological basis (e.g. a sort of homeostasis of the overall activity of the brain) is matter for future research. However, our findings may be relevant for the construction of brain models to reproduce the critical behavior of the brain.

References

1. Chialvo, D.R.: Critical brain networks. Physica A 340(4), 756–765 (2004)
2. Lo, C.C., Bartsch, R.P., Ivanov, P.C.: Asymmetry and basic pathways in sleep-stage transitions. EPL 102, 10008 (2013)
3. Schreiber, T.: Measuring Information Transfer. Phys. Rev. Lett. 85, 461–464 (2000)
4. Barnett, L., Barrett, A.B., Seth, A.K.: Granger Causality and Transfer Entropy are Equivalent for Gaussian Variables. Phys Rev. Lett. 103, 238701 (2009)
5. Roudi, Y., Tyrcha, J., Hertz, J.: Ising model for neural data: Model quality and approximate methods for extracting functional connectivity. Phys. Rev. E 79(5), 051915 (2009)
6. Smirnov, D.A., Bezruchko, B.P.: Phys Rev. E 79, 046204 (2009)
7. Faes, L., Nollo, G., Chon, K.H.: Assessment of Granger causality by nonlinear model identification: Application to short-term cardiovascular variability. Annals of Biomedical Engineering 36, 381–295 (2008)
8. Marinazzo, D., Pellicoro, M., Stramaglia, S.: Kernel Method for Nonlinear Granger Causality. Phys. Rev. Lett. 100, 144103 (2008)
9. Marinazzo, D., Pellicoro, M., Stramaglia, S.: Kernel-Granger causality and the analysis of dynamical networks. Phys. Rev. E 77, 056215 (2008)
10. Marinazzo, D., Liao, W., Chen, H., Stramaglia, S.: Nonlinear Connectivity by Granger causality. NeuroImage 58, 330–338 (2011)
11. Friston, K.J.: Functional and effective connectivity: A review. Brain Connectivity 1, 13 (2011)
12. Marinazzo, D., Wu, G., Pellicoro, M., Angelini, L., Stramaglia, S.: Information Flow in Networks and the Law of Diminishing Marginal Returns: Evidence from Modeling and Human Electroencephalographic Recordings. PLoS ONE 7(9), e45026 (2012)
13. Kawasaki, K.: Diffusion Constants near the Critical Point for Time-Dependent Ising Models. I. Phys. Rev. 145, 224–230 (1966)
14. Broyd, S.J., Demanuele, C., Debener, S., Helps, S.K., James, C.J., Sonuga-Barke, E.J.S.: Default-mode brain dysfunction in mental disorders: A systematic review. Neuroscience and Biobehavioral Reviews 33(3), 279–296 (2009)
15. Sanz Leon, P., Knock, S.A., Woodman, M.M., Domide, L., Mersmann, J., McIntosh, A.R., Jirsa, V.: The Virtual Brain: A simulator of primate brain network dynamics. Front. Neuroinform. (2013), doi:10.3389/fninf.2013.00010
16. Hagmann, P., Cammoun, L., Gigandet, X., Meuli, R., Honey, C.J., Weeden, V.J., Sporns, O.: Mapping the Structural Core of Human Cerebral Cortex. PLoS Biology 6(7), e159 (2008)

Deviation from Criticality in Functional Biological Networks

Tom Lorimer, Florian Gomez, and Ruedi Stoop

Institute of Neuroinformatics,
University of Zurich and ETH Zurich,
Winterthurerstrasse 190, 8057 Zurich, Switzerland
{lorimert,fgomez,ruedi}@ini.phys.ethz.ch

Abstract. Claims based on power laws that cognition occurs in a critical state often rely on the assumption that the network observables studied are observables of cognition, however this relationship to function is not clear. Our novel approach to investigate this problem is instead to consider functional output during (goal-directed) pre-copulatory courtship of *Drosophila melanogaster*, which we study as a complex network. This courtship body language, expressed through a symbolic dynamics, has previously been shown to be situation specific and grammatically complex; here, we show that the networks underlying it deviate from a scale-free structure when recursive grammars are included. This structural deviation is modelled by a simple network growth algorithm which adds internal edge saturation to the preferential attachment paradigm. From this, we suggest that a critical state may not be compatible with higher level cognition.

Keywords: Criticality, cognition, complex networks, network growth.

1 Introduction

Understanding brain function is now firmly established as one of the great challenges in contemporary physics. In the last decade, the term "criticality" has taken a strong position in this literature, where it is even suggested to be essential for cognition [1]. Recasting cognition in terms of critical phase transitions is as attractive as it is controversial: this is familiar territory to physicists, but the onus of demonstrating a critical state from small noisy datasets is problematic, and beyond claims of optimal information transfer and dynamic range, it is unclear how a critical state should be related to goal-directed function. Here we present new and contrary evidence, and discuss it in terms of a hidden assumption behind brain criticality: that the observables claimed to be power-law distributed are the ones that give rise to cognitive function.

Power law distributed observables are necessary but not sufficient to infer a critical state. Great statistical care must be taken when testing for power laws, and generally this has been done since appropriate methods were outlined in Ref. [2]. However, particularly in the case of small datasets, the significant role of truncation effects is still often overlooked (for appropriate methods in this case, see [3]).

V.M. Mladenov and P.C. Ivanov (Eds.): NDES 2014, CCIS 438, pp. 309–316, 2014.

Beyond evidence supporting the existence of a power law, claims of a critical state in cognition also refer to deviations from power law behaviour either side of the claimed critical state. Experimental limitations hinder this line of argument: there is seldom sufficient resolution of an order parameter to observe the neccessary cusp at the proposed critical control parameter value (e.g. [4]), and isolated experimental cases which deviate from a power law serve only to demonstrate that non-critical behaviour is also possible (e.g. [5]). We will not focus on whether there is sufficient evidence for criticality; instead we ask whether the critical state is useful in cognition, or simply a resting state from which the brain deviates to achieve cognition.

Appealing to intuition, it may seem that the availability of structure at all scales in a critical state should be useful for complex and adaptive function, however the evidence supporting this intuition is questionable. Claims that a critical state enables optimal information propagation, storage, and dynamic range in neural networks are largely based on artificial network simulations, so are strongly dependent on the networks and measurements used. For instance, in Ref. [6] it was claimed that "reservoir computing" systems are able to perform calculations on time series when the "reservoir" is near a critical state, though the success of this is contingent on the feed-forward inputs and outputs inherent in such a system. In another example, a measure of information propagation was defined in Ref. [7] based on avalanche size distributions - a claimed critical order parameter observed in biological neural networks - but then when noise was added (as is always the case in biological systems) the measure was no longer maximised in the critical state. Claims that dynamic range is maximal in the critical state seem intuitive, though it is unclear to us how the sensitivity range can be fruitfully translated into output range. Indeed, for mammalian cochleae, the correct sensitivity profile can only be obtained for coupled Hopf oscillators by tuning away from the critical bifurcation point (see e.g. [8] for a single element, and [9] for the networked context). The relationship between power law observations in biological networks and their functional output is not clear. The first invasive multielectrode array observations of power law distributed neuronal avalanches in rat cortex slices [10] were from *in vitro* recordings (thus with no clear functional relevance), but even *in vivo*, sizes of neuronal avalanches [11] or locations of high activity clusters [12] do not have clearly defined functional outputs. Power-law like bursts of cortical activity were also observed on a more macroscopic scale in sleep experiments [13,14]. The issue of functional relevance is avoided in this paper by working instead with functional output itself: the body language actions of *Drosophila melanogaster* during its pre-copulatory courtship [15,16] provides an opportunity to code and study distinct, relevant, neural states, in terms of a clearly defined and finite set of fundamental actions. Sequences of these fundamental actions naturally define a network, which gives a glimpse of the topology of transitions between relevant neural states expressable by the system. Further, these sequences have been shown to convey important and specific information with a broad range of grammatical sophistication [15,16,17].

2 Networks Underlying *Drosophila* Pre-copulatory Courtship (DPC)

From visual inspection of high speed video courtship recordings, DPC was decomposed into 37 fundamental actions: body language acts which are non-overlapping, occur on timescales well beyond the limit of neural refractory periods, do not generally have inherent physical restrictions on their ordering, and for which further decomposition yields no further information about the courtship sequence [15]. Thus the actions performed by each individual *Drosophila* during courtship pairing can be represented as a finite sequence $S_a = \{a_1, a_2, ..., a_m\}$, where each $a_i \in S_a$ is one of the 37 fundamental actions.

The undirected topological network underlying such a DPC sequence has vertices defined by the actions present, and edges defined by adjacency of actions (vertices) in the sequence. i.e., vertex set $V = \{v_i : v_i \in S_a\}$ and edge set $E = \{(v_i, v_j) : v_i = a_n, v_j = a_{n+1}; a_n, a_{n+1} \in S_a\}$. Generally we deal with networks composed from more than one sequence, which are defined in the usual graph union sense. This topological, undirected representation avoids introducing bias due to incomplete data, while still elucidating the structure on which the sequences were (necessarily) produced by some network walk. The rules governing this walk are not considered here; only the network structure on which they operate. The structure of these functional networks is a fundamental observable of the system during goal-directed (towards copulation) behaviour. If the system is in a critical state, this network structure should be statistically consistent with a scale-free degree distribution.

In all, 10 fundamental action sequence types were recorded [15], each corresponding to one pairing-protagonist combination. For instance, the paring of *male* and *mature virgin female* drosophila gives the two sequence types (protagonist vs. antagonist): *male* vs. *mature virgin female*; and *mature virgin female* vs. *male*. 3 main pairings give 6 of the sequence types: *male* paired with each of *immature virgin female*, *mature virgin female* and *mated female*. Further pairing of males with a genetic mutation, here termed *fruitless*, with each of (normal) *male* and *mature virgin female* give a further 4 sequence types (a total of 10). In this paper, the sequence types are studied in 3 groups: *male behaviour*; *female behaviour*; and *male and female behaviour* (Table 1). We use *behaviour* in the sense defined in [15,16], so that the *female behaviour* includes the gender switching observed in *male* vs. *fruitless* (however sequences produced by *fruitless* as protagonist are excluded).

Truncated Power Law Fitting. The degree distribution of these DPC networks is the key structural observable, which, if the system is in a critical state, could be expected to follow a (truncated) power law. When fitting truncated power laws to these degree distributions, we follow closely the method outlined in [3] (which in turn closely follows [2]). Maximum likelihood estimation of the discrete truncated power law exponent is used to define a fitted distribution, from which 1000 surrogate datasets are sampled. Each of these surrogate datasets is

Table 1. Definitions of the DPC sequence groups. Here "females" comprises: *immature virgin female*; *mature virgin female*; *mated female*. See text for explanation of *female behaviour* group.

group name	included sequence types
male behaviour	*male* vs. females
female behaviour	females vs. *male* *male* vs. *frutiless* *mature virgin female* vs. *fruitless*
male and female behaviour	sequences in *male behaviour* sequences in *female behaviour*

then also fit by maximum likelihood estimation, and their Kolmogorov-Smirnov (KS) distances are compared to the KS distance of the original data to its fit. A "p-value" is then calculated as the fraction of surrogate datasets with a larger KS distance (a worse fit) than the original data, i.e. the probability that the data are better fit by a power law than surrogate data: a measure of goodness of fit. Of course, a high p-value is not sufficient to demonstrate that the data follow a power law (since this was an assumption in the test), but only to indicate that a power law is feasible. A low p-value on the other hand is sufficient for rejection of a power law, though the selection of a rejection threshold is rather arbitrary.

Due to the small size of the networks, the lower and upper truncation bounds were selected as the minimum and maximum degrees present, except in the *male and female behaviour* case, where the second largest degree was selected, to avoid a large gap in the degree distribution. Due to the small size of the dataset, we plot results as the complement of the cumulative distribution $P(k)$, called the *survival function*, $SF(k) = 1 - P(k)$. Readers may expect that a power law should appear as a straight line in log-log space, however we emphasise that for a truncated survival function this is not the case, as the distribution must reach zero at the upper bound, so deviation from the non-truncated power law line occurs as this upper bound is approached. Over a small range, power law and exponential distributions are very similar; we also tested exponential distributions on our data and obtained very similar fits and p-values (not shown).

3 Results, Modeling, Discussion

Fitting truncated power laws to the degree distributions of the 3 group networks reveals a distinct difference in structure between them. The *female behaviour* network is consistent with a scale-free network (Fig. 1a), while the other two deviate sharply from a power-law degree distribution at high degree (Fig. 1b,c). We do not claim that the *female behaviour* network is scale free; for such small networks, there is little difference between power law and exponential fits. However we do claim that the *male behaviour* network and the *male and female behaviour* network are inconsistent with a scale-free structure. This observation is particularly interesting in light of an earlier result [17] which showed not only that DPC language often exhibits recursion (making it to the best of our knowledge the only known animal example of a Context-Free or Context-Sensitive language in the Chomsky hierarchy) but that this recursive property is predominantly exhibited in male DPC. It seems possible that the inclusion of such high level grammar precludes a scale-free network structure.

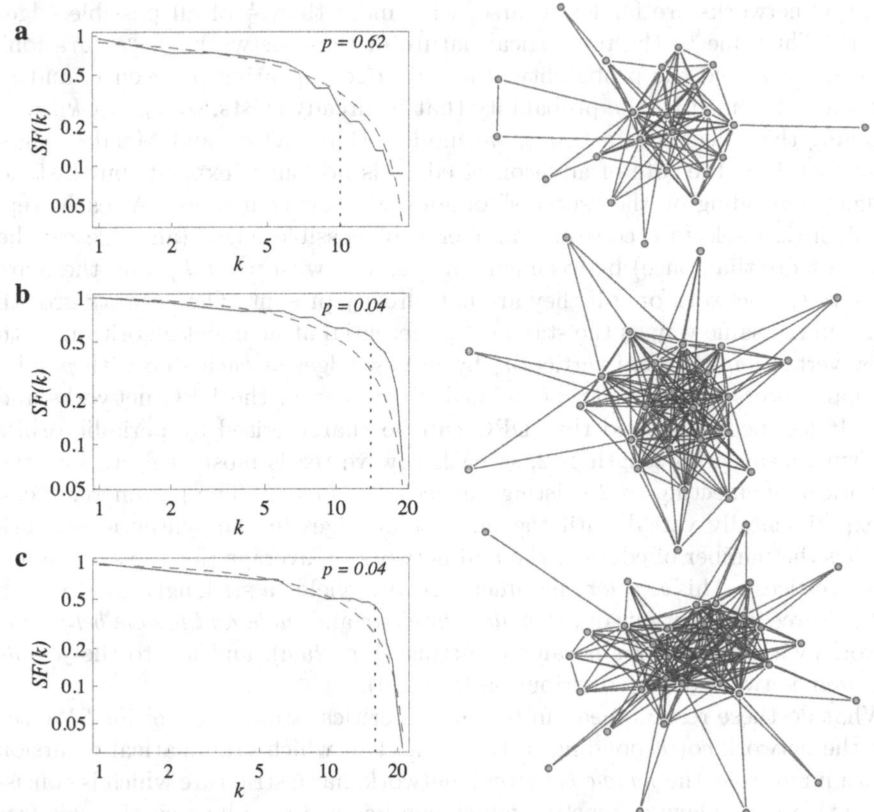

Fig. 1. Left panel: log-log plots of truncated power law fits (*dashed lines*) to network degree distribution survival functions (*solid lines*) showing KS difference location (*dotted lines*). Right panel: corresponding networks. **a** *female behaviour*: $p = 0.62$; power law exponent 0.37; degree bounds $k_{min} = 1$, $k_{max} = 19$; KS difference 0.10. **b** *male behaviour*: $p = 0.04$ (rejected); power law exponent 0.001 (minimum permitted); degree bounds $k_{min} = 2$, $k_{max} = 20$; KS difference 0.22. **c.** *male and female behaviour*: $p = 0.04$ (rejected); power law exponent 0.076; degree bounds $k_{min} = 1$, $k_{max} = 22$; KS difference 0.18.

The evidence given above does not stand alone as an example of deviation from scale-free structure in language networks. Ferrer i Cancho and Solé [18] showed that Zipf's famous empirical "law" for the frequency of occurrence of words in English breaks into two scaling regimes for sufficiently large *corpi*, and this result was reproduced by a simple network growth algorithm for the undirected topological word network ("word web") by Dorogovtsev and Mendes [19]. This original algorithm in Ref. [19] added an internal linking process to the standard preferential attachment growth algorithm (from Ref. [20]) whereby "internal" edges between vertices v_i and v_j were added with probability $p \propto k_i k_j$, where k denotes vertex degree. However in contrast to the English word webs, the DPC networks are far less sparse, with more than $\frac{1}{3}$ of all possible edges present. Thus due to the topological nature of these networks, a "saturation" effect appears, and the probability of a new edge appearing between v_i and v_j is strongly modified by the probability that it already exists, so $p_{new} \not\propto k_i k_j$.

Taking this effect into account, we modify Dorogovtsev and Mendes' algorithm such that the rate of addition of edges is no longer explicit, but instead implicit, depending on the "success" of adding a new connection. At each step, our algorithm selects a constant number r of possible edges (an edge can be selected more than once) between existing vertices with $p \propto k_i k_j$, and these are added to the network only if they are not already present. The network growth occurs in the same way as the standard preferential attachment algorithm, with a new vertex joined to old vertices v_i by m new edges at each step with $p \propto k_i$. The parameters r and m can be defined directly from the DPC networks and data. It has been observed that DPC can be characterised by periodic orbits [15], with mean orbit length > 2, so each new vertex is most likely to join the network by connecting to 2 existing vertices, i.e. $m = 2$. The parameter r can be experimentally varied until the number of edges in the synthetic network matches the number of edges in the real network on average (for the same number of vertices). This *a priori* parameter setting yields a strikingly good match to the degree distributions of the *male behaviour* and *male and female behaviour* networks without need for parameter fitting (Fig. 2a,c), and not to the *female behaviour* network degree distribution (Fig. 2d).

What do these results mean in terms of a critical state in cognition? We see that the network corresponding to the dataset in which grammatical recursion plays a minor role, the *female behaviour* network, has a structure which is consistent with a power law; a structure which can be generated by a single scale-free process. With the introduction of significant recursion however, the power law fit breaks down, and instead a network growth algorithm which includes a second process (internal preferential linking with saturation, breaking the scale-free preferential attachment process), is able to match these network structures (but not the scale-free one) without fitting. Recursion is essential for higher level abstract reasoning [17], and requires a secondary "book keeping" function to keep track of the references between elements. It appears from the above evidence that such a second process is neccessary for the *male behaviour*, and incompatible with a critical state. It would seem then that this type of higher level cognitive process is also incompatible with a critical state.

In conclusion, we have presented a clear example of how power laws may not highlight essential higher functional processes. Power laws observed by other

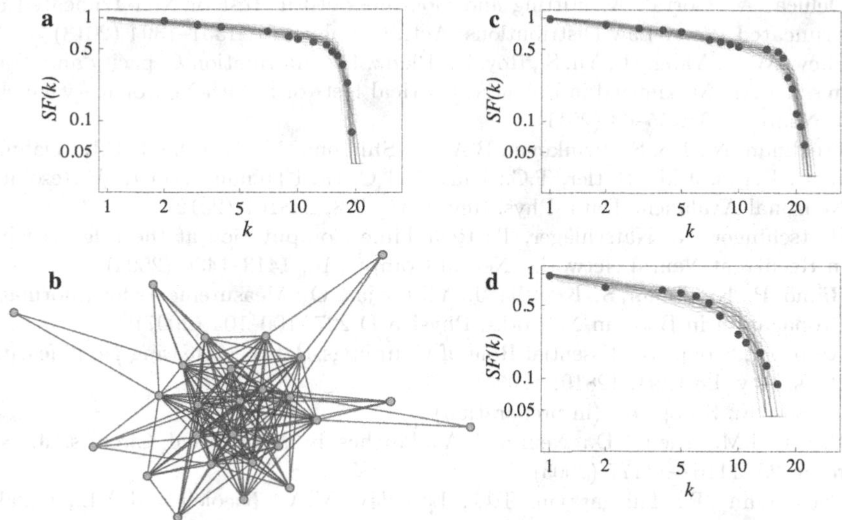

Fig. 2. Saturation growth algorithm match to data. Log-log plots of degree distribution survival functions: 100 algorithm realisations (*lines*); data (*dots*, plotted only for degrees present in the network). **a** *male behaviour.* **b** Example synthetic network matching *male behaviour* network. **c** *male and female behaviour.* **d** *female behaviour*, showing that this secondary process breaks the fit as discussed in the text. Parameters m, r as mentioned in text.

authors during brain function, need not be power laws of functional states themselves. Even "functional networks" (e.g. [12]) are only defined by activity correlations between brain regions. At the most fundamental level, claims of brain criticality assume that the observables chosen are active and relevant in cognition. It is not obvious that this should be the case. The DPC networks by contrast are networks *of* function, during specific goal directed behaviour. We suggest that this "network of function" approach may exert a crucial influence in the brain criticality debate.

Acknowledgments. The authors acknowledge the generous support of the Swiss National Science Foundation SNF, and the Swiss Federal Commission for Scholarships for Foreign Students FCS.

References

1. Chialvo, D.R.: Emergent Complex Neural Dynamics. Nature Phys. 6, 744–750 (2010)
2. Clauset, A., Shalizi, C.R., Newman, M.E.J.: Power-Law Distributions in Empirical Data. SIAM Rev. 51, 661–703 (2009)

3. Deluca, A., Corral, A.: Fitting and Goodness-of-Fit Test of Non-Truncated and Truncated Power-Law Distributions. Acta Geophys. 61, 1351–1394 (2013)
4. Shew, W.L., Yang, H., Yu, S., Roy, R., Plenz, D.: Information Capacity and Transmission Are Maximized in Balanced Cortical Networks with Neuronal Avalanches. J. Neurosci. 31, 55–63 (2011)
5. Friedman, N., Ito, S., Brinkman, B.A.W., Shimono, M., DeVille, R.E.L., Dahmen, K.A., Beggs, J.M., Butler, T.C.: Universal Critical Dynamics in High Resolution Neuronal Avalanche Data. Phys. Rev. Lett. 108, 208102 (2012)
6. Bertschinger, N., Natschläger, T.: Real-Time Computation at the Edge of Chaos in Recurrent Neural Networks. Neural Comput 16, 1413–1436 (2004)
7. Rämö, P., Kauffman, S., Kesselia, J., Yli-harjaa, O.: Measurements for Information Propagation in Boolean Networks. Physica D 227, 100–104 (2007)
8. Kern, A., Stoop, R.: Essential Role of Couplings Between Hearing Nonlinearities. Phys. Rev. Lett. 91, 128101 (2003)
9. Gomez, F., Stoop, R.: (in preparation)
10. Beggs, J.M., Plenz, D.: Neuronal Avalanches in Neocortical Circuits. J. Neurosci. 23, 11167–11177 (2003)
11. Petermann, T., Thiagarajan, T.C., Lebedev, M.A., Nicolelis, M.A.L., Chialvo, D.R., Plenz, D.: Spontaneous Cortical Activity in Awake Monkeys Composed of Neuronal Avalanches. Proc. Natl. Acad. Sci. U.S.A. 106, 15921–15926 (2009)
12. Eguíluz, V.M., Chialvo, D.R., Cecchi, G.A., Baliki, M., Apkarian, A.V.: Scale-Free Brain Functional Networks. Phys. Rev. Lett. 94, 018102 (2005)
13. Lo, C.-C., Nunes Amaral, L.A., Havlin, S., Ivanov, P.Ch., Penzel, T., Peter, J.-H., Stanley, H.E.: Dynamics of Sleep-Wake Transitions During Sleep. Europhys. Lett. 57, 625–631 (2002)
14. Lo, C.-C., Chou, T., Penzel, T., Scammell, T.E., Strecker, R.E., Stanley, H.E., Ivanov, P.Ch.: Common Scale-Invariant Patterns of Sleep-Wake Transitions Across Mammalian Species. Proc. Natl. Acad. Sci. U.S.A. 101, 17545–17548
15. Stoop, R., Arthur Jr., B.I.: Periodic Orbit Analysis Demonstrates Genetic Constraints, Variability, and Switching in Drosophila Courtship Behavior. Chaos 18, 023123 (2008)
16. Stoop, R., Joller, J.: Mesocopic Comparison of Complex Networks Based on Periodic Orbits. Chaos 21, 016112 (2011)
17. Stoop, R., Nüesch, P., Stoop, R.L., Bunimovich, L.A.: At Grammatical Faculty of Language, Flies Outsmart Men. PLoS ONE 8, e70284 (2013)
18. Ferrer i Cancho, R., Solé, R.V.: Two Regimes in the Frequency of Words and the Origins of Complex Lexicons: Zipf's Law Revisited. J. Quant. Linguist. 8, 165–173 (2001)
19. Dorogovtsev, S.N., Mendes, J.F.F.: Language as an Evolving Word Web. Proc. R. Soc. Lond. B 268, 2603–2606 (2001)
20. Barabási, A.-L., Albert, R.: Emergence of Scaling in Random Networks. Science 286, 509–512 (1999)

Cardiodynamic Complexity: Electrocardiographic Characterization of Arrhythmic Foci

Oriol Pont and Binbin Xu

Inria Bordeaux – Sud-Ouest

Abstract. The electrical activity of the heart consists of nonlinear interactions emerging as a complex system. As such, proper characterization of arrhythmias and arrhythmogenic areas requires nonlinear analysis methods. Electrocardiographic imaging provides a full spatiotemporal picture of the electric potential on the human epicardium. Rhythm reflects the connection topology of the pacemaker cells driving it. Hence, characterizing the attractors as nonlinear, effective dynamics can capture the key parameters without imposing any particular microscopic model on the empirical signals. A dynamic phase-space reconstruction from an appropriate embedding can be made robust and numerically stable with the presented method. We show that both the phase-space descriptors and those of the *a priori* unrelated singularity analysis are able to highlight the arrhythmogenic areas on cases of atrial fibrillation.

Keywords: Wavelets, nonlinear analysis, heartbeat dynamics, embedding methods, multifractal analysis.

1 Introduction and State of the Art

The usual approach to electrocardiography (ECG) consists of qualitatively characterizing typical shapes of potential signals, often labeled as *waves*. Some characteristic signatures of such shapes are specific to certain arrhythmias and this is principally what makes them a standard diagnostic tool. Alternatively, but in a quite similar way, the statistical approach tries to highlight specificities that are not evident in the electrical signals themselves but in their distribution functions or some measure derived from them. Statistical methods for processing cardiac signals have had a notable success in several areas, e.g., providing measures of complexity of a certain arrhythmia. Additionally, they output quantitative measures, they provide confidence intervals and, sometimes, we know robust implementations for them.

Standard statistical methods do not always respect the signal invariances. This fact motivates looking for fractal [9] or multifractal [7,6] approaches to heart rate variability, which do capture the multiscale structure of the process. More interestingly, multifractal analysis reconnects the statistical features of the signal with a geometrical interpretation linked to the effective macroscopic transfer of information in the signal [12,13,10,11,15].

V.M. Mladenov and P.C. Ivanov (Eds.): NDES 2014, CCIS 438, pp. 317–324, 2014.

Taking a different path, embedding methods can also reconstruct the effective attractor of the system that produced the signal. This effect reinforces the possibilities for cross-validating the obtained parameters as well as their physical interpretation [17,16]. This makes an interesting duality of approaches, with the one based on a multiscale hierarchy and the other on reconstructing a chaotic dynamics. Since the analysis is done at the effective level (i.e., independently of the microscopic descriptors), it is unclear whether the assumption of chaos suffices [3]. Actually, *chaotic* cardiac electrophysiological models coexist with *stochastic* ones. A stochastic component can advantageously account for the empirical spectral signature [8] despite that, at a general level, the observability of the *deterministic vs. stochastic* distinction can always be challenged [18].

Atrial Fibrillation. (AF), the most common form of cardiac arrhythmia, consists in the chaotic operation, electrical activation and pumping of the atria. In some circumstances, it can induce life-threatening complications such as inducing a heart failure or forming blood clots that lead to stroke. In cases where medication is impossible or ineffective, a successful treatment consists in radiofrequency ablation of the endocardial tissue to ease an appropriate electrical conduction. In case of paroxysmal AF, Haïssaguerre et al. have shown [5] that for 80 % of patients, electrical insulation of the pulmonary veins allows the patient to regain a normal heart rhythm, but in persistent or permanent AF, the location of problematic areas remains difficult and is still an open problem. In persistent AF, the arrhythmogenic foci can be anywhere in the atria and so need a well focused ablation.

The paper has the following structure: section 2 introduces the basic methods used for processing the cardiac electrical signals. In section 3 we show our results in the identification of dynamical regimes. Finally, in section 4 we discuss the results and present the conclusions of our study.

2 Statistical Methods Based on Nonlinear Dynamics

The electrical activity of the heart is often described as a complex system. Complexity can have multiple interpretations varying in both nature and extent. Nevertheless, the effect in all cases is that the behaviour of the system as a whole is an emergent behaviour, which could not be derived from separately considering the microscopic mechanisms (be them at cell level or even at molecular level). This non-separability of the different scales involved is a consequence of nonlinear interactions or, in other words, expressing the effect at a given level requires more than just the sum of effects at lower levels. This is what makes global synchronization possible, also can amplify microscopic fluctuations to perturb the whole regime [6].

Emergence of chaos becomes even more important in complex arrhythmic regimes, where linearized descriptors fail to provide meaningful parameterizations except only possibly for very short time windows, or microscopic space or parametric scales. In this context, any appropriate processing methodology must be nonlinear in nature. Singularity analysis provides a robust framework that identifies dynamical transition fronts and information content [12,14,10] which is useful for cardiac electric potential signals [15,17].

2.1 Singularity Analysis

The degree of singularity/regularity of a given point in a signal tells how rare is the signal at this point and therefore how much information it contains. A local expansion of the signal around this point has a leading order that dominates at the local neighbourhood (short distances, small position perturbations). This leading order is a scale parameter raised to an exponent, which is not necessarily integer. A signal s has a (fractional) singularity exponent h at point $\boldsymbol{x}$ if

$$\mathcal{T}_{\Psi}\mu(\boldsymbol{x},r) = \alpha_{\Psi}(\boldsymbol{x})\, r^{h(\boldsymbol{x})} + o\left(r^{h(\boldsymbol{x})}\right) \qquad (r \to 0), \tag{1}$$

with $\mathcal{T}_{\Psi}\mu(\boldsymbol{x},r) = \int_{\mathbb{R}^d} \mathrm{d}\mu(\boldsymbol{x}')\, \Psi\left((\boldsymbol{x}-\boldsymbol{x}')/r\right)$ as the wavelet-projected measure μ at scale r and Ψ as a certain wavelet kernel. The measure is differentially defined: $\mathrm{d}\mu(\boldsymbol{x}) = \|\nabla s\|(\boldsymbol{x})\, \mathrm{d}\boldsymbol{x}$.

Now, the sole requirements of being deterministic, linear, isotropic and translational invariant permit to define a local reconstruction kernel [11]. Actually, this minimal-assumption reconstruction identifies a *reduced* signal which is reconstructed only from the orientation of the signal on its most singular points [15]. The actual signal s and its reduced counterpart r are related through a complex but slow-changing modulation called *source field* [15], which is defined as the Radon-Nykodym derivative between their respective measures: $\mathrm{d}\mu_s/\mathrm{d}\mu_r(\boldsymbol{x},t)$.

2.2 Dynamical Attractor Reconstruction

Time series evolution is mapped to an object embedded in a phase space in abstract coordinates. m independent observations construct an mD phase space, as per the embedding theorem [3]. Most advantageously, this mapping preserves the topological properties and thus we can properly call it a *phase space reconstruction* even when the physically meaningful variables are not directly available. The dimension m is the least one that embeds the dynamics (which is twice plus one the Minkowski dimension of its attractor set). With appropriate filtering, the method can be made robust and well adapted to empirical signals. The result is a compact dynamical description that characterizes complexity degree and information distribution [17].

Time Lags. τ are the shortest for which the m coordinates do not mutually interfere. For a time series $\boldsymbol{x}(t)$, phase space reconstruction is represented as:

$$\mathcal{X}(\tau,m) = [\boldsymbol{x}(t), \boldsymbol{x}(t+\tau), \boldsymbol{x}(t+2\tau), \cdots, \boldsymbol{x}(t+(m-1)\tau)], \tag{2}$$

where m is the embedding dimension and τ is the time lag. The actual system dimensionality m is generally unknown *a priori* from empirical data. Furthermore, high-dimensional attractors require extremely long time series to obtain any meaningful reconstruction of them. Essentially, τ determines the correlation structure. If τ is too small, the trajectories of $\boldsymbol{x}(t)$ and $\boldsymbol{x}(t+\tau)$ are redundant, and if τ is too large, the two coordinates become statistically disconnected and chaotic evolution prevents retrieving the actual attractor. Since dependencies are evaluated on the ensemble of coordinates, τ and m affect each other.

Both singularity analysis and phase-space reconstruction give a geometric perspective to the system dynamics, which is only partially observed from electrographic variables. When applicable, both methods provide accurate measures of complexity. In both cases, all the structural complexity of the system is abstractly represented as means of effective dynamics. In practical terms, singularity analysis follows an approach that is radically different from the one of a local phase-space reconstruction. Time lag fluctuations of reconstruction correlate with atrial fibrillation episodes. Correspondingly, the dynamical changes implied from singularity analysis also highlight atrial fibrillation in a local way, as described in [15] and presented as well in [17,16].

3 Results on Analyzing Atrial Fibrillation

To show the presented methods in practice, we present the analyses made on electrocardiographic maps made on cases of AF at Bordeaux Haut-Leveque Hospital. First, heart geometry is extracted from a tomographic scan. A 252-electrode vest measures the electric potential around the torso and these measures are mapped on the epicardial surface. This inverse problem is essentially ill-posed and requires extensive regularization. Therefore, it is important that the analysis methods are robust to perturbations and regularization artefacts [4].

- Patient 1 is a 67-year-old man with persistent atrial fibrillation (sustained episodes) for 13 months. With the atria dilated by the fibrillation (dilated cardiomyopathy) and had the pulmonary veins already insulated on a previous, unsuccessful operation. The cardiologist further ablated several areas, finally turning the fibrillation to a still arrhythmic but milder atrial tachycardia, which remained sustained even with antiarrhythmic drugs. Ablated areas, in order: inferior left atrium, left atrial appendage, left atrial anterior wall, right pulmonary vein, atrial septum, right atrial appendage, lower right atrium, cavotricuspid isthmus, mitral isthmus and atrial roof. Fibrillation cycle length significantly decreased after the left appendage ablation and fibrillation turned to tachycardia after the roof ablation.
- Patient 2 is a 42-year-old man with persistent atrial fibrillation for 1 month. Dilated cardiomiopathy. During the operation, the cardiologist targeted the ablation first on the pulmonary veins, unsuccessfully, and then the left atrial appendage, which fully stopped the fibrillation.

The singularity-analysis approach here presented is equivalent to the source-field concept presented in [17] for a setting similar to the one presented here. While time-evolving, the mapping series are too short to extract a tachogram from them and so we processed the potential signal itself, whose fluctuations dynamically evolve – an effect usually related to *fractionation*. Exponents are computed in space-time domain, with the typical propagation speed regularizing the time coordinate. As expected, exponents are quite regular everywhere, but especially at the fibrillation foci (Fig. 1), for both patients.

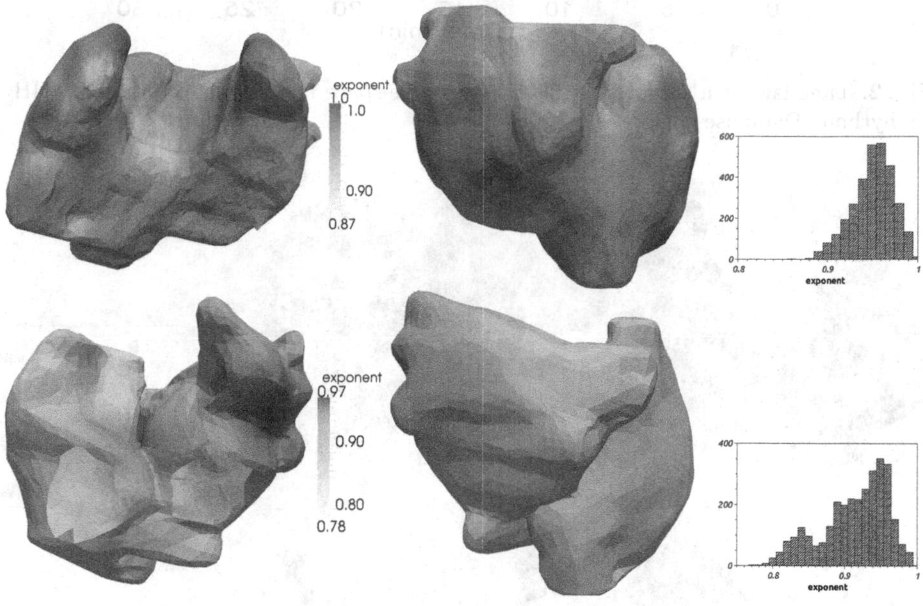

Fig. 1. Exponent maps and histogram for Patient 1 (top) and Patient 2 (bottom). Mapping projected on the atria: anterior view (left) and posterior view(right).

In our previous approach to phase-space reconstruction, presented in [17], we processed data from the MIT-BIH Arrhythmia Database [2]. These consist of ECG measurements under different arrhythmic regimes for patients of Boston's Beth Israel Hospital. Remarkably, ECGs have been hand annotated at the beat level by multiple experts independently, and these annotations have been audited and verified by the full community of the database users since 1980. Case #217 of MIT-BIH Arrhythmia Database displays a fragmented electrogram with many intermittent episodes of atrial flutter and atrial fibrillation. Embedding dimensions do not evolve significantly, but time lags reveal a strong correlation of their fluctuations with the arrhythmic episodes, as seen at Fig. 2 [17]. This looks consistent with reported attenuation patterns in spiral-wave simulations [1].

Unlike singularity exponents, time lags are dimensional. This means that either in terms of samples or time units, they are not invariant: changes in anatomy, regulation mechanisms, pulse rate and action-potential duration affect them. Nevertheless, even if their typical values are not much meaningful, their relative proportions seem to sharply indicate which areas are driving the fibrillation.

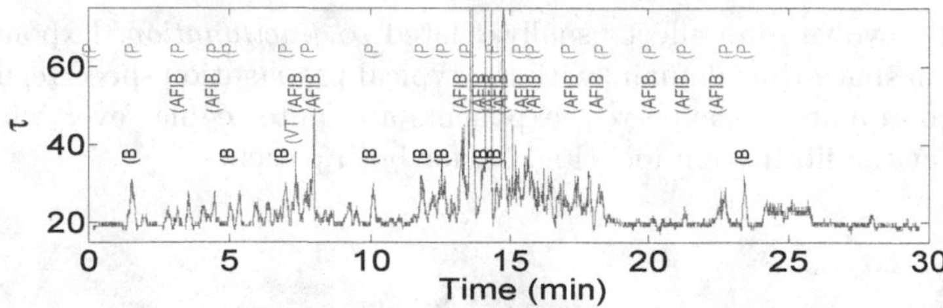

Fig. 2. Time lag τ (in samples, at 360Hz) of phase-space reconstruction for MIT-BIH Arrhythmia Database case #217

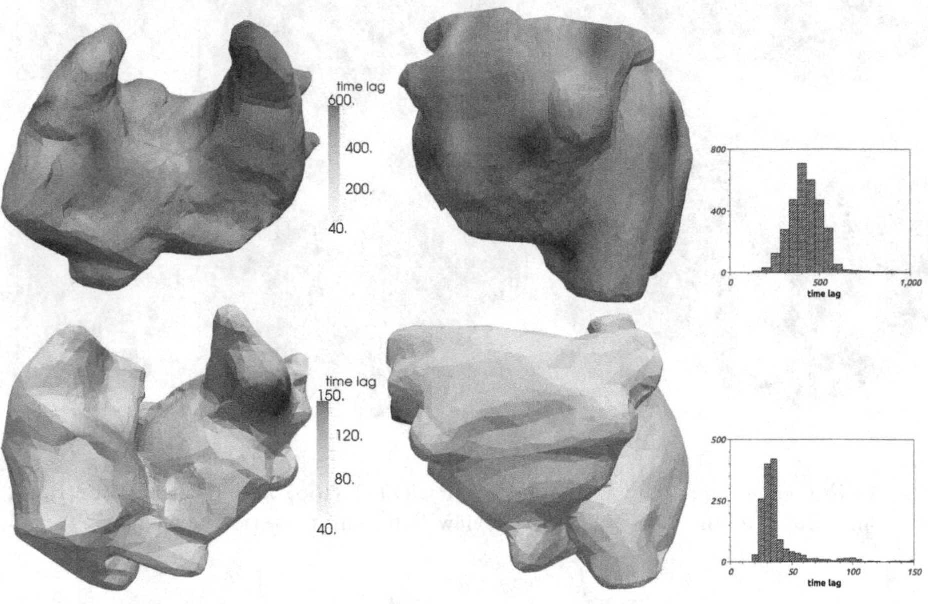

Fig. 3. Time lag τ (in samples, at 1kHz) of phase-space reconstruction before ablation for Patient 1 (top) and Patient 2 (bottom), arranged as in Fig. 1. Fibrillation of Patient 1 just turned to tachycardia, so it was not fully resolved. Complex areas are widely distributed and also lag values. We can notice how the method highlights the key areas: left appendage and roof. Patient 2, a much simpler case of persistent AF, reached the healthy sinus rhythm after ablation of the left appendage, which we see sharply marked.

For the case here, as we can observe (Fig. 3), time lag maps highlight the fibrillation drivers too.

4 Discussion and Conclusions

Nonlinear analysis provides methods for characterizing cardiac dynamics. Singularity analysis and phase-space reconstruction are physically meaningful complexity measures with minimal assumptions on the underlying interactions. They are based on effective descriptions derived from first principles, and as a consequence, parameters are robustly estimated. We have validated this approach on ECGs and electrocardiographic maps [17].

Key parameters vary infrequently and exhibit sharp transitions, which show where information concentrates and correspond to actual dynamical regime changes. Singularity exponents sift a simple fast dynamics from its slow modulation [15]. In space domain, extreme values highlight arrhythmogenic areas. We observe a correspondence of time lag fluctuations of phase-space reconstructions with atrial fibrillation episodes in the same way as with the dynamical changes coming from singularity exponents. This characterization of information transitions could be used in the regularization of inverse-problem mapping of electrocardiographic epicardial maps. Furthermore, this opens the way for improved model-independent complexity descriptors to be used in non-invasive, automatic diagnosis support and ablation guide for electrical insulation therapy, in cases of arrhythmias such as atrial flutter and fibrillation [15,17].

The robustness of the approach can be corroborated by comparing it with an independent analysis method. In that sense, we have performed phase-space reconstructions of the chaotic signals by means of the embedding theorem. We observe a correspondence of time lag fluctuations of such reconstructions with atrial fibrillation episodes in the same way as with the dynamical changes coming from singularity exponents. These results open the way for improved model-independent complexity descriptors to be used as non-invasive diagnosis support or operation guide in cases of cardiac arrhythmias, particularly atrial fibrillation.

Acknowledgements. B. Xu is financially supported by the French IHU in cardiac rhythmology *L'Institut de rythmologie et modélisation cardiaque* LIRYC and the regional council of Aquitania. We thank CHU Bordeaux - Haut-Leveque hospital, IHU LIRYC and CardioInsight Technologies for the epicardial electrocardiographic maps.

References

1. de la Casa, M.A., de La Rubia, F.J., Ivanov, P.C.: Patterns of spiral wave attenuation by low-frequency periodic planar fronts. Chaos: An Interdisciplinary Journal of Nonlinear Science 17(1), 015109 (2007)
2. Goldberger, A.L., Amaral, L.A.N., Glass, L., Hausdorff, J.M., Ivanov, P.C., Mark, R.G., Mietus, J.E., Moody, G.B., Peng, C.K., Stanley, H.E.: Physiobank, physiotoolkit, and physionet: Components of a new research resource for complex physiologic signals. Circulation 101(23), e215–e220 (2000)
3. Govindan, R.B., Narayanan, K., Gopinathan, M.S.: On the evidence of deterministic chaos in ecg: Surrogate and predictability analysis. Chaos 8(2), 495–502 (1998)
4. Haissaguerre, M., Hocini, M., Shah, A.J., Derval, N., Sacher, F., Jais, P., Dubois, R.: Noninvasive panoramic mapping of human atrial fibrillation mechanisms: A feasibility report. J. Cardiovasc. Electrophysiol. 24(6), 711–717 (2013)

5. Haïssaguerre, M., Jaïs, P., Shah, D.C., Takahashi, A., Hocini, M., Quiniou, G., Garrigue, S., Le Mouroux, A., Le Métayer, P., Clémenty, J.: Spontaneous initiation of atrial fibrillation by ectopic beats originating in the pulmonary veins. New England Journal of Medicine 339(10), 659–666 (1998)
6. Ivanov, P., Amaral, L., Goldberger, A., Havlin, S., Rosenblum, M., Struzik, Z., Stanley, H.: Multifractality in human heartbeat dynamics. Nature 399, 461–465 (1999)
7. Ivanov, P., Rosenblum, M., Peng, C.K., Mietus, J., Havlin, S., Stanley, H., Goldberger, A.: Scaling behaviour of heartbet intervals obtained by wavelet-based time-series analysis. Nature 383, 323–327 (1996)
8. Ivanov, P.C., Amaral, L.A.N., Goldberger, A.L., Stanley, H.E.: Stochastic feedback and the regulation of biological rhythms. Europhys. Lett. 43(4), 363–368 (1998)
9. Peng, C.K., Mietus, J., Hausdorff, J.M., Havlin, S., Stanley, H.E., Goldberger, A.L.: Long-range anticorrelations and non-gaussian behavior of the heartbeat. Phys. Rev. Lett. 70(9), 1343–1346 (1993)
10. Pont, O., Turiel, A., Perez-Vicente, C.: Empirical evidences of a common multifractal signature in economic, biological and physical systems. Physica A 388(10), 2025–2035 (2009)
11. Pont, O., Turiel, A., Perez-Vicente, C.: On optimal wavelet bases for the realization of microcanonical cascade processes. Int. J. Wavelets Multi., IJWMIP 9(1), 35–61 (2011)
12. Pont, O., Turiel, A., Perez-Vicente, C.J.: Application of the microcanonical multifractal formalism to monofractal systems. Physical Review E 74, 061110–061123 (2006)
13. Pont, O., Turiel, A., Perez-Vicente, C.J.: Description, modeling and forecasting of data with optimal wavelets. Journal of Economic Interaction and Coordination 4(1), 39–54 (2009)
14. Pont, O., Turiel, A., Yahia, H.: Singularity analysis of digital signals through the evaluation of their unpredictable point manifold. International Journal of Computer Mathematics 90(8), 1693–1707 (2013)
15. Pont, O., Haissaguerre, M., Yahia, H., Derval, N., Hocini, M.: Microcanonical processing methodology for ECG and intracardial potential: application to atrial fibrillation. Transactions on Mass-Data Analysis of Images and Signals 3(1), 15–34 (2011)
16. Pont, O., Xu, B.: Characterizing Complexity of Atrial Arrhythmias through Effective Dynamics from Electric Potential Measures. In: Computing in Cardiology, CinC 2013. Saragossa, Spain (2013)
17. Pont, O., Yahia, H., Xu, B.: Arrhythmic dynamics from singularity analysis of electrocardiographic maps. In: EMBC 2013. 35th Annual International Conference of the IEEE Engineering in Medicine and Biology Society. IEEE EMBS, Osaka (2013)
18. Werndl, C.: Are deterministic descriptions and indeterministic descriptions observationally equivalent? Studies in History and Philosophy of Science Part B: Studies in History and Philosophy of Modern Physics 40(3), 232–242 (2009)

Stochastic Modeling of Excitable Dynamics: Improved Langevin Model for Mesoscopic Channel Noise

Igor Goychuk

Institute for Physics and Astronomy, University of Potsdam
Karl-Liebknecht-Str. 24/25, 14476 Potsdam-Golm, Germany
igoychuk@uni-potsdam.de
http://www.physik.uni-augsburg.de/~igor/

Abstract. Influence of mesoscopic channel noise on excitable dynamics of living cells became a hot subject within the last decade, and the traditional biophysical models of neuronal dynamics such as Hodgkin-Huxley model have been generalized to incorporate such effects. There still exists but a controversy on how to do it in a proper and computationally efficient way. Here we introduce an improved Langevin description of stochastic Hodgkin-Huxley dynamics with natural boundary conditions for gating variables. It consistently describes the channel noise variance in a good agreement with discrete state model. Moreover, we show by comparison with our improved Langevin model that two earlier Langevin models by Fox and Lu also work excellently starting from several hundreds of ion channels upon imposing numerically reflecting boundary conditions for gating variables.

Keywords: Excitable dynamics, ion channels, mesoscopic noise, Langevin description.

1 Introduction

Hodgkin-Huxley (HH) model of neuronal excitability [1] provides a milestone for biophysical understanding of information processing in living systems [2] in terms of electrical spikes mediated by ionic currents through voltage-dependent membrane pores made by ion channel proteins. One considers the cell membrane as an insulator with specific electrical capacitance C_m per unit of area, which is perforated by ionic channels providing generally voltage-dependent parallel ionic pathways with specific conductances G_i per unit of area for various sorts of ion channels. This yields the following equation for transmembrane electrical potential difference V

$$C_m \frac{dV}{dt} + G_{\mathrm{K}}(n)(V - E_{\mathrm{K}}) + G_{\mathrm{Na}}(m,h)(V - E_{\mathrm{Na}}) + G_{\mathrm{L}}(V - E_L) = I_{\mathrm{ext}} \,. \quad (1)$$

Here, three ionic currents are taken into account, sodium Na, potassium K and unspecific leakage current (mainly due to chloride ions). This is nothing else

V.M. Mladenov and P.C. Ivanov (Eds.): NDES 2014, CCIS 438, pp. 325–332, 2014.

the Kirchhoff current law, which takes into account the ionic and capacitance currents, as well as an external current $I_{\rm ext}$ which can stimulate electrical excitations. This equation reflects assumption on Ohmic conductance of completely open ion channels with E_i being the reversal or Nernst potentials. They emerge due to the difference of ionic concentrations inside and outside of the excitable cell, which are kept approximately constant by the work of ionic pumps, which is not considered explicitely. Nonlinearity comes from the open-shut gating dynamics of sodium and potassium channels. The corresponding specific conductances

$$\begin{aligned} G_{\rm K}(n) &= g_{\rm K}^{\rm max} n^4(V,t), \\ G_{\rm Na}(m,h) &= g_{\rm Na}^{\rm max} m^3(V,t)h(V,t), \end{aligned} \tag{2}$$

depend on three voltage-dependent gating variables, n, m, and h, where $n(t)$ is the probability of one gate of potassium channel to be open (more precisely the fraction of open gates), m corresponds to one activation gate of sodium channel, and h is the fraction of closed sodium inactivation gates. One assumes four independent identical gates for potassium channel, hence its opening probability is n^4, as well as three activation and one inactivation gate for the sodium channel. Hence, m^3h is the fraction of open sodium channels. The maximal conductances $g_{\rm K}^{\rm max}$ and $g_{\rm Na}^{\rm max}$ can be expressed via the unitary conductances $g_{i,0}$ of single ion channels as $g_i^{\rm max} = g_{i,0}\rho_i$, where ρ_i is the membrane density of the ion channels of sort i. The gating dynamics is in turn described by the relaxation kinetics

$$\frac{d}{dt}x = \alpha_x(V)\,(1-x) - \beta_x(V)\,x, \quad x = m, h, n\,, \tag{3}$$

with voltage-dependent rates

$$\alpha_m(V) = \frac{0.1\,(V+40)}{1-\exp[-(V+40)/10]},\ \beta_m(V) = 4\,\exp[-(V+65)/18]\,, \tag{4}$$

$$\alpha_h(V) = 0.07\,\exp[-(V+65)/20],\ \beta_h(V) = \{1+\exp[-(V+35)/10]\}^{-1}, \tag{5}$$

$$\alpha_n(V) = \frac{0.01\,(V+55)}{1-\exp[-(V+55)/10]},\ \beta_n(V) = 0.125\,\exp[-(V+65)/80]\,. \tag{6}$$

Here the voltage is measured in millivolts and rates in inverse milliseconds. Other classical parameters of HH model suitable to describe excitable dynamics of squid giant axon are: $C_m = 1\,\mu{\rm F/cm^2}$, $E_{\rm Na} = 50\,{\rm mV}$, $E_{\rm K} = -77\,{\rm mV}$, $E_{\rm L} = -54.4\,{\rm mV}$, $G_{\rm L} = 0.3\,{\rm mS/cm^2}$, $g_{\rm K}^{\rm max} = 36\,{\rm mS/cm^2}$, $g_{\rm Na}^{\rm max} = 120\,{\rm mS/cm^2}$.

The set of four coupled nonlinear differential equations defined by (1)- (6) presents a milestone in biophysics and neuroscience because of its very clear and insightful physical background. In the same spirit, one can build up various other conductance-based biophysical models starting from the pertinent molecular background and following to the bottom-up approach. However, it assumes macroscopically large numbers of ion channels in neglecting completely the mesoscopic channel noise effects. The number of ion channels in any cell is, however, finite, and the corresponding channel noise can be substantial [2]. Especially, one confronts with this question by considering the spatial spike propagation among approximately piece-wise isopotential membrane clusters of ion channels [3].

2 Stochastic Hodgkin-Huxley Equations

How to take stochastic dynamics of ion channels within the physical framework of HH model into account is subject of numerous studies [2, 4, 5]. The most rigorous way is to consider the variable population of open ion channels as a birth-and-death process [6]. Consider for simplicity a population of N independent two-state ion channels (one gate only) with opening rate α and closing rate β (constant under voltage clamp). Each ion channel fluctuates dichotomously between the closed state with zero conductance and the open state having unitary conductance g_0. For such two-state Markovian channels, the stationary probability of opening is $p_0 = \alpha/(\alpha + \beta)$ and the averaged conductance is $\langle g(t)\rangle = p_0 g_0$. The number of open channels n is binomially distributed with probability $P_N^{\rm st}(n) = p_0^n(1 - p_0)^{N-n}N!/(n!(N - n)!)$, average $\langle n\rangle = p_0 N$, and variance $\langle(n - \langle n\rangle)^2\rangle = Np_0(1 - p_0) = N\alpha\beta/(\alpha + \beta)^2$. For sufficiently large $N \geq 100$, we introduce quasi-continuous variable $0 \leq x(t) = n(t)/N \leq 1$, with smoothened binomial probability density $p_N^{\rm st}(x) = Np^{xN}(1-p)^{N(1-x)}N!/(\Gamma(1+xN)\Gamma(1 + (1 - x)N))$. Use of approximate Stirling formula $n! \approx (n/e)^n$ yields

$$p_N^{\rm st}(x) \approx C_N(\alpha,\beta)\left(\frac{\alpha}{x}\right)^{Nx}\left(\frac{\beta}{1-x}\right)^{N(1-x)}, \tag{7}$$

where $C_N(\alpha,\beta)$ is a normalization constant. We are looking for the best diffusional (continuous) approximation for discrete state birth-and-death process defined by the master equation

$$\dot{P}_N(n) = F(n-1)P_N(n-1) + B(n+1)P_N(n+1) - [F(n) + B(n)]P_N(n) \tag{8}$$

for $1 \leq n \leq N - 1$, with forward rate $F(n) = \alpha(N - n)$ and backward rate $B(n) = \beta n$, complemented by the boundary conditions

$$\dot{P}_N(0) = B(1)P_N(1) - F(0)P_N(0), \tag{9}$$

$$\dot{P}_N(N) = F(N-1)P_N(N-1) - B(N)P_N(N)\,. \tag{10}$$

2.1 Diffusional Approximations for Birth-and-Death Process

Kramers-Moyal Expansion and Standard Diffusional Approximation. A standard way to obtain diffusional approximation for $p_N(x) := P_N(xN)/\Delta x$ ($\Delta x = 1/N$) with rates $f(x) := F(xN)\Delta x$, $b(x) := B(xN)\Delta x$ is to do the Kramers-Moyal expansion [6, 7], like $p_N(x+\Delta x) \approx p_N(x) + (\partial p_N(x)/\partial x)\Delta x + (\partial^2 p_N(x)/\partial x^2)(\Delta x)^2/2$, $f(x+\Delta x) \approx f(x) + (df(x)/dx)\Delta x + (d^2 f(x)/dx^2)(\Delta x)^2/2$, to the second order. This yields the Fokker-Planck equation

$$\frac{\partial}{\partial t}p(x,t) = -\frac{\partial}{\partial x}[f(x) - b(x)]p(x,t) + \frac{\partial^2}{\partial x^2}D_{\rm KM}(x)p(x,t) \tag{11}$$

with diffusion coefficient $D_{\rm KM}(x) = [f(x)+b(x)]/(2N)$. This Fokker-Planck equation corresponds to the Langevin equation

$$\dot{x} = f(x) - b(x) + \sqrt{2D_{\rm KM}(x)}\xi(t), \tag{12}$$

where $\xi(t)$ is white Gaussian noise of unit intensity, $\langle \xi(t)\xi(t')\rangle = \delta(t-t')$, in pre-point, or Ito interpretation [8]. This equation is quite general for any one-dimensional birth-and-death process within this standard diffusional approximation. For the considered population of ion channels,

$$\dot{x} = \alpha(1-x) - \beta x + \sqrt{[\alpha(1-x)+\beta x]/N}\xi(t). \tag{13}$$

This is stochastic equation for a gating variable in the stochastic generalization of Hodgkin-Huxley equations by Fox and Lu [5]. It replaces Eq. (3) with corresponding voltage-dependent $\alpha_x(V)$, $\beta_x(V)$, and $N = N_{\rm Na} = \rho_{\rm Na}S$ for m and h, or $N = N_{\rm K} = \rho_{\rm K}S$ for the variable n. S is the area of membrane patch, and $\rho_{\rm Na} = 60\mu m^{-2}$, $\rho_{\rm K} = 18\mu m^{-2}$ within HH model [4]. Clearly, in the limit $N \to \infty$ the channel noise vanishes, restoring the deterministic HH model. We name this model the second model by Fox and Lu (Fox-Lu 2) in application to stochastic HH dynamics.

Linear Noise Approximation. The further approximation (Fox-Lu 1 within stochastic HH model) is obtained by $D_{\rm KM}(x) \to D_{\rm KM}(x_{\rm eq}) = const$, where $x_{\rm eq}$ is equilibrium point of deterministic dynamics, $f(x_{\rm eq}) = b(x_{\rm eq})$. It corresponds to the so-called $1/\Omega$ expansion with linear additive noise approximation advocated by van Kampen [6]. Then, with $x_{\rm eq} = p_0 = \alpha/(\alpha+\beta)$ Eq. (13) reduces to

$$\dot{x} = \alpha(1-x) - \beta x + \sqrt{2\alpha\beta/[N(\alpha+\beta)]}\xi(t). \tag{14}$$

Diffusional Approximation with Natural Boundaries. The both diffusional approximations are not quite satisfactory because they do not guarantee the boundary conditions in a natural way. As a result, for a sufficiently small opening probability $p_0 \ll 1$, and not sufficiently large number of channels the negative values, $x < 0$, become possible with appreciable probabilities $p(x,t) > 0$. Likewise, the larger than one values, $x > 1$, are also possible when the opening probability p_0 is close to one. However, this deficiency can easily be corrected numerically by imposing reflecting boundary conditions at $x = 0$ and $x = 1$ in stochastic simulations. With this correction, Langevin approximation of stochastic HH dynamics is widely used [9–11, 3]. However, it is not quite clear if this procedure indeed delivers the correct results [12]. To clarify the issue, we consider a different diffusional approximation with natural reflecting boundaries which naturally bound stochastic dynamics to the interval $0 \le x \le 1$.

For this, we first demand that the diffusional approximation is consistent with the stationary distribution of birth-and-death process, which can be expressed as $P_N^{\rm st}(n) = \exp[-\Phi(n)]P_N^{\rm st}(0)$ in terms of a pseudo-potential $\Phi(n) = -\sum_{n=1}^{N} \ln [F(n-1)/B(n)]$ [6]. Hence, in the continuous limit, $p_N^{\rm st}(x) \propto \exp[-N\phi(x)]$, with pseudo-potential $\phi(x) = -\int_0^x \ln[f(x')/b(x')]\,dx' = \ln(1-x) - x\ln(\alpha(1-x)/(x\beta))$. This indeed yields the probability density (7). The corresponding Fokker-Planck equation must read

$$\frac{\partial}{\partial t}p(x,t) = \frac{\partial}{\partial x}\left(D(x)e^{-N\phi(x)}\frac{\partial}{\partial x}e^{N\phi(x)}p(x,t)\right) \tag{15}$$

$$= \frac{\partial}{\partial x}ND(x)\phi_x'(x)p(x,t) + \frac{\partial}{\partial x}D(x)\frac{\partial}{\partial x}p(x,t) \tag{16}$$

with

$$ND(x)\phi_x'(x) = b(x) - f(x) \tag{17}$$

in order to be also consistent with the deterministic limit $N \to \infty$. The last equation fixes the diffusion coefficient as

$$D(x) = \frac{1}{N}\frac{f(x) - b(x)}{\ln[f(x)/b(x)]}\,. \tag{18}$$

The Langevin equation which corresponds to this best diffusional approximation of the birth-and-death processes [7, 13] reads

$$\dot{x} = f(x) - b(x) + \sqrt{2D(x)}\xi(t), \tag{19}$$

in the post-point, or Klimontovich-Hänggi interpretation [7]. In the standard Ito interpretation suitable for integration with stochastic Euler algorithm [8] the corresponding Langevin equation becomes

$$\dot{x} = f(x) - b(x) + D_x'(x) + \sqrt{2D(x)}\xi(t) \tag{20}$$

with spurious drift $D_x'(x)$. In application to stochastic dynamics of one gating variable it reads

$$\dot{x} = \alpha(1-x) - \beta x + D_x'(x) + \sqrt{2D(x)}\xi(t)\,. \tag{21}$$

with

$$D(x) = \frac{1}{N}\frac{\alpha(1-x) - \beta x}{\ln[\alpha(1-x)/(\beta x)]}\,. \tag{22}$$

Replacing with such equations the stochastic equations for gating variables in the standard Langevin variant of stochastic Hodgkin-Huxley equations we obtain the improved Langevin description of mesoscopic channel noise, with natural boundaries because $D(0) = D(1) = 0$, i.e. the channel noise (and the probability flux) vanishes exactly at the reflecting boundaries, in the theory. Nevertheless, in numerical algorithm one must yet additionally secure such boundaries for any *finite* integration time step δt. Notice also that near the equilibrium point with $|f(x) - b(x)| \ll f(x) + b(x)$, $D(x) \approx D_{\rm KM}(x)$, and the standard diffusional approximation is almost restored, almost, if to neglect the spurious drift correction $D_x'(x)$, which still remains within the Ito interpretation.

We test the best diffusional approximation for a gating variable against the earlier Langevin descriptions with reflecting boundary conditions implemented numerically. For this we use stochastic Euler algorithm with time step $\delta t = 0.001$ for several values of N and the simulation software XPPAUT [14]. The results are shown for $\alpha = 1$ and $\beta = 9$ with $p_0 = 0.1$ in Fig. 1 for $N = 100$ (a) and $N = 10$ (b). As a big surprise, the simplest linear noise approximation actually seems to work best, if only to implement reflecting boundary conditions. For $N = 100$, it reproduces well the still somewhat skewed binomial distribution with the exact mean $\langle x \rangle = 0.1$ and standard deviation $\langle \Delta x^2 \rangle^{1/2} = 0.03$. Even for $N = 10$, it gives the mean closer to the correct value of 0.1 within the discrete state model. However, the variance then deviates from the theoretical value $\langle \Delta x^2 \rangle^{1/2} \approx 0.095$ larger than within two other approximations. For a sufficiently large $N = 1000$ (not shown), all three diffusional approximations give practically identical results, within the statistical errors of simulations. Surprisingly, all three work reasonably well even for $N = 10$! However, such a performance is *a priori* not guaranteed for stochastic nonlinear dynamics with voltage-dependent $\alpha(V(t))$ and $\beta(V(t))$. In fact, for a multistable dynamics the best diffusional approximation is generically expected [13] to operate essentially better.

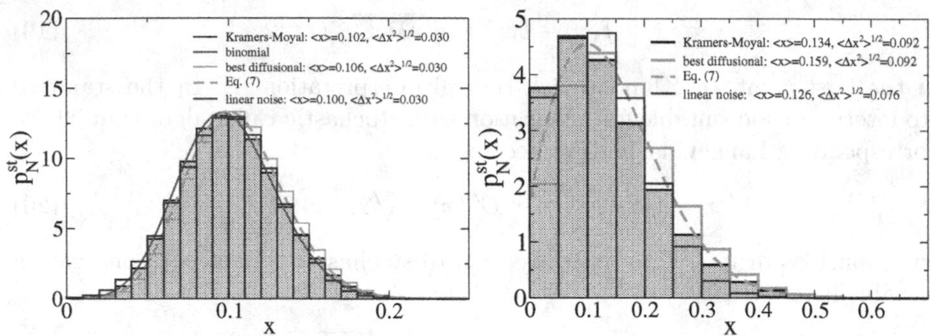

Fig. 1. Stationary distributions of gating variable x for two ensembles of ion channels with $\alpha = 1$ and $\beta = 9$, (a) $N = 100$ and (b) $N = 10$. Numerics are compared with binomial distribution (a) and distribution (7) for the best diffusional approximation.

We compare three different Langevin descriptions of stochastic HH dynamics in Fig. 2, for two different membrane patches. Here, the interspike interval distributions are presented, together with the corresponding mean, $\langle \tau \rangle$, standard deviation, $\langle [\tau - \langle \tau \rangle]^2 \rangle^{1/2}$, and the relative standard variation, or the coefficient of variation, $C_V = \langle [\tau - \langle \tau \rangle]^2 \rangle^{1/2} / \langle \tau \rangle$, which measures the spike coherence. For $S = 10\ \mu m^2$, all three approximations agree well. However, for $S = 1\ \mu m^2$ the discrepancies become apparent, and we prefer our improved Langevin description on general theoretical grounds.

The coefficient of variation C_V, calculated within our Langevin variant of stochastic HH model, is plotted *vs.* the patch size S in Fig. 3. It displays a typical

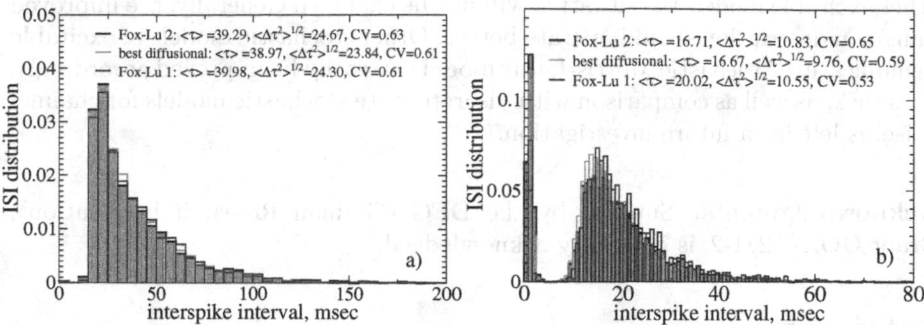

Fig. 2. Interspike time interval distribution for self-excitable dynamics, $I_{\mathrm{ext}} = 0$, due to the channels noise for two membrane patches: (a) $S = 10\mu m^2$ ($N_{\mathrm{Na}} = 600$, $N_{\mathrm{K}} = 180$), and (b) $S = 1\mu m^2$ ($N_{\mathrm{Na}} = 60$, $N_{\mathrm{K}} = 18$)

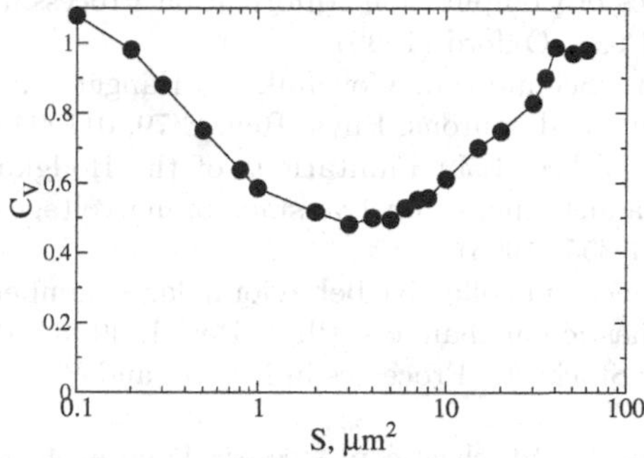

Fig. 3. Coefficient of variation versus the membrane patch size within our variant of stochastic HH model. Self-excitable dynamics, $I_{\mathrm{ext}} = 0$.

coherence resonance [15] behavior revealed earlier within stochastic HH models in [9, 16] as a system-size coherence resonance. There exists an optimal patch size (optimal number of ion channels) with most coherent stochastic dynamics due to internal mesoscopic noise.

2.2 Summary and Conclusions

In this paper, we presented the best diffusional Langevin approximation for excitable cell dynamics within stochastic Hodgkin-Huxley model, with natural boundary conditions for the channel noise implemented. It has clear theoretical advantages over the standard diffusional approximation in the case of transitions induced by mesoscopic noise as discussed for bistable birth-and-death processes long time ago [13]. However, within stochastic HH model for a sufficiently large number of ion channels, the standard diffusional approximations were shown to work also very good. Hence, this work confirmes the validity of the previous work done within the Langevin approximations of stochastic HH dynamics, for a sufficently large number of channels. This does not mean, however, that for other excitable models the situation will not be changed. Generally, the improved Langevin description should operate better. Other stochastic models of excitable dynamics, e.g. stochastic Morris-Lecar model can easily be improved accordingly. This task, as well as comparison with discrete state stochastic models for channel noise, is left for a future investigation.

Acknowledgments. Support by the DFG (German Research Foundation), Grant GO 2052/1-2, is gratefully acknowledged.

References

1. Hodgkin, A.L., Huxley, A.F.: A quantitative description of membrane current and its application to conduction and excitation in nerve. J. Physiol. 117, 500–544 (1952)
2. Koch, C.: Biophysics of Computation, Information Processing in Single Neurons. Oxford University Press, Oxford (1999)
3. Ochab-Marcinek, A., Schmid, G., Goychuk, I., Hänggi, P.: Noise-assisted spike propagation in myelinated neurons. Phys. Rev. E 79, 011904 (2009)
4. Strassberg, A.F., DeFelice, L.J.: Limitations of the Hodgkin-Huxley formalism: Effects of single channel kinetics on transmembrane voltage dynamics. Neuronal Computation 5, 843–855 (1993)
5. Fox, R.F., Lu, Y.: Emergent collective behavior in large numbers of globally coupled independently stochastic ion channels. Phys. Rev. E 49, 3421–3431 (1994)
6. Van Kampen, N.G.: Stochastic Processes in Physics and Chemistry. North Holland, Amsterdam (1981)
7. Hänggi, P., Thomas, H.: Stochastic processes: Time evolution, symmetries, and linear response. Phys. Rep. 88, 207–319 (1982)
8. Gard, T.C.: Introduction to Stochastic Differential Equations. Marcel Dekker, New York (1988)
9. Schmid, G., Goychuk, I., Hänggi, P.: Stochastic resonance as a collective property of ion channel assemblies. Europhys. Lett. 56, 22–28 (2001)
10. Schmid, G., Goychuk, I., Hänggi, P.: Effect of channel block on the spiking activity of excitable membranes in a stochastic Hodgkin-Huxley model. Phys. Biol. 1, 61–66 (2004)
11. Schmid, G., Goychuk, I., Hänggi, P.: Capacitance fluctuations causing channel noise reduction in stochastic Hodgkin-Huxley systems. Phys. Biol. 3, 248–254 (2006)
12. Schmid, G., Goychuk, I., Hänggi, P., Zeng, S., Jung, P.: Stochastic resonance and optimal clustering for assemblies of ion channels. Fluct. Noise Lett. 4, L33–L42 (2004)
13. Hänggi, P., Grabert, H., Talkner, P., Thomas, H.: Bistable systems: Master equation versus Fokker-Planck modeling. Phys. Rev. A 29, 371–378 (1984)
14. Ermentrout, B.: Simulating, Analyzing, and Animating Dynamical Systems: A Guide to XPPAUT for Researchers and Students. SIAM (2002)
15. Pikovsky, A.S., Kurths, J.: Coherence resonance in a noise-driven excitable system. Phys. Rev. Lett. 78, 775–778 (1997)
16. Jung, P., Shuai, J.W.: Optimal sizes of ion channel clusters. Europhys. Lett. 56, 29–35 (2001)

Typical Dynamics of Bifurcating Neurons with Double Base Signal Inputs

Yusaku Yanase, Shota Kirikawa, and Toshimichi Saito

Hosei University, Koganei, Tokyo, 184-8584 Japan
tsaito@hosei.ac.jp

Abstract. This paper studies the bifurcating neuron whose base signal consists of fundamental and third harmonic components. The BN can exhibit a variety of bifurcation phenomena which are impossible in the case of single component. For example, periodic spike-train in the single component is changed into co-existence states of chaotic and periodic spike-trains in the double components. Using the mapping procedure, such phenomena are analyzed precisely. Presenting a simple test circuit, typical phenomena are investigated experimentally.

Keywords: Chaos, bifurcation, spiking neurons.

1 Introduction

The bifurcating neuron (BN [1]-[5]) is a simple switched dynamical system inspired by spiking neurons [6]. Repeating integrate-and-fire behavior between a threshold and periodic base signal, the BN can output a variety of chaotic/ periodic spike-trains. The dynamics can be analyzed by a one-dimensional spike-phase map (SPM). As the shape of base signal varies, the shape of the SPM (i.e., the dynamics of BN) varies. In previous works of the BN, sinusoidal and triangular base signals have been mainly used [1]-[4]. Applications of the BN are many and include signal processing, ultra wide band communications and neural-prosthesis [7]-[10]. Analysis of the BN is important not only as a nonlinear dynamical system [11] but also for engineering applications.

This paper studies the BN whose base signal ($B(t)$) consists of fundamental ($B_1(t)$) and third harmonic waveforms ($B_3(t)$). If $B(t) = B_1(t)$ or $B(t) = B_3(t)$ (single component), the BN exhibits well-known phenomena such as period doubling bifurcation to chaos [2]. However, if $B(t) = B_1(t) + B_3(t)$ (double components), the BN can exhibit a variety of bifurcation phenomena which are impossible in the case of single component. Note that the superposition theorem is not satisfied in the BN. Since general analysis the bifurcation phenomena is hard, this paper considers basic phenomena. For example, periodic spike-train in the single component is changed into co-existence states of chaotic and periodic spike-trains in the double components. The BN exhibits either spike-train depending on the initial value. Also, chaotic spike-train in the single component is changed into periodic spike-train in the double components. Using the Pmap,

V.M. Mladenov and P.C. Ivanov (Eds.): NDES 2014, CCIS 438, pp. 333–340, 2014.

such phenomena are analyzed precisely. Presenting a simple test circuit, typical phenomena are investigated experimentally. Note that the case of double components is the first step to consider the case of many components whose analysis is very hard [5].

2 Circuit Model of Bifurcating Neuron

Fig. 1 (a) shows a circuit model of the bifurcating neuron (BN). Below a positive threshold V_T, the capacitor voltage v increases by integrating a constant current $I > 0$. If v_1 reaches V_T, the comparator triggers a monstable multi-vibrator (MM) to output a spike $Y(t) = E$. The spike closes a switch SW and v is reset to a periodic base signal $B(t) < V_T$ with period T. Repeating the integrate-and-fire dynamics. the BN outputs a spike train $Y(t)$. We regards the base signal $B(t)$ as an input of BN. As shape of the $B(t)$ varies, the BN can exhibit various chaotic/periodic phenomena. In this paper, we consider the third harmonics:

$$B(t) = B_1(t) + B_3(t),\ B_1(t) = -K_1 \sin \omega t,\ B_3(t) = -K_3 \sin 3\omega t,\ T \equiv \frac{2\pi}{\omega} \quad (1)$$

If $K_1 \neq 0$ and $K_3 = 0$ then the base signal is fundamental component only and the BN exhibits period-doubling bifurcation of chaos as suggested in [1]-[3]. If $K_1 = 0$ and $K_3 \neq 0$ then the base signal is third harmonics only and the BN exhibits similar phenomena to the case of fundamental component. The problem is what happens for $K_1 \neq 0$ and $K_3 \neq 0$ (Fig. 2). It goes without saying that the theorem of superposition is not valid in this nonlinear system. We have fabricated a breadboard prototype and have performed laboratory experiments. The base signal is given by the function generator AFG-2005. The BN exhibits various interesting phenomena and we show two examples.

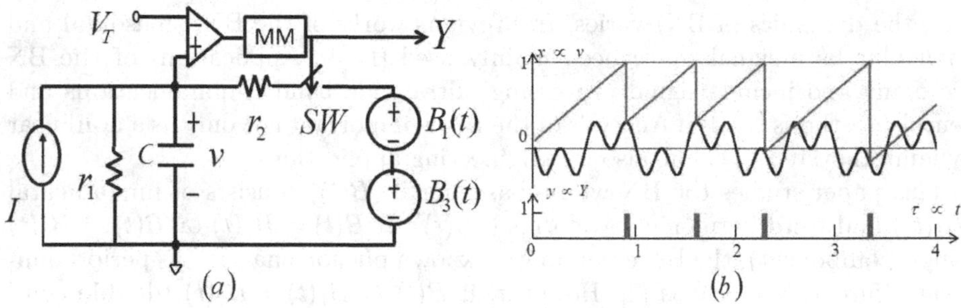

Fig. 1. Bifurcation neuron (a) Circuit model, (b) Waveform

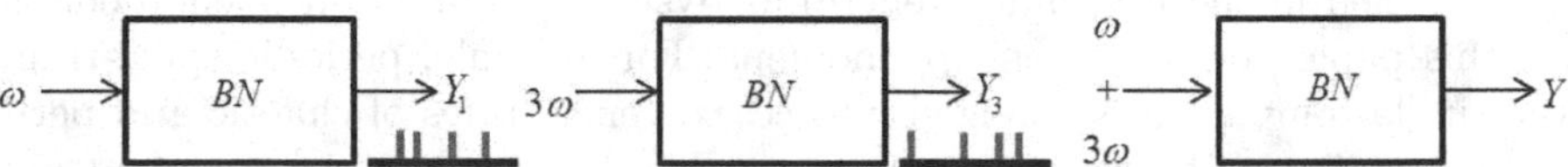

Fig. 2. Bifurcating neuron with double base inputs

Fig. 3 shows the first example. The BN exhibits periodic waveform for $B(t) = B_1(t)$ (fundamental waveform only) or $B(t) = B_3(t)$ (third harmonics only). Note that a spike signal $Y = E$ is generated when v is reset to the base and the sequence of reset signals corresponds to the spike-train. However, if $B(t) = B_1(t)+B_3(t)$ then the BN exhibits chaos as shown in Fig. 3 (d). This phenomenon is referred to as "order + order = chaos". The periodic waveform can coexist with chaos as shown in the figure. The BN exhibits either chaos or periodic waveform depending on the initial value.

Fig. 4 shows the second example. The BN exhibits chaos for $B(t) = B_1(t)$ or $B(t) = B_3(t)$. However, if $B(t) = B_1(t) + B_3(t)$ then the BN exhibits periodic waveform. This phenomenon is referred to as "chaos +chaos = order".

3 Circuit Equation and Return Map

We derive the return map and analyze the typical phenomena. For simplicity, the inner resistors are ignored ($r_1 \to \infty$, $r_2 \to 0$) and the switching is assumed to be ideal: v_1 is reset instantaneously without delay. The circuit dynamics is described by

$$\begin{cases} C\dfrac{dv}{dt} = I, & Y(t) = -E \text{ for } v(t) < V_T \\ v(t_+) = B(t_+), & Y(t_+) = E \text{ if } v(t) = V_T \end{cases} \tag{2}$$

Using dimensionless variables and parameters:

$$\tau = \frac{t}{T}, \; x = \frac{v}{V_T}, \; \dot{x} = \frac{dx}{d\tau}, \; y = \frac{Y+E}{2}, \; k_1 = \frac{K_1}{V_T}, \; k_3 = \frac{K_3}{V_T}, \; s = \frac{IT}{CV_T} \tag{3}$$

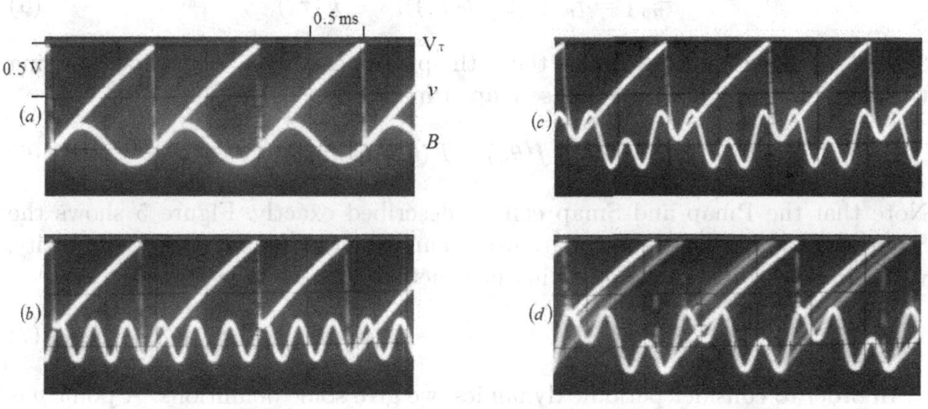

Fig. 3. Waveforms of "order + order = chaos", horizontal = t[0.5ms/div.], vertical = v[0.5V/div.], $C \doteq 0.022[\mu\text{F}]$, $I \doteq 0.033[\text{mA}]$, $r_1 \doteq 121.1[\text{k}\Omega]$, $r_2 \doteq 0.45[\text{k}\Omega]$, $V_T \doteq 1[\text{V}]$, $T \doteq 1[\text{ms}]$, ($s \doteq 1.05$), $\omega \doteq 1[\text{kHz}]$. (a) Periodic waveform for $(K_1, K_3) \doteq (0.17, 0)$, (b) Periodic waveform for $(K_1, K_3) \doteq (0, 0.17)$, (c) & (c') Co-existing chaotic and periodic waveforms for $(K_1, K_3) \doteq (0.17, 0.17)$.

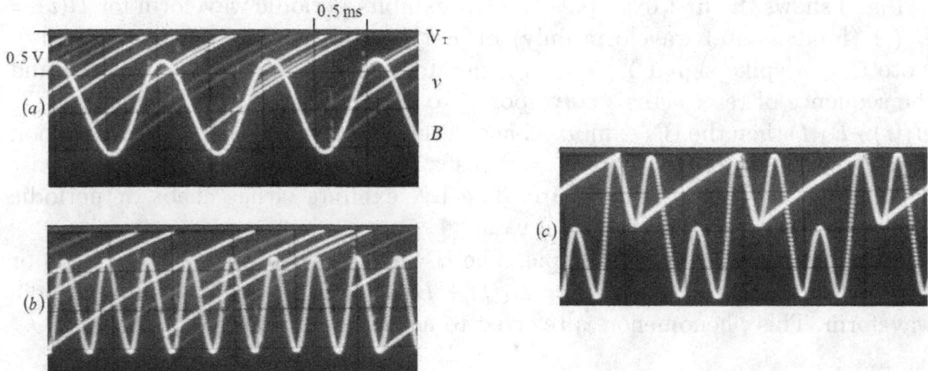

Fig. 4. Waveforms of "chaos + chaos = order". horizontal = t[0.5ms/div.], vertical = v[0.5V/div.], $C \doteq 0.022[\mu\text{F}]$, $I \doteq 0.033$[mA], $r_1 \doteq 121.1[\text{k}\Omega]$, $r_2 \doteq 0.45[\text{k}\Omega]$, $V_T \doteq 1$[V], $T \doteq 1$[ms], ($s \doteq 1.05$), $\omega \doteq 1$[kHz]. (a) Chaotic waveform for $(K_1, K_3) \doteq (0.58, 0)$, (b) Chaotic waveform for $(K_1, K_3) \doteq (0, 0.58)$, (c) Periodic waveform for $(K_1, K_3) \doteq (0.58, 0.58)$.

Eqs. (2) and (1) are transformed into

$$\begin{cases} \dot{x} = s, & y(\tau) = 0 \quad \text{for } x < 1 \\ x(\tau_+) = b(\tau_+), & y(\tau_+) = 1 \text{ if } x(\tau) = 1 \end{cases} \quad b(\tau) = -k_1 \sin 2\pi\tau - k_3 \sin 6\pi\tau \qquad (4)$$

Let τ_n denote the n-th spike-position (firing moment). Since τ_{n+1} is determined by τ_n, we can define the spike position map (Smap):

$$\tau_{n+1} = \tau_n + (1 - b(\tau_n))/s \equiv F(\tau_n), \qquad (5)$$

Since $b(\tau)$ is period 1, we define the n-th spike-phase by $\theta_n = \tau_n \bmod 1$. Using this, we can define the spike-phase map (Pmap):

$$\theta_{n+1} = f(\theta_n) = F(\theta_n) \bmod 1 \qquad (6)$$

Note that the Pmap and Smap can be described exactly. Figure 5 shows the Smap and Pmap. Using the Pmap, we can analyze the dynamics. For simplicity, we consider the BN in the following parameter subspace hereafter:

$$P_s \equiv \{(s, k_1, k_3) \mid s = 1,\ b(\tau) = -k_1 \sin 2\pi\tau - k_3 \sin 6\pi\tau < 1,\ \forall\tau\} \qquad (7)$$

In order to consider periodic dynamics, we give some definitions. A point p is said to be a period point (PEP) with period k if $f^k(p) = p$ and $f^k(p) \neq p$ for $0 < l < k$. A PEP with period 1 is referred to as a fixed point. A PEP with period

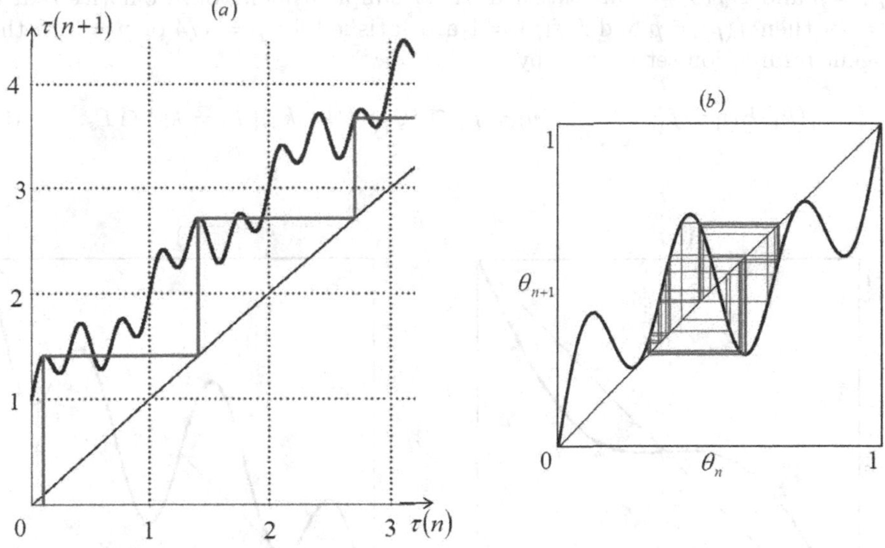

Fig. 5. Skipe-position map (Smap, (a)) and spike-phase map (Pmap, (b))

k is said to be stable, critical and unstable for initial value if $Df^k(p) < 1, Df^k(p) = 1, and Df^k(p) > 1$, respectively. where $Df(p)$ is the slope of f at p. A sequence of PEPs $(f_1(p), \cdots, f_1^k)$ is said to be a periodic orbit (PEO).

In order to characterize chaos, we use the Lyapunov exponent of the Pmap:

$$\Lambda = \frac{1}{M}\sum_{n=1}^{M} \ln|Df(\theta(n))|, \;\; Df(\theta) = 1 + 2\pi k_1 \sin 2\pi\theta + 6\pi k_3 \sin 6\pi\theta \qquad (8)$$

Convergence of Λ is confirmed for $M = 10^4$. Usually, $\Lambda > 0$ indicates chaos [11]. Figure 6 shows Pmaps which correspond to Fig. 3 of "order + order = chaos". Stable fixed point of Pmap (a) corresponds to periodic waveform of BN with fundamental component ($B(t) = B_1(t)$) and stable periodic point with period 2 of Pmap (b) corresponds to periodic waveform of BN with third harmonics $B(t) = B_3(t)$. In the Pmap for double input case $b(\tau) = b_1(\tau)+b_3(\tau)$, stable fixed point (c) and chaotic orbit (d) coexist. These correspond to periodic and chaotic waveforms in Fig. 3 (c) and (d). The Pmap and BN exhibit either of periodic or chaotic phenomenon depending on the initial value. Figure 7 shows Pmaps which correspond to Fig. 4 of "chaos + chaos = order". Chaotic orbits of Pmaps (a) and (b) correspond to chaotic waveforms of BN with single component. Stable fixed point of Pmap (c) corresponds to periodic waveform of BN with double components. This stable fixed point p is born by the tangent bifurcation at which

$f(p) = p$ and $Df(p) = 1$ are satisfied. After simple calculation, we notice that if $k_1 = k_3$ then $f(p) = p$ and $Df(p) = 1$ are satisfied for $p = 1/4$ or $p = 3/4$: the tangent bifurcation set is given by

$$\{(k_1, k_3)|\ Df(p) = 1,\ f(p) = p\} \cap P_s = \{(k_1, k_3)|\ k_1 = k_3\} \cap P_s. \quad (9)$$

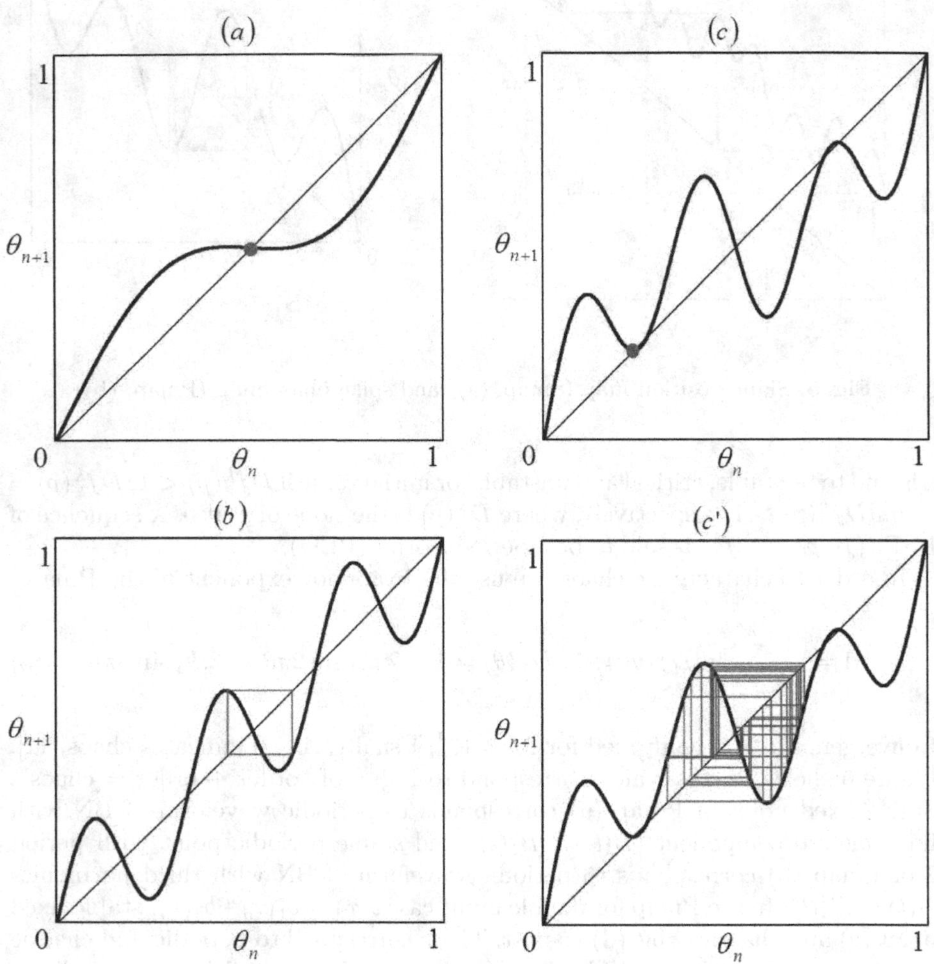

Fig. 6. Pmaps of "order + order = chaos" (a) stable fixed point for $(k_1, k_3) = (0.17, 0)$, $(\Lambda \doteq -1.17)$ (b) stable periodic point with period 2 for $(k_1, k_3) = (0, 0.18)$, $(\Lambda \doteq -0.11)$ (c) stable fixed point for $(k_1, k_3) = (0.17, 0.18)$, $(\Lambda \doteq -1.30)$ (c') chaos for $(k_1, k_3) = (0.17, 0.18)$, $(\Lambda \doteq 0.32)$

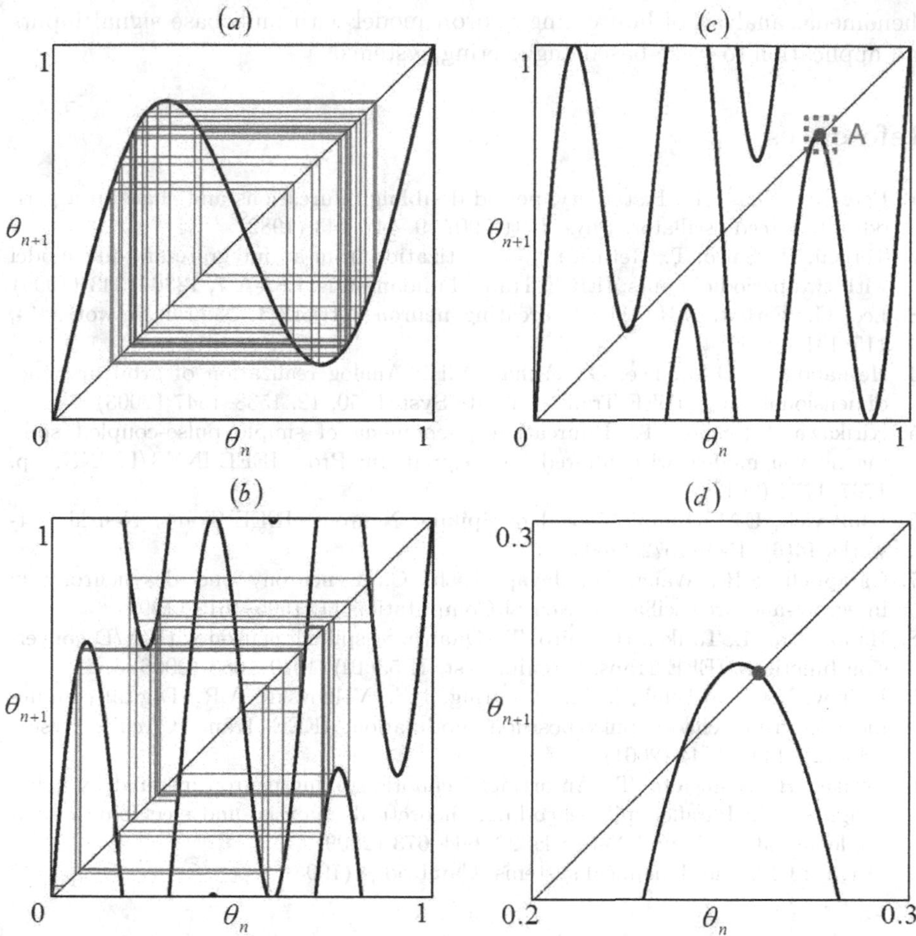

Fig. 7. Pmaps of "chaos + chaos = order" (a) chaos for $(k_1, k_3) = (0.58, 0)$, $(\Lambda \doteq 0.17)$ (b) chaos for $(k_1, k_3) = (0, 0.59)$, $(\Lambda \doteq 0.0.79)$ (c) stable fixed point for $(k_1, k_3) = (0.58, 0.59)$, $(\Lambda \doteq -0.04)$. (c') Enlargement of a part A of (c)

4 Conclusions

We have studied bifurcating neuron circuit that have a variety of spike-train dynamics. The base signal consists of the double inputs (fundamental and third harmonic components). The BN can exhibit a variety of phenomena and typical examples are shown: "order + order = chaos" and "chaos + chaos = order". Applying the mapping procedure, the phenomena are analyzed precisely. Presenting a simple test circuit, typical phenomena are observed experimentally(CCO, OOC). The dynamics can be integrated into the Pmap and basic bifurcation phenomena are analyzed. Future problems include detailed analysis of bifurcation phenomena, analysis of bifurcating neuron model with muli base signal inputs and application to spike-based engineering systems.

References

1. Perez, R., Glass, L.: Bistability, period doubling bifurcations and chaos in a periodically forced oscillator. Phys. Lett., 90A 9, 441–443 (1982)
2. Torikai, H., Saito, T.: Return map quantization from an integrate-and-fire model with two periodic inputs: IEICE Trans. Fundamntals, E82-A 7, 1336–1343 (1999)
3. Lee, G., Farhat, N.H.: The bifurcating neuron network 1. Neural Networks 14, 115–131 (2001)
4. Hernandez, E.D.M., Lee, G., Farhat, N.H.: Analog realization of arbitrary one-dimensional maps. IEEE Trans. Circuits Syst. I, 50, 12, 1538–1547 (2003)
5. Kirikawa, S., Saito, T.: Bifurcation phenomena of simple pulse-coupled spiking neuron models with filtered base signal. In: Proc. IEEE-INNS/IJCNN, pp. 1767–1774 (2013)
6. Izhikevich, E.M.: Simple Model of Spiking Neurons. IEEE Trans. Neural Networks 14(6), 1569–1572 (2003)
7. Campbell, S.R., Wang, D., Jayaprakash, C.: Synchrony and desynchrony in integrate-and-fire oscillators. Neural Computation 11, 1595–1619 (1999)
8. Hamanaka, H., Torikai, H., Saito, T.: Quantized spiking neuron with A/D conversion functions. IEEE Trans. Circuits Syst. II 53(10), 1049–1053 (2006)
9. Rulkov, N.F., Sushchik, M.M., Tsimring, L.S., Volkovskii, A.R.: Digital communication using chaotic-pulse-position modulation. IEEE Trans. Circuits Systs., I 48(12), 1436–1444 (2001)
10. Torikai, H., Nishigami, T.: An artificial chaotic spiking neuron inspired by spiral ganglion cell: Parallel spike encoding, theoretical analysis, and electronic circuit implementation. Neural Networks 22, 664–673 (2009)
11. Ott, E.: Chaos in dynamical systems, Cambridge (1993)

On the Mechanisms for Formation of Segmented Waves in Active Media

Andrey Polezhaev and Maria Borina

P.N. Lebedev Physical Institute of the Russian Academy of Sciences Russia, Leninskiy prosp., 53, 119991 Moscow, Russia

Abstract. We suggest three possible mechanisms for formation of segmented waves and spirals. These structures were observed in the Belousov-Zhabotinsky reaction dispersed in a water-in-oil aerosol microemulsion. The first mechanism is caused by interaction of two coupled subsystems, one of which is excitable, and the other one has Turing instability depending on the parameters. It is shown that, segmented spirals evolve from ordinary smooth spirals as a result of the transverse Turing instability. We demonstrate that depending on the properties of subsystems different segmented spirals emerge. For the second mechanism we suggest "splitting" of the traveling wave in the vicinity of the bifurcation point of codimension-2, where the boundaries of the Turing and wave instabilities intersect. Finally we show that segmented waves can emerge in some simple two-component reaction-diffusion models having more than one steady state, particularly in a FitzHugh-Nagumo model.

Keywords: Active media, autowaves, segmentation.

1 Introduction

Chemical waves discovered in the Belousov-Zhabotinsky (BZ) reaction is a vivid example of spatial-temporal patterns arising in active media. They were discovered more that 40 years ago [1] and for long time remained the only type of patterns which were observed experimentally. A different class of patterns, stationary Turing structures [2], were found experimentally only 20 years later in another chemical system, the chlorite-iodide-malonic acid reaction [3]. At present several experimental laboratory systems are known which alow to observe and study formation of complex spital-temporal patterns [4,5,6]. Perhaps one of the most "plentiful" of them appeared to be the the BZ system dispersed in aerosol OT (AOT = sodium bis(ethylhexyl) sulfosuccinate) water in oil microemulsion (BZ-AOT system) [7] . This system made it possible to observe new types of waves and patterns [8]. In particular for the first time the waves were obtained which in the course of propagation split into segments of certain scale – the so-called dashed and segmented waves (see Fig. 1). They were observed in the freshly made microemulsion and disappeared after 2-3 hours [9,10]. Later by a proper choice of parameters it appeared possible to obtain a more long-living

V.M. Mladenov and P.C. Ivanov (Eds.): NDES 2014, CCIS 438, pp. 341–348, 2014.

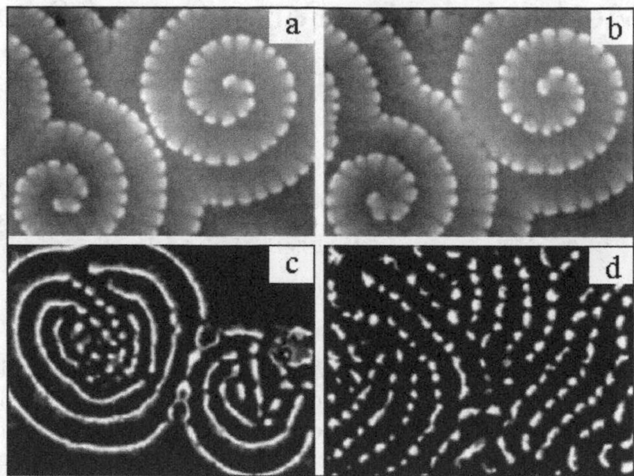

Fig. 1. Segmented spirals (a, b) and dashed waves (c,d) in BZ-AOT system. Time interval between (a) and (b) is 66 seconds and between (c) and (d) – 0.5 hour. The scale of the first pair of images – $3.72 \times 4.82mm^2$ and of the second pair – $2.54 \times 1.88mm^2$. (From [9,10]).

regime [11]. Segmented waves were also discovered in a chlorine dioxide-iodine-malonic acid reaction [12] and in a reaction-diffusion convection system [13,14].

The aim of the present investigation is to analyze the possible mechanisms of segmented waves formation. We suggest three variants: interaction of two coupled subsystems, one of which is excitable and the other one has Turing instability; "splitting" of a traveling wave in the vicinity of the bifurcation point of codimension-2, where the boundaries of the Turing and wave instabilities intersect; emergence of segmented waves in two-component reaction-diffusion models having more than one steady state, one of which being excitable, and the other one displaying pseudo-Turing instability.

2 Interaction of Excitability and Turing Instability

We consider the case of a spatially-distributed system which is a combination of two subsystems: one corresponding to the excitable medium and the other potentially (for corresponding parameters) having Turing instability. The first subsystem parametrically influences the second one transferring it into unstable state. For the first subsystem we have chosen the FitzHugh-Nagumo model with the single excitable steady state which for appropriate initial conditions generates a spiral wave. For the second subsystem we explored either Brusselator or another FitzHugh-Nagumo model with one of the parameters depending on the variables of the first subsystem.

Numerical experiments were carried out in a bounded domain with zero-flux boundary conditions using the method of alternating directions (details of the numeric scheme are given in [15]). Below we present some results of our simulations.

2.1 FitzHugh-Nagumo Model Plus Brusselator

Let us consider the following set of equations:

$$\begin{aligned} \dot{u} &= u - \tfrac{u^3}{3} - v + D_u \Delta u \,, \\ \dot{v} &= \varepsilon(u - \gamma v + \delta) \,, \\ \dot{x} &= a - (b(u) + 1)\, x + x^2 y + \Delta x \,, \\ \dot{y} &= b(u) x - x^2 y + D_y \Delta y \,. \end{aligned} \tag{1}$$

One of the parameters b of the second pair of equations corresponding to Brusselator depends on the variable u in the following way: $b(u) = \begin{cases} b_c + ku, & u \geq 0 \\ b_c, & u < 0 \end{cases}$, where b_c is the value of the parameter b corresponding to the Turing bifurcation. (We considered also the case when the second parameter a depended on the first subsystem and obtained similar results.)

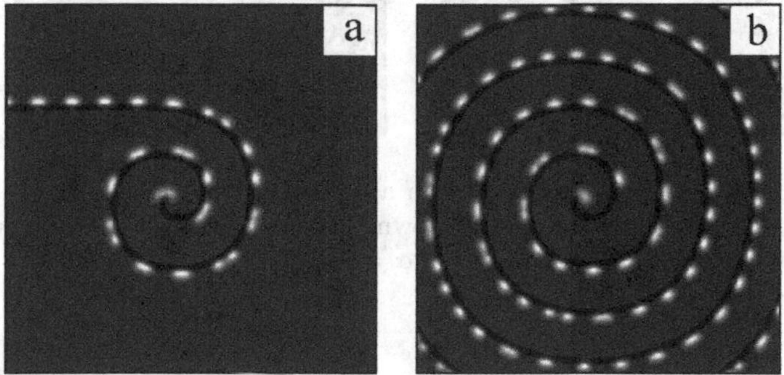

Fig. 2. Formation of a segmented wave in a square spatial domain in the model (1) (variable x, here and further light color corresponds to higher concentration). Two moments of time are shown: (a) $t = 120$, (b) $t = 300$. The parameters used: $\varepsilon = 0.09, \gamma = 0.5, \delta = 0.7, a = 2, D_u = 0.1, D_y = 100, b_c = 1.25, k = 2$. Size of the domain: 200×200.

The results of numerical simulations of the model (1) are presented in Fig. 2. For initial conditions a segment of a plane wave with a free end for the (u, v) subsystem and a uniform randomly perturbed state for the (x, y) subsystem were used. One can see that due to Turing instability a smooth wave developing in the first subsystem splits into segments.

2.2 Two FitzHugh-Nagumo Models

Now we consider a model being the combination of two FitzHugh-Nagumo models:

$$\begin{aligned} \dot{u} &= u - \tfrac{u^3}{3} - v + D_u \Delta u \,, \\ \dot{v} &= \varepsilon(u - \gamma v + \delta) \,, \\ \dot{x} &= \alpha(x - \tfrac{x^3}{3} - y + I(u) + D_x \Delta x) \,, \\ \dot{y} &= \alpha(x - \gamma y + \delta_1 + \Delta y) \,, \end{aligned} \tag{2}$$

where $I(u) = \begin{cases} I_c + ku, & u \geq 0 \\ I_c, & u < 0 \end{cases}$, I_c is the value of the parameter I corresponding to the Turing bifurcation. Fig. 3 illustrates the formation of a segmented wave for this model.

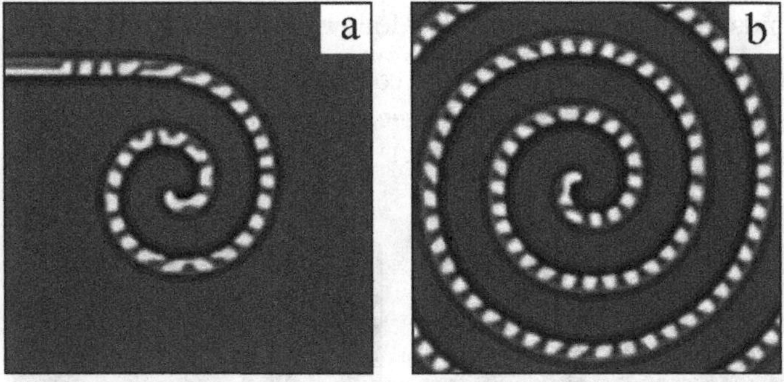

Fig. 3. Formation of a segmented wave in a square spatial domain in the model (2) (variable x). Two moments of time are shown: (a) $t = 75$, (b) $t = 200$. The parameters used: $\varepsilon = 0.09, \gamma = 0.5, \delta = 0.7, \delta_1 = 0.5, \alpha = 5, , D_u = 0.1, D_x = 0.05, I_c = 0.1, k = 0.03$. Size of the domain: 100×100.

These and other numerical simulations demonstrate that depending on the properties of the interacting subsystems and values of the parameters used this mechanism is able to produce a variety of segmented waves differing in shape and size of segments.

3 Traveling Wave Segmentation Near the Turing and Wave Bifurcations Codimension-2 Point

In the vicinity of the point of intersection in parametric space of the boundaries of Turing and wave bifurcations a wave can be presented in the form of superposition of corresponding modes: $u(r,t) = A_T^{ik_T r} + A_L^{i(\omega t + k_W r)} + A_R^{i(\omega t - k_W r)} + c.c.$ Here A_T is the complex amplitude of the mode corresponding to the wave vector k_T, which became unstable due to the Turing bifurcation; A_R and A_L are complex amplitudes of modes corresponding to equal in magnitude but opposite in direction wave vectors $\pm k_W$ and frequency ω, which became unstable due to the wave bifurcation.

Assuming that both bifurcations are supercritical, we can describe the dynamics of interacting modes by the following amplitude equations [16]:

$$\begin{aligned}\dot{A}_T &= \eta_T A_T - |A_T|^2 A_T - h_{TW}(|A_L|^2 + |A_R|^2)A_T\ ,\\ \dot{A}_L &= \eta_W A_L - (1 - ic_1)|A_L|^2 A_L - h_{LR}(1 - ic_2)|A_R|^2 A_L\\ &\quad - h_{WT}(1 - ic_3)|A_T|^2 A_L\ ,\\ \dot{A}_R &= \eta_W A_R - (1 - ic_1)|A_R|^2 A_R - h_{LR}(1 - ic_2)|A_L|^2 A_R\\ &\quad - h_{WT}(1 - ic_3)|A_T|^2 A_R\ .\end{aligned} \tag{3}$$

η_T and η_W are bifurcation parameters for Turing and wave bifurcations correspondingly; h_{TW}, h_{WT}, h_{LR} describe strength of competition between modes; c_1, c_2 and c_3 specify ratios between imaginary and real parts of the coefficients for the corresponding cubic terms.

Depending on the parameters the set of equations (3) has several stable stationary solutions corresponding to different regimes: stationary Turing patterns, propagating waves and a mixed regime, when both a stationary pattern and a traveling wave coexist (see [16]). The latter is just the case when a Turing pattern is superimposed on a traveling wave resulting in its "splitting" into segments. The emerging spatial-temporal pattern – a train of traveling "splitted" waves – is shown in Fig. 4(a). We extract one of the waves and show it in two moments of time to demonstrate its motion (Fig. 4(b,c)).

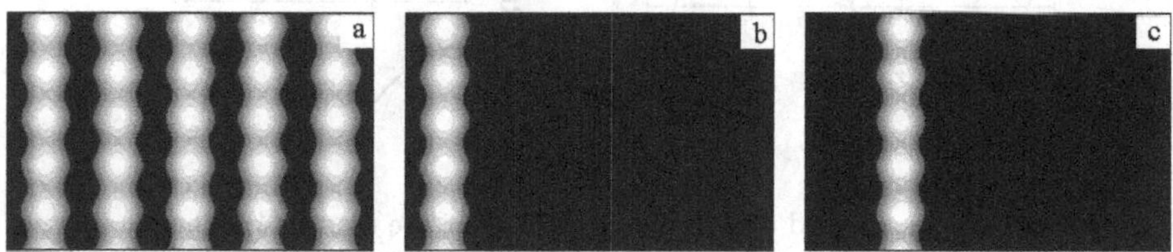

Fig. 4. (a) A mixed regime in Eq. (3) corresponding to segmented traveling waves in a rectangular spatial domain. One of the waves is extracted and shown for $t = 5$ (b) and $t = 11$ (c). Size of the domain: 250×80.

4 Segmented Waves in Two-Component Reaction-Diffusion Models with Multiple Steady States

This mechanism was suggested by Vanag et al. [9]. The authors explain segmentation of the wave by interaction of the excitable state and the state which demonstrates the so-called pseudo-Turing instability. (Interval of unstable modes includes the zero mode corresponding to uniform oscillations.) In numeric experiments they obtained segmented waves in the Gray-Scott model and in Oregonator [9,10].

We attempted to obtain a segmented wave in the FitzHugh-Nagumo model in the case when it has three stationary states:

$$\begin{aligned} \dot{u} &= u - \tfrac{u^3}{3} - v + \triangle u \,, \\ \dot{v} &= \varepsilon(\rho u - v + \delta) + D\triangle v \,. \end{aligned} \tag{4}$$

Analysis of these equations has shown that in the case of constant values of parameters it is impossible to meet the conditions of both excitability (small ε), on the one side, and pseudo-Turing instability, on the other side. It becomes possible if one of the parameters, say ρ, is not constant: $\rho(u) = \begin{cases} \rho_1, & u \geq 0 \\ \rho_2, & u < 0 \end{cases}$. However the range of parameters enabling segmented wave formation appears to be very narrow. The matter is that the critical point responsible for excitability is to be very close to one extremum of the zero isocline $\dot{u} = 0$ on the (u, v) plane while both other critical points should be near the other extremum, what poses strict restrictions on parameters. The phase portrait of zero isoclines of equations (4) is shown in Fig. 5(a). Dispersion curves for the model (4), linearized near the stationary point responsible for pseudo-Turing instability, are presented in Fig. 5(b).

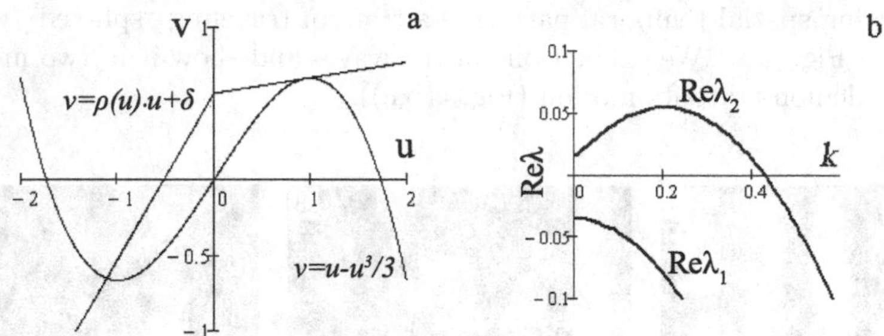

Fig. 5. Zero isoclines (a) and dispersion curves (b) for the model (4). The parameters used: $\varepsilon = 0.09,\ \delta = 0.568,\ \rho_1 = 1.1,\ \rho_2 = 0.1016,\ D = 1.2$.

A segmented wave formed in the model (4) is shown in Fig. 6. It should be noted that the wave is short-living and in the course of time splits into segments and disappears.

5 Discussion

In the present paper we considered three possible mechanisms for formation of segmented waves. Patterns of this kind were discovered in spatially-distributed Belousov-Zhabotinsky reaction dispersed in a water-in-oil aerosol OT microemulsion.

The first mechanism is caused by interaction of two coupled subsystems, one of which is excitable, and the other one has Turing instability depending on the

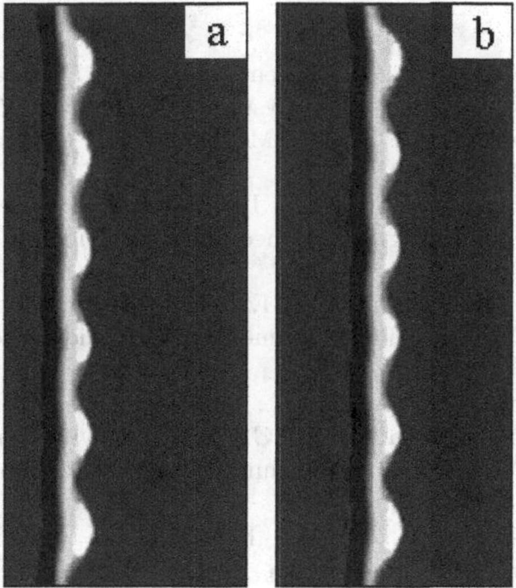

Fig. 6. A segmented wave in the FitzHugh-Nagumo model (4) at the moments of time: (a) $t = 50$, (b) $t = 60$. The parameters used: $\varepsilon = 0.09$, $\delta = 0.568$, $\rho_1 = 1.1$, $\rho_2 = 0.1016$, $D = 1.2$. Size of the domain: 65×200.

parameters. It is shown that, segmented spirals evolve from ordinary smooth spirals as a result of the transverse Turing instability. For the second mechanism we suggest "splitting" of the traveling wave in the vicinity of the bifurcation point of codimension-2, where the boundaries of the Turing and wave instabilities intersect. Finally we show that segmented waves can emerge in two-component reaction-diffusion models having more than one steady state, particularly in a FitzHugh-Nagumo model.

In our opinion the most probable candidate for explanation of the observed patterns is the first mechanism. First of all, it is the only mechanism which is robust and is realized in a broad range of parameters. The system can be either near bifurcation (as it takes place in the case of "splitting" of a wave near codimension-2 bifurcation point) or far from it. Besides, patterns that arise according to this mechanism are stable and long-living, contrary to the segmented waves that are formed due to the third mechanism. Finally, this mechanism depending on the parameters demonstrates a great variety of segmented waves differing in shape and size of segments.

Acknowledgments. This work was partly supported by grant from RFBR No. 14-01-00196.

References

1. Zaikin, A.N., Zhabotinsky, A.M.: Concentration Wave Propagation in Two Dimensional Liquid–Phase Self–Oscillating System. Nature 225, 535–537 (1970)
2. Turring, A.M.: The Chemical Basis of Morphogenesis. Philos. Trans. R. Soc. London, Ser. B 237, 37–72 (1952)
3. Castets, V., Dulos, E., Boissonade, J., De Kepper, P.: Experimental Evidence of a Sustained Standing Turing-type Nonequilibrium Chemical Pattern. Phys. Rev. Lett. 64, 2953 (1990)
4. Lengyel, I., Rabai, G., Epstein, I.R.: Experimental and Modeling Study of Oscillations in the Chlorine Dioxide-Iodine-Malonic Acid Reaction. J. Am. Chem. Soc. 112, 9104–9110 (1990)
5. Orban, M., De Kepper, P., Epstein, I.R.: An Iodine-Free Chlorite-Based Oscillator: The Chlorite-Thiosulfate Reaction in a C.S.T.R. J. Phys. Chem. 86, 431–432 (1982)
6. Mikhailov, A.S., Ertl, G.: Nonequilibrium Structures in Condensed Systems. Science 272, 1596–1597 (1996)
7. Vanag, V.K., Epstein, I.R.: Pattern Formation in a Tunable Medium: The Belousov-Zhabotinsky Reaction in an Aerosol OT Microemulsion. Phys. Rev. Lett. 87, 228301 (2001)
8. Vanag, V.K.: Waves and Patterns in Reaction-Diffusion Systems. Belousov–Zhabotinsky Reaction in Water-in-Oil Microemulsions. Phys. Usp. 47, 923–941 (2004)
9. Vanag, V.K., Epstein, I.R.: Dash Waves in a Reaction-Diffusion System. Phys. Rev. Lett. 90, 098301 (2003)
10. Vanag, V.K., Epstein, I.R.: Segmented Spiral Waves in a Reaction-Diffusion System. Proc. Natl. Acad. Sci. USA 100, 14635–14638 (2003)
11. Carballido-Landeira, J., Berenstein, I., Taboada, P., Mosquera, V., Vanag, V.K., Epstein, I.R., Perez-Villar, V., Munuzuri, A.P.: Long-Lasting Dashed Waves in a Reactive Microemulsion. Phys. Chem. Chem. Phys. 10, 1094–1096 (2008)
12. Yang, L., Berenstein, I., Epstein, I.R.: Segmented Waves from a Spatiotemporal Transverse Wave Instability. Phys. Rev. Lett. 95, 038303 (2005)
13. Rossi, F., Liveri, M.L.T.: Chemical Self-Organization in Self-Assembling Biomimetic Systems. Ecological Modelling 220, 1857–1864 (2009)
14. Rossi, F., Budroni, M.A., Marchettini, N., Carballido-Landeira, J.: Segmented Waves in a Reaction-Diffusion-Convection System. Chaos 22, 037109 (2012)
15. Borina, M., Yu., P.A.A.: Diffusion Instability in a Three Component Model of the Reaction-Diffusion Type. Kompjuternye Issledovanija i Modelirovanie 3, 135–146 (2011) (in Russian)
16. Nicola, E.M.: Interfaces between Competing Patterns in Reaction-diffusion Systems with Nonlocal Coupling. Dissertation, Dresden (2001)

Single Optical Fiber Probe Spectroscopy for Detection of Dysplastic Tissues

Murat Canpolat[1], Tuba Denkçeken[1], Ayşe Akman-Karakaş[2], Erkan Alpsoy[2], Recai Tuncer[3], Mahmut Akyüz[3], Mehmet Baykara[4], Selçuk Yücel[4], Ibrahim Başsorgun[5], M. Akif Çiftçioğlu[5], Güzide Ayşe Gökhan[5], Elif İnanç-Gürer[5], Elif Peştereli[5], and Şeyda Karaveli[5]

[1] Biomedical Optic Research Unit, Department of Biophysics, Faculty of Medicine, Akdeniz University, Antalya, Turkey
[2] Department of Dermatology and Venereology, Faculty of Medicine, Akdeniz University, Antalya, Turkey
[3] Department of Neurosurgery, Faculty of Medicine, Akdeniz University, Antalya, Turkey
[4] Department of Urology, Faculty of Medicine, Akdeniz University, Antalya, Turkey
[5] Department of Pathology, Faculty of Medicine, Akdeniz University, Antalya, Turkey

Abstract. A single optical fiber probe with a core diameter of 100 μm was used to deliver visible light to tissue and from tissue to a spectrometer to diagnose cancerous tissues of different organs. Measurements were performed on brain, skin, cervix and prostate tumor specimens and surrounding normal tissues ex-vivo. The spectral slopes were positive for normal tissues and negative for the tumors. Signs of the spectral slopes were used as a discrimination parameter to differentiate tumor from normal tissues for the four organ tissues. Sensitivity and specificity of the system in differentiation between tumors from normal tissues were 93% and %100 for brain, 87% and 85% for skin, 93.7% and 46.1% for cervix and 98% and 100% for prostate respectively.

Keywords: Single-scattering spectroscopy, Cancer, Diagnosis, Non-invasive, Real-time.

1 Introduction

Incidence of cancer increases worldwide. Non-invasive and real-time cancer diagnosis is a key factor for use a medical device for screening proposes and early diagnosis. Early diagnosis of cancerous tissue reduces mortality significantly. Most of the cancer incidences start at mucosal cell layers of body cavity. Therefore, diagnosing cancerous cells in mucosal layers may increase probability of early diagnosis.

Several groups have used visible or near infrared (VIS-NIR) spectroscopy to diagnose cancerous tissues [1,2]. In diagnosis of cancerous tissue, employing a spatially resolved steady-state diffuse reflectance VIS-NIR spectroscopy [3], provide information about scattering and absorption coefficients of tissue. Reflectance spectroscopy has also been used to diagnose cancerous tissues by several groups [4-6]. In reflectance spectroscopy studies, difference between the spectra acquired from non-cancerous and cancerous tissues were used for the diagnosis.

V.M. Mladenov and P.C. Ivanov (Eds.): NDES 2014, CCIS 438, pp. 349–354, 2014.

Angular distribution of single scattered laser light from normal and cancerous cell in saline has been measured using a goniometer system and showed that angular distribution of the single scattered (phase function) light for cancerous cells is different than normal cells. The reason of the difference between the phase functions are morphological alteration of cancerous cells such as increased ratio of nuclei to cell volume [7]. Then, spectrum of single-scattering light from back reflection geometry from in-vivo measurement was acquired using and optical fiber probe consists of two adjacent optical fibers were used as source and detector. Diffuse component of the back reflected light estimated using diffusion approximation and subtracted from the detected signal in order to get singly scattered component of the spectrum [8]. Changes in the polarization of the back reflected light was also used to detect single scattered light rather than diffused light to diagnose cancerous tissues [9]. The developed methods were employed to diagnose cancerous tissues on different organs successfully.

In order to get information about microstructure of turbid medium such as tissue, single optical fiber probe was used to deliver visible light to and from turbid media VIS-NIR light was delivered to tissue phantoms using a single fiber optical fiber probe and back reflected light was detected with the same fiber. Since core diameter of the optical fiber is 100 μm, probability of detection of multiply scattered photons by the same fiber is small, therefore most of the detect light are singly scattered. It has been shown by observing Mie oscillation on the spectrum acquired from mono-sized polystyrene microspheres. The modality has been named elastic light single-scattering spectroscopy (ELSSS) [10].

In the presented study, ELSSS was used to detect cancerous tissues from brain, skin, cervix and prostate ex-vivo. ELSSS measurements were performed both in-vivo and ex-vivo on skin tissues. Spectroscopic measurements on brain, cervix and prostate tissues were performed ex-vivo.

2 Material and Methods

2.1 Spectroscopic System

ELSSS of tissues were acquired using a system consisting of a spectrometer (USB2000 with OOIBase32TM Platinum Spectrometer Operating Software, Ocean Optics, Tampa, FL), a tungsten halogen white light source, a single-fiber optical probe, and a laptop computer. The single-fiber optical probe was used for both delivery and detection of white light to and from the tissue as illustrated in Fig. 1. The single-fiber optical probe was a 1x2 fiber optical coupler with a splitting ratio of 50%. One proximal end of the coupler was connected to the light source and the other was connected to the spectrometer. The diameter of the distal end of the probe's fiber was 100 μm with an NA of 0.22; which was used to gently touch the tissue samples during the experiments. A detailed description of the system is given elsewhere [10]. The measured spectra were corrected for the wavelength dependence of system components and specular reflection. The corrected spectrum is

$$R(\lambda) = \frac{R(\lambda)_s - R(\lambda)_{bg}}{R(\lambda)_c - R(\lambda)_{bg}} \tag{1}$$

Where, $R(\lambda)_s$ is a spectrum of the tissue, $R(\lambda)_c$ is a spectrum of Spectralon (Labsphere, Inc.) in water and $R(\lambda)_{bg}$ is a background spectrum taken from pure water in a black container. Spectral data were processed in real time and the corrected spectra were visualized on the computer screen. The single- fiber optical probe was used to record a spectrum from 10% polystyrene micro-spheres with a diameter of 2 0.02 μm dispersed in pure water to test whether the probe could detect singly-scattered photons from the tissue phantom. The observation of Mie oscillation from the spectrum of mono-dispersed micro-spheres confirmed that the ELSSS system detects single-scattered photons from turbid media rather than diffused photons.

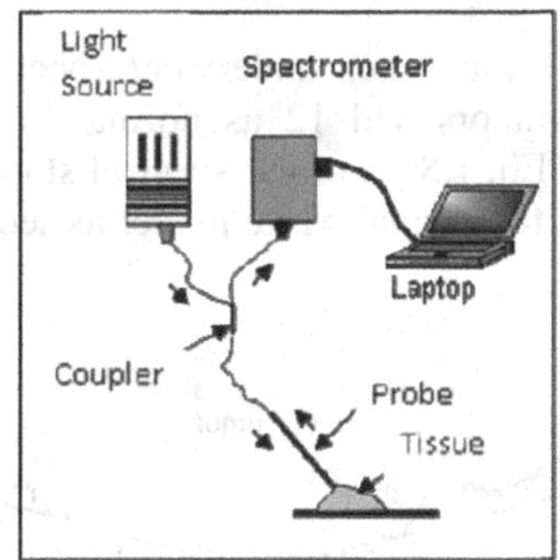

Fig. 1. Elastic light scattering system

2.2 Patients

The clinical studies were conducted at Akdeniz University Hospital with the approval of the Akdeniz University Institutional Review Board.

ELSSS spectra were acquired from 41 brain tissue specimens ex-vivo. After the excision, each tissue sample was placed on a black plastic sheet to prevent back-reflection from the holder during the spectral data acquisition. Then, all the tissue samples were sent for histopaology examination [11].

In diagnosis of skin cancer, 23 consecutive patients with 33 lesions were enrolled in the study [12]. ELSSS data were acquired before and after the excision of the lesions. Before the excision, spectra were acquired on the lesions and normal appearing tissue approximately 1 cm away from the lesions. The spectra acquired from the normal appearing skin tissue were used as a control group.

In detection of cervical cancer, the study was conducted on two groups of the patients. In the first group, after the colposcopical examination, surgical biopsy was performed and samples of one to four pathologic colposcopic locations were removed from the cervix for permanent pathology examination. ELSSS spectra were acquired from 95 cervix tissue specimens of 60 patients. In the second group, just after hysterectomy, ELSSS spectra were acquired on the transformation zone of 10 normal cervix tissues to obtain average spectra of normal cervical tissue to form a negative control group [13].

Patients undergoing open radical prostatectomy at Akdeniz University Urology Department were recruited for the study. The prostate tissues of 12 patients serially sectioned at 3-mm intervals in a plane perpendicular to the prostatic urethra, and each of them is placed in separate labeled small pathology cassettes to prepare the tissue samples for the pathologic examination. Before the pathologic examination, ELSSS data were acquired on different locations of 108 prostate specimens in the pathology cassettes by gently placing the single-fiber optical probe on each prostate specimen in the pathology cassettes. Then, the pathology cassettes were transported to the histopathology room. Elastic light single-scattering spectra and pathology results were compared to find a correlation between the spectra and the tissue pathologies.

3 Results

In the brain tumor study, 41 total sites underwent spectroscopic analysis, 29 were classified histologically as tumors and 12 as normal brain tissue. All the ELSSS spectra were corrected using Eq. 1.Sign of the spectral slopes in the wavelength range of 450-750 nm were positive for normal white matter as seen in Fig. 2.

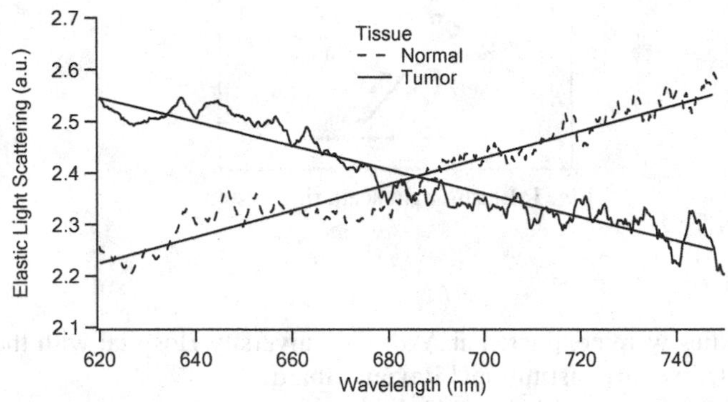

Fig. 2. Spectral slope is positive for normal and negative for tumor tissues.

Sign of the spectral slopes were used as a discrimination parameter between normal and tumor brain tissues. Discrimination of normal white matter brain tissue from tumors using the sign of the spectral slopes results in a sensitivity of 93% and a specificity of 100%, but the ELSSS system is unable to differentiate grey matter from tumors based on the sign of the spectral slopes.

In skin malignancy diagnosis, total 28 lesions underwent spectroscopic analysis. Histopathological examination classified 9 as basal cell carcinoma (BCC), 4 as melanoma, 2 as squamous cell carcinoma (SCC), and 13 as benign skin tissues. Average slopes of the spectra taken on BCC, Melanoma, SCC and benign tissue were calculated. There was a significant difference between the corrected spectra of the malignant (BCC, SSC, and melanoma) and benign skin tissues. Spectral slopes of malign tissues were negative and benign tissue specimens were positive. Discrimination of benign tissue from malignant tissue using the sign of the spectral slopes results in a sensitivity of 87% and a specificity of 85%.

Histopathology classification of the cervical tissue was following; 43 cervicities, 36 low-grade squamous intraepithelial lesions (LSIL), 16 high-grade squamous intraepithelial lesions (HSIL) and 10 normal. ELSSS defined 15 out of the 16 HSIL biopsy samples as positive and only one of them was misclassified as non-HSIL.

Ten spectral measurements in the negative control group were taken on normal smear just after hysterectomy and all of them were classified as normal by the ELSSS. LSIL were accepted as normal and 10 of them were classified as normal and 26 as HSIL of the total 36 tissue samples. Spectral measurements were taken on 43 cervicitis tissues and 21 of them were classified as negative and 22 as HSIL. Based on the above results, sensitivity and specificity of the system are 92.5% and 47.5% respectively in differentiation HSIL tissues from other pathologies and normal cervix tissues.

In differentiation of cancerous prostate tissues from non-cancerous by ELSSS system, all 22 benign prostate tissues correctly defined as normal and 84 tumor tissue specimens defined as tumor out of 86. Only two tumor tissue specimens misclassified as normal. Sensitivity and specificity of ELSSS system in differentiation cancerous from normal prostate tissue are 98% and 100% respectively.

4 Discussion

Light scatters in a medium due to heterogeneity in the index of refraction of the medium. The index of refraction for cell membranes is greater than the index of refraction for extra-cellular liquid, which causes scattering of light at cell membrane. Light also scatters inside a cell at the nucleus, mitochondria and other organelles due to difference in the index of refraction between intracellular compartments and the surrounding cytoplasm. There are several techniques to detect single-scattering component of back reflected light from a turbid medium. One of the techniques is using two adjacent optical fibers, one of the fibers is used to deliver light to tissue and other one is used to detect back reflected light [8]. Diffuse component of the back reflected light is modeled and subtracted from the detected signal in order to obtain single-scattering component of the spectrum. In the modeling, absorption and scattering coefficients should be known. Estimating the coefficients requires additional measurements. Another technique to obtain single-scattering component of the backreflected light is polarization light scattering spectroscopy [9]. The system set up is complicated and expensive.

Single fiber optical probe with a core diameter of 100 μm was used to deliver white light to and from tissue. Single optical fiber probe mostly detects singly scattered light from tissue rather than diffused light. Since nuclei of cancerous cells are larger than nuclei of non-cancerous cells, spectra were taken on cancerous tissues are different than spectra of non-cancerous tissues. As seen in Fig.2, signs of the spectral slope are positive for normal tissues and negative for cancerous tissues. Hence, sign of the spectral slopes were used as a diagnostic parameter in differentiation between cancerous and non-cancerous tissues. First time single optical fiber probe was used to detect single-scattering spectrum of light from tissues and sign of the spectral slopes in the wavelength range of 620-750 nm were used as a diagnostic parameter to detect cancerous tissues.

Each spectrum of the ELSSS system provides information about the tissue under the optical fiber probe. Where, interrogated area of the tissue is limited by surface area of the optical fiber probe in the each measurement. Average propped depth of the tissue is 500 μm. Sign of the spectral slopes were used as a parameter to discriminate benign tissue from malign tissue. Therefore, ELSSS is not an imaging system as a confocal microscope or optical coherence tomography (OCT) and only can be used to differentiate between benign and malign tissue.

The single fiber optic probe ELSSS system has the potential to differentiate between malign and benign lesions in-vivo and real-time. The ELSSS system may allow surgeons to surgical margins in real-time either on the biopsy specimen or in the surgical cavity. The ELSSS system cannot replace pathology, because it does not provide information about tissue pathology. It can be used only to differentiate malign tissue from benign tissue in real time. Therefore, it has potential to be used as a device in operating room to detect positive surgical margins in real time.

References

1. Marks, R.: An overview of skin cancers. Incidence and causation. Cancer 75, 607–612 (1995)
2. http://www.nccc-online.org/
3. Farell, T.J., Patterson, M.S.: A diffusion theory model of spatially resolved, steady-state diffuse reflectance for the noninvasive determination of tissue optical properties. Med. Phys. 19, 879–888 (1992)
4. A'amar, O.M., Ley, R.D., Bigio, I.J.: Comparison between ultraviolet-visible and near-infrared elastic scattering spectroscopy of chemically induced melanomas in an animal model. J. Biomed. Opt. 9(6), 1320–1326 (2004)
5. Mirabal, N.Y., Chang, S.K., Atkinson, E.N., et al.: Reflectance spectroscopy for in vivo detection of cervical precancer. J. Biomed. Opt. 7, 587–594 (2009)
6. Drezek, R., Guillaud, M., Collier, T., et al.: Light scattering from cervical cells throughout neoplastic preogression: Influence of nuclear morphology, DNA content, and chromatin texture. J. Biomed. Opt. 8, 7–16 (2003)
7. Mourant, J.R., Freyer, J.P., Hielsher, A.H., et al.: Mechanism of light scattering from biological cells relevant to noninvasive optical-tissue diagnostics. Appl. Opt. 37(16), 3586–3593 (1998)
8. Perelman, L.T., Backman, V., Wallace, M., et al.: Observation of periodic Fine Structure in Reflectance from Biological Tissue: A New Technique for Measuring Nuclear size distribution. Phys. Rev. Lett. 80, 627–630 (1998)
9. Backman, V., Gurjar, R.: Polarized Light Scattering Spectroscopy for Quantitative Measurement of Epithelial Cellular Structures in Situ. IEEE J. Sel. Top. Quantum Electron 5, 1019–1026 (1999)
10. Canpolat, M., Mourant, J.R.: Particle Size Analysis of Turbid Media with a Single Optical Fiber in Contact with the Medium to Deliver and Detect white Light. Appl. Opt. 40(22), 3792–3799 (2001)
11. Canpolat, M., Akyüz, M., Gökhan, G.A., et al.: Intra –operative brain tumor detection using elastic light single-scattering spectroscopy: A feasibility study. J. Biomed. Optic 14(5), 054021 (2009)
12. Canpolat, M., Akman-Karakas, A., Gokhan-Ocak, G.A., et al.: Diagnosis and Demarcation of Skin Malignancy Using Elastic Light Single-Scattering Spectroscopy: A Pilot Study. Dermatologic Surgery 38(2), 215–223 (2012)
13. Denkçeken, T., Şimşek, T., Erdoğan, G., et al.: Elastic light single-scattering spectroscopy for the detection of cervical precancerous Ex-vivo. IEEE Trans. Biomed. Eng. 60(1), 123–127 (2013)

A Hierarchical Coding-Window Model of Parkinson's Disease

Andres Daniela Sabrina[1,2,3,*], Gomez Florian[1], Cerquetti Daniel[2], Merello Marcelo[2], and Stoop Ruedi[1]

[1] Institute of Neuroinformatics, ETH and UZH Zurich, Winterthurerstrasse 190, 8057, Zurich, Switzerland

[2] Institute for Neurological Research Raul Carrea, Fleni Institute, Movement Disorders Section, Buenos Aires, Argentine

[3] Society in Science, The Branco-Weiss Fellowship, administered by ETH, Zurich, Switzerland

dandres@ini.uzh.ch

Abstract. Parkinson's disease is an ongoing challenge to theoretical neuroscience and to medical treatment. During the evolution of the disease, neurodegeneration leads to physiological and anatomical changes that affect the neuronal discharge of the Basal Ganglia to an extent that impairs normal behavioral patterns. To investigate this problem, single Globus Pallidus pars interna (GPi) neurons of the 6-OHDA rat model of Parkinson's disease were extracellularly recorded at different degrees of alertness and compared to non-Parkinson control neurons. A structure function analysis of these data revealed that the temporal range of rate-coded information in GPi was substantially reduced in the Parkinson animal-model, suggesting that a dominance of small neighborhood dynamics could be the hallmark of Parkinson's disease. A mathematical-model of the GPi circuit, where the small neighborhood coupling is expressed in terms of a diffusion constant, corroborates this interpretation.

Keywords: Parkinson's disease, neuronal code, structure function, diffusive coupling.

For the coding of neuronal information, the precise relative timing of neuronal firing within the neuronal ensemble is an attractive scheme. However, also the average discharge rate of neurons or of neuronal ensembles has its advantage: clearly, such a code would be more robust, in particular with respect to noise that we find ubiquitous in the cortex [1, 2]. The disadvantage of rate coding is a smaller maximal information content and an increased latency. The discharge time-window across which a rate is calculated has necessarily two limiting time scales: a time beyond which further spikes no longer modify the rate-coded information (long-term correlations, however, might be present nonetheless) and a short time scale given by the finite resolution beyond which two spikes can no longer be distinguished from a single one. As different environmental conditions will induce different information scales on the neural system, it can be expected that different coding systems co-exist, but have their distinguished relevance for different time-scales.

* Corresponding author.

V.M. Mladenov and P.C. Ivanov (Eds.): NDES 2014, CCIS 438, pp. 355–362, 2014.

Parkinson's disease (PD) is characterized by a conflict between different intrinsic neuronal firing time scales [3]. Identifying different states of a system according to its different scaling behaviors can be a key to the study of complex systems based on time series analysis [4]. To this goal, different approaches have been successfully applied. For example, the evaluation of multifractal properties has been used in the cardiovascular system, where it allowed distinguishing between healthy and pathological conditions [5]. Another well-tested approach is the structure function analysis [6-8]. Comparing structure functions with autocorrelation analysis, both offer essentially the same information, but structure functions have been shown to be more robust to the presence of drift, low frequency noise and short time series [9]. Also, the advantage of studying a family of structure functions is that geometrical properties of the neuronal discharge could be captured in terms of corresponding scaling exponents [10]. We apply the structure function approach to interspike interval (ISI) time series, where $I(j)$ is the jth interspike interval and $\Delta I(\tau) = I(j+\tau) - I(j)$ is the difference between successive intervals, separated by an index increment $\tau \in N+$. The structure function $S_q(\tau)$ is defined as

$$S_q(\tau) = \left\langle \left| \Delta I(\tau)^q \right| \right\rangle \tag{1}$$

where $\langle\cdot\rangle$ accounts for the statistical average over the time series and q is a real number. The scaling behavior of S_q is then characterized by the power law relationship

$$S_q(\tau) \sim \tau^{\zeta(q)} . \tag{2}$$

For a stationary process with independent increments $\zeta(q) = 0$, which expresses that the mean correlation between successive events does not depend on the event index [11]. Monofractal, non-intermittent time series imply $\zeta(q) = const$, whereas multifractal behavior is characterized by $\zeta(q)'' < 0$. The zero-slope regime ($\zeta(q) = 0$) of the structure function is of particular interest, since it marks the temporal scale across which only random processes are at work. For neuronal signals, this regime precludes coding schemes other than a rate code.

We applied the structure function analysis to data from single neurons of the Globus Pallidus pars interna (GPi). Spontaneous neuronal firing was recorded in an animal model of PD (6-hydroxydopamine - 6OHDA - partial lesion model of Sprague-Dawley adult rats) and a control group. Experiments were performed at two conditions, firstly with the animals under deep chloral-hydrate anesthesia and following at full alertness, both in relaxed, head-restrained conditions. We analyzed 52 neuronal recordings in total, 27 of the PD group (13 under anesthesia and 14 in the awake state) and 25 of the control group (13 under anesthesia, 12 awake). Details of the procedures of the animal experiments and recording technique can be found in the supplementary material [12]. Signals were processed off-line; spikes were extracted and classified using standard spike-sorting [13]. Raw neuronal recordings were around 60 minutes long and the time series extracted included roughly 10000 events (14731 ± 7210 ISI, mean ± SD). Interspike intervals (ISI) time series were constructed and structure functions of increasing order were calculated for each time series. For the structure function analysis, we varied τ between 1 and 2000. For simplicity, q was restricted to the set {1,2,...,6}. The behavior of the structure functions was checked to be consistent for $1 \leq q \leq 6$ and since they show a low dependence on order $q \leq 6$, the detailed analysis will be restricted to order one.

In the majority of the neurons studied, three distinct regimes of the temporal structure function emerge (separated by index-points s_1 and s_2, cf. Fig. 1). Roughly, breakpoint s_1 is defined as the locus where the initially ascending behavior changes abruptly into an essentially constant state, the end of which is indicated by breakpoint s_2. More precisely, for the location of the breakpoints, $\log(S_q(\tau))$ was plotted vs. $\log(\tau)$ and a first-order function was fitted to it. The main part of regime II was obtained by maximizing the range over which the fitted function would have absolute slope less than exp (-3), with data scattered around it having a standard error of the same size. This area was then extended to meet with the monotonous behavior manifestly present at smaller and larger scales. In only 7 out of the 52 cases studied (13%), regime I presented a descending instead of an ascending behavior, meaning that there aren't any small-scale processes with memory affecting the time series at a scale larger than unity (ISI). Such variants were also encountered during our model simulations, and thus appear to describe a natural phenomenon (see below).In the case of the long-term correlations regime III, an ascending behavior was found in 40 out of 52 neurons studied (77%). This regime reflects the presence of memory processes at long time-scales. However, its study is limited by the finite length of the time series. Therefore, a descending behavior of regime III could reflect the presence of higher-scale oscillations, and the distinction between the implications of an ascending or a descending behavior of regime III is less clear.

Functionally, regime II indicates the window where temporal rate-coding is expected. This regime is substantially reduced in the PD group at all alertness levels. In 9 out of 27 recorded PD neurons, a shortened regime II was found, while in the remaining cases regime II had fully disappeared (see for example Fig. 1, panel B). The interpretation that the disappeared windows are extreme cases of shortened windows was corroborated by a subsequent simulation study (see below). To translate our findings from the index space underlying the structure function to the time-axis, we multiplied the indices s_1 and s_2 by the average ISI duration of the time series, from which we obtained the corresponding characteristic time-breakpoints t_1 and t_2. Using time as the abscissa, under anesthesia, PD neurons show a substantially increased t_1 (on average 38 sec. vs. 1 sec. on average, p=0.07, two sample t-test). At full alertness, t_2 is more than eight times smaller in PD neurons compared to the control group (4 sec. vs. 35 sec. on average, p=0.01). Regarding the anaesthesia-alertness transition, whereas for the control group it is characterized by the increase of t_2 (from 8 to 35 sec., p=0.01), the extension of the long-range correlation regime III is the main feature in the PD group.

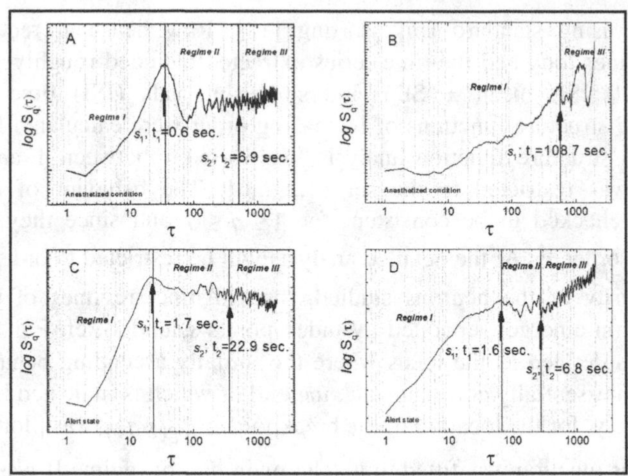

Fig. 1. Order-1 structure functions of the four experimental groups (characteristic examples). A), B) Deep anesthesia condition, control and PD neuron, respectively. C), D) Full alertness condition, control and PD neuron, respectively. Smoothing over 20 data points was applied to the original data. Arrows: times t_1 and t_2 corresponding to breakpoints *s1* and *s2*. Control neurons show a clear zero-slope scale-range (regime II) at both alertness conditions, prolonged after the transition to full alertness (larger t_2). PD neurons not always show such a region. Under deep anesthesia, PD neurons have an increased t_1 compared to the control group, with a temporal structure disruption at long scales. At full alertness PD neurons show an extended long-scale correlations window (regime III).

Our interpretation of these findings in the network context is the following. Under deep anesthesia, the most marked difference between PD and control neurons is a significant prolongation of regime I of the temporal structure function. Since in this state external input is minimum, the observed differences between the two groups can be attributed to a differing small-range network structure, playing a stronger role in PD than in the control group. To corroborate this interpretation, we performed a simulation study. In this network, the small-range interactions were modeled by a diffusive coupling among the GPi cells. For the comparison of the PD vs. the control case, the identical network architecture was used. To recover the experimentally measured data, in addition to a stronger diffusion, also differing excitatory / inhibitory input levels were chosen (see below), in agreement with the hallmark of PD in the BG network [3]. In the simulation, each neuron had the form (Rulkov [14])

$$x_{i,n+1} = f(x_{i,n}, y_{i,n} + \beta_{i,n}) \tag{3}$$

$$y_{i,n} = y_{i,n} - \mu(x_{i,n} + 1) + \mu\sigma + \mu\sigma_{i,n} \,, \tag{4}$$

where the index n indicates the iteration step, and where function f is given by

$$f(x_n, y) = \begin{cases} \alpha/(1-x_n)+y, & x_n \le 0 \\ \alpha + y, & 0 < x_n < \alpha + y \quad and \quad x_{n-1} \le 0 \\ -1, & x_n \ge \alpha + y \quad or \quad x_{n-1} > 0. \end{cases} \tag{5}$$

External input is modeled by

$$\sigma = \sigma_u + I \ , \tag{6}$$

where σ_u represents the initial excitability of each isolated neuron. In total, 101 GPi neurons were implemented, aligned on a ring structure. The Subthalamic (STN) and Striatal (Str) inputs to the GPi are modeled as excitatory and inhibitory inputs respectively, and the spatial distribution of both inputs is close to the available histological data [15]: Input to GPi is mediated by 101 STN axons, each of which sends collaterals to 10 neighboring cells using identical synaptic weights ($w_{STN\text{-}GPi}$=0.1). Str input to the GPi is mediated by 101 axons producing 10 collaterals each: one central connection to a GPi neuron with a high synaptic weight ($w_{Str\text{-}GPi\text{-}I}$=0.9) and 9 connections to adjacent cells with a lower weight ($w_{Str\text{-}GPi\text{-}II}$=0.01). Inputs were modeled by uniformly distributed random numbers from the unit interval, multiplied by the amplitude A_e, for excitatory input or by amplitude A_i, for inhibitory input, respectively. The distinction between anesthetized and alert conditis is modeled by changed input amplitudes. Whereas the anesthetized condition was modeled by A_e=1.5/25 and A_i=-1.2/-24.5, the alert condition was modeled by A_e=2/50 and A_i=-1.5/-48.5 (control / PD cases, respectively). In this way, the two conditions were characterized by slightly different A_e/A_i ratios of 1.25/1.02 (anesthesia) and 1.33/1.03 (alertness) for the control / PD case, respectively. These values were chosen in order to reproduce best our experimental measurements, but they also have physiological justification (cf. [3] and references contained).

The coupling from each neuron i to its neighbors is described by the following equations (cf. [14], where we set β_e and σ_e to 1):

$$\beta_{i,n} = g_{ji}\beta^e(x_{j,n} - x_{i,n}) \tag{7}$$

$$\sigma_{i,n} = g_{ji}\sigma^e(x_{j,n} - x_{i,n}) \ . \tag{8}$$

The local dependence of the coupling on the neighbor order j is implemented by

$$g_{ij} = \frac{D}{|(i - j)|^2} \ . \tag{9}$$

For every neuron, the parameter values α=4.5 and μ=0.001 were used. To account for variability in initial neuronal excitation, σ_u was drawn uniformly from [0.05, 0.15]. The value of α was chosen to reproduce our experimental burst length, and the slight variations in σ_u helped including in the model the subtle transitions from a bursting to a spiking regime that were observed during our experiments (for a detailed description of the parameter space of Rulkov neurons, see ref. [14]; for details on the bursting behavior of our measured neurons, see ref. [12]). As was stated above, the main difference between PD and control cases was a different choice of diffusion strength (*D*=0.3 in the PD vs. *D*=0.01 in the control case), in addition to the different input levels implemented. Once a stationary regime had been achieved, the system was iterated for 180,000 time steps and spikes were extracted to obtain ISI time series. On these data, the same structure function analysis as for the experimental data was applied.

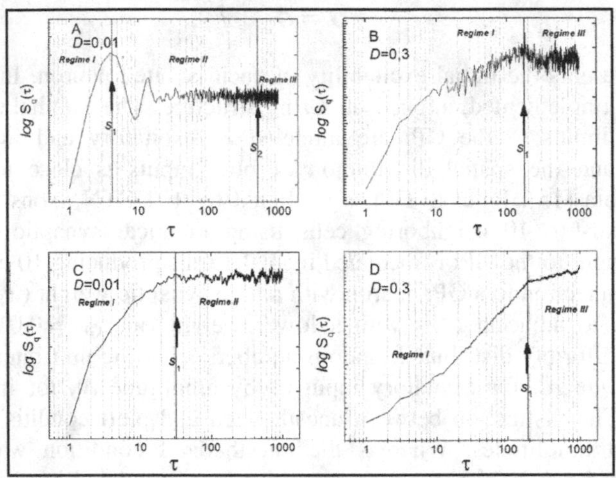

Fig. 2. Modeling results (characteristic examples, corresponding to Fig. 2). The main qualitative features measured are reproduced by changing the strength of diffusive coupling and the excitatory/inhibitory input levels (see text).

The simulations that are fully based on the available physiological data, show substantial agreement with our experiments (Fig. 2). The variants of behavior observed in our experiments were also encountered in our model. In the model, not all neurons evolve in exactly the same subset of the parameter space, because of the initial variability introduced. This fact explains the slight differences observed. Regarding the duration of regime I, in our simulations s_1 depends on two variables: the frequency of discharge has a direct relation with s_1, while the coupling strength *D* shows an inverse relationship with it. In this context, we propose that the increased coupling strength may act as an adaptive response to the primarily increased frequency in the parkinsonian GPi. From this, our conclusion is that diffusive coupling is a key feature shaping the temporal structure of the neuronal discharge, and that its pathologic increase might lead to the abnormalities observed in PD. Moreover, we propose that this increase could explain the excessive correlation found in the BG [16-20]. Experimental evidence from other works supports this interpretation: under PD, the coupling between adjacent Striatal cells has been shown to be significantly increased [21-22]; a similar effect might also take place in the GPi.

At all alertness levels, the rate-coding window is extremely shortened for PD neurons. Under anesthesia, small-range dynamics dominates the neuronal firing in

PD, while at full alertness, almost all the PD dynamics is captured by long-term correlations. This makes the PD system extremely sensitive to noise, since all fluctuations of the environmental input on temporal scales longer than a characteristic time (our time scale t_2) trigger long-time correlations in the system. In the control group, on the contrary, the transition to full alertness is characterized by an extension of the rate-coding window. This is a cell-correlate of de-synchronizing neural populations reported from electroencephalographic studies during this transition. In this way, healthy neurons are robust to noise over larger time- and frequency-scales than PD neurons, and this difference is further augmented towards full alertness.

Our findings are relevant to the application of deep brain stimulation (DBS) therapy. DBS consists in chronically stimulating a selected BG target with a train of high frequency electric pulses, which may successfully alleviate the symptoms of PD and also of other neurological conditions, such as dystonia, Tourette-syndrome and essential tremor. However, the mechanism of action of DBS has remained to be unknown, and therefore we currently lack a rational basis for the selection of the optimal stimulation parameters (train frequency, pulse amplitude and width) and the specific stimulation target (GPi and STN are preferred ones). On this basis, the outcome of DBS therapy is hard to predict in individual patients. Understanding the mechanisms that underlie the coding of information in the BG and how they are altered by PD is a key issue for understanding how PD neurons respond to DBS. For the first time we showed that several coding-schemes with characteristic time-scales are present in the BG, and that they are altered in PD due to excessively strong diffusive coupling (for experimental evidence from human PD patients at full alertness, cf. [23]). We suggest that successful DBS not only breaks the pathologic correlations between PD neurons, but also restores the rate-coding window at the proper time-scale, which in turn re-establishes a robust coding of information in the BG.

Acknowledgments. We acknowledge the support by the technical personnel at the laboratories of the Center for Applied Neurological Research, Fleni Institute, Buenos Aires, Argentina.

References

1. Stein, R.B., Roderich Gossen, E.R., Jones, K.E.: Neuronal variability: Noise or part of the signal? Nat. Rev. Neurosci. 6, 389–397 (2005)
2. Shadlen, M.N., Newsome, W.T.: Noise, neural codes and cortical organization. Curr. Opin. Neurobiol. 4(4), 569–579 (1994)
3. Obeso, J.A., Rodríguez-Oroz, M.C., Rodríguez, M., Lanciego, J.L., Artieda, J., Gonzalo, N., et al.: Pathophysiology of the basal ganglia in Parkinson's disease. Trends Neurosci. 23(suppl.), S8–S19 (2000)
4. Hu, K., Ivanov, P.C., Chen, Z., Carpena, P., Stanley, E.: Effect of trends on detrended fluctuation analysis. Phys Rev. E 64(1), 011114 (2001)
5. Ivanov, P.C., Nunes Amaral, L.A., Goldberger, A.L., Havlin, S., Rosenblum, M.G., Struzik, Z.R., Stanley, E.: Multifractality in human heartbeat dynamics. Nature 399(6735), 461–465 (1999)

6. Romano, G.P., Antonia, R.A., Zhou, T.: Evaluation of LDA temporal and spatial velocity structure functions in a low Reynolds number turbulent channel flow. Exp. Fluids 27, 368–377 (1999)
7. Frisch, U.: Turbulence: The legacy of A.N. Kolmogorov. University Press, Cambridge (1996)
8. Benzi, R., Ciliberto, S., Tripiccione, R., Baudet, C., Massaioli, F., Succi, S.: Extended self-similarity in turbulent flows. Phys. Rev. E 48(1), R29–R32 (1993)
9. Schuls-DuBois, E.O., Rehberg, I.: Structure function in Lieu of correlation function. Appl. Phys. 24, 323–329 (1981)
10. Lin, D.C., Hughson, R.L.: Modeling heart rate variability in healthy humans: A turbulence analogy. Phys. Rev. Lett. 86, 1650–1653 (2000)
11. Vainshtein, S.I., Sreenivasan, K.R., Pierrehumbert, R.T., Kashyap, V., Juneja, A.: Scaling exponents for turbulence and other random processes and their relationships with multifractal structure. Phys. Rev. E 50(3), 1823–1835 (1994)
12. Andres, D.S., Gomez, F., Cerquetti, D., Merello, M., Stoop, R.: A hierarchical coding-window model of Parkinson's disease (2014), `http://arxiv.org/abs/1307.6028`
13. Quian Quiroga, R., Nadasdy, Z., Ben-Shaul, Y.: Unsupervised spike detection and sorting with wavelets and superparamagnetic clustering. Neural Comput. 16, 1661–1687 (2004)
14. Rulkov, N.F.: Modeling of spiking-bursting neural behavior using two-dimensional map. Phys. Rev. E 65, 041922 (2002)
15. Difiglia, M., Rafols, J.A.: Synaptic organization of the Globus Pallidus. J. Electron Microsc. Tech. 10, 247–263 (1988)
16. Gatev, P., Darbin, O., Wichmann, T.: Oscillations in the basal ganglia under normal conditions and in movement disorders. Movement Disord 21(10), 1566–1577 (2010)
17. Dostrovsky, J., Bergman, H.: Oscillatory activity in the basal ganglia – relationship to normal physiology and pathophysiology. Brain 127, 721–722 (2004)
18. Weinberger, M., Mahant, N., Hutchison, W.D., Lozano, A.M., Moro, E., Hodaie, M., Lang, A.E., Dostrovsky, J.O.: Beta oscillatory activity in the subthalamic nucleus and its relation to dopaminergic response in Parkinson's disease. J. Neurophysiol. 96, 3248–3256 (2006)
19. Levy, R., Dostrovsky, J.O., Lang, A.E., Sime, E., Hutchison, W.D., Lozano, A.M.: Effects of Apomorphine on subthalamic nucleus and Globus Pallidus Internus neurons in patients with Parkinson's disease. J. Neurophysiol. 86, 249–260 (2001)
20. Kühn, A.A., Kempf, F., Brücke, C., Doyle, L.G., Martinez-Torres, I., et al.: High-frequency stimulation of the subthalamic nucleus suppresses oscillatory b activity in patients with Parkinson's disease in parallel with improvement in motor performance. J. Neurosci. 28(24), 6165–6173 (2008)
21. Cepeda, C., Walsh, J.P., Hull, C.D., Howard, S.G., Buchwald, N.A., Levine, M.S.: Dye-coupling in the Neostriatum of the rat: I. Modulation by dopamine-depleting lesions. Synapse 4, 229–237 (1989)
22. Onn, S.P., Grace, A.A.: Dye-coupling between rat striatal neurons recorded in vivo: Compartmental organization and modulation by dopamine. J. Neurophysiol. 71(5), 1917–1934 (1994)
23. Andres, D.S., Cerquetti, D.F., Merello, M.: Turbulence in Globus pallidum neurons in patients with Parkinson's disease: Exponential decay of the power spectrum. J. Neurosci. Meth. 197(1), 14–20 (2011)

Using Coherence for Robust Online Brain-Computer Interface (BCI) Control

Martin Spüler[1], Wolfgang Rosenstiel[1], and Martin Bogdan[1,2]

[1] Wilhelm-Schickard-Institute for Computer Science, University of Tübingen, Germany

[2] Computer Engineering, University of Leipzig, Germany

Abstract. A Brain-Computer Interface (BCI) enables the user to control a computer by brain activity only. In this paper we investigated the use of different brain connectivity methods to control a Magnetoencephalography (MEG)-based Brain-Computer Interface (BCI). We compared the use of coherence, phase synchronisation and a widely used method for spectral power estimation and found coherence to be a more robust feature extraction method, when using the BCI over a longer time interval across sessions. To validate these results we implemented an online BCI system using coherence and could show that coherence also performed more robust in an online setting than traditional methods.

Keywords: Brain connectivity; Brain-Computer Interface (BCI); Coherence; Magnetoencephalogrpahy (MEG); non-stationarity.

1 Introduction

A Brain-Computer Interface (BCI) provides a user with the means to control a computer by pure brain activity [1]. Its main purpose is to restore communication in people who have lost voluntary muscle control due to neurodegenerative diseases or traumatic brain injuries. The basic principle of a BCI relies on the user being able to voluntarily alter his brain activity. These changes in the recorded brain activity can be detected and used as an input signal for a computer. There are different signal acquisition techniques that allow to measure the brain activity of a user. While Electroencephalography (EEG) is the most commonly used method for recording brain activity, we focus on Magnetoencephalography (MEG) in this paper. While MEG has been shown to work well with BCI [2], it is rarely used, which may be attributed to the lack of portability and its high costs. On the bright side, MEG offers a higher spatial resolution and more information in the higher frequency range above 40 Hz compared to EEG [3], since the magnetic field is not distorted by skull and scalp.

Regardless of the underlying recording technique, the power spectrum is the feature that is most commonly used in motor imagery BCIs [4]. If a user imagines a one-sided hand movement, a power decrease in the mu rhythm (8-13 Hz) can be detected over the contralateral motor cortex. Methods for estimating brain connectivity like coherence or phase synchronisation have been shown as

V.M. Mladenov and P.C. Ivanov (Eds.): NDES 2014, CCIS 438, pp. 363–370, 2014.

alternatives for detection of motor imagery [5]. While the power spectrum only yields information about the local brain activity, connectivity methods allow to capture the dynamics between different brain regions and thereby may give additional information that could be used to detect the user's intention.

That connectivity methods can also be used as a feature extraction method for classification in BCI has been shown in several publications with EEG [6,7,8,9], but so far only Bensch et al. [10] have investigated the use of connectivity methods for MEG-based BCIs.

More importantly, none of them has addressed the robustness of those connectivity methods for feature extraction. The robustness of the features used is important because of the non-stationary nature of the recorded brain signals, which is still a critical issue in current state-of-the-art Brain-Computer Interfacing [11]. This non-stationarity especially is a problem when a classifier trained on data of a previous session is used for classification in a current session, which is often referred to as the session-transfer problem. While adaptation is one way to counter non-stationarity [12], the use of more robust features is another way to alleviate the problem of non-stationarity. In this paper we evaluate connectivity methods like coherence and Phase Locking Value and show how robust they are across sessions in an online MEG-based BCI.

2 Methods

In this chapter, we will first describe the three different methods for feature extraction that were evaluated in this paper. Afterwards we will describe the data and methods we used for the offline analysis of MEG data and the online experiment.

2.1 Feature Extraction Methods

To better classify the brain signals, feature extraction methods can be used to extract specific features of the signals, which allow for a better classification.

In the following, we introduce three different feature extraction methods, that were used and evaluated in this paper. In the brackets, the corresponding abbreviation is given, which is used later in this paper to identify the different methods.

Power Spectral Density Estimated by Autoregressive Model (AR). Estimating the power spectrum by means of an autoregressive model is a commonly used method for feature extraction in EEG- and MEG-based BCIs [2]. In this paper we used the Burg algorithm [13] for estimating the coefficients for the autoregressive model with a model order of 16.

Coherence (COH). Given two signals $s_x(t)$ and $s_y(t)$, the coherence [7] can be derived from the cross-spectrum $S_{xy}(f)$ of the two time-series:

$$S_{xy}(f) = \langle X_x(f) X_y^*(f) \rangle \tag{1}$$

with $X_x(f)$ being the Fourier transform of $s_x(t)$ and $X_x^*(f)$ being the complex conjugate of the Fourier transform $s_x(t)$. The expectation operator is denoted by $\langle\rangle$. The complex coherence can be obtained by normalizing the cross-spectrum with the two spectra of the corresponding signals:

$$C_{xy}(f) = \frac{S_{xy}(f)}{\sqrt{S_{xx}(f)S_{yy}(f)}} \tag{2}$$

By taking the absolute value of the complex coherence, the coherence can be calculated:

$$Coh_{xy}(f) = |C_{xy}(f)| \tag{3}$$

To obtain a scalar value, the average coherence in a specified frequency band or the whole frequency range can be calculated. The result is a value between 0 and 1 with 0 meaning that the frequency components of both signals are not correlated.

Phase Locking Value (PLV). Given two signals $s_x(t)$, $s_y(t)$ and their corresponding phases $\varphi_x(t)$, $\varphi_y(t)$ the Phase Locking Value (PLV) [7] describes the stability between $\varphi_x(t)$ and $\varphi_y(t)$. It can be computed across subsequent time samples with:

$$PLV = |\langle e^{j(\varphi_x(t)-\varphi_y(t))}\rangle| \tag{4}$$

where $\langle\rangle$ denotes the expectation operator. Therewith the PLV equals the absolute mean over all $e^{j(\varphi_x(t)-\varphi_y(t))}$ in one window.

For a randomly distributed phase difference (between $[0, 2\pi]$), the PLV will be 0, while it will be 1 for a constant phase difference (phase synchronisation). The Hilbert transform can be employed to compute the instantaneous phase $\varphi(t)$. For computation of the PLV we used the improved algorithm published in [6].

2.2 Offline Analysis of MEG Data

Data and Tasks. To evaluate the use of the different methods, we performed an offline analysis on data recorded for a previous study [14]. In this study 10 subjects performed different mental tasks in two sessions. In each session 51 trials were recorded per task. Recording was done with a 275-channel whole-head MEG-system (VSM MedTech Ltd.) at a sampling rate of 586 Hz. Since it was shown in [14] that *right hand* and *subtraction* were the two mental tasks that could be classified with the highest accuracy we concentrated on data from these two tasks. In the task *right hand* the subject had to imagine a right hand movement and in the *subtraction* task the subject had to do subtractions by choosing a random number (around 100) and subtract 7 repeatedly until the end of the trial. Each trial lasted 4.05 seconds with about 6 seconds of break between the trials. Instructions were given on a screen and a fixation cross was displayed during trials to minimize eye movement.

Signal Processing and Classification. Before classification, the signals were filtered and resampled to 200 Hz. For spatial filtering a small Laplacian derivation was applied. To reduce the number of channels we only used the 185 inner channels, since the outer channels are supposed to hold little task-related information and are more influenced by possible artifacts. After the preprocessing, we used the three different feature extraction methods AR, COH and PLV as previously described. To reduce the computational load in favor of later online applications of the BCI, we wanted to use the same default parameters for each session and each subject. To estimate the best default parameters for each feature extraction method we tested different frequency ranges by cross-validation and selected the frequency range that gave the best accuracy for each method. For the AR method we found the range of 1 to 40 Hz in 2 Hz bins to yield the highest accuracies, which is why AR was used with this frequency range. For coherence the frequency range of 2.5 Hz to 15 Hz was chosen, while the frequency range was not limited for PLV. Before classification we did a feature selection based on r^2-values [15] and used the 1000 features with the highest r^2-values for classification. For classification we chose a Support Vector Machine (SVM) [16], using the LibSVM [17] implementation with standard parameters (RBF-Kernel, $C = 1$, $\gamma = 0.001$). SVM was used for classification, since it was shown to be superior to other classification methods on BCI data [18].

To estimate classification accuracies, we used two different approaches. The first one was a a 100x2-fold cross-validation where the data from session 1 and session 2 are mixed together. The data is then randomly permuted and partitioned in 2 blocks with equal size. Each block is used for training the classifier once and tested on the other block. This procedure is repeated 100 times with different permutations and the total accuracies are averaged.

For the second approach we used the first session for training the classifier and the second session to test the classifier and evaluate its accuracy (referred to as S1S2-validation later). Since the data is randomly mixed for the cross-validation, the influence of non-stationarity on the classification accuracy is very small. The contrary is the case for the S1S2-validation. Thereby the comparison of both validation-techniques gives us a method to assess the robustness of the tested feature extraction methods. The number of 2 blocks for the cross-validation was chosen to have exactly the same amount of data in the training and testset as we have in the S1S2-validation.

2.3 Online Experiment with MEG

To validate the results from the offline analysis of the MEG data (will be shown in the results section), we performed an online experiment, where we recruited 10 subjects (mean age 28.2 ±2.6, 5 female, 5 male) with only 2 of them having previous experience with a motor imagery BCI. The study was approved by the local ethics committee of the Medical Faculty at the University of Tübingen and written consent was obtained from all subjects. Each subject participated in two sessions. The first session was used to record training data without feedback. The second session was used for testing and feedback was given to the subject.

Recording was done with a 275-channel whole-head MEG-system (VSM MedTech Ltd.) at a sampling rate of 586 Hz. During measurement, the head position relative to the MEG sensors was continuously monitored and saved along with the MEG data. Due to a slightly different technical setup we had to use a different block size than during the recording of the data used for the offline analysis. This resulted in slightly different time intervals for trials and breaks. First a cue was given, what task should be performed next. Then a symbol was displayed to indicate that the trial starts in less than a second. During the trial, the subject should perform the mental task and a fixation cross was displayed to the subject to minimize eye movements. After the trial a 3 second break followed. If feedback was given online, this was done after the trial in the first 2 seconds of the relax period. The classification for the online feedback was done using coherence as feature with the same preprocessing and classification methods as described previously. Similar to the offline analysis, only the 185 inner channels were used in the online experiment. The calculation of the different feature exctraction methods for 1000 features and 185 channels took on average 103 ms for AR, 123 ms for COH and 394 ms for PLV.

To compare the online results using COH, we also did a simulated online experiment with AR and PLV as feature. To appropriately simulate the online case, all parameters and data were exactly the same as in the online experiment, except the different feature extraction methods that were tested. For comparability, we used the same default parameters as for the previous offline analysis.

3 Results

3.1 Offline Analysis of MEG Data

The results for the 100x2 cross-validation are shown in table 1. When using AR as feature, an average accuracy of 87.3 % (±7.6 %) was achieved. When using COH and PLV, an average accuracy of 87.0 % (±9.7 %) and 87.4 % (±8.3 %), respectively, was achieved.

In contrast to the cross-validation, COH and PLV perform better than AR in the S1S2-validation, in which the classifier was trained with data from session 1 and tested with data from session 2 (see table 1). While an average accuracy of 79.2 % (±9.9 %) was reached when using AR, using COH and PLV for classification resulted in an average accuracy of 84.8 % (±9.5 %) and 84.2 % (±7.1 %), respectively.

3.2 Online Experiment with MEG

The results from the online experiment as well as the simulations using AR are shown in table 2. At first it is notable, that subjects BI and BJ were not able to control the BCI and achieved an accuracy around or below chance level (50 %). During the online experiment, in which COH was used for feature extraction, an average accuracy of 73.0 % (±16.4 %) was achieved. When subjects BI and BJ

Table 1. Classification accuracies with different feature extraction methods obtained by a 100x2-fold cross-validation (on the left) and when using session 1 for training and session 2 for testing (S1S2-validation; on the right)

100x2-fold cross-validation				S1S2-validation			
subject	AR	COH	PLV	subject	AR	COH	PLV
AA	87.2 %	75.9 %	75.4 %	AA	79.4 %	74.5 %	74.5 %
AB	83.6 %	90.5 %	90.9 %	AB	82.3 %	87.2 %	88.2 %
AC	87.4 %	88.2 %	86.8 %	AC	74.5 %	91.1 %	84.3 %
AD	86.1 %	71.0 %	79.1 %	AD	83.3 %	69.6 %	83.3 %
AE	89.7 %	93.2 %	92.7 %	AE	68.6 %	88.2 %	78.4 %
AF	84.0 %	91.8 %	90.3 %	AF	64.7 %	81.3 %	87.2 %
AG	93.0 %	93.0 %	92.3 %	AG	93.1 %	93.1 %	88.2 %
AH	96.2 %	97.4 %	97.2 %	AH	76.4 %	97.0 %	90.2 %
AI	69.6 %	73.7 %	74.2 %	AI	73.5 %	73.5 %	72.5 %
AJ	95.8 %	95.1 %	94.9 %	AJ	96.0 %	92.1 %	95.1 %
mean	87.3 %	87.0 %	87.4 %	mean	79.2 %	84.8 %	84.2 %

are disregarded, due to their inability to control the BCI, an average accuracy of 79.3 % (±10.8 %) was achieved for COH while AR resulted in an average accuracy of 70.8 % (±12.9 %).

When pooling the results of the 10 subjects from the offline analysis with the results of the 10 subjects from the online MEG experiment, a one-sided Wilcoxon signed rank test shows that COH and PLV both perform significantly better ($p < 0.05$) than AR.

Table 2. Online accuracies with coherence (COH) as feature, simulated online accuracies with power spectrum (AR) and PLV as feature and days between session 1 and session 2. Since subjects BI and BJ did not achieve significant BCI control, also the average accuracy without these subjects is given.

subject	AR	COH	PLV	days between S1 and S2
BA	82.5 %	88.0 %	82.5 %	5
BB	51.0 %	87.5 %	71.0 %	2
BC	84.0 %	87.5 %	90.5 %	4
BD	80.0 %	88.0 %	81.0 %	6
BE	52.5 %	76.0 %	63.5 %	5
BF	76.5 %	62.5 %	71.0 %	6
BG	67.5 %	64.0 %	61.0 %	2
BH	72.5 %	81.0 %	69.0 %	1
BI	50.0 %	52.0 %	59.0 %	1
BJ	50.0 %	43.5 %	46.5 %	0
mean	66.7 %	73.0 %	69.5 %	3.2
mean (without BI, BJ)	70.8 %	79.3 %	73.7 %	3.9

4 Discussion and Conclusion

This study aimed at evaluating the robustness of different feature extraction methods in light of the session-transfer problem. To do this, we analyzed MEG data offline using a cross-validation and a S1S2-validation. The difference between the results obtained by cross-validation and the results obtained by S1S2-validation gives an estimate of the performance-drop due to the session-transfer. Because of the session-transfer, there is a covariate shift [19] of the data during the S1S2-validation, but not during the cross-validation. Therefore, one can use the difference between the cross-validation results and the results obtained by S1S2-validation as a measure for the robustness of a feature in regard of the session-transfer problem. Using this method to assess the robustness of the feature extraction methods, we could show in an offline analysis on MEG data that the connectivity measures coherence (COH) and PLV seem to be more robust features than the power spectrum (AR).

To validate these results, we performed an online experiment, where COH was used as feature for online feedback. Using the data from the online experiment, we also simulated the online scenario with AR and PLV as feature extraction methods and could validate the results from the offline analysis that connectivity measures are more robust features on MEG data.

For the online experiment it is also noticeable that 2 subjects did achieve online accuracies below 60 %, which is not sufficient for communication. This finding agrees with earlier results by Vidaurre et al. [20], who state that BCI control does not work for 15 % to 30 % of the subjects.

As a conclusion, we have shown coherence and PLV to be more robust features in an MEG-based BCI, that help to alleviate the session-transfer problem. We have also demonstrated that coherence can be utilized for online BCI control. Further studies are needed to also evaluate the robustness of those brain connectivity methods for EEG-based BCI and to investigate the reasons for connectivity methods being affected less by non-stationarity.

Acknowledgments. This study was granted by the *Deutsche Forschungsgemeinschaft* (DFG, Grant RO 1030/15-1, KOMEG).

References

1. Wolpaw, J.R., Birbaumer, N., McFarland, D.J., Pfurtscheller, G., Vaughan, T.M.: Brain-computer interfaces for communication and control. Clinical Neurophysiology 113(6), 767–791 (2002)
2. Mellinger, J., Schalk, G., Braun, C., Preissl, H., Rosenstiel, W., Birbaumer, N., Kübler, A.: An MEG-based brain-computer interface (BCI). NeuroImage 36(3), 581–593 (2007)
3. Kaiser, J., Walker, F., Leiberg, S., Lutzenberger, W.: Cortical oscillatory activity during spatial echoic memory. European Journal of Neuroscience 21(2), 587–590 (2005)

4. Pfurtscheller, G., Neuper, C.: Motor imagery and direct brain-computer communication. Proceedings of the IEEE 89(7), 1123–1134 (2001)
5. Spiegler, A., Graimann, B., Pfurtscheller, G.: Phase coupling between different motor areas during tongue-movement imagery. Neuroscience Letters 369(1), 50–54 (2004)
6. Gysels, E., Celka, P.: Phase synchronization for the recognition of mental tasks in a brain-computer interface. IEEE Transactions on Neural Systems and Rehabilitation Engineering 12(4), 406–415 (2004)
7. Brunner, C., Scherer, R., Graimann, B., Supp, G., Pfurtscheller, G.: Online Control of a Brain-Computer Interface Using Phase Synchronization. IEEE Transactions on Biomedical Engineering 53(12), 2501–2506 (2006)
8. Wei, Q., Wang, Y., Gao, X., Gao, S.: Amplitude and phase coupling measures for feature extraction in an EEG-based brain-computer interface. Journal of Neural Engineering 4(2), 120 (2007)
9. Hamner, B., Leeb, R., Tavella, M., del Millan, J.R.: Phase-based features for motor imagery brain-computer interfaces. In: Conf. Proc. IEEE Eng. Med. Biol. Soc., pp. 2578–2581 (2011)
10. Bensch, M., Bogdan, M., Rosenstiel, W.: Phase Synchronization in MEG for Brain-Computer Interfaces. In: Proceedings of the 3rd Int. Brain-Computer Interface Workshop, Graz, pp. 18–19 (September 2006)
11. Krusienski, D.J., Grosse-Wentrup, M., Galn, F., Coyle, D., Miller, K.J., Forney, E., Anderson, C.W.: Critical issues in state-of-the-art brain-computer interface signal processing. Journal of Neural Engineering 8(2), 025002 (2011)
12. Spüler, M., Rosenstiel, W., Bogdan, M.: Principal component based covariate shift adaption to reduce non-stationarity in a meg-based brain-computer interface. EURASIP Journal on Advances in Signal Processing 2012(1), 129 (2012)
13. Priestley, M.B.: Spectral analysis and time series. Academic Press, London (1981)
14. Bensch, M., Mellinger, J., Bogdan, M., Rosenstiel, W.: A multiclass BCI using MEG. In: Proceedings of the 4th Int. Brain-Computer Interface Workshop, Graz, pp. 191–196 (September 2008)
15. Spüler, M., Rosenstiel, W., Bogdan, M.: A fast feature selection method for high-dimensional MEG BCI data. In: Proceedings of the 5th Int. Brain-Computer Interface Conference, Graz, pp. 24–27 (September 2011)
16. Vapnik, V.N.: Statistical Learning Theory, 1st edn. Wiley-Interscience (September 1998)
17. Chang, C.-C., Lin, C.-J.: LIBSVM: A library for support vector machines (2001), Software available at `http://www.csie.ntu.edu.tw/~cjlin/libsvm`
18. Lotte, F., Congedo, M., Lécuyer, A., Lamarche, F., Arnaldi, B., et al.: A review of classification algorithms for eeg-based brain-computer interfaces. Journal of Neural Engineering, 4 (2007)
19. Sugiyama, M., Krauledat, M., Müller, K.-R.: Covariate shift adaptation by importance weighted cross validation. J. Mach. Learn. Res. 8, 985–1005 (2007)
20. Vidaurre, C., Blankertz, B.: Towards a cure for bci illiteracy. Brain Topography 23, 194–198 (2010), doi:10.1007/s10548-009-0121-6

Applying Nonlinear Techniques for an Automatic Speech Recognition System

Daniela Şchiopu

Petroleum-Gas University of Ploieşti, Department of Informatics, Information Technology, Mathematics and Physics,
Blvd. Bucureşti 39, 100680 Ploieşti, Romania
daniela_schiopu@yahoo.com

Abstract. The variability of the speech signal is a challenging problem which the field of automatic speech recognition (ASR) has to deal with. And when the speech is continuous and the goal is the execution of the uttered commands by a machine, the provocation is more significantly. This paper evaluates the use of different nonlinear techniques for improving performance on a given task. An overview of the field of ASR and main techniques is given, with an emphasis on practical implementation issues.

Keywords: Automatic speech recognition, hidden Markov models, control system, statistical methods.

1 Introduction

The goal of getting a machine to understand fluent spoken speech, and in particular to recognize the spoken words uttered by a human and execute them, has been pursuing by the researchers for more than 60 years. And although major advances have been made in this area, the same challenges still remain, because of the problematic of the speech itself. Some of the factors that make automatic speech recognition quite difficult are: natural speech is continuously and sometime may contain disfluencies, different pronunciations of the same speaker in different context, the effect of coarticulation, large vocabularies are often confusable, recorded speech is variable over surroundings' acoustics, background noise, microphone and channel characteristics, speech recognition task. All these factors make speech recognition by machine to be a challenging problem.

In spite of the variability of speech, the advantages of automatic speech recognition are tremendous: cost reduction (replacing humans performing different tasks with automated machines is one of the earliest goals for speech recognition systems), new opportunities (speech recognition enables applications like automatic call processing, embedded control systems, translation and dictation, e-learning).

The progress in automatic speech recognition has made for different spoken languages (English, French, Japanese, German, Swedish, Spanish, etc.) and the technologies and methods used for this task were discussed in the literature (see e.g. [1], [2], [3], [4], [5]). Important achievements in the development of automatic speech

V.M. Mladenov and P.C. Ivanov (Eds.): NDES 2014, CCIS 438, pp. 371–378, 2014.

recognition systems were made for the Romanian language too (recent research can be found in [6], [7], [8], [9]). The study described in this paper is focused on building an automatic speech recognition system for continuous Romanian speech, applying different nonlinear techniques.

The paper is organized as follows. Section 2 presents the field of automatic speech recognition and reviews the main techniques used in the study. Section 3 describes how these techniques were implemented in a proposed speech recognition system and the results obtained. The last section is dedicated to the conclusions of this paper.

2 Overview of the ASR Framework

2.1 The General Architecture of an ASR System

Speech recognition, known as automatic speech recognition or spoken language recognition or speech-to-text conversion is essentially the ability of machine to identify as input a message (acoustical signal) uttered by a human, the goal being the obtaining of a string of words as output.

Speech is a highly variable signal, characterized by many parameters. But the sources of speech variability are manifold. The speech signal doesn't transmit only the message (linguistic information), but information about the speaker himself: age, gender, regional and social origin, emotional state and even his identity. Speaker uniqueness results from a complex combination of physiological and cultural aspects.

The goal of the field of ASR is to develop techniques and systems, so the performance of the machine ability to be improved, to reach that of humans and to allow the speech to become a viable computer interaction.

Fig. 1 shows a proposed general representation of an ASR system (used in Section 3). The main steps followed by a recognizer are:

(1) signal processing and feature extraction;
(2) the pattern recognition or the decoder, which might have three components: acoustic model, phonetic model and language model.

Speech recognition process consists of two main parts:

- the recognition is made based on acoustical parameters extracted from the speech signal and not using the uttered message directly;
- the recognition is realized based on the three models developed independently.

The role of the acoustic model is to estimate the likelihood of uttering a speech message, given a sequence of words. The acoustic units used by this model can be words (especially when it is an isolated spoken word recognizer) or sub-lexical units, such as phonemes or senones [10]. In order to implement the acoustic model, there are used hidden Markov models (HMMs) [11].

The language model is designed to estimate the probabilities of all sequences of words in the search space, helping the system to decide for the sequence with the highest probability (whether that sequence is a valid sentence in the language).

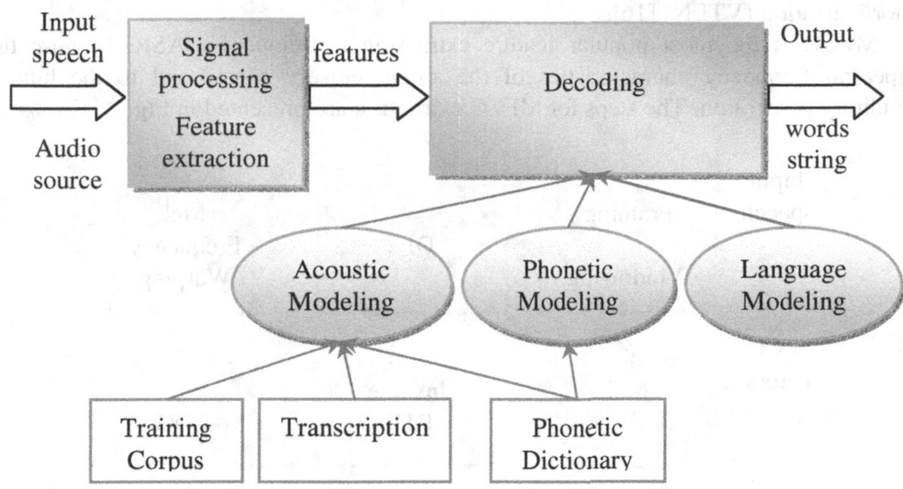

Fig. 1. The general architecture of the proposed recognizer

In continuous speech recognition systems with large vocabulary, there are used statistical language models, *n*-gram models. They are built using large text corpora adapted to the domain of the spoken messages which must be decoded.

In continuous or isolated word ASR systems with reduced vocabulary, there are used finite state grammars as language models.

The role of phonetic model is to connect acoustic model (which models the generation of the sounds) to the language model (which models how the words succeed). The phonetic model might be a phonetic dictionary (pronunciation dictionary), associating the appropriate sequence of phonemes to each word from the vocabulary (how to pronounce that word). Phonetic dictionary contains the correspondence between writing and phonetic form of words from a certain language.

2.2 Speech Analysis Techniques

The non-stationary nature of the speech signal requires definitions of its parameters, based on its local properties. Estimation of these parameters (features) can be done with short-time analysis methods. These features can have applications starting with the delimitations of the portions corresponding to the speech signal (energy and power) till recognition itself.

Feature extraction has an important role in the speech recognition process, because this module must identify the characteristics of the uttered message from the others. A vast range of feature extraction techniques for the speech recognition task are successfully used, some of these techniques being: mel-frequency cepstral coefficients (MFCC) [12], linear predictive coding coefficients (LPC) [13], perceptual linear prediction coefficients (PLP) [14], as well as normalizations via cepstral mean substraction (CMS) [15], relative spectral filtering (RASTA), vocal tract length normalization (VTLN) [16].

MFCC is the most popular feature extraction technique for ASR, because the spectral frequency characteristics of the signal closely correspond to the human auditory perception. The steps for MFCC extraction are presented in Fig. 2 [17].

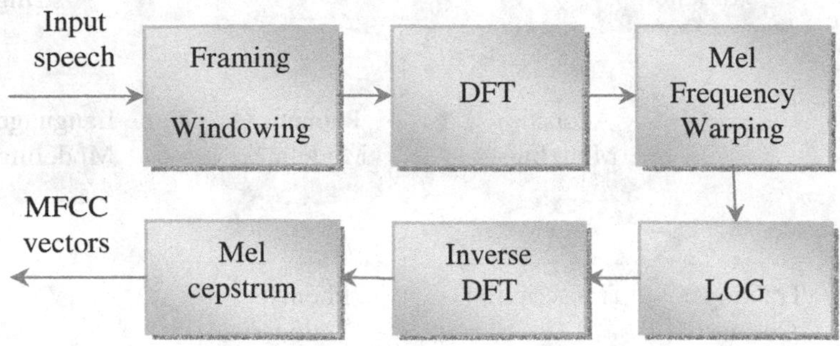

Fig. 2. MFCC feature extraction

2.3 Techniques Used in ASR

There are three approaches of ASR techniques classification: acoustic-phonetic methods, pattern recognition methods and knowledge-based methods.

The first category of methods consists of the following steps: spectral analysis of speech, segmentation and labeling the speech, resulting in a phoneme lattice, and determination whether a certain word or a sequence of words is valid according to the first two steps.

The pattern recognition methods include template based approach and stochastic approach, although the last one is the most popular and suitable. Many methods from this category are used, such as hidden Markov models (HMMs) [11], support vector machines (SVM), dynamic time warping (DTW), vector quantization (VQ).

Hidden Markov models were described for the first time in 1972 by Baum [18], subsequently they have been expanded to speech recognition, and thus they became the predominant approach in the area of ASR, because of their ability to manage the variability of speech signal. A HMM is a temporal stochastic model, in which the state of changing process is described by a single discrete random variable. The possible values of the variable are the possible states of the modeled world. HMMs underlie the acoustic modeling, which may express the transformation of speech into text as follows: given an observation sequence *O*, find the words sequence *W* which has the maximum probability *P*(*W*|*O*), as in (1):

$$\hat{\mathrm{W}} = \underset{W}{\arg\max} P(W \mid O) = \underset{W}{\arg\max} \frac{P(W)P(O \mid W)}{P(O)} = \underset{W}{\arg\max} P(W)P(O \mid W)\,. \quad (1)$$

Knowledge-based methods for ASR include artificial neural networks (ANNs). ANNs consists of a large number of simple processing units (neurons), which send each other's performance through a network of excitatory weights. ANNs provide different types of learning methods, like supervised, unsupervised or reinforced learning.

3 Experimental System

In this section, we present an ASR system based on HMMs and MFCC. The general architecture is the one proposed in Section 2.

The goal of the system is the uttering a few commands (Romanian words) to a robot (machine) for executing them. Thus, the recognizer must handle the following words: *start* (start), *stop* (stop), *întoarce* (turn), *roteşte* (rotate), *stânga* (left), *dreapta* (right), *treizeci* (30), *şaizeci* (60), *nouăzeci* (90), *de* (of), *grade* (degrees), *sus* (up), *jos* (down), *prinde* (catch), *lasă* (leave). The selection of these words was not random. We chose these vocabulary as every word means an action which a robot (for example, a robotic arm) must execute it. The words *start* and *stop* were chosen to the detriment of other two Romanian words (*porneşte* and *opreşte*), because we observed in another experiments [19] that these words are acoustically very similar and for this reason the word *opreşte* was replaced very often with the other one.

Example of typical inputs might be: "*start întoarce stânga treizeci de grade stop*" ("start turn left 30 degrees stop"), where *start* and *stop* mark the beginning and the end of the action of the robot. The speech is continuous, with very short pauses between the words. Because these sequences of words might be used for controlling a robot, we were interested in the situation that a command is misunderstood.

As in general architecture (Fig. 1), we build the training corpus and the textual transcriptions. For this purpose, we used voices of several speakers (11 females and 6 males speakers), each of them uttered a sequence of words one or many times.

Every speech data was converted into a set of feature vectors. For each frame, we extracted 39 coefficients: 12 MFCC parameters along with the first and the second derivatives and their log-energy.

For each of the acoustic events (the words from the vocabulary and silence – *sil*), we defined an HMM. The topology of the five HMMs is presented in Fig. 3.

We built a grammar definition language for specifying a simple task grammar. This grammar is a finite state grammar and consists of a set of productions (rules)

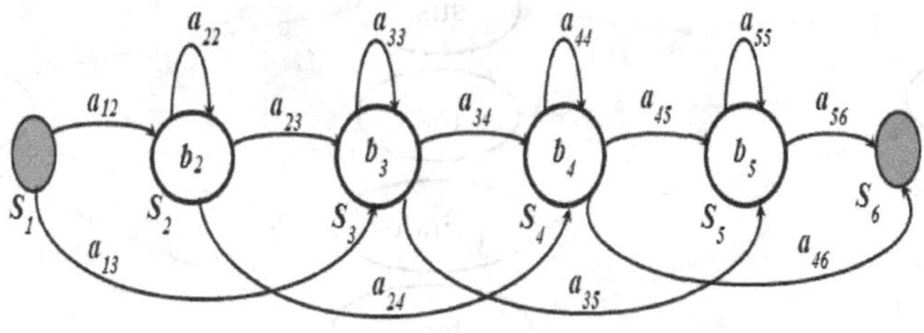

Fig. 3. HMM topology

for the syntactic variables (non-terminals) followed by a regular expression describing the words to recognize.

The grammar in a Chomsky form [22] might be:

```
S • <start> Action <stop>
Action • <sus> | <jos> | <prinde> | <lasa> | Turn
<de> <grade> | Rotate <de> <grade>
Turn • <intoarce> Left_Right Number
Rotate •<roteste> Left_Right Number
Left_Right • <stanga> | <dreapta>
Number • <treizeci> | <saizeci> |<nouazeci>
```

where the non-terminals are: *S, Action, Turn, Rotate, Left_Right* and *Number*, and the terminals are the words which the system has to recognize them. This grammar is a context-free grammar (CFG).

The grammar in HTK (Hidden Markov Model Toolkit) [21] form, it might be:

```
$action1 = sus | jos | prinde | lasa;
$action2 = intoarce | roteste ;
$action3 = stanga | dreapta ;
$action4 = treizeci | saizeci | nouazeci ;
({START_SIL} [ start <$action1> stop | <$action2>
<$action3> <$action4> de grade stop ] {END_SIL})
```

In Fig. 4 we present a flow diagram for this grammar.

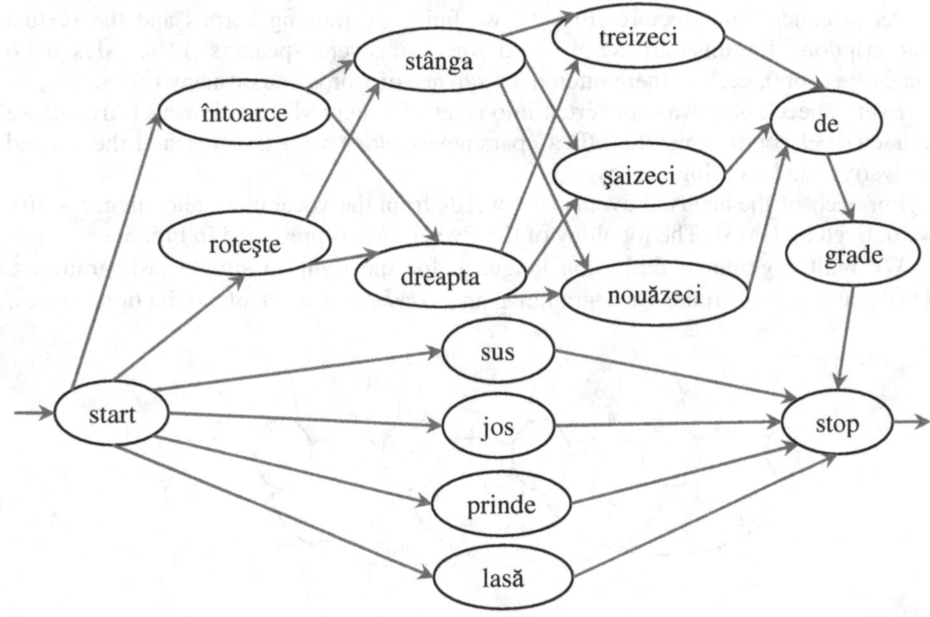

Fig. 4. The flow diagram for the grammar

A HMM (considered equivalent to a stochastic regular grammar) can be modeled more efficiently by a stochastic context-free grammar (SCFG) [23]. A SCFG is a CFG which assigned a probability to each production. In our case, we considered the grammar as a SCFG with the same probability for each production corresponding to a non-terminal.

Evaluating the performance of the system, each speech signal from the test corpus (which consists of the utterances from 3 female and 2 male speakers) is converted first into a set of acoustic vectors. The result is then processed with Viterbi algorithm [20] which compares the signal with the HMMs in the recognition module. The overall results are provided by HTK [21]. Word recognition rate in average for this experimental system was 78.92%.

4 Conclusions and Future Work

This paper evaluates how nonlinear techniques, such as MFCC feature extractions and HMM stochastic technique, can be used in speech recognition control systems.

We have presented an HMM-based continuous speech recognition system for the Romanian language with a reduced vocabulary.

The system estimates the parameters of a set of HMMs using a training speech database and the associated transcription of the speech data (manual labeling of the files containing speech).

Comparing the present system with previous experiments [19], where the speech was isolated and the vocabulary contained only four Romanian words, this system is using continuous voice, and the choice of the words was more inspired, because there were a few words which can be confused.

The results demonstrate the efficiency of the proposed system and we consider that the system presented in this paper can be implemented in automatic control area for human-machine interaction, robots manipulation through voice commands, and many others.

As future developments, we are going to enhance the performance of this system by increasing the training corpus and applying discriminative training methods, and extend the present system to a control system for controlling a robot by voice commands.

References

1. Rabiner, L.R., Juang, B.-H.: Speech Recognition: Statistical Methods. Encyclopedia of Language & Linguistics, 1–18 (2006)
2. Anusuya, M.A., Katti, S.K.: Speech Recognition by Machine: A Review. International Journal of Computer Science and Information Security 6(3), 181–205 (2009)
3. Gold, B., Morgan, N., Ellis, D.: Speech and Audio Signal Processing. Processing and Perception of Speech and Music. John Wiley and Sons (2011)
4. Desai, N., Dhameliya, K., Desai, V.: Feature Extraction and Classification Techniques for Speech Recognition: A Review. International Journal of Emerging Technology and Advanced Engineering 13(12), 367–371 (2013)
5. Baker, J.K., Deng, L., Glass, J., Khudanpur, S., Lee, C.H., Morgan, N., O'Shaughnessy, D.: Developments and Directions in Speech Recognition and Understanding, Part 1. IEEE Signal Processing Magazine 26(3), 75–80 (2009)

6. Chivu, C.: Romanian Continuous Speech Recognition Applied to Automatic Controlled Systems. Annals of the Oradea University, Fascicle of Management and Technological Engineering VI (XVI), 728–735 (2007)
7. Dumitru, C.-O., Gavăţ, I.: Progress in Speech Recognition for Romanian Language, Advances in Robotics, Automation and Control. In: Aramburo, J., Trevino, A.R. (eds.) InTech (2008)
8. Teodorescu, H.-N.: AI Tools for Speech Analysis Applied to the Romanian Language. In: Proceedings of the 4th European Computing Conference, pp. 272–279 (2010)
9. Buzo, A., Cucu, H., Burileanu, C.: Text Spotting in Large Speech Databases for Under-Resources Languages. In: Proceedings of SpeD (2013)
10. Hwang, M., Huang, X.: Subphonetic Modeling with Markov States - Senone. In: Proceedings ICASSP, vol. 1, pp. 33–36 (1992)
11. Rabiner, L.R.: A Tutorial on Hidden Markov Models and Selected Applications in Speech Recognition. Proceedings of the IEEE 77(2), 257–286 (1989)
12. Davis, S., Mermelstein, P.: Comparison of Parametric Representations for Monosyllabic Word Recognition in Continuously Spoken Sentences. IEEE Trans. Acoust. Speech Signal Processing 28(4), 357–366 (1980)
13. Hermansky, H.: Perceptual Linear Predictive (PLP) Analysis of Speech. J. Acoust. Soc. Amer. 87(4), 1738–1752 (1990)
14. Rosenberg, A.E., Lee, C.H., Soong, F.K.: Cepstral Channel Normalization Techniques for HMM-Based Speaker Verification. In: Proceedings of IEEE ICASSP, pp. 1835–1838 (1994)
15. Hermansky, H., Morgan, N.: RASTA Processing of Speech. IEEE Trans. Speech Audio Processing 2(4), 578–589 (1994)
16. Eide, E., Gish, H.: A Parametric Approach to Vocal Tract Length Normalization. In: Proceedings of IEEE ICASSP, pp. 346–349 (1996)
17. Prabhakar, O.P., Sahu, N.K.: A Survey on: Voice Command Recognition Technique. International Journal of Advanced Research in Computer Science and Software Engineering 3(5) (2013)
18. Baum, L.E.: An Inequality and Associated Maximization Technique in Statistical Estimation of Probabilistic Functions of Markov Processes. Inequalities 3, 1–8 (1972)
19. Şchiopu, D.: Using Statistical Methods in a Speech Recognition System for Romanian Language. In: Proceedings of 12th IFAC/IEEE International Conference on Programmable Devices and Embedded Systems (PDeS), Velke Karlovice, Czech Republic, pp. 252–256 (2013)
20. Viterbi, A.J.: Error Bounds for Convolutional Codes and an Asymptotically Optimum Decoding Algorithm. IEEE Transactions on Information Theory 13(2), 260–269 (1967)
21. Hidden Markov Model Toolkit, `http://htk.eng.cam.ac.uk/`
22. Chomsky, N.: On Certain Formal Properties of Grammars. Information and Control 2, 137–167 (1959)
23. Lari, K., Young, S.J.: The Estimation of Stochastic Context-Free Grammars using the Inside-Outside Algorithm. Computer Speech and Language 4, 35–56 (1990)

The Challange of Clustering Flow Cytometry Data from Phytoplankton in Lakes

Stefan Glüge[1], Francesco Pomati[2], Carlo Albert[2], Peter Kauf[1], and Thomas Ott[1]

[1] ZHAW Zurich University of Applied Sciences, Switzerland
[2] Swiss Federal Institute of Aquatic Science and Technology (EAWAG), Switzerland

Abstract. Flow cytometry (FC) devices count and measure cells in fluids in an automated procedure. In this paper we present our work in progress on the clustering of FC data. We compare standard clustering algorithms such as K-means, Ward's clustering, etc., to the more advanced approach of sequential superparamagnetic clustering (SSC). We found Ward's hierarchical clustering to perform best regarding internal cluster validation measures, while SSC yielded the best results based on the visual inspection of the clustering results.

Keywords: Clustering, Sequential Superparamagnetic Clustering, Flow Cytometry, Phytoplankton.

1 Introduction

Flow cytometry (FC) devices basically count/measure cells in fluids. Usually, cells pass a single wavelength laser beam and emit a specific optical signal depending on their structure and fluorescence. The area of FC applications is wide, e.g. monitoring of industrial processes [8], aquatic system monitoring [6], and medical research [4].

At the Swiss Federal Institute of Aquatic Science and Technology (Eawag) a FC approach is applied to survey phytoplankton occurrences in lakes [6]. The focus is on understanding and predicting the effects of environmental change on biodiversity. One mayor issue is the need to introduce a trait-based analysis using the characteristics of individual phenotypes rather than species. Thereby one would create a measures of functional diversity, which is a predictor of ecosystem functioning across a range of communities.

In this paper we present our work in progress on the clustering of FC data, collected in the pre-alpin Lake Zurich (Switzerland). We compare several standard clustering algorithms (e.g. K-means) to the more advanced approach of SSC [5].

The data poses several interesting challenges, that is, a large amount of unlabeled samples, clusters of different shapes and densities, and an unknown number of classes/categories. Further, the data contains a significant number of outliers.

Our results induce that SSC is a promising approach to the automatic analysis of the complex FC data. Even though, for the application in field studies some labeled data is needed for a more reliable evaluation of the clustering results.

V.M. Mladenov and P.C. Ivanov (Eds.): NDES 2014, CCIS 438, pp. 379–386, 2014.

2 Methods

2.1 Flow Cytometry Data

The phytoplankton dataset used in this study consists of water samples collected from April to December 2009 in monthly intervals in Lake Zurich at 14 different water depths from $0, \ldots, 135$m. For flow-cytometry analysis 50ml of sampled water were fixed with a filter-sterilized solution of paraformaldehyde and glutaraldehyde (0.01 and 0.1% final concentration, pH 7). A scanning flow-cytometer from Cytobuoy[1] was used for counting and characterization of phytoplankton particles. It allows the analysis of pulse-signals providing 54 descriptors of 3D structure and fluorescence (FL) profile for each particle [7]. Raw Cytobuoy data were visually inspected for the distribution of FL signals in order to set threshold levels to extract FL particles (phytoplankton) with a size larger than 1µm.

2.2 Data Preprocessing

The whole data set consists of 73055 samples, each represented by 54 features (cf. Sec. 2.1). At first the raw data was centered and normalised to its standard deviation.

Some of the features are highly correlated, or anti-correlated respectively. Therefore, we reduced the feature space by applying a principle component analysis, ending up with 20 principle components (PCs) representing 90.75% of the variance in the data. For a comparison based on visual inspection we also applied the clustering algorithms on the first 3 PCs of the data. Further, clustering was done on 5% of the data (3653 samples) that where chosen randomly from the complete data set. We had to limit ourself on this subset due to constraints on memory and computation time.

Figure 1 shows the first three PCs of the selected 5% of the data to give an impression of the clustering problem we face. Looking on the data one can identify at least three different clusters. There is a small cluster (C1) on the left at PC1 ≈ 2 and PC2 ≈ -2 separated quite well, another big dense cluster (C2) at the bottom lying more or less in the PC2-PC3–plane, and one big cluster (C3) in the middle of the plot surrounded by a rather unspecific halo that was considered to be noise. The clusters where labelled manually to create values that allowed us an external cluster validation. However, these labels are prone to errors, especially in the regions of the edges the labelling is rather subjectively.

2.3 Clustering Algorithms

For our comparative study we considered four standard clustering algorithms that are usually used in FC data analysis (cf. [1]), namely K-means, Partitioning Around Medoids (PAM), model based clustering (mclust), and Ward's hierarchical clustering (ward). Further, we applied SSC, which is a more recent and advanced approach [5].

[1] Woerden, the Netherlands; `http://www.cytobuoy.com`

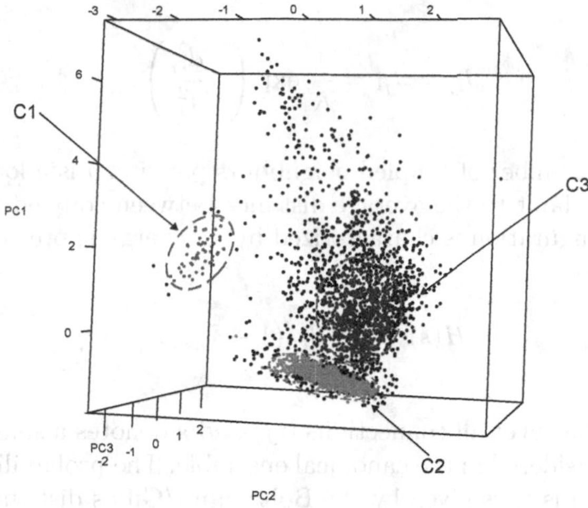

Fig. 1. 5% of the flow cytometry (FC) data projected on its first three principle components (PCs). At least three clusters can be identified by visual inspection.

K-means clustering is a method of vector quantization. It aims to partition n observations into k clusters in which each observation belongs to the cluster with the nearest mean, serving as a prototype of the cluster.

PAM is quite similar to K-means. Both algorithms break the datasets into groups, and both try to minimize the error. However, PAM works with Medoids, that are an entity of the dataset that represent the group in which it is inserted, while K-means works with Centroids [2].

mclust is based on distribution models. We applied Gaussian mixture models using the expectation-maximization algorithm. To obtain a hard clustering, samples were assigned to the Gaussian distribution they most likely belong to.

ward is a hierarchical cluster method. The criterion for choosing the pair of clusters to merge at each step is based on the minimum of some inter-cluster distance measure. Ward used the minimum variance method which minimizes the total within-cluster variance [9].

SSC is based on superparamagnetic clustering enhanced by a sequential procedure to select the most natural clusters [5].

Superparamagnetic Clustering can be description via Potts spin models. For N samples to be clustered with pairwise affinities d_{ij}, an inhomogeneous grid of Potts spins is constructed in the following way: each sample i is represented by one site of the grid with Potts spin variable $s_i \in \{1, \ldots, q\}$. q is typically set to 10 or 20. Each spin is symmetrically coupled to its k nearest neighbours. The coupling strength J_{ij} is a decreasing function of d_{ij},

$$J_{ij} = J_{ji} = \frac{1}{\hat{K}} \exp\left(\frac{-d_{ij}^2}{2a^2}\right). \tag{1}$$

$\hat{K}$ is the average number of coupled neighbours per site. a is a local length scale which is set by default to the average distance between coupled spins.

Each spin configuration is characterized by an energy expressed by the Potts spin Hamiltonian

$$H(s) = \sum_{(ij)} J_{ij} \left(1 - \delta_{s_i s_j}\right), \tag{2}$$

where the sum runs over all connections (ij) and s denotes a spin configuration. The system is considered in the canonical ensemble. The probability for a certain spin configuration is thus given by the Boltzmann/Gibbs distribution

$$p(s) = \frac{1}{Z} \exp\left(\frac{-H(s)}{T}\right), \tag{3}$$

where the partition function $Z = Z(T)$ serves as a normalization factor.

At a given temperature T, clusters are identified with the help of the pair correlation: two points i and j belong to the same cluster if the pair correlation

$$G_{ij} = \sum_{s} p(s) \delta_{s_i s_j} \tag{4}$$

exceeds a given threshold Θ.

Clear clusters express themselves as regions of order that are stable over a substantial range of T. The idea is to choose the clusters that have the largest T-range extensions (denoted by T_{cl}). Consequently, T-stability s_T of a cluster is defined as

$$s_T = \frac{T_{\text{cl}}}{T_{\max}}, \tag{5}$$

where $T_{\max}$ is the temperature of the paramagnetic transition. Hence, s_T represents the stability of a cluster in relation to the stability of the whole set. However, the most natural clusters may not be the most stable ones. Different densities, shapes, and sizes of the clusters result in different ranges of temperature where they occur. Consequently, often natural clusters emerge only for short T-ranges when dense superclusters break-up at higher temperatures. The sequential procedure overcomes this problem by reclustering the most stable cluster in terms of s_T with readjusted weights. That means, the most stable cluster is clustered in two new separate sessions, and the connectivity and weights are independently redetermined for each set according to the criterion of k nearest neighbours and (1).

2.4 Cluster Validation Measures

As we worked on originally unlabeled data we relied on internal cluster validation measures. These were Connectivity and the Dunn Index because they are well known and widely used in practice.

Unfortunately, internal validation measures usually have the drawback that they can identify only well separated hyper sphere shaped clusters. This is because the indices measure the variance of the clusters around some representative points [3].

In our case the clusters are of arbitrary shape and may not have representative centre point. Therefore, we evaluated the clustering results concerning our expected result (cf. Fig. 1) using the Jaccard coefficient. Further, two qualitative criteria were evaluated, that are:

- QC1 are the *three* clusters in Fig. 1 separated (yes/no)
- QC2 is the background noise separated (yes/no)

Connectivity measures to what extent observations are placed in the same cluster as their nearest neighbours in the data space. The connectivity has a value between 0 and ∞, and should be minimized.

Dunn Index is the ratio of the smallest distance between observations not in the same cluster to the largest intra-cluster distance. The Dunn Index has a value between 0 and ∞, and should be maximized.

Jaccard Coefficient is defined as: $J_{\text{coeff}} = \frac{n_{11}}{n_{11}+n_{10}+n_{01}}$, where n_{11} is the number of observation pairs where both observations are in both clusterings, n_{10} is the number of pairs where the observations are in the first clustering but not the second, and n_{01} is the number of pairs where the observations are in the second clustering but not the first. The index takes values between 0 and 1, and is maximized when both clusterings are identical.

For K-means, PAM, mclust and Ward's clustering we evaluated the clustering according to Connectivity and Dunn Index for $k = 2 \ldots 20$ clusters. Then the best results where evaluated according to the qualitative criteria and the Jaccard Coefficient. For SSC the number of clusters to be found cannot be set directly, but is controlled by several parameters. The parameters were set in trail-and-error manner to achieve the most reasonable result according to visual inspection.

3 Results

Table 1 summarises results for the different clustering algorithms. The evaluation according to the Jaccard coefficient and qualitative criteria (QC1,QC2) was done on the clustering result that provided the best Dunn Index for each method.

Figure 2 shows the results of Ward's clustering which are the best according to the internal cluster validation measures. In Fig. 2a only two clusters are

Table 1. Results for the different clustering algorithms for data samples using 3/20 PCs. Connectivity and Dunn Index are shown together with the number of clusters k. The cluster results with the highest Dunn Index are also evaluated according to Jaccard coefficient and two qualitative criteria QC1 (C1-C3 found) and QC2 (background noise detected).

	PCs	Dunn Index / k	Connectivity / k	Jaccard	QC1	QC2
K-means	3	0.013 / 3	123.00 / 2	0.520	no	no
	20	0.047 / 4	260.90 / 2	0.554	no	no
PAM	3	0.009 / 3	109.36 / 2	0.568	no	no
	20	0.041 / 4	287.76 / 2	0.561	no	no
mclust	3	0.006 / 3	166.84 / 2	0.631	no	yes
	20	0.029 / 2	857.70 / 2	0.574	no	no
ward	3	0.012 / 5	40.52 / 2	0.580	no	no
	20	0.052 / 5	148.97 / 2	0.556	yes	no
SSC	3	0.003 / 4	299.69 / 2	0.642	yes	yes
	20	0.002 / 3	1249.15 / 3	0.389	no	yes

found. These clusters are separated quite well, which leads to a low connectivity. However, we would expect at least four clusters (C1-C3 plus background noise). In Fig. 2b C1-C3 are found while especially C1 dose not separate very well. Additionally, C3 and background are subdivided in 3 separate clusters.

Finally, Fig. 3 shows the result of SSC which performed best regarding the Jaccard coeffients and our qualitative clustering measures QC1 and QC2. All cluster C1-C3 are identified while the remaining points are grouped together in one cluster that we consider to be noise. Obviously it is hard make a clear decision whether some specific point should be part of a cluster or rather is noise.

4 Discussion

Our study shows that the FC data presents an interesting challenge for cluster algorithms. We have to deal with a large amount of unlabeled data. The clusters C1-C3 are of different shapes and densities, and the data contains a significant number of outliers.

We can conclude that Ward's hierarchical clustering performs best regarding the internal cluster validation measures. However, those measures are of limited significance as they work only on separated hyper sphere shaped clusters (cf. Fig. 2). Therefore, we introduced qualitative measures based on the visual inspection of the clustering results and compared the clustreing against our expected clustering using the Jaccard coefficient. According to these handcrafted, but reasonable, measures SSC yielded the best results (cf. Fig. 3).

The Jaccard coefficient of the model based clustering (mclust) with three PCs is quite high with $J_{\text{coeff}} = 0.631$ (cf. Tab. 1). Cluster C1 was not found by the algorithm. As this cluster is small (cf. Fig. 1), it was of limited influence on the Jaccard coefficient. However, the difference to the expected clustering is rather

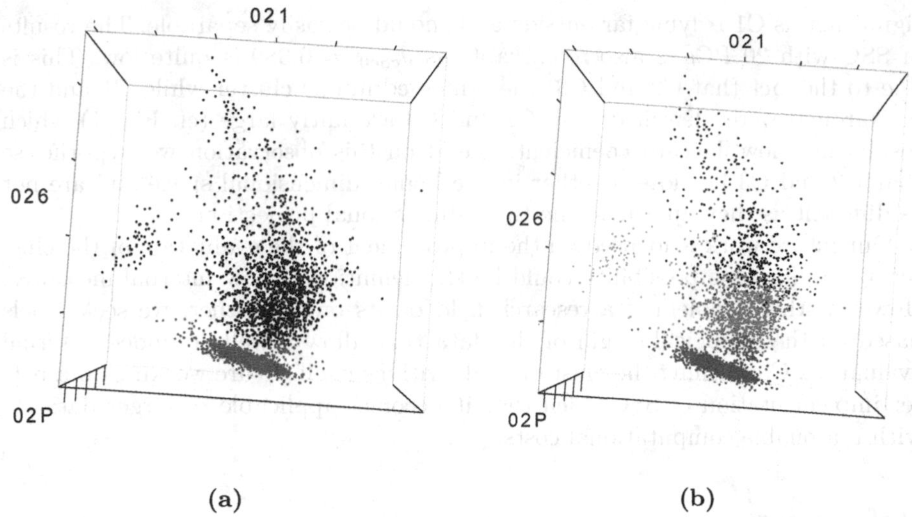

(a) (b)

Fig. 2. Ward's clustering result with best Connectivity 40.52 (a) and best Dunn Index 0.052 (b)

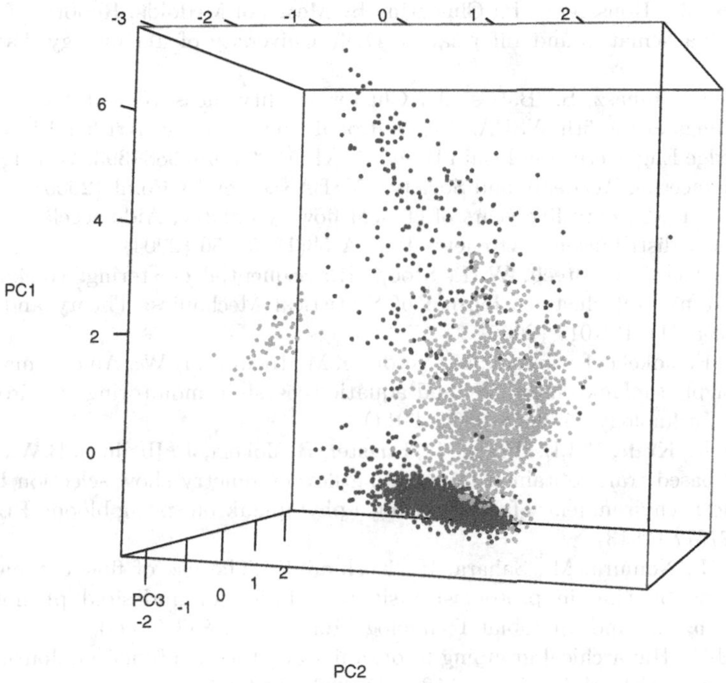

Fig. 3. SSC clustering result of flow cytometry (FC) data based on the first three principle components

significant as C1 is lying far outside and should be easily separable. The results of SSC with 20 PCs is also remarkable as $J_{\mathrm{coeff}} = 0.389$ is quite low. This is due to the fact that C2 and C3 where merged in one cluster while C1 and the background were identified well. C2 and C3 are fairly large (cf. Fig. 1) which results in a low Jaccard coefficient. Based on this observation we hypothesise that C2 and C3 lie close together in the higher dimensional space and are not as different as they appear in the three dimensional projection.

Our future work is focused on the improvement of the evaluation of the clustering results. One possibility could be the definition of new internal measures. However, this Problem is a research field on its own. Further, we seek labels based on the biological origin of the data that allow a well-grounded external evaluation. Concerning the clustering algorithms itself we are working on a better implementation of SSC, such that it becomes applicable to larger datasets with reasonable computational costs.

References

1. Boddy, L., Wilkins, M.F., Morris, C.W.: Pattern recognition in flow cytometry. Cytometry 44(3), 195–209 (2001)
2. Kaufman, L., Rousseeuw, P.: Clustering by Means of Medoids. Reports of the Faculty of Mathematics and Informatics. Delft University of Technology, Fac., Univ. (1987)
3. Legány, C., Juhász, S., Babos, A.: Cluster validity measurement techniques. In: Proceedings of the 5th WSEAS International Conference on Artificial Intelligence, Knowledge Engineering and Data Bases, AIKED 2006, pp. 388–393. World Scientific and Engineering Academy and Society (WSEAS), Stevens Point (2006)
4. Mandy, F.F.: Twenty five years of clinical flow cytometry: Aids accelerated global instrument distribution. Cytometry Part A 58(1), 55–56 (2004)
5. Ott, T., Kern, A., Steeb, W.H., Stoop, R.: Sequential clustering: tracking down the most natural clusters. Journal of Statistical Mechanics: Theory and Experiment 2005(11), P11014 (2005)
6. Pomati, F., Jokela, J., Simona, M., Veronesi, M., Ibelings, B.W.: An automated platform for phytoplankton ecology and aquatic ecosystem monitoring. Environmental Science Technology 45, 9658–9665 (2011)
7. Pomati, F., Kraft, N.J.B., Posch, T., Eugster, B., Jokela, J., Ibelings, B.W.: Individual cell based traits obtained by scanning flow-cytometry show selection by biotic and abiotic environmental factors during a phytoplankton spring bloom. PLoS ONE 8(8), e71677 (2013)
8. Urano, N., Nomura, M., Sahara, H., Koshino, S.: The use of flow cytometry and small-scale brewing in protoplast fusion: Exclusion of undesired phenotypes in yeasts. Enzyme and Microbial Technology 16(10), 839–843 (1994)
9. Ward, J.H.: Hierarchical grouping to optimize an objective function. Journal of the American Statistical Association 58(301), 236–244 (1963)

Air Quality Forecasting by Using Nonlinear Modeling Methods

Elia Georgiana Dragomir[1] and Mihaela Oprea[2]

University Petroleum-Gas of Ploiesti,
[1]Department of Information Technology, Mathematics and Physics
[2]Department of Automation, Computers and Electronics
Bd. Bucuresti Nr. 39, Ploiesti, 100680, Romania

Abstract. The air quality forecasting became an important problem to be solved by the environmental monitoring systems in order to prevent the occurance of severe air pollution episodes, or, at least, to reduce their effects on human health. Several parameters can influence the result of the air quality forecasting: air pollutants concentrations, meteorological parameters, seasonality, time, and other factors specific, for example, to the geographical area. Due to the high complexity of this problem, the main class of methods that could solve it in a more proper way is given by the nonlinear methods that could capture the nonlinearity of the air pollution phenomena. The paper presents the use of some nonlinear modeling methods for air quality forecasting: artificial neural networks (ANNs) and support vector machines (SVMs). A comparative study among these methods, ANN, SVM and a Linear Regression (LR) method is also discussed.

1 Introduction

Air pollution is a highly nonlinear phenomenon determined by physical and chemical processes that arise in the atmosphere at spatial and temporal scales. The understanding of the dependencies and relationships between various factors and parameters involved in these dynamic processes is very important in order to make more accurate real time forecasts of air quality, and thus, to reduce the effects of severe air pollution episodes on the human health. The main air pollutants are carbon monoxide (CO), carbone dioxide (CO_2), nitrogen monoxide (NO), nitrogen dioxide (NO_2) and nitrogen oxides (NO_x), suspended particulates (particulate matters - PM: respirable PM_{10}, and fine $PM_{2.5}$), sulfur dioxide (SO_2), ozone (O_3), lead (Pb), volatile organic compounds (VOC) etc.

Several parameters can influence the result of the air quality forecasting: the air pollutants concentrations, the meteorological parameters (air temperature, wind speed, wind direction, relative humidity, atmospheric pressure etc), seasonality, time, and other factors specific, for example, to the geographical and industrial area. Due to the high complexity of the air quality forecasting problem [1], the main class of methods that could solve it in a more proper way is given by the nonlinear modeling methods that capture more accurately the nonlinearity of the air pollution phenomena. Some

V.M. Mladenov and P.C. Ivanov (Eds.): NDES 2014, CCIS 438, pp. 387–394, 2014.

research works that were recently reported in the literature highlight the importance of using nonlinear methods (see e.g. [2], [3], [4], [5]).

In this paper it is presented a study of applying nonlinear modeling methods (such as artificial neural networks and support vector machines) and a linear regression method for the air quality forecasting in the area of Ploiesti, Romania, an industrial city where the petrochemical and chemical industries are the main contributors to the air pollution.

2 The Air Quality Forecasting

Starting from the air quality forecasting problem formulation, we present the main approaches proposed in the literature, and briefly describe two nonlinear methods, the feed forward artificial neural networks and support vector machines.

2.1 Problem Formulation

The problem of air quality forecasting can be formulated by equation (1).

$$y = f(x_1, x_2, \ldots, x_n) \tag{1}$$

where y represents the forecasted air quality parameter or a vector of air quality parameters (e.g. in the case of time series forecasting where the forecasting window is greater than 1, i.e. more that one step ahead are forecasted – as e.g. two hourly forecasts or five daily forecasts), and x_i, $i = 1 \div n$, represents the factors that can influence the forecasted parameter. The forecasted air quality parameter can be the air quality index (AQI) or the common air quality index (CAQI) or the concentration of a certain air pollutant (past measurements); The analyzed parameters can be the values of the air pollutants concentrations, the meteorological factors (such as wind speed, air temperature, relative humidity etc), seasonality, time, other factors (e.g. factors depending on the geographical and industrial area where the forecasting is performed). The function f is a nonlinear function. The forecasting problem solving finds an approximation of the f function.

Based on historical data, it can be forecasted the value of the forecasted air quality parameter for short-term or longer-term. Usually, the air quality forecasting can be made hourly, daily, monthly or even yearly, when large historical databases are used.

2.2 Forecasting Approaches

Several types of approaches were developed or adapted to solve the forecasting problem (see e.g. [6], [7], [8]). We can identify two main classes: (a) deterministic approaches, based on numerical models that represent physical and chemical processes involved in the air pollution phenomenon, and (b) data-driven approaches that apply statistical and artificial intelligence methods (including computational intelligence techniques). The deterministic approaches use a more complete model of the phenomenon, and thus are computationally demanding and less useful in real time systems, having the main advantage of highly forecasting accuracy. The data-driven approaches use a more incomplete model of the phenomenon with approximations and patterns of forecasting knowledge embedded in the historical datasets. Another viewpoint on the forecasting approaches classification is whether they apply a nonlinear modeling by keeping the nonlinearity of the phenomenon or a linear modeling by simplifying it. Under this last classification we can identify linear modeling methods (e.g. Linear Regression - LR, Multiple Linear Regression – MLR), and nonlinear modeling methods (e.g. artificial neural networks - ANNs, support vector machines - SVM, Adaptive Neuro-Fuzzy Inference Systems - ANFIS).

Linear modeling methods do not always capture the complex relationships that exist in the air monitoring data, especially for the air quality forecasting problems. On contrary, nonlinear modeling methods could do it more properly.

2.2.1 Artificial Neural Networks

Artificial neural networks are biologically inspired modeling tools for complex processes and phenomena. They can solve noisy problems and their computational power is given by their ability to estimate missing values. An ANN can model complex nonlinear phenomenon like air pollution [9]. It can be trained to successfully approximate any continuous function [10], by using supervised learning. The artificial neural networks were first proposed for the adaptive control of nonlinear dynamical systems in [11]. They are well suited for modeling nonlinear problems such as the air quality forecasting [12]. Moreover, the adaptive nature of the ANNs enhances their ability for time series forecasting. Therefore, an artificial neural network can learn the *f* function from equation (1).

An ANN is composed by a set of interconnected processing elements (artificial neurons units) that are organized under a certain topology that provide the type of the artificial neural network. Examples of ANN types are: multilayer perceptron (MLP, named also feed forward ANN), Hopfield, Boltzmann etc. In our study we have used the MLP model, i.e. a feed forward artificial neural network (FFANN) with an input layer, one or more hidden layers and an output layer. The generic architecture of an air quality forecasting FFANN is shown in Figure 1.

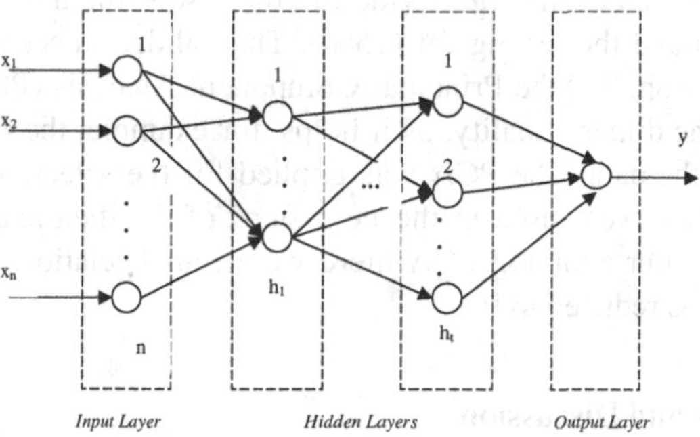

Fig. 1. The generic architecture of an air quality forecasting FFANN

2.2.2 Support Vector Machines

Support vector machines are binary classifiers. They are a kernal-based learning method from statistical learning theory that uses an implicit mapping of the input data into a higher dimensional feature space defined by the kernel function. A linear hyperplane is derived as a decision function for nonlinear problems such as the forecasting problem. A back-transformation is applied in the nonlinear space. The choice of kernel influences drastically the performance of the SVMs depending on the problem considered [13]. In the case of air quality forecasting SVM has yielded very good generalization performance. This domain typically has involved high-dimensional input space, and the good performance is also related to the fact that SVMs learning ability can be independent of the feature space's dimensionality.

3 A Comparative Study

Several experiments were conducted in order to forecast the air quality index based on the previously recorded data for the most important air pollution parameters recorded by the Romanian National Air Quality Monitoring Network [14] for the Ploiesti city. Hourly air pollutants concentrations are measured and recorded in the databases. We have made a comparative study between three forecasting methods: FFANN, SVM and LR.

3.1 Data Set

For this study we have used hourly recorded data for all air quality parameters monitored in the Ploiesti city at three stations (PH-2, PH-3 and PH-6) – air pollutants concentrations for the air pollutants (such as CO, SO_2, NO, NO_2, NO_x, PM, O_3 etc) and the meteorological parameters, over a period of 3 years from January 2009 to December 2011. The data were preprocessed in order to be used by the Weka software [15], a data mining software tool that contains a collection of visualization tools and algorithms for data analysis and predictive modeling. We have used the dataset for all the forecasting methods that were experimented, ANN, SVM and LR.

The dataset used by the ANN was divided in three sets: the training set (50%), the validation set (25%) and the testing set (25%). The validation set was used to avoid over-fitting. We have applied the Principal Component Analysis (PCA) to the dataset in order to change the dimensionality, as it helps, for example, the ANN to find more easily patterns into the data. The PCA was applied for the selection of the most important parameters that were used in the next steps of the data analysis from 22 air quality parameters. After applying PCA there were found relations among these and the inputs number was reduced to 9.

3.2 Experiments and Discussion

We have performed three experiments, corresponding to each forecasting method: ANN, SVM and LR, by using the same dataset and the Weka software tool.

Experiment 1 (FFANN)

During the first experiment we have used a FFANN for AQI forecasting. The best topology of the FFANN was determined by experiments. Different values of the nodes' number on the input and hidden layers, of the number from the hidden layers, for the learning rate and momentum, for the training epochs were tested. Figure 2 presents the best topology that was found containing 9 nodes on the input layer corresponding to the ones determined by the PCA and 1 unit on the output layer for the air quality index, and 2 hidden layers each one with 10 units. The validation method that was used is a 10-fold cross validation approach known as repeated random sub-sampling validation [16]. The cross validation accuracy is given by the root mean squared error.

The main data pre-processing steps involved in this experiment were the use of an interpolation method to calculate missing values, and the normalization of the input data. This last pre-processing step was realized by using a hyperbolic tangent sigmoid transfer function applied to the hidden layers. In the output layer it is applied a linear transfer function.

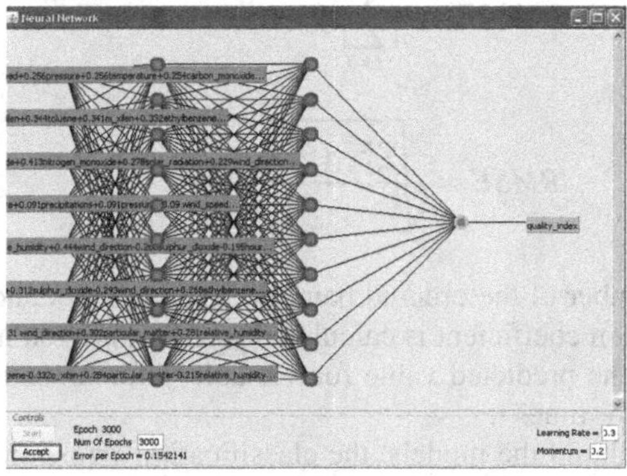

Fig. 2. The FFANN topology

Experiment 2 (SVM)

In order to establish which is the most appropriate combination of parameters and the type of kernel used there were designed different models. Several kernels are available in Weka for learning: the homogeneous polynomial kernel together with the linear kernel, by setting the polynomial exponent to 1 and the Gaussian kernel [17]. Two major models were designed: one with a polynomial kernel and one with Gaussian kernel. The results show that if the parameter C is higher or equal than 1.0, there are no visible effects on the model. If the values of parameter C are lower than 1.0, there it is a correctly classified instances percent that reflects a larger number of misclassifications i. e C=0.8 for the first model and C=1.0 for the second one.

Experiment 3 (LR)

For the third experiment Linear Regression has been used. LR is a statistical method for predicting the value of a dependent variable y, based on the value of an independent variable x. In our study case the air quality index represents the variable y and the other database parameters are the x variables. There were also tested two types of attribute selection method: M5 method and Greedy method. For both selection methods the results are the same.

Comparative analysis

The performance of the three forecasting models (ANN, SVM and LR) was measured by the following statistical criteria parameters: correlation coefficient (R), given by equation (2); the mean absolute error (MAE), given by equation (3); and the root mean squared error (RMSE), given by equation (4).

$$r_{xy} = \frac{\sum_{i=1}^{n} x_i y_i - \frac{(\sum_{i=1}^{n} x_i)(\sum_{i=1}^{n} y_i)}{n}}{\sqrt{[(\sum_{i=1}^{n} x_i^2 - \frac{(\sum_{i=1}^{n} x_i\ )^2}{n_x})(\sum_{i=1}^{n} y_i^2 - \frac{(\sum_{i=1}^{n} y_i\ )^2}{n_y})]}} \tag{2}$$

$$MAE = \frac{1}{n}\sum_{i=1}^{n} |\hat{y}_i - y_i| \tag{3}$$

$$RMSE = \sqrt{\frac{\sum_{t=1}^{n}(\hat{y}_t - y_t)^2}{n}} \tag{4}$$

where n is the number of the ordered pairs in the sample, x and y are the variables between the correlation coefficient is calculated (e.g. x=nitrogen monoxide and y= air quality index), $\acute{y}_t$ is the predicted value for variable y for times t, y_t is the measured value for variable y for times t.

The time taken to build the models, the classification errors along with other parameters are synthesized in table 1.

Table 1 points out that even that, the time taken by the LR classifiers to build model is the smallest i. e 0.03s, the analysis of another three parameters i. e R, MAE, RMSE revealed that the model formed by FFANN classifier is better. The best values of MAE and RMSE are obtained for the nonlinear methods, i. e FFANN and SVM (figure 3). The mean absolute error for the FFANN model is also more efficient comparative to the value of 0.9361 of the same statistics measure for the LR model. It is also seen that the performance of FFANN and SVMs classifiers are very good and almost the same, thus, sustaining the idea that nonlinear data can be successfully modeled with FFANN and SVMs.

The comparative study of these results revealed that, using the same dataset, a model bases on the FFANN technique is a better classifier for the new instances. Figure 4, 5 and 6 present the plots of the air quality index predicted and the actual one for each method. The graphics for FANN and SVMs reveal the strong relation between the predicted and the actual values compared with the week relation reflected in the LR results in figure 6.

The experimental results showed that nonlinear methods can be successfully applied in the air quality forecasting, and could provide a good decision support tool for the atmospheric environmental management.

Table 1. The Experimental Results

	ANN	SVMs		Linear Regression
		Kernel= a polynomial function	Kernel = a RBF function	
Correctly Classified Instances	99.8320 %	99.7731 %	99.2027 %	
Incorrectly Classified Instances	0.20%	0.2269 %	0.7973 %	
Correlation coefficient	*0.9983*	0.9972	0.9901	0.6261
Mean absolute error	*0.2031*	0.2042	0.2043	0.9361
Root mean squared error	*0.3001*	0.3013	0.3012	1.1928
Time taken to build model	482.08 s	11.7 s	109.99 s	0.03 s

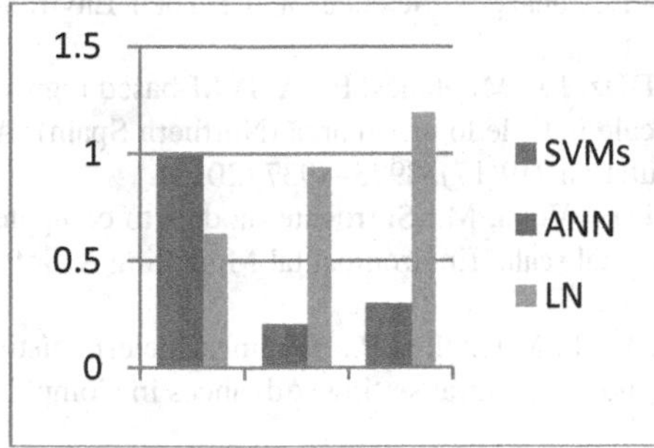

Fig. 3. The comparative analysis of the R, MSE and RMSE parameters for the three models

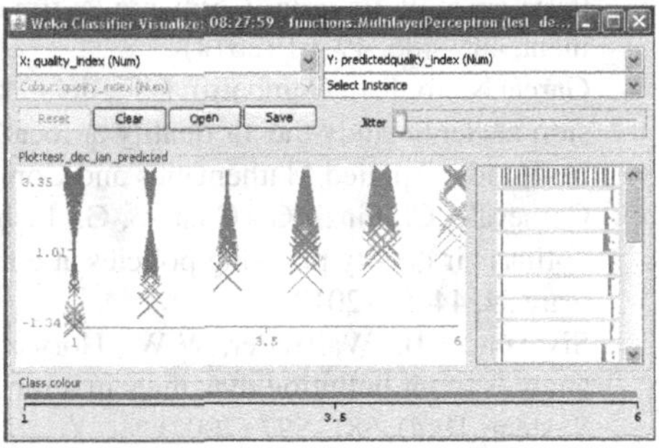

Fig. 4. The air quality index predicted by the FFANNs

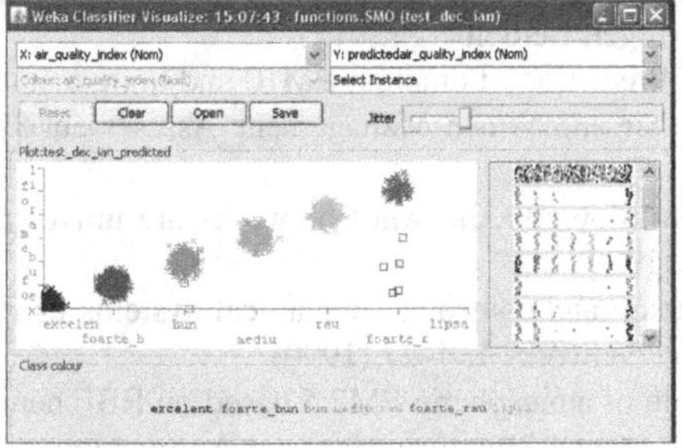

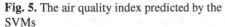

Fig. 5. The air quality index predicted by the SVMs

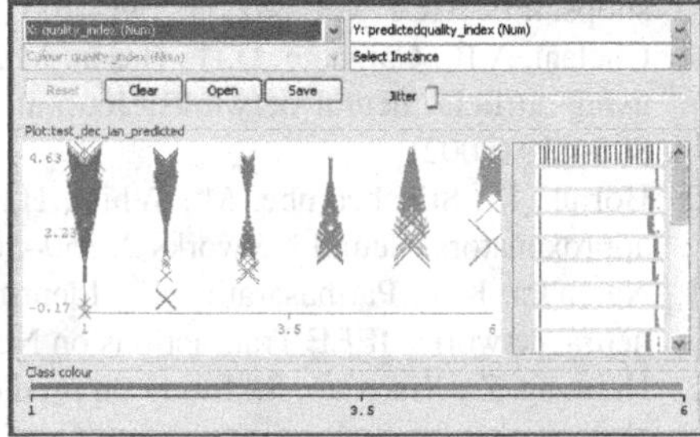

Fig. 6. The air quality index predicted by the LR

4 Conclusion and Future Work

The paper presented a case study on using nonlinear modeling methods such as ANNs and SVM for the air quality forecasting problems. The highly nonlinear nature of the air pollution phenomenon was analyzed with these methods. A comparative study that was conducted in Ploiesti, a city with petrochemical and chemical predominant industries, revealed a very good performance of the nonlinear modeling methods that proved to be useful tools for air quality forecasting. These methods can be applied with success to other domains of applications (computer networks, electronics, physics, biology, economics, environmental engineering etc), where nonlinear phenomena (dynamics) occur.

References

1. Moussiopoulos, N. (ed.): Air Quality in Cities. Springer, Berlin (2003)
2. Singh, K.P., Gupta, S., Kumar, A., Shukla, S.P.: Linear and nonlinear modeling approaches for urban air quality prediction. Science of the Total Environment 426, 244–255 (2012), doi:10.1016.j.scitotenv.2012.03.076
3. Breitner, S., Wolf, K., Devlin, R.B., Diaz-Sanchez, D., Peters, A., Schneider, A.: Short-term effects of air temperature on mortality and effect modification by air pollution in three cities of Bavaria, Germany: A time-series analysis. Science of the Total Environment, 485–486, 49–61 (2014)
4. García Nieto, P.J., Combarro, E.F., del Coz Díaz, J.J., Montañés, E.: A SVM-based regression model to study the air quality at local scale in Oviedo urban area (Northern Spain): A case study. Applied Mathematics and Computation 219(17), 8923–8937 (2013)
5. Carnevale, C., Finzi, G., Guariso, G., Pisoni, E., Volta, M.: Surrogate models to compute optimal air quality planning policies at a regional scale. Environmental Modelling & Software 34, 44–50 (2012)
6. Sivakumar, B., Wallender, W.W., Horwath, W.R., Mitchell, J.P.: Nonlinear deterministic analysis of air pollution dynamics in a rural and agricultural setting. Advances in Complex Systems 10(4), 581–597 (2007)
7. Mesin, L., Orione, F., Pasero, E.: Nonlinear adaptive filtering to forecasting pollution, Adaptive Filtering Applications. In: Garcia, L. (ed.) InTech (2011)
8. Kyriakidis, I., Karatzas, K., Papadourakis, G., Ware, A., Kukkonen, J.: Investigation and Forecasting of the Common Air Quality Index in Thessaloniki, Greece. In: Iliadis, L., Maglogiannis, I., Papadopoulos, H., Karatzas, K., Sioutas, S. (eds.) AIAI 2012 Workshops. IFIP AICT, vol. 382, pp. 390–400. Springer, Heidelberg (2012)
9. Chelani, A.B., Gajghate, D.G., Hasan, M.Z.: Prediction of ambient PM10 and toxic metals using artificial neural networks. Journal of Air and Waste Management Assessment 52, 805–810 (2002)
10. Hornik, K., Stinchcombe, M., White, H.: Multilayer feedforward networks are universal approximators. Neural Networks 2, 359–366 (1989)
11. Narendra, K.S., Parthasarathy, K.: Identification and control of dynamical systems using neural networks. IEEE Transactions on Neural Networks 1, 4–27 (1990)
12. Haiming, Z., Xiaoxiao, S.: Study on prediction of atmospheric PM2.5 based on RBF neural networks. In: Proceedings of Int. Conf. on Digital Manufacturing and Automation, pp. 1287–1289 (2013)
13. Hamel, L.: Knowledge Discovery With SVMs. John Wiley& Sons (2009)
14. Romanian National Air Quality Monitoring Network: `http://www.calitateaer.ro`
15. Weka: `http://www.cs.waikato.ac.nz/ml/weka/`
16. Vapnik, V.: The Nature of Statistical Learning Theory. Springer, Verlag (1995)
17. Boser, B.E., Guyon, I.M., Vapnik, V.N.: A Training Algorithm for Optimal Margin Classifiers. In: Proceedings of the 5th Annual Workshop on Comp. Learning Theory, COLT 1992, pp. 144–152. ACM Press, Pittsburgh (1992)

Bifurcation Analysis of Time Delayed Ecological Model

Tibor Kmet[1] and Maria Kmetova[2]

[1] Constantine the Philosopher University, Department of Informatics,
Tr. A. Hlinku 1, 949 74 Nitra, Slovakia
tkmet@ukf.sk
http://www.ukf.sk

[2] Constantine the Philosopher University, Department of Mathematics,
Tr. A. Hlinku 1, 949 74 Nitra, Slovakia
mkmetova@ukf.sk

Abstract. We consider a simple ecological model consists of phosphorus, algea and zooplankton. The model is described by a system of delay partial differential equations. The stability analysis of spatially constant equilibria and some numerical simulations are given. It is shown that Hopf bifurcation may occur depending on the time delay.

Keywords: Ecosystem model, partial differential equations with discrete time delay, stability analysis of equilibria, Hopf bifurcation.

1 Introduction

The purpose of this paper is to investigate a simplified ecosystem model. The model consists of phosphorus (x_1) as a limiting nutrient for growth of algae (x_2) and zooplankton (x_3). Similar models of n species of microorganisms competing exploitatively for one, two or more growth-limiting nutrients are used to study continuous culture of microorganisms in the chemostat under constant conditions without the presence of predators (see [1, 6, 15–17, 22–24]) . The detail description of chemostat and its general theory is given in [20]. The simple ecosystem model is described by a system of partial differential equations with discrete time delay, which is a generalization of the models presented in [2, 9–13].

We examine the stability analysis of the equilibria. Some numerical simulations are also given under constant environmental conditions. It is shown that Hopf bifurcation may occur depending on the filtration rate and time delay, respectively.

This paper is organized as follows. In Section 2, we present a description of the model. Section 3, studies the existence and stability of the spatially constant equilibria of the model. Some numerical results are also given. Conclusions are being presented in Section 4.

V.M. Mladenov and P.C. Ivanov (Eds.): NDES 2014, CCIS 438, pp. 395–402, 2014.

2 Description of the Model

The model is described by a system of three delay partial differential equations. It is derived from the models of the series AQUAMOD [12, 18, 19], and modified by the inclusion diffusion and discrete time delay τ of food uptake by Daphnia.

The following system of partial functional differential equations with discrete time delay is proposed as a model of simple ecosystem:

$$\frac{\partial x_i(p,t)}{\partial t} = D_i \frac{\partial^2 x_i(p,t)}{\partial p^2} + F_i\left(x(p,t), x(p,t-\tau), t\right), \; i = 1, \ldots, 3, \tag{1}$$

where

$$F_1\left(x(p,t), x(p,t-\tau), t\right) = a_7(a_1 - x_1(p,t)) - x_1(p,t)\frac{p_2 x_2(p,t)}{x_1(p,t) + s_2} + r_2 x_2(p,t) + x_3(p,t) C_2 x_2(p,t)\left(1 - \frac{d_2}{a_4 + x_2(p,t)}\right),$$

$$F_2\left(x(p,t), x(p,t-\tau), t\right) = \frac{p_2 x_1(p,t) x_2(p,t)}{x_1(p,t) + s_2} - r_2 x_2(p,t) - E_2 x_2(p,t) x_3(p,t) - d_1 x_2(p,t) + a_2 a_7,$$

$$F_3\left(x(p,t), x(p,t-\tau), t\right) = a_6 d_2 x_3(p,t-\tau)\frac{a_8 E_2 x_2(p,t-\tau)}{a_4 + x_2(p,t-\tau)} - a_5 x_3(p,t) + a_7 a_3,$$

with Neumann boundary condition

$$\frac{\partial x_i}{\partial p}(0,t) = \frac{\partial x_i}{\partial p}(1,t) = 0 \tag{2}$$

and initial conditions

$$x_i\,(p,t) = \phi_i\,(p,t) \geq 0, \; 0 \leq p \leq 1, \; t \in \langle -\tau, 0\rangle, \; i = 1, \ldots, 3. \tag{3}$$

Here t denotes the time, p represents the spatial location, D_i are the diffusion coefficients and $x_i(t,p)$, i=1,...,3 are the concentration of phosphorus, algae and zooplankton, respectively at time t and in spatial location p. The constant τ stands for the discrete time delay in uptake of algal species by Daphnia.

3 Stability Analysis of Spatially Constant Equilibria

In this section we investigate the existence and stability of equilibria. One parameter analysis of existence and stability of the equilibria of (1) is carried out using zooplankton filtration rate a_8 as bifurcation parameters [14], respectively. We consider the equilibrium solutions $\bar{x}$ to exist only if they lie in the nonnegative cone.

Next we try to find all the different equilibia of system (1). To simplify our consideration suppose that $a_2 = 0$. The system has four possible equilibria. Now these equilibria are given by

(i) the trivial equilibrium point $\bar{x}^{00} = (a_1, 0, 0, 0, 0, 0)$,
(ii) the algae free equilibrium points $\bar{x}^{0} = (a_1, 0, 0, 0, 0, \frac{a_6 a_7}{a_5})$,
(iii) the Daphnia free equilibrium points $\bar{x}^{20} = (\bar{x}_{12}, \bar{x}_2, 0)$, where

$$\bar{x}_{12} = \frac{s_2(r_2 + d_1)}{p_2 - r_2 - d_1},$$
$$\bar{x}_2 = \frac{a_7(a_1 - \bar{x}_{12})}{\frac{p_2 \bar{x}_{12}}{\bar{x}_{12} + s_2} - r_2},$$

(iv) the "interior" equilibria $\bar{x}$ with $\bar{x}_1,\ \bar{x}_2,\ \bar{x}_3 > 0$.

System (1) for a given set of parameters has a positive spatially constant equilibrium $\bar{x} = (\bar{x}_1, \bar{x}_2, \bar{x}_3)$, with

$$\bar{x}_1 = \frac{-B + \sqrt{B^2 - 4AC}}{2A},$$
$$\bar{x}_2 = \frac{a_5 a_4}{a_6 d_2 a_8 E_2 - a_5},$$
$$\bar{x}_3 = \frac{\frac{p_2 \bar{x}_1}{\bar{x}_1 + s_2} - r_2 - d_1}{E_2},$$

where $\bar{x}_1$ is a positive root of the quadratic equation $Ax_1^2 + Bx_1 + C = 0$, where

$$A = a_7 E_2,$$
$$B = E_2 a_7 s_2 + \frac{a_5(p_2 + d_1 + r_2)}{a_6} - E_2 a_7 a_1 - \frac{E_2 a_6 d_2 a_5 (a_8 d_1 + p_2 - r_2 + a_8 r_2 - a_8 p_2)}{a_6(a_5 - E_2 a_6 d_2 a_8)},$$
$$C = s_2 \left(E_2 a_7 a_1 + \frac{E_2 a_6 d_2 a_5 (a_8 d_1 - r_2 + a_8 r_2)}{a_6(a_5 - E_2 a_6 d_2 a_8)} + \frac{a_5 d_1 + a_5 r_2}{a_6} \right).$$

The characteristic value λ is a solution of the following characteristic equation:

$$H(\lambda) \equiv J(\bar{x}) - \lambda I - k^2 \pi^2 D = 0,$$

where

$$J(\bar{x}) = \begin{pmatrix} A_{11} & A_{12} & A_{16} \\ A_{21} & 0 & A_{26} \\ 0 & A_{62} e^{-\lambda \tau} & -a_5 + A_{66} e^{-\lambda \tau} \end{pmatrix}.$$

The stability of spatially constant equilibrium depends on the stability of linearization around $\bar{x}$ [4, 5, 25]. The spatially constant equilibrium $\bar{x}$ is asymptotically stable if and only if all roots of the characteristic equation have negative real parts. The characteristic equation can be written in the form

$$\lambda^3 + A_1 \lambda^2 + A_2 \lambda + A_3 = e^{-\lambda \tau}(B_1 \lambda^2 + B_2 \lambda + B_3), \tag{4}$$

where
$A_1 = (k\pi)^2(D_1 + D_2 + D_3) - A_{11} + a_5,$

$A_2 = (k\pi)^4(D_1D_2 + D_1D_3 + D_2D_3) + (k\pi)^2(D_2a_5 - D_2A_{11} - D_3A_{11} + D_1a_5) - A_{12}A_{21} - A_{11}a_5,$
$A_3 = (k\pi)^6 D_1D_2D_3 + (k\pi)^4(D_1D_2a_5 - D_2D_3A_{11}) - (k\pi)^2(D_2A_{11}a_5 + D_3A_{12}A_{21}) - A_{12}A_{21}a_5,$
$B_1 = A_{66},$
$B_2 = (k\pi)^2(D_2A_{66} + D_1A_{66}) - A_{11}A_{66} + A_{26}A_{62},$
$B_3 = (k\pi)^4 D_1D_2A_{66} + (k\pi)^2(-D_2A_{11}A_{66} + D_1A_{26}A_{62}) - A_{21}A_{12}A_{66} + A_{62}A_{21}A_{16} - A_{26}A_{62}A_{11}.$
Now for $\tau = 0$ characteristic equation (4) reduced to:

$$\lambda^3 + (A_1 - B_1)\lambda^2 + (A_2 - B_2)\lambda + A_3 - B_3 = 0. \tag{5}$$

It is convenient to use the Routh-Hurwitz criterion (see [3]) to test the stability properties of (5). By this criterion the real parts of the eigenvalues are negative if and only if $A_1 - B_1,\ A_3 - B_3 > 0$ and $(A_1 - B_1)(A_2 - B_2) > A_3 - B_3$. By straightforward computation we get

$$\begin{aligned}
&A_1 - B_1 = (D_1 + D_2 + D_3)k^2 - A_{11},\\
&A_3 - B_3 = D_1D_2D_3k^6 - A_{11}D_2D_3k^4 - (A_{12}A_{21}D_3 + A_{26}A_{62}D_1)k^3 +\\
&A_{11}A_{26}A_{62} - A_{16}A_{21}A_{62},\\
&(A_1 - B_1)(A_2 - B_2) - (A_3 - B_3) =\\
&(D_1^2D_2^2 + D_1D_3 + D_1^2D_2 + 2D_1D_2D_3 + D_1D_3^2 + D_2^2D_3 + D_2D_3^2)(k\pi)^6 +\\
&(-A_{11}D_2^2 - 2A_{11}D_2D_3 - 2A_{11}D_1D_2 - A_{11}D_3^2 - 2A_{11}D_1D_3)(k\pi)^4 +\\
&(A_{11}^2D_2 + A_{11}^2D_3 - A_{12}A_{21}D_1 - A_{12}A_{21}D_2 - A_{26}A_{62}D_2 - A_{26}A_{62}D_3)(k\pi)^2 +\\
&A_{11}A_{12}A_{21} + A_{16}A_{21}A_{62}.
\end{aligned}$$

Here, of course, all of $A_{11},\ A_{12},\ A_{16},\ A_{21},\ A_{26},\ A_{62}$ and A_{66} are function of a_8 since the equilibrium $\bar{x}$ depends on a_8, furthermore $A_{11},\ A_{26} < 0$ and $A_{16},\ A_{21},\ A_{62} > 0$. If $A_{12} < 0$ and $A_{11}A_{26}A_{62} > A_{16}A_{21}A_{62}$ then $(A_1 - B_1),\ (A_3 - B_3),\ (A_1 - B_1)(A_2 - B_2) - (A_3 - B_3) > 0$. Summarizing the above discussion, we can now obtain the next proposition.

Proposition 1. *Assume that $\tau = 0$ and there exist $a_{9_{min}},\ a_{9_{max}} > 0$ such that for all $a_8 \in (a_{8_{min}}, a_{8_{max}})$ a spatially constant positive equilibrium $\bar{x}$ exists. If $A_{12} < 0$ and $A_{11}A_{26}A_{62} > A_{16}A_{21}A_{62}$, then the spatially constant positive equilibrium $\bar{x}$ of the system (1) is locally asymptotically stable.*

Let us analyze the system (1) in the presence of the discrete time delay $\tau > 0$ around its interior equilibrium $\bar{x}$. We set $\lambda = \alpha + i\beta$ and substituting into (4) similarly as in [7], [8], [21] we obtain the following equations :

$$\begin{aligned}
&\alpha^3 - 3\alpha\beta^2 + A_1(\alpha^2 - \beta^2) + A_2\alpha + A_3 =\\
&\quad e^{-\alpha\tau}\left[\left\{B_1\left(\alpha^2 - \beta^2\right) + B_2\alpha + B_3\right\}\cos(\beta\tau) + \left\{2B_1\alpha\beta + B_2\beta\right\}\sin(\beta\tau)\right], \quad (6)\\
&3\alpha^2\beta - \beta^3 + 2A_1\alpha\beta + A_2\beta =\\
&\quad e^{-\alpha\tau}\left[\left\{B_1\left(\alpha^2 - \beta^2\right) + B_2\alpha + B_3\right\}\sin(\beta\tau) - \left\{2B_1\alpha\beta + B_2\beta\right\}\cos(\beta\tau)\right].
\end{aligned}$$

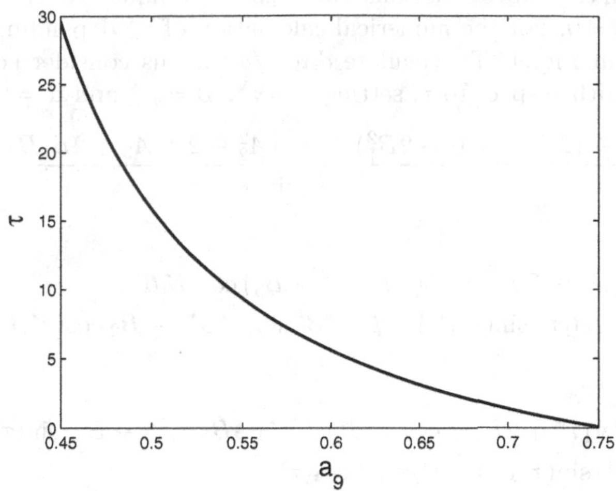

Fig. 1. Numerical calculation of the bifurcation discrete time delay τ depending on a_8

Let τ^* be such that $\alpha(\tau^*) = 0$. On substituting this into Eqs. (6), they reduce to

$$\begin{aligned} -A_1\beta^2 + A_3 &= \left\{-B_1\beta^2 + B_3\right\}\cos(\beta\tau) + B_2\beta\sin(\beta\tau), \\ -\beta^3 + A_2\beta &= \left\{-B_1\beta^2 + B_3\right\}\sin(\beta\tau) - B_2\beta\cos(\beta\tau). \end{aligned} \tag{7}$$

Squaring and adding the equations (7) and simplifying we obtain an equation for β of the form:

$$\beta^6 + (A_1^2 - 2A_2 - B_2^2)\beta^4 + (A_2^2 - 2A_1A_3 + 2B_1B_3 - B_2^2)\beta^2 + A_3^2 - B_3^2 = 0. \tag{8}$$

Let β^* be a simple largest positive root of equation (8). We now show that for $\beta = \beta^*$ there is a τ^* such that $\alpha(\tau^*) = 0$. The equation (7) can be written in the form:

$$\begin{aligned} Q_1\cos(\beta\tau) + Q_2\sin(\beta\tau) &= Q_3, \\ -Q_2\cos(\beta\tau) + Q_1\sin(\beta\tau) &= Q_4, \end{aligned}$$

where $Q_3^2 + Q_4^2 = Q_1^2 + Q_2^2 = Q_5^2$. The equations

$$\begin{aligned} Q_5\cos(\delta) &= Q_1, \\ Q_5\sin(\delta) &= Q_2 \end{aligned}$$

determine a unique $\delta \in \langle 0, 2\pi)$ and with this δ we have

$$\begin{aligned} Q_5\cos(\beta^*\tau^* - \delta) &= Q_3, \\ Q_5\sin(\beta^*\tau^* - \delta) &= Q_4. \end{aligned}$$

As it follows from the above equations, there exists a unique $\tau^* \in \langle \delta/\beta^*, (2\pi+\delta)/\beta^*)$ for which $\alpha(\tau^*) = 0$. For the numerical calculation of τ^* depending on a_8. The result is shown in Fig. 1. To calculate $d\alpha(\tau)/d\tau$ let us consider equations (6). Differentiating with respect to τ, setting $\tau = \tau^*$, $\beta = \beta^*$ and $\alpha = 0$ we get

$$\frac{d\alpha(\tau^*)}{d\tau} = \frac{3\beta^{*6} + (2A_1^2 - 4A_2 - 2B_1^2)\beta^{*4} + (A_2^2 - 2A_1A_3 + 2B_1B_3 - B_2^2)\beta^{*2}}{h_1^2 + h_2^2},$$

where

$$h_1 = A_2 - 3\beta^{*2} + \tau^*(-B_1\beta^{*2} + B_3)\cos(\tau^*\beta^*) - \\ 2B_1\tau^*\sin(\tau^*\beta^*) + B_2\tau^*\beta^*\sin(\tau^*\beta^*) - B_2\cos(\tau^*\beta^*)$$

and

$$h_2 = -2A_1\tau^* + 2B_1\tau^*\cos(\tau^*\beta^*) + \tau^*(-B_1\beta^{**}2 + B_3)\sin(\tau^*\beta^*) \\ -B_2\sin(\tau^*\beta^*) - B_2\tau^*\beta^*\cos(\tau^*\beta^{)}.$$

Let $\eta = \beta^2$, then we can reduce equation (8) to

$$\phi(\eta) = \\ \eta^3 + (A_1^2 - 2A_2 - B_2^2)\eta^2 + (A_2^2 - 2A_1A_3 + 2B_1B_3 - B_2^2)\eta + A_3^2 - B_3^2. \quad (9)$$

Since β^* is the largest positive simple root of equation (9), we have

$$\frac{d\phi(\eta^*)}{d\eta} > 0.$$

Hence

$$\frac{d\alpha(\tau^*)}{d\tau} = \frac{\beta^{*2}\frac{d\phi(\eta^*)}{d\eta}}{h_1^2 + h_2^2} > 0.$$

Now we state the following Lemma due to Gopalsamy [4]:

Lemma 1. *The equilibrium point $\bar{x}$ is locally asymptotically stable for all $\tau > 0$ if and only if $\bar{x}$ is locally asymptotically stable for $\tau = 0$ and there is no purely imaginary root of the characteristic Eq. (4).*

Clearly, if Eq. (9) has no positive root, then the interior equilibrium $\bar{x}$ is locally asymptotically stable for all $\tau > 0$. We obtain the spatially constant positive equilibrium $\bar{x}$, furthermore assumptions of Proposition 4 are satisfied, i.e. $A_1 - B_1$, $A_3 - B_3$, $(A_1 - B_1)(A_2 - B_2) - A_3 - B_3 > 0$ for all $a_8 \in \langle 0.45, 0.75\rangle$. For $k = 0$, there exist a simple largest positive root β^* of the equation (8) and a $\tau^* > 0$ such that $\pm\beta^* i$ are simple eigenvalues of the Eq. (4) and all other eigenvalues have negative real parts.

For $k > 0$, there are no real roots of the Eq. (8). According to Lemma 1 all eigenvalues of the Eq. (4) have negative real parts. Hence $\frac{d\alpha(\tau^*)}{d\tau} > 0$, we can apply Hopf bifurcation theorem [25] to show that the system (1) has a family of periodic solutions bifurcating from the spatially constant interior equilibrium $\bar{x}$, when τ is near τ^*. For numerical solutions see Fig. 2.

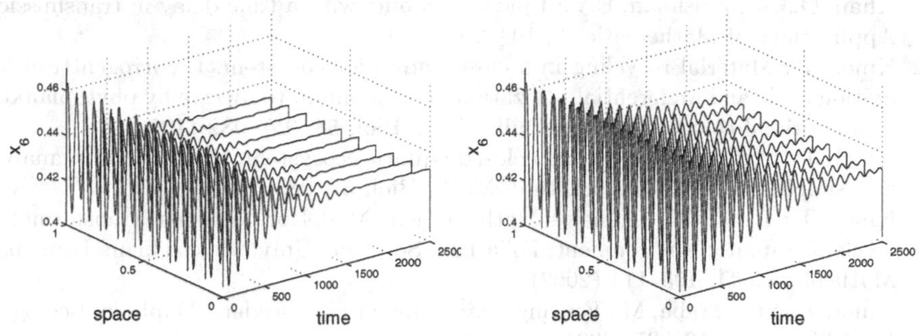

Fig. 2. Numerical solution of system (1) for $a_8 = 0.5$ and $\tau = 14.88$ and $\tau = 16.88$

4 Conclusion

We considered a simple ecological model consisting of phosphorus as a limiting nutrient for growth of algae and Daphnia. The model is described by a system of three partial functional differential equations with a discrete time delay. The dynamical behaviour of the system is analyzed. The partial functional differential equations model exhibited a much more complicated dynamical behavior than that by the corresponding partial differential equations model. It is evident from the results that the discrete time delay in the uptake of algae by Daphnia can destabilize an otherwise a stable equilibrium, and also a simple Hopf bifurcation can occur when the time delay passes transversally through its critical value.

Acknowledgment. This paper was written under the scientific project number KEGA 010UJS-4/2014 and VEGA 1/0699/12.

References

1. Busenberg, S., Kumar, S.K., Austin, P., Wake, G.: The Dynamics of a Model of a Plankton-Nutrient Interaction. Bull. of Math. Biol. 52, 677–696 (1990)
2. Ellermeyer, S.F.: Competition in the Chemostat: Global Asymptotic Behavior of a Model with Delayed Response in Growth. SIAM J. Appl. Math. 54, 456–465 (1994)
3. Gantmacher, F.R.: The theory of matrices, New York, Chelsea (1959)
4. Gopalsamy, K.: Stability and Oscillation in Delay Different Equations in Population Dynamics. Kluwer Academic Publisher, Boston (1992)
5. Hale, L.S., Verduyn Lunel, S.M.: Introduction to Functional Differential Equations. Springer, New York (1993)
6. Hsu, S.B., Hubebell, S., Waltman, P.: A Mathematical Theory for Single-nutrient Competition in Continuous Cultures of Micro-organisms. SIAM J. Appl. Math. 32, 366–383 (1977)
7. Jana, S., Chakraborty, M., Chakraborty, K., Kar, T.K.: Global stability and bifurcation of time delayed prey-predator system incorporating prey refuge. Mathematics and Computers in Simulation 85, 57–77 (2012)

8. Khan, Q.J.A., Krishnan, E.V.: Epidemic model with a time delay in transmission. Applications of Mathematics 3, 193–203 (2003)
9. Kmet, T.: Material recycling in a closed aquatic ecosystem. I. Nitrogen transformation cycle and preferential utilization of ammonium to nitrate by phytoplankton as an optimal control problem. Bull. Math. Biol. 58, 957–982 (1996)
10. Kmet, T.: Material recycling in a closed aquatic ecosystem. II. Bifurcation analysis of a simple food-chain model. Bull. Math. Biol. 58, 983–1002 (1996)
11. Kmet, T.: Kmet, T.: Diffusive Mathematical Model of Nitrogen Transformation Cycle in Aquatic Environment: Folia Fac. Sci. Nat. Univ. Masarykiane Brunensis, Mathematica 11, 105–114 (2002)
12. Kmet, T., Straskraba, M.: Feeding adaptation of filter feeders: Daphnia. Ecological Modelling 178, 313–327 (2004)
13. Kmet, T.: Model of Nitrogen Transformation Cycle. Mathematical and Computer Modelling 44, 124–137 (2006)
14. Kmet, T., Kmetova, M.: Modelling and Simulation of Filter Adaptation by Daphnia. In: Burczynski, T., Kolodziej, J., Byrski, A., Carvalho, M. (eds.) ECMS 2011, Krakow, pp. 122–128 (2011)
15. Li, B., Wolkowicz, G.S.K., Kuang, Y.: Global asymptotic behaviour of a chemostat move with two perfectly complementary resources and distributed delay. SIAM J. App. Math. 60, 2058–2086 (2000)
16. Li, B., Smith, H.: How many species can two essential resources support? SIAM J. App. Math. 62, 336–366 (2001)
17. Pan, S., Wan, S.: Traveling wave fronts a delayed population model of Daphnia magna. Applied Math. and Computation 215, 1118–1123 (2009)
18. Radtke, E., Straskraba, M.: Self-optimization in a phytoplankton model. Ecological Modelling 9, 247–268 (1982)
19. Straskraba, M., Gnauck, P.: Freshwater Ecosystems, Modelling and Simulation, Developments in Environmental Modelling. Elsevier, Amsterdam (1985)
20. Smith, H.L., Waltman, P.: The Theory of the Chemostat, Dynamics of Microbial Competition. Cambridge Univ. Press, Cambridge (1995)
21. Su, Y., Wei, J., Shi, J.: Hopf bifurcations in a reactiondiffusion population model with delay effect. J. Differential Equations 247, 1156–1184 (2009)
22. Wolkowicz, G.S.K., Xia, H.: Global Asymptotic Behavior of a Chemostat Model with Discrete Delays. SIAM J. Appl. Math. 57, 1019–1043 (1997)
23. Wolkowicz, G.S.K., Xia, H., Ruan, S.: Competition in the Chemostat: A Distributed Delay Model and its Global Asymptotic Behavior. SIAM J. Appl. Math. 57, 1281–1310 (1997)
24. Wolkowicz, G.S.K., Lu, Z.: Global dynamics of a mathematical model of competition in the chemostat: general response function and differential death rates. SIAM J. Appl. Math. 52, 222–233 (1992)
25. Wu, J.: Theory and Applications of Partial Functional Differential Equations. Applied Math. Sciences, vol. 119. Springer, New York (1996)

Author Index